Insecticides - Impact and Benefits of Its Use for Humanity

Edited by Ramón Eduardo Rebolledo Ranz

Published in London, United Kingdom

IntechOpen

Supporting open minds since 2005

Insecticides - Impact and Benefits of Its Use for Humanity
http://dx.doi.org/10.5772/intechopen.95712
Edited by Ramón Eduardo Rebolledo Ranz

Contributors
Samira Khodayari, Nayereh Hamedi, Lawrence N. Nkosikhona Malinga, Mark D. Laing, Ahmed Boughdad, Nahid Haouache, Soukaina El Asri, Adil Asfers, Abdelhadi Ait Houssa, Bouchra Tazi, Taruna Kaura, Abhishek Mewara, Nikita Sharma, Valéria Verebová, Jana Staničová, Alberto Angioni, Nicola Arru, Mohamed Abdel-Raheem, Onyeka Kingsley Nwosu, Ayibapreye John, Muzafar Riyaz, Gulzar Ahmed Rather, Saqib Hassan, Mrinalini Kumari, Shyam Babu Sah, Atul Srivastava, Subhashini, U. Shamjana, Tony Grace, S.A. Dwivedi, V.K. Sonawane, T.R. Pandit, Abu Hazafa, Muhammad Naeem, Muhammad Murad, Muhammad Umer Masood, Shahid Bilal, Muhammad Nasir Khan, Qasim Farooq, Muhammad Omer Iqbal, Muhammad Shakir, Nambangia Justin Okolle, Ekwa Yawa Monono, Amungwa Ivan Tabikam, Mambo Stephania Kinge, Magwell Pierre Fils Rodrigue, Enrique Mejías, Tatiana Garrido, Haci Ahmet Deveci, Gökhan Nur, Abdulsamed Kükürt, İnan Kaya, Müge Mavioğlu Kaya, Ömer Faruk Başer, Ayla Deveci, Mahmut Karapehlivan, Volkan Gelen, Shrawan Kumar Sahani, Vikas Kumar, Subhajit Pal, Kenneth Ojotogba Achema, Charity Jumai Alhassan, Rodrigo Rosario Cruz, Delia Ines Domínguez García, Saul Lopez Silva, Fernando Rosario Dominguez, Sushil Nyaupane, Tapan Kumar Hath, Biswajit Patra

First published in London, United Kingdom, 2022 by IntechOpen
IntechOpen is the global imprint of INTECHOPEN LIMITED, registered in England and Wales, registration number: 11086078, 5 Princes Gate Court, London, SW7 2QJ, United Kingdom
Printed in Croatia

British Library Cataloguing-in-Publication Data
A catalogue record for this book is available from the British Library

Additional hard and PDF copies can be obtained from orders@intechopen.com

Insecticides - Impact and Benefits of Its Use for Humanity
Edited by Ramón Eduardo Rebolledo Ranz
p. cm.
Print ISBN 978-1-83969-026-6
Online ISBN 978-1-83969-027-3
eBook (PDF) ISBN 978-1-83969-028-0

For EU product safety concerns:
IN TECH d.o.o., Prolaz Marije Krucifikse Kozulić 3, 51000 Rijeka, Croatia, info@intechopen.com or visit our website at intechopen.com.

Meet the editor

Dr. Ramón Rebolledo Ranz obtained a Ph.D. in Agricultural Engineering with a mention in Plant Protection from the Polytechnic University of Madrid, Spain in 1994. He has worked in the Faculty of Agricultural and Forestry Sciences, Universidad de La Frontera, Chile since 1986. He has published more than ninety scientific articles in national and international journals, two books, five book chapters, and has edited four books in his specialty. He has presented more than 100 scientific papers at national and international conferences and has directed more than 90 theses, both undergraduate and postgraduate. He has been a responsible investigator in sixteen research projects and has participated in several others as a co-investigator. He is an active member of several scientific societies and has given several conferences in the areas of pest control, biodiversity, and beekeeping. For years, he has been the scientific coordinator of the National Beekeeping Network of Chile. He has organized more than twenty-five national and international scientific conferences. He is also a consultant for companies in the private sector on the subject of pests and beekeeping. He is the creator and curator of the Entomological Museum, Faculty of Agricultural and Forestry Sciences, Universidad de La Frontera. Dr. Ranz has served as a chair in zoology, pest control, agricultural entomology, and environmental entomology at the undergraduate and postgraduate levels.

Contents

Preface

This book provides a comprehensive overview of insecticides, including information on their adverse effects on the environment, the resistance of pests to insecticides, accumulation in food, and regulations and rules in different countries regarding the application of insecticides.

It includes twenty-one chapters written by authors with extensive national and international experience. They address such topics as the use and impact of insecticides in Cameron and Nepal, how insecticides stop working after application, the biochemistry of insecticides, the role of microbial insecticides in pest control, biological pest control versus insecticides, control of tetranychids, the effect of insecticides on natural enemies, integrated management of ticks in cattle, pest resistance to insecticides, and much more.

This book fills a gap in the literature on pest control and is a useful resource for undergraduate and graduate students in agriculture, forestry, veterinary science, and environmental science.

Dr. Ramón Eduardo Rebolledo Ranz
Universidad de La Frontera,
Temuco, Chile

Section 1

General Aspects on Insecticides

Chapter 1

Insecticide Use and Application in Cameroon

Nambangia Justin Okolle, Ekwa Yawa Monono, Amungwa Ivan Tabikam, Mambo Stephania Kinge and Magwell Pierre Fils Rodrique

Abstract

Cameroon has a diverse natural environment with five agro-ecological zones that support the cultivation of many crops. The country relies mainly on agriculture, with main export crops such as cocoa, coffee, cotton and banana. The agricultural sector occupies an average of about 47 million hectares of land with different production system practices such as mono cropping, mixed cropping, intercropping etc. Biotic factors are major constraints. These biotic constraints are seriously hindering the crop production, resulting to pre-and post-harvest losses and lower yields. These pests are mainly from the arthropod, vertebrate, mollusk, weeds and nematode groups and disease-causing pathogen such as fungi, viruses and bacteria. With these constraints famers in developing world like Cameroon are forced to look for ways to control these pests and pathogens which lead to the use of numerous preventive and curative techniques including the use of insecticides. Using insecticides is not against the law but the application methods and the supply routes carried out by farmers and traders respectively might be bad. This is why this chapter reviews the insecticides supply routes, registration procedure and registered insecticides, insecticidal application with its malpractices while looking at its intoxication as well as the alternatives to the use of synthetic insecticides in Cameroon and make possible recommendations to promote judicious use of insecticides in Cameroon.

Keywords: registration, malpractices, intoxications, synthetic chemical alternatives

1. Introduction

The purpose of this research was to gather relevant information concerning the sources, practices and intoxications resulting from the use of insecticides in Cameroon. It was also to collect information on different options that can replace synthetic insectides and finally to give recommendations that will help foster best practices and minimize cost of production to users as well as minimize effects to human health and the environment.

To meet our purpose, we carried out mainly desk study that allowed us to collect, analyze and summarize information from different sources such as internet search engines (mainly www.google.com), library of the Institute of Agricultural

Research for Development—IRAD and that of the African Research Centre on Bananas and Plantains—CARBAP), consultation of some national experts especially those at the Phytosanitory Department of the Ministry of Agriculture & Rural Development (MINADER), and Whatsapp groups of related professionals. The information collected were mainly primary research (peer-reviewed manuscripts published in journals) as well as scientific reports or articles in conference proceedings or annual reports. We also carried out secondary literature review from news bulletins, magazines, and books. Review focus on all these sources was on the following aspects; the agricultural sector, sources of insecticides, registration procedure & list of registered insecticides, insecticide application methods & associated Malpractices, different intoxications or poisonings resulting from malpractices, and alternatives to the use of synthetic insecticides.

2. The agriculture sector and agro-ecological zones

2.1 Geographical location and regions

Cameroon is a country located at the crossroads of West and Central Africa with a surface land size of 475,440 km^2 [1] and a population of about 22.71 million inhabitants. It is bordered by Nigeria, Chad, Central African Republic, Equatorial Guinea, Gabon and Republic of Congo to the West, Northeast, East, and South respectively. Cameroon lies on the Bight of Bonny coastline, which is part of the Gulf of Guinea and the Atlantic Ocean [1]. Cameroon is called '*Africa in miniature*' because it is characterized by a richly diversified natural environment such as mountains, desert, rain forest, savannah grassland and oceanland. The country consists of three main natural regions [2].

- The southern forest (including the regions of Centre, East, Littoral, South and South West) is situated in the maritime and equatorial zones

- The western highlands charatristize (covering the regions of West and North West)

- The Sudano-sahelian north (covering Adamawa, North and Far North) (**Figure 1**).

2.2 Agro-ecological zones

The natural regions in Cameroon are divided into five agro-ecological zones (**Table 1** and **Figure 1**) each characterized by dominant physical, climatic, and vegetative features.

2.3 Importance and production systems

The economy of Cameroon relies mainly on agriculture, with main export crops such as cocoa, coffee, cotton and banana. On an average about 47 million hectares of Cameroon's land are used for the agricultural sector [2]. It estimated to be less than 5% of the entire territory; moreover, the exploitation of forestry, mining and fisheries represent an additional contribution to the economy of the country. Agriculture in Cameroon is currently employing about 70% of its workforce and providing 44% of its gross domestic product and 30% of its export revenue [1]. The different production systems practiced in the country are:

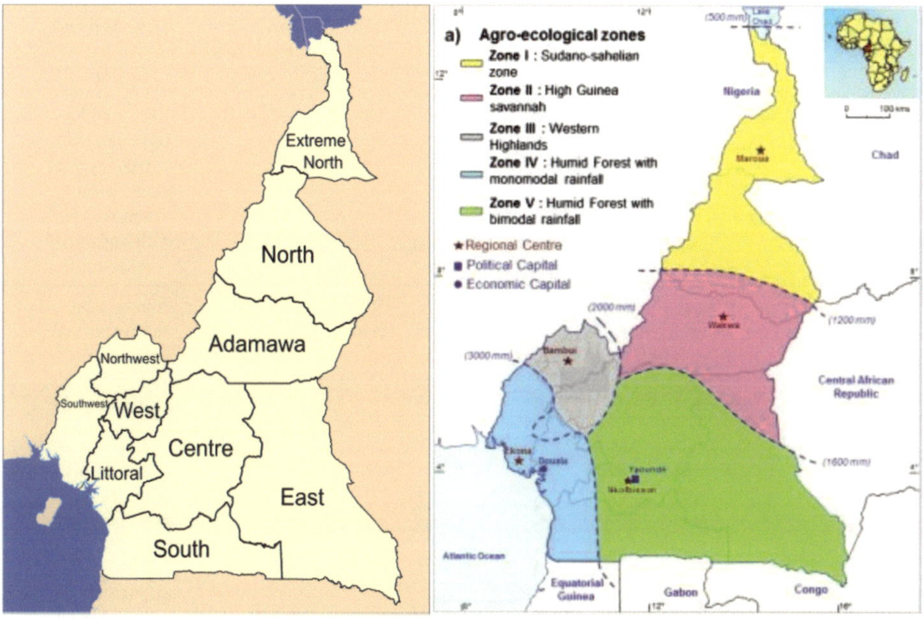

Figure 1.
Different regions (a) and agroecological zones (b) in Cameroon.

2.3.1 Dry farming and terracing

These are commonly practiced in the western highlands of Cameroon. Terracing is practice across steep-slop banks on the western highlands with main crops cultivated such as plantains, yams, and potatoes. Dry farming system is practiced along the few fertile gentle slopes and in localities of undulating reliefs [3].

2.3.2 Intercropping and mixed cropping

These are practiced mainly by smallholder farmers. The practices differ from one agro-ecological zone to another but many characteristics of these practices are the same nationwide. Intercropping/mixed farming is mostly carried out on small farm sizes, generally those <2 hectares. Most of the intercrop mixes contain one leguminous crop. Crops intercrops are selected based on their importance for household consumption and market. While there are cases where food crops are intercropped with cash crops, the practice is predominantly carried out by food crop farmers [4]. Some of the farmers grow more than two crops on the same land during one growing season, especially those cultivating on a land size <1 hectares [3]. Some of such farms also have one or more livestock.

2.3.3 Crop rotation

This is the practice of planting different crops sequentially on the same land to reduce the spread and rapid attack of crops by pests and diseases. The crops also differ from one agro-ecological zone to another.

2.3.4 Mono cropping

This is commonly practiced by large scale farmers >5 hectares. This cropping system is mostly industrial agricultural institutes such as CDC, PAMOL,

Agro-ecological zones	Regions	Altitude (m)	Rainy days/ year	Rainy months/ year	Rainfall (mm)	Mean annual temperature (range)	Main crop and animal production
Sudano-Sahelian	North and Far North	250–500	90–120	3–5	500-900	28°C (7.7)	Maize, millet-sorghum, rice, cowpea, soybean, onion, sesame, fruits, cotton, cattle and small ruminants
Sudano-Guinean (high Guinea savannah)	Adamawa	500–1500	110–150	7	1500–1800	23°C (6.4)	Maize, yam, cassava, sweet potatoes, rice, cotton, cattle, pig, small ruminants, poultry birds
Western highlands	West and North West	1500–2500	175–220	7–9	1800–2400	21°C (2.2)	Maize, beans, potatoes, rice, sweet potatoes, vegetables, coffee, pig, poultry, cattle, small ruminants, fisheries
Humid forest (monomodal rainfall)	Littoral and South West	0–500	180–240	9–12	2000–11,000	26°C (2.8)	Banana, plantain, cassava, cocoyam, sweet potatoes, maize, vegetables, cocoa, coffee, oil palm, rubber, fruits, poultry, pig, poultry birds, small ruminants, fisheries
Humid forest (bimodal rainfall)	Centre, East and South	400–1000	125–175	7–9	1500–2000	25°C (2.4)	Plantain, cassava, banana, maize, cocoyam, sweet potatoes, cocoa, oil palm, rubber, coffee, maize, cocoa, oil palm, fruits, poultry, pig, fisheries, small ruminants

Table 1.
Agro-ecological of Cameroon characteristics, geographical features, rainfall patterns and major crops cultivated and animal species reared (source: authors compilation).

SOCAPALM, Tole tea etc. where they cultivate crops like palms, banana, tea, rubber etc.

2.4 Main constraints in the agricultural sector

Like most developing countries, the agricultural sector in the country faces lots of constraints; (i) poor farm-to-market roads, (ii) insufficient access to finance, (iii) most are small farms still using rudimentary tools, (iv) most of the smallholder farmers have not been trained on good agricultural practices, (v) insufficient links to market and market information, (vi) insufficient and low standard transformation/processing of commodities, (vii) pests and diseases are a major constraint, (viii) most of the smallholder farmers misusing agrochemicals.

Biotic constraints (pests and pathogens) are seriously hindering the crop production, resulting to pre-harvest losses, post-harvest losses and lower yields. Arthropod pests especially insects are the most common and serious group causing damage to crops. Most damaging insect Orders are Coleoptera, Lepidoptera, Hemiptera, and Diptera. Most mentioned group of disease-causing agents are fungi and viruses. Severity of the pests and diseases varies with the crop and crops hardest hit by diseases are cocoa, tomatoes, banana/plantain, onions, sorghum, maize, pineapple, cucumbers, pepper and water melons while those hardest hit by pests are cocoa, tomatoes, maize, groundnuts, bananas/plantains, sorghum, millet, cotton, pineapple, coffee and cucumbers.

The different agro-ecological zones are faced with biotic and abiotic constraints that affect the agricultural productivity. Pests are mainly from the arthropod, vertebrate, mollusk, weeds and nematode groups of living things. Disease-causing pathogens are mainly fungi, viruses and bacteria. For cocoa, black pod; capsids; rodents and primates reported as the main pests/diseases. Corm borer weevils, mealy bugs, aphids, snails, black sigatoka, banana bunchy top disease and Banana streak virus for bananas/plantains; blight, fruit flies, white flies, aphids and mole crickets for tomatoes, peppers, and African nightshade; fall army worm, stem borers, snails, rust and maize weevil for maize; boll worm for cotton; bruchids for cowpea; thrips and mildew for onions, leaf miners for oil palm, grasshoppers, caterpillars, aphids and whiteflies also attack lots of traditional African vegetables.

3. Sources of insecticides and purpose

3.1 Definition

Insecticides are agrochemicals in the pesticide family used to control insects by killing them or preventing the establishment or proliferation of those considered harmful. They play an important role in agriculture and public health by improving the yield and productivity of crops caused by pests and by reducing the rate of vector-borne diseases by killing or affecting growth and development of vectors such as mosquitoes, houseflies, tse tse flies, sand flies, cockroaches, etc. They are pesticides formulated essentially for repelling, killing, harming or mitigating insects from crops and other agri-food sources such as ranches, poultry farms, etc. Insecticides work differently based on their mode of actions; some disrupt the nervous system, whereas others damage the exoskeletons, others repel or control them. Insecticides application helps in managing and mitigating insects; thereby guaranteeing crop protection and preventing yield loss, they are the main weapons against insects in Africa. Insecticides are widely used in Cameroon by farmers and traders to protect their plants and products during production and post-harvest storage [5].

3.2 Insecticide supply routes

There are two main supply routes in Cameroon—a legal and an illegal supply chain. Pesticides (including insecticides) are imported mainly from France (30.9%), Switzerland (14.0%), Norway (5.7%) and the USA (5.1%). Other countries (such as Germany, China, and India) supply 16.6% while 27.7% were of unidentified sources. Pesticides are imported by local companies to large distributors and then distributed to chemical retailers (95% of all pesticides) or to farmers' societies (5%) [6].

Cameroon does not currently produce any insecticides for crop protection, timber protection and public health uses, but repackaging is practiced by some distributors [7]. Consequently, all national demand for insecticides is met by imports. In Cameroon, the quantities of pesticides imported are undergoing a clear evolution; from 960 tons in 2015 to 1163 tons in 2019 [8].

3.3 Main dealers of insecticides

There are about 13 recognized large companies that carry out the majority of insecticide and other agrochemical imports into Cameroon although the major companies are: FIMEX International SA, AGROCHEM, ARYSTA Life Science, JACO, AFRICAWARE, ADER, BASE-F, Syngenta and YARA (Ministry of Agriculture & Rural Development). These main private companies operating in the phytosanitary business are mostly grouped within the association CropLife Cameroon, an international association of pesticide companies whose mission is to promote the availability of quality pesticides within the country. CropLife's advocacy role involves financial support and regular sensitisation of pesticide import and distribution companies to conduct their activities in a manner that takes into account national and international regulatory requirements in order to ensure the promotion of appropriate health, safety and environmental protection measures for all those who may be directly or indirectly affected by their activities.

The responsibility of the distributor is his/her obligation to answer for the act of making available to the end user plant protection products that have safety and quality standards. Law No. 2003/003 of 21 April 2003 on plant protection stipulates that the distributor must verify that the plant protection products are registered or have a valid Provisional Sales Authorization (PSA) in Cameroon. In order to meet this expectation, he must confirm that the product of standard (formulation, mode of action and type or family of products, he must be able to distinguish the products, which means he must know how to read the label and has adequate training). The regulations stipulate that the technician in charge of pesticide distribution must have been trained by an approved agricultural training institution. Although the distribution of phytosanitary products is an economic activity, the technician must be able to give technical advice to his clients.

3.4 Distribution of insecticides in Cameroon

The distribution of pesticides is subject to strict rules and the distributor must be registered in the MINADER database. Article 24:(1) of Law No. 2003/003 of 21 April 2003 on plant protection stipulates that the marketing of plant protection products in bulk or on display is prohibited. The same applies to the possession of obsolete plant protection products [9]. Small scale distributors or retailers buy from the large companies. Some of these large companies put in place measures that allow the retailers to be registered before being allowed to get products. Sellers mainly buy their supplies from large authorized distributors. In addition to vendors, permanent retailers operate agricultural input sales outlets in the main markets of Cameroon. In addition to permanent retailers, hawkers do sell in several markets at once, traveling from one market to another.

In 2017, a pilot study on the harmful effects of agricultural pesticides on human health and the environment in some regions of Cameroon, carried out in five different zones representing the main agricultural production basins of the 5 agro-ecological zones indicated that although all the retailers claim to get their supplies preferentially from approved distributors, the majority of pesticide sellers found in the markets do not have any sales authorization as required by the Law and are not listed in MINADER's data base. This creates an informal market for agricultural pesticides (including insecticides). Some of them sell in bulk and on stalls. This creates a situation where unregistered and expired products are found in the markets (**Figure 2**) [10].

3.5 Importance of insecticide

Insecticides are widely used in the country by chemical retailers, farmers' organizations, farmers and post-harvest traders to protect their plants and products during production and post-harvest storage. The importance of plant protection products in agriculture is justified by their impact on increasing crop yields (cereals, banana and plantain, pineapple, tomato, rubber, cocoa, wood, vegetable and fruit crops, cotton, sugarcane, carrots, rice, oil palm, maize, coffee, logs, stocks, cabbage) by an average of 30–40%, reduction of damage caused by pests and diseases which can reach 30–50% of losses in the field or after harvest [11, 12].

Insecticides are applied to the environment to reduce the population and damage caused by insect pests below a level that cannot cause economic damage. They therefore play an important role in agriculture and public health. In Cameroon, public health, insect pests and/or vectors of importance include mosquitoes, house flies, cockroaches, bed bugs, midges. Insecticides are also used to control urban pests such as beetles and termites destroying ceilings and other furniture of houses.

(a) (b) (c)

Figure 2.
Photo of a typical store where pesticides are sold. (a) and (b) Pictures of pesticides of retailers salers. (c) Picture of typical store where pesticides are sold (source: Pouokam, 2016; Cameroon Tribune 2021).

Figure 3.
Sample damage caused by pests and diseases: photos of bed bug and its effects on humans, fall army worm on maize leaves, cocoa black pod disease and mold on maize cobs.

The use of insecticides has an advantage in terms of economic efficiency and improved human health and welfare (**Figure 3**).

4. Registration procedure and analysis of registered insecticides

4.1 Registration procedure

4.1.1 Legal and institutional framework

In Cameroon, laws exist that guide the distribution and appropriate use of agrochemicals especially synthetic pesticides. These laws concern the use or misuse of pesticides and align with certain international conventions (e.g. Stockholm, Rotterdam, International Plant Protection) of which Cameroon is a signatory. One important law is No. 2003/003 of April 2003 concerning phytosanitory measures or crop protection practices. According to this law, pesticides are substances or group of substances (example insecticides) used to destroy or control below threshold levels crop pests, disease vectors, species that are undesirable to plants and animals or negatively affect the entire value chain of agricultural products. Pesticides can cause harm to humans, animals or the environment if not properly used or disposed.

Also, some pesticides residues or phytosanitary products may accumulate in organisms after repeated applications andthis may cause diseases and subsequently death. For this reason, laws and decrees are usually published to regulate the sale, storage and use of all phytosanitary products that enter the Cameroonian markets. In relation to this, some products or active ingredients that are identified as toxic by the competent authorities have been banned and removed from the market. This Law documents the principles and rules governing plant protection in Cameroon. Generally, pest control is carried out through (i) the development, adoption and adaptation of standards, (ii) the prevention and fight against pests of plants and plant products, (iii) theuse of pesticides that are safe to human and animal health and for the environment, (iv) the dissemination and popularization of appropriate techniques for plant protection, (v) control of the import and export of pesticides.

Furthermore, chemical treatments are applied in accordance with good agricultural practices issued by the competent authority in order to protect human and animal health and protect the environment from hazards arising from the presence or accumulation of pesticide residues. Any natural or legal person wishing to perform phytosanitary treatments in a professional capacity must first be approved by the competent authority. In addition, only registered pesticides or those with a provisional sales authorization must be imported, distributed, packaged or used in Cameroon. All plants, plant products, soil or growing medium, bodies and biological pest control products are subject to: phytosanitary inspection regardless of their place of production, multiplication and storage and their mode of transport; control during their manufacture, import, export, packaging, distribution and use.

4.1.2 Procedure for registration in ECCAS and Cameroon

The procedure for registration of pesticides in Cameroon is almost the same for the countries in the Economic Community of Central African States (ECCAS) region although there might be some slight differences. This procedure has been endorsed by the Central African Pesticides Committee (CPAC) in collaboration with all the ECCAS countries [13]. The procedure involves three main steps:

Step 1: Submission of a complete registration application file to the Permanent Secretariat of the Central African Pesticides Registration Committee (CPAC) together with payment of an examination fee. The file is then forwarded to experts for examination.

Step 2: After examination of the file, CPAC may decide to either (i) register the pesticide in Central Africa region for 10 years; (ii) grant a Provisional Sale Authorization (PSA) for a two-year period pending further studies; (iii) retain the file under study pending additional information or (iv) refuse to register the pesticide. A registered pesticide is issued a unique number that is valid for all CPAC member states.

Step 3: The CPAC Permanent Secretariat transmits the results of the deliberation to the applicant and to the member states, and publishes the list of registrations and PSA in CPAC periodical.

The pesticide registration application file comprises all information necessary to assess the efficiency of the pesticide and the potential hazards that such a pesticide might pose to humans, non-target organisms and the Central African environment as a whole. It includes all information on the identification and the physico-chemical properties of the product and the active ingredient, toxicology, effects on the environment and wildlife, the residues as well as information on the safety measures on the use of the product [13]. The file includes the following items submitted in French or English:

- An application for the registration of a commercial product;
- A specification sheet;
- A technical package;
- An analytical file;
- A toxicology file;
- The original label or scale model;
- A reference sample of the active ingredient(s) contained in the commercial product and a sample of the commercial product;
- A registration certificate in the country of origin.

The registration criteria comprise of:

- an administrative information (name and address of applicant, patent holder, manufacturer of formulation and manufacturer of active ingredients);
- identity of the formulation (brand name of the formulation, names and proportion of active ingredients, etc.);
- identity of the active ingredients (ISO, purity, proportions of additives, etc.);
- intended uses (type of pesticide, target crops, countries with similar ecologies where the formulation is registered).

The files include physico-chemical, biological efficacy, analytical, toxicology, environmental, residue and packaging and labeling files. The files shall comprise only abstracts of these studies. The complete studies is made available to CPAC on request.

Labeling of pesticide containers is designed as a means of attaining a high level of communication between the pesticide dealer and the user. Therefore, it should be clear and concise and should contain fundamental data for the use of pesticide in complete safety and with guaranteed efficiency throughout its life span. The label should describe the content, present a clear visible indication of the hazard, direction for the sound use of the content, name and address of manufacturer as well as manufacture and expiry dates. Additionally, a specification sheet or technical notice should be enclosed to supplement information on the description of active ingredients, direction for use and necessary precautions.

The importation, sale and use of pesticides in Cameroon are regulated by Law No. 2003/003 of 21st of April 2003 regarding phytosanitary protection, particularly in section 1 of chapter III of the law. Here, it is clearly stated that only registered phytosanitary products or products that have a Provisional Sale Authorization (PSA) could be imported, distributed, conditioned or used in Cameroon. These products are supposed to be marketed and used only in their original packaging material. In addition, equipments used for the application of pesticides are supposed to respect specific norms. To this, their production, importation and distribution in Cameroon are regulated by the law. Interested persons are supposed to submit an application file for certification of the equipment to the National Commission for Homologation of Pesticides and the Certification of phytosanitary equipment (CNPHCAT) and must pay an evaluation fee. Registration is a process at the end of which a competent authority approves the importation, distribution and use of a product after results of scientific analysis indicate that the product is effective, does not present any risks or danger to humans, animals or the environment when used as recommended.

The regulation binding the registration of pesticides in Cameroon is the same for the countries in the ECCAS. This regulation was put in place by CPAC in collaboration with all the ECCAS. The regulation clearly indicates that a pesticide may not be homologated unless its formulation conforms to the following criteria:

- It is sufficiently effective against the target organism; has no phytotoxic effect, it is not harmful to humans and wildlife not initially targeted and has no negative effects on the environment.
- It has acceptable biological efficacy.
- Established experimental and analytical methods can determine the components, impurities and residues of the pesticide.
- Maximum residue limits for agricultural products intended for human consumption and subject to homologation.

Where most of the above criteria are respected, a Provisional Sale Authorization (PSA) is granted, which will be valid for a limited period of 2 years nonrenewable.

Registration of a product involves the following:

- Chemical analysis of a sample of the product conducted in an accredited laboratory;
- Biological efficacy tests conducted by a research institute during one or two cropping seasons;
- Pre extension tests conducted by plant protection services of the Ministry of Agriculture and Rural Development over at least one cropping season;
- Combined tests of bio-efficacy and pre extension for at least one cropping season.

Individuals or group of persons who intend to submit a phytosanitary product for registration are supposed to deposit an application file to the National Commission for Homologation of Pesticides and the Certification of phytosanitary equipment (CNPHCAT) and must pay an examination fee. This commission is created in MINADER but includes one or two members from other government ministries such as Ministries of Scientific Research and Innovation, Higher Education, Public

Health, Animal Husbandry and Fisheries, Environmental Protection and Nature Protection among others. The chair of the commission is the Minister of Agriculture and Rural Development as stated in chapter IV of the Prime Ministerial Decree no. 2005/0772/PM of 06 April 2005. This decree also provides details on procedure for submission of a phytosanitary product for homologation in its chapter II.

4.2 Analysis of registered insecticides

The list of pesticides registered in Cameroon as of 04 March 2021 by the National Commission for the Homologation of Phytosanitary Products and Certification of Treatment Equipment housed within the Ministry of Agriculture and Rural Development for the control of crop and wood pests and for public health uses is estimated at nearly 900 pesticides, 90% of which are pesticides and 10% growth regulators [14, 15]. Of this list, insecticides represent about 34% of the total registered pesticides. It is subject to periodic renewal, but provides a framework that is binding on all at the national level. The toxicological class of insecticides according to the World Health Organization classification, indicate that 2% are in Class Ia (extremely dangerous) for use in food storage; 3% of registered insecticides are in Class Ib (very dangerous), including insecticides and nematicides used to treat cotton, tomatoes, plantains, vegetables; 32% in Class II (moderately dangerous); and 63% in Class III (slightly dangerous) [14].

Of the 311 insecticides registered by March 2021, mostare registered for use on cotton (24.4%), tomatoes (22.5%), cocoa (21.9%), and public health (13.2%) (**Figure 4**). **Table 2** shows the different sectors on which insecticides are registered, major group of insecticides and their percentages, target pests and examples of main active ingredients.

From **Table 3**, 11 commercial insecticides are registered with characteristics of highly hazardous pesticides (HHPs) comprising 07 active ingredients according to World Health Organization. These HHPs are registered only for banana/plantain, woods, cotton, stored products, and tomato. Cocoa with 68 registered insecticides has no active ingredient which is HHP.

Only four biocontrol agents have been registered (*Bacillus thuringiensis kurstaki*, *B. subtilis*, emamectine benzoate, and a nuclear polyhedrovirus). These microbial biocontrol agents are to be used mainly on caterpillars infesting cabbage and maize as well as for public health. So far, no botanical insecticide has been registered in Cameroon although several are used by farmers—including neem aqueous extracts,

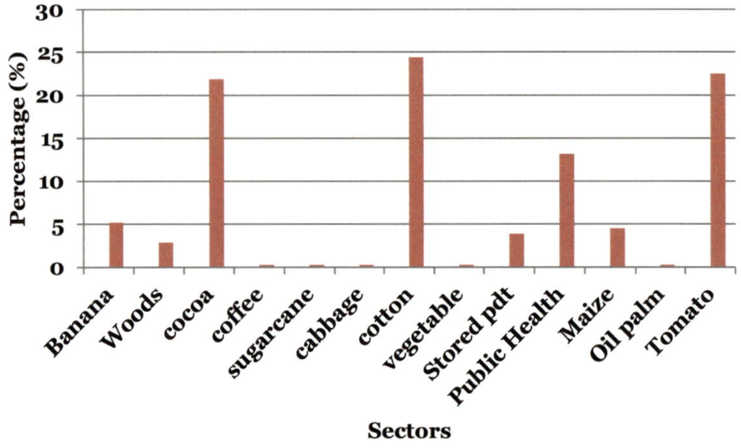

Figure 4.
Distribution of insecticides registered in Cameroon.

Sector	Target pest	Major groups of insecticide	Percentage of the group	Example of most registered active ingredient of the group
Banana/ plantain	Corm borer weevils Nematodes	Neonicotynoid	37.5 (6/16)	Imidaclopride Thiametoxam
Woods	Xylophage insects	Different types mainly combinations	—	Thiacloprid Boric acid Magnesuim phosphide
Cocoa	mirids	Neonicotynoid	80.9 (55/68)	Imidaclopride Thiametoxam Acetamiprid
Coffee	Coffee berry borers	Neoticotynoid	100 (1/1)	Imidacloprid Lamda cyahothrine
Sugarcane	Termites	Pheny pyrazole	100 (1/1)	Fipronil
Cabbage	Caterpillars	Microbial	100 (1/1)	*Bacillus thuriengensis* var. *kurstaki*
Cotton	Boll worm	Ivermectine Pyrethroid	60.5 (46/76)	Emamectine benzoate Cypermethrine
Vegetable crop	White flies Thrips		100 (1/1)	Abamectine
Stored crops	Store pests	Organophosphates in combination	83.3 (10/12)	Pyrimifos-methyl Permithrine Aluminum phosphide
Public health	Mosquitoes Midges	Pyrethroid	70.7 (29/41)	Transfluthrine Lamda cyahalothrine Alpha cypermethrine
Maize	Fall army worm Stem borers	Ivermectine	42.9 (6/14)	Emamectine benzoate
Oil palm	Leaf miner	Organophosphafe	100 (1/1)	Thioryclam hydrogenoxalate
Tomato	Fruit flies White flies	Pyrethroid	77.3 (58/75)	Cypermethrine Lamda cyahalothrine

Table 2.
Target pests and major group of insecticides and active ingredients registered for different sectors in Cameroon.

neem oil, wood ash, pepper extracts, Piper spp. powder, and water extracts of tobacco and hemp.

5. Insecticide application methods and associated malpractices

5.1 Application methods

The mode of application of insecticides is very important for optimizing their functions especially as less than 1% of these applied insecticides get to their target organisms [16, 17]. The over-reliance on chemical pesticides in combating pests throughout the world cannot be emphasized, yet it remains the most efficient, cheap and most accessible control mechanism in controlling pests [18]. Danquah et al. [19] reports cases of organochlorine insecticides residues in Cameroon, Ghana and India within vegetable cultivation, water resources and soil sediments which results largely due to application malpractices.

Sector	Commercial name	Active ingredients	Toxicological class*
Banana/plantain	Counter 15 FC	Terbufos 150 g/kg	Ia
	Moking 10 G	Ethoprofos 100 g/kg	Ib
	Vykingran 10 G	Oxamyl 50 g/kg	Ib
Woods	Magtoxin	Magnesium phosphide 660 g/kg	Ib
Cotton	Almec 036 EC	Abamectine 36 g/L	Ib
Stored products	Aladin	Aluminum phosphide 56%	Ia
	Magtoxin	Magnesium phosphide 660 g/kg	Ib
	Phostoxin	Aluminum phosphide 56%	Ia
Tomato	Abamet 18 EC	Abamectine 18 g/L	Ib
	Tetrakill 20 EC	Abamectine 20 g/L	Ib
	Furadent Super 10 G	Oxamyl 10%	Ib

**According to World Health Organization.*

Table 3.
List of highly hazardous insecticides registered in Cameroon.

5.1.1 Current insecticide application methods widely used in Cameroon

The desired results of any insecticides can be achieved based on the appropriate methods used and also respecting the time factor. Methods of application varies depending on the nature of the insecticide, their formulation, the soil characteristics, the pests and the availability of clean water [20]. The applications of insecticides is basically for seed treatment, soil treatment or foliar treatment. Insecticides come either in solid, liquid or powder forms, and it is these forms that dictate the methods and appliances used.

5.1.1.1 Sprayers, broomsticks and leaves

Numerous studies show that in Cameroon and Africa, more than 70% of farmers use sprayers which ranges from knapsack, to motorized knapsack sprayers and others in applying pesticides [19, 21, 22]. It is the most widely used method in the application of insecticides that comes in both powder and liquid forms requiring mixing with water before usage.

As observed with most small scale farmers, the mixing of the insecticides with water is carried out with plastic buckets ranging from 10 to 15 liters while others use their knapsacks, fetching water from nearby streams and using teaspoons, table-spoons or corks of bottles as their measuring units. These are all incorrect estimation measures that lead to either over-dosage or under-dosage of insecticides.

As reported by Christos et al. [22] some resource-poor farmers often improvise by using brooms, mesh and leaves on a wooden short stick in sprinkling pesticides mixed in buckets (**Figure** 5).

5.1.1.2 Dusting

This is the application of powder insecticides without diluting in water, usually using a duster [22]. In Cameroon very few farmers actually know and use dusters in the application of insecticides. From observation and sampling a handful of tomatoes farmers around Buea, just one in ten farmers know about the dusting techniques and also these few could not afford the duster. These farmers improvise by

Figure 5.
Using manual knapsack sprayer on sweet potatoes and fuel-powered sprayer on maize at Ekona, South West Region, Cameroon.

mixing these insecticides with water and spray. Some farmers apply these powder insecticides with hands upon wearing gloves, without any protective equipment especially for their nostrils and mouth. This practice leads to either over-dosing or under-dosing of plants often resulting in environmental contamination and pollution. Also, farmers have been reported using their bare hands to mix wood ash and terbufos and thereafter apply the mixture on leaves of maize and vegetables such as African nightshade (*Solanum* spp).

5.1.1.3 Granular application

These are insecticides designed in the forms of solid granules which are made to last longer with the gradual release of the active ingredients slowly upon contact with the soil or roots of plants. In Cameroon more than 70% of the farmers interviewed apply granular pesticides with their hands without any protective, with some using empty tomato tins in spreading them upon the soil.

Most often, these tomato tins serve as their measuring instrument, and depending on the plants, some apply a tin per plantain or banana but for smaller plants like tomatoes, they use a tin for 5–10 plants.

5.2 Malpractices linked to the application methods

5.2.1 Inappropriate application appliances

Besides using sprayers, sprinklers and dusters, using of broomsticks and bundles of leaves and brushes attached to sticks and dipping in a bucket of mixed insecticides is the usual practice in Cameroon and in most other African countries [19, 21]. This leads to ineffective application of insecticides resulting in wastage and environmental pollution.

5.2.2 Application without personal protective equipment (PPE)

In Cameroon, applying pesticides without all the required PPEs is a common practice by a majority of small-scale farmers especially as no monitoring mechanism is in place to enforce usage of PPEs. In a studies conducted in Buea by Christos et al. [21] about 76.4% tomatoes farmer use inappropriate or no PPEs during the application of insecticides, while Asongwe et al. [23] shows that 95% of farmers in Bamenda Municipality do not protect themselves during pesticide applications. Rugalema and Mnyone [18] reported scenarios of insecticides sprinkling at homes with the hands.

5.2.3 Incorrect measurement and different mixtures

Farmers in Cameroon use teaspoons, corks of beer bottles, or corks of containers, together with empty liters as their measuring units for insecticides. Using these listed items as their measuring units most often results in inappropriate measurement of pesticides that automatically results in environmental pollution and possible the contamination of farmers themselves.

Usually water is fetched in any nearby stream or river irrespective of the source or quality and more recently farmers have dug wells where they fetch water for watering their crops especially during the dry seasons with the same buckets used in mixing the insecticide.

Different concentration of different insecticides with varying active ingredients are usually mixed and used in managing insecticides, especially when the pests affecting the plants is unknown. According to Tarla et al. [24], during the rainy seasons farmers in Foumbot apply pesticides as often as twice a week and when the rains are at their peak, their fields are treated thrice a weak. These frequency of treatment overloads the environment and compromises on the quality of the plant with residues.

5.3 Factors affecting Insecticides malpractices

It is without doubts that pesticides have greatly increased crop productivity and simultaneously contaminated and polluted our soils and water resources, and also affecting our health. The technology used for their application remains hugely remote and archaic in most parts of Africa, particularly Cameroon, and it is the cost for the massive wastage and unnecessary environmental pollution. Some of the main factors affecting insecticides application malpractices include;

5.3.1 Knowledge of farmers

Many of the malpractices surrounding the safe usage and handling of pesticides is due to the fact that a majority of farmers including vendors have not had any formal training or technical support on the judicious use and safe handling of pesticides; Sonchieu and Blandine et al. [25, 26] pointed out that educational levels is the factor affecting knowledge, attitude and overall practices about pesticides usage.

5.3.2 Equipment cost and maintenance

Generally, spraying equipments in use are often in poor conditions due to lack of proper maintenance and the cost of buying original of these equipment or their parts is expensive for majority of these farmers. As reported by Abhilash and Singh [27] a large number of farmers never change sealing washers in their equipment and this is equally true for Cameroon, where most farmers share equipment without any maintenance knowledge on these equipment besides controlling the nozzles for faster application which does not necessarily lead to efficiency.

5.3.3 Repackaging or no labeling/or labeling in different languages

In Cameroon, using of pesticides without labels is a practice by vendors for selling banned products [21, 26]. In some cases, repackaging of pesticides in small quantities is to meet customer's demand due to the high cost in purchasing the entire container that is usually more than the amount required, thereby resulting to packaging in smaller containers without labels.

There are also products that are package in the Chinese and French, with little or no English language making it difficult for the farmers to follow the instructions and without proper guidance on usage there is bound to be malpractice.

5.3.4 Weather conditions

In order to maximize the best results possible from pest control mechanisms, accurate timing in pesticides application is paramount [27]. Unfortunately, Cameroon does not have reliable meteorological stations to provide farmers with information related to weather conditions and this situation has been compounded by the changing climates. This makes it difficult for small-scale farmers to accurately apply insecticides in a timely manner that will be most beneficial for plant growth. According to Balasha and Nsele [28], about 60% of farmers apply insecticides during the early mornings and late evenings from about 4–6 pm, with a handful applying irrespective of the time as it is the case observed in Wotutu area where farmers apply more pesticides after the rainfall had washed the previous pesticides about an hour after application.

5.3.5 Inappropriate concepts and economic pressure

In Cameroon, farming is looked upon as the last resort for many people especially cultivation of vegetables like tomatoes, cucumbers, and watermelons so as to make some fast profit. Yet in order to achieve optimum results these respective endeavors it requires mastery, so most Cameroonians believe more utilization of pesticides and chemicals automatically translates to high yields; thereby overreliance and abusive use of pesticides is practiced, as it is the case reported by Tarla et al. [24] that pesticides use does not always adhere to recommended safe practices [29].

Any desired results with regards to pesticides can only be guaranteed by properly respecting the guidelines of the producer and timing too is of the essence. Unfortunately application methods still depends on many variables such as the nature of the insecticide, its formulation, pests to be controlled, application site, water availability and training of the farmer, without of any these factors in place malpractice is most likely to occur.

6. Intoxications resulting from malpractices

Intoxications or poisonings resulting from insecticides are mainly the result of malpractices at the level of agrochemical companies, retailers, and farmers [24, 30]. Highest poisoning cases reported for Littoral and followed by South West Region. These are then followed by North West, West and Centre Regions while the least is from the Northern Regions (Adamawa, North and Extreme North). Highest cases from Littoral and South West is not surprising because very large cocoa and banana plantations are found in these Regions and these two crops are noted for their intensive use of pesticides especially insecticides and fungicides. In addition, there are also many vegetable farmers who cultivate to meet the demands of Douala—the economic capital found in the Littoral Region and near to the South West Region. To have high yields and high quality crops, these farmers use much insecticides and therefore possibilities of misuse and ultimately intoxications.

The crops most implicated for these poisonings are tomato, cotton, banana, cabbage, lettuce, onion, African nightshade, and cocoa while the most implicated insecticides are terbufos (Class Ia), ethoprofos (Class Ia), cypermethrine (Class II & III), chlorpyriphos-ethyl (Class II), combination of lambda cyhalothrine & imidacloprid (Class II). Although the signs/symptoms of intoxications vary with the different implicated active ingredients, the common signs and symptoms of acute intoxications reported are vomiting, abdominal pain, diarrhea, skin irritations, headaches, blurred visions, convulsions as well as rare cases of immediate death, most of which are accidental and occupational. The most common routes of exposure is the skin and respiratory tract. Most of the health personnel have no training or knowledge on recognizing and managing intoxications. From their responses, except for one case of suicide attempt, all cases of intoxications mentioned in the health centres or hospitals are accidental.

7. Alternatives to the use of synthetic insecticides

Apart from their cost, pesticides have a significant impact on human health, wildlife (terrestrial and aquatic) and the environment (soil and water pollution, etc.). Pesticides are often of poor quality or illegal, due to the low level of control of sales and distribution channels, and are generally misused because farmers are poorly trained in their use and have little knowledge of pest biology and ecology. Cameroon produces several agricultural commodities for export and domestic consumption hence this has increased the rate of application of inputs such as pesticide to help protect the crops from pests and diseases. Chemicals continue to be the main tools to protect crops during field and storage in spite of the deleterious effects as a result of malpractices.

Generally, the pest management practices reported by researchers and farmers include mainly use of synthetic pesticides and their alternatives such as cultural practices, botanical pesticides, use of biological control agents and the use of resistant/tolerant varieties. Of all the alternative pest management techniques, cultural practices and use of plants with insecticidal properties are the most common. Use of botanical insecticides is very common for cocoa, tomatoes, and stored products (e.g. maize, beans, cowpea). Sanitation, manual weeding, use of traps, and pruning are the most common cultural practices for field crops while use of containers with tight-fitting lids or bags is most common practice to protect stored products such as grains of maize, beans and cowpea.

In Cameroon, farmers depend on the use of synthetic insecticides to reduce crop losses due to insect pest, increase crop production, and urge to meet up with the high demand for agricultural products [25, 31]. However, over dependence and inappropriate applications of synthetic insecticides has led to harmful effects on the environment and human health [32]. These effects are numerous such as disruption of the normal functioning of the ecosystem, toxicity for a wide range of non-target organisms including humans and a high tendency to accumulate in the environment [33]. In addition to these, there is also the problem of pest ecological backlashies such as resurgence, replacement, and resistance. Most insecticides easily become ineffective due to the development of resistance by the insect pest. In order to minimize or completely eliminate the challenges posed by these chemicals, many researchers have carried out a number of researches in an attempt to find sustainable alternatives to replace synthetic insecticides [32]. Numerous efforts have been undertaken by IRAD, CARBAP, Universities and some International organizations like IITA and CIRAD to search and evaluate efficacies and effectiveness of alternatives to synthetic insecticides for the management of insect pests. Alternatives to synthetic insecticides include techniques such as:

- Crop rotation
- Host plant resistance
- Physical and mechanical control
- Botanical insecticides
- Biological control agents

7.1 Cultural control methods

Cultural techniques for pest management involves the manipulation of the environment or implementation of preventive practices with the aim to reduce pest population and their damages [32].

7.1.1 Intercropping

This is the practice of increasing crop diversity by growing more than one plant species in close proximity in a field to overcome insect pest outbreaks associated with monocultures [34]. Tanyi et al. [35] used this system of intercropping beans and maize together with use of extracts of Piper spp. to manage fall army worm in the Buea municipality. Most smallholder farmers have diverse cropping systems and therefore do not have much insect pest problem.

7.1.2 The use of Insect resistant plant varieties

Although the use of synthetic pesticides is usually effective, most pesticides easily become ineffective due to the development of resistance by the weevils or insects [32]. Cultivation of plant varieties which are resistance to insect attack minimizes the need for insecticide applications [32, 34]. For cocoa, resistant/tolerant varieties used are IMC60, Catongo Trinitario, Playa Alta2, SIC5, SNK614.

For tomato, use of improved or tolerant varieties such as Dona F1, Heinz 1370, Fline Mecline and Mobo line. For bananas/plantains, use of tolerant varieties such as CRBP 039, FHIA 21. For cotton, use of Bt-Cotton, use of tolerant/resistant varieties such as NGT115, SARC-1-57-2, KVX-165-14-1, LORI, IT97K-556-6-IITA. The use of resistant/tolerant varieties such as the local variety VYA-Cowpea.

7.2 Biological control methods

Biological control is an environmentally sound and effective means of reducing or mitigating pests and their effects without having to spray the plant with chemical insecticides but rather through the use of natural enemies. Biological control agents of insect pests are in three groups namely; predators, parasitoids and pathogens. There are three primary methods of using biological control agents in pest control programs which are the importation, augmentation and conservation methods.

7.2.1 Entomopathogens

Although not widely applied on agro-ecosystems in the country, some entomopathogens have been evaluated on research stations/sites. *Trichoderma* spp. has been used in the biological control of cocoa brown rot caused by *Phytophtora infestans* while endophytes have been evaluated for crop protection purposes [36–38]. Pathogenic organisms such as entomopathogenic fungi (*Beauveria bassiana* and *Metarhizium anisopliae*), entomopathogenic nematodes (*Steinernema* spp. and *Heterorhabditis* spp) and endophytes (nonpathogenic *Fusarium* spp) have been tested against *Cosmopolites sordidus* (banana borer weevil) [32]. After the application of these biological agents, results revealed that entomopathogenic fungi and nematodes are effective in controlling adult weevils while endophytes effectively control the immature stages of the pest mainly at the level of the laboratory/shade houses [32]. Emamactine benzoate and oxymatrine are used for stem borer and fall army worm (caterpillars) management on maize while sex pheromones are sometimes used to capture and kill adult moths. For cotton, emmamectine benzoate, spinosad are used to control the cotton bollworms (*Helicoverpa armigera*). *Metarrhizium anisoplae* used on flower bud thrips on cowpea.

7.2.2 Predators and parasitiods

In banana/plantain farms, some farmers encourage the trap-jaw ant (*Odontomachus* spp) as they are convinced these have excellent searchability and feed on eggs, larvae and pupae of *C. sordidus*. In the Northern regions, where majority of the cotton farms are found, they encourage and promote survival of natural enemies: parasitoids such as *Pediobius vigne*, *Trichogramma*, *Schixopyrannus* and predatory mites used for the management of caterpillars on cotton. The weaver ants (*Oecophylla longinoda*) are usually conserved by some cocoa farmers to help reduce the population of cocoa capsids (mirids).

7.3 Plants with insecticidal properties

Several plants with insecticidal properties (mortality, repellence, development disruptor, feeding and oviposition deterrence) have been identified and their efficacies tested in the laboratory and fields. A total of 29 plant species reported as plants having some pesticidal properties and use for one or more different pests and diseases. Based on journal publications on insecticidal plants in the country, there are 29 plant species belonging to 21 different plant families of which Fabaceae

(19.1%), Amaryllidaceace (19.1%), and Meliaceae (14.3%) were the most reported. Most common parts used are seeds and leaves mainly as extracts or dry powders. These insecticidal plants are usually applied in the form of ash, essential oils, powder formulations and aqueous or alcohol extracts.

7.3.1 Ash

Ash is one of the most common product used by smallholder farmers especially in the rural and peri-urban areas. The ash collected from household kitchens or burnt wood [39]. Some use only the dry wood ash while others mix fine dry soil with wood ash. Some others mix their ash with conventional insecticides such as Mocap (ethoprofos or terbufos), cypermethrine or kerosene. Some mix the ash with water or kerosene and used as sprays [39].

Most farmers in Buea Sub-Division, South West Region, Cameroon believe that applying a combination of terbuphos (Counter 10G®) and ash (from rubber plants wood and oil palm bunch residue ash) as seed treatments helps to protect suckers from the banana weevil (*Cosmopolites sordidus*) and stimulate plant growth better than their sole forms [40]. This mixture is mostly applied on soil around the corms or put in the planting holes before planting the plants. Sometimes ash of plant parts are used as insecticide such as the leaves of *Cupressus arizonica*, *Eucalyptus grandis*, *Ocimum gratissimum* and root ash of *Vetiveria zizanioides* against *Sitophilus zeamais* (Coleoptera: Curculionidae) [41]. The ash of *E. grandis* and *O. gratissimum* at the rate of 0.25 g/25 g maize grains significantly reduced the number of emerged weevils but if *E. grandis* ash is increase at 20 g/2 kg grains it significantly reduced grain weight loss and protected grains for 6 months without adversely affecting the germination of the seeds [41].

7.3.2 Essential oils

Essential oils from so many plant parts have been use as insecticide in Sub Saharan Africa especially in Cameroon. The essential oils are extracted from the seeds, leaves, bark and roots of the plant. Crude essential oils of *Lippia rugosa* and *Hyptis spicigera* are the most promising for *Tribolium castaneum* because of their efficacy on the other life stages. They are more efficient, with 100% mortality, on larvae at early stages and young adults [42]. Oben et al. [39] reported that essential oil from dried seed powder of *Piper guineense* has potentials for development as an organic insecticide against *Sitophilus oryzae* L. and other pests of stored grains. This is because *Piper* carene, copaene, α-caryophyllene or β- caryophyllene with insecticidal propertie [39].

The insecticidal properties of formulations based on *Ocimum gratissimum* essential oil and montmorillonite clay against the maize weevil (*Sitophilus zeamais*). The formulations based on essential oils adsorbed on modified clays is considered as alternatives to synthetic insecticides for use in storage and protection of maize grains from *Sitophilus zeamais* The insecticidal potentials of essential oils of *Chenopodium ambrosioides* and *Cupressus sempervirens* are very effective low persistence against *S. zeamais* on stored maize because persistence of both oils dropped to zero within 2 weeks [43]. In some localities in Cameroon, essential oils of pericarps of ripe fruits from *Citrus aurantifolia, C. limon*, *C. sinensis* and *C. reticulata* are used for the development of natural biocides [44]. *Azadirachta indica* (neem) seeds have been widely used in the northern part of Cameroon as essential oil. *A. indica* seed oils and powders from sun-dried kernels, shade-dried kernels, sun-dried seeds and shade-dried seeds seed oil was more active towards *Callosobruchus maculatus* on cowpea seeds and *Sitophilus zeamais* on maize grains [45].

Essential oils of *Lippia adoensis* leaves have already shown good insecticidal efficacy on *Sitophilus zeamais*. The essential oil of young leaves killed more weevils than that of old leaves, but old leaves were more repellent (II ≤ percent repellency ≤ V) than young leaves (0 ≤ percent repellency ≤ I). With its persistence levels of 98.47% after 96 h, the essential oil of old leaves more persistent than that of young leaves (94.66% after 96 h). The efficacy of essential oil from dry leaves of *Callistemon viminalis* (Myrtaceae) against *Acanthoscelides obtectus* (Say) (Coleoptera; Bruchidae) a major pest of *Phaseolus vulgaris* of stored beans in Cameroon was evaluated and the results revealed that at the end of the first day of exposure, the highest concentrations of essential oil applied to the beans (0.40 µl/g) and to the filter paper discs (0.251 $\mu l/cm^2$) caused mortality rates of 72.6% and 80%, respectively. These rates increased to 97.5% and 100% respectively after 4 days of exposure.

Essential oils from the leaves of *Ocimum canum* Sims (camphorated basil) and *Ocimum basilicum* L. (sweet basil) caused 100% mortality on *Anopheles funestus* adults at a concentration of 200 ppm for *O. canum* and 250 ppm for *O. basilicum*.

7.3.3 Powder formulation

In Cameroon some of the method of application is through powder formulation where the plant parts are dried then later grind and may be sieve or not depending on the user. These powders are sometimes mixed with other materials such as wood ash or other plants. The application is done by band placement or broadcasting depending on the farmer. For example in the western highlands of Cameroon plant materials consisting of leaves, seed and/or roots; *Cupressus arizonica*, *Eucalyptus grandis*, *Ocimum gratissimum*, *Vetiveria zizanioides, Balanites aegyptiaca, Lophira lanceolata, Hemizygia welwitschii, Plectranthus glandulosus, Laggera pterodonta* and *Azadirachta in*dica are considered as alternatives to synthetic insecticides for the protection of stored maize grains against *Sitophilus zeamais* [43, 45–47] Dried ground leaves/seed powders of some spices (*Syzygium aromaticum, Piper guineense, Aframomum citratum* and *Ocimum basilicum*), leave/seed powder of *Hemizygia welwitschii* and/or *Plectranthus glandulosus, Azadirachta indica* have an insecticidal potential against the cowpea weevil (*Callosobruchus maculatus*), common bean weevil and maize, thus small-scale farmers have been using them to protect their stored-product [45, 46, 48, 49]. In the Far North region of Cameroon, powders of *Hyptis spicigera* (Lamiaceae), *A. indica* (Meliaceae) and *Vepris heterophylla* (Rutaceae) are mostly used as insecticides single or associated with food during storage in combination with two or three others plants to control the red flour weevil (*Tribolium castaneum*) [50]. The flours from *Phaseolus sativum* and *Phaseolus vulgaris* seeds are also used to protect sorghum grains against the attack of *Sitophilus oryzae* [51].

7.3.4 Aqueous extact

Azadirachta indica (neem) fruit reduces insects' pest in the heading and maturation stages of rice plant [52]. The *aqueous leaves, seeds, roots (aqueous extracts) of Azadirachta indica, Boswellia dalzielii, M. anisopliae, Lippia rugosa, Annona senegalensis and Jatropha curcas* and their combinations are considered as potential natural insecticide in the management of thrips population on *Vigna unguiculata* in the fields. This would increase *Vigna unguiculata* yield and free environmental pollution from synthetic insecticides [53]. Aqueous extracts of *Cannabis sativa, Guibourtia tessannii, Erythrophleum ivorense, Thevetia peruviana, Azadiratchta indica, Ceiba pentandra, Pachyelasma tessmanii, Nicotinia tabacum,* wood ash, *Chnopodium ambrosoides, Lobelia columnaris, Carica papaya, Urtica dioica, Grotolania juncea,* Pepper, as well as smoke from burnt flowers of oil palm and cocoa husks are used in cocoa farms to manage capsids (mirids) and caterpillar infestations. *Cupressus benthanmii, Vetiveria zizaniodes, Piper guineense, Tithonia diversifolia, Mucuna cochinchinensis,* garlic, onion, ginger used on tomatoes to reduce population of fruit flies, leaf miners, white flies and caterpillars. *Afromomum melegueta, Piper guineense,* and *A. indica* extracts with greater effectiveness against *C. sordidus*, mealybugs and aphids infesting *Musa* spp. [32]. Extracts of three Cupressus species (*C. macrocarpa, C. sempervirens and C. arizonica*), as well as *Chenopodium ambrosioides* have been shown to have larvicidal and repellent properties against the *Anopheles gambiae* in Cameroon [54, 55].

7.3.5 Plant emulsions

The insecticidal properties of detergent-oil (mixture of liquid washing detergent and vegetable oil) emulsions of four plants (*Lantana camara, Allium sativum, Coffea arabica,* and *Jatropha curcas*) are used to control mealy bug infestation in banana and plantain field [40].

7.4 Integrated pest management

Integrated Pest Management (IPM) is a concept that is widely used or documented in diverse literature related to agriculture, environment and pesticides. This concept is becoming more and more popular as people become more and more conscious of the detrimental effects of pesticides and other chemicals used in agriculture, forestry and public health. According to Norris et al. [56], IPM is defined as a decision support system for the selection and use of pest control tactics singly or harmoniously coordinated into a management strategy, based on cost-benefit analyses that take into account the interests of and impacts on producers, society, and the environment. Considering this definition IPM can simply be defined as the use of all available pest management techniques or options that are compatible, cost-effective and environment-friendly.

From several discussions (especially with farmers and agriculture officers and researchers), literature and rapid observational assessments, it is clear that on the part of most farmers in the rural areas and a few in urban/peri-urban areas they are applying IPM unknowingly. These farmers have been working for long on the same pieces of land and/or on the same crops. They therefore have some wealth of information about their agro-ecosystems (the behavior of the crops, the different pests and their population dynamics, etc.). With this knowledge, most of these farmers have learnt to apply two or more techniques for managing the pest or pest complexes in their farms. They commonly apply combinations of cultural, physical and chemical options. Those who are flexible have (with the help of agriculture extension workers and researchers) added a genetic component (the use of improved and/or tolerant/resistant varieties) to their management strategy.

In addition, most farmers that are closer to research stations and the large agro-industrial plantations are having extra advantage by considering them as their role models. The farmers cultivate the same crops and put every effort to implement the same IPM strategy that these research stations and the plantations use in their farms. Apart from these hands-on experiences, the farmers (some of which are workers in these research stations and plantations) benefit from the few seminars/workshops concerning IPM for specific crops and good agricultural practices or best management practices. Such seminars/workshops are very common for crops such as cocoa, bananas, maize, cotton, and vegetables in peri-urban zones or in areas where pesticide abuse and misuse are common practices.

For the agriculture technicians/extensionists, research technicians, agric/crop protection students, and research assistants/researchers, from time to time they benefit from seminars/workshops/trainings on best management options and good agricultural practices for specific crops or pests. Most of these trainings/workshops are usually organized and sponsored by the Ministries of Agriculture and Health and/or international organizations/research centers such as IITA, CIRAD, IRAD, World Vegetable Centre, FAO, and GIZ. These organizations have played an important role in the implementation of Farmers' Field Schools (FFSs) in some rural areas of the country. For instance, GIZ in collaboration with the Delegations of Agriculture in the South West, Centre, South and Littoral Regions have organized and sponsored FFSs in the major cocoa producing areas. In these FFSs, farmers are taught and followed up in the farms on good agricultural practices and on using of pesticides only when absolutely necessary.

In addition, texts of certain decrees give some information on how biological control and IPM should be carried out. According to these texts, IPM is to be carried out with the following aim in mind; (i) reduce dependency on pesticides, (ii) effective control on the use of pesticides, (iii) reduce the risks resulting from abusive and inappropriate use of pesticides.

8. Recommendations to promote judicious use of insecticides

- The decision to select a particular insecticide should be based on an assessment of the risks and benefits and the potential hazard to public health and the environment.

- Legislation to control and regulate the manufacture, importation, distribution and sale of Insecticides should be strict. Only registered and recommended products should be used.

- Urgent need for government regulatory authorities to monitor the use of agrochemicals in the country, to strengthen controls for effective implementation of pesticide bans and to implement rigorous control of obsolete pesticide stocks in Cameroon.

- Raise awareness of the need for proper monitoring of insecticides in the country and develop strategies to reduce insecticide residues in food, and measures should be taken by regulatory authorities to manage the country's stock of obsolete pesticides and regulate the use of agrochemicals in the country.

- Training and intensive monitoring of farmers so as to effectively monitor the application of insecticides by the insecticides units.

- Promoting safe and proper use of insecticides by producers and the general public.

- The national pesticide registration committee should limit the registration of many commercial products with the same active ingredients and concentrations. In addition, it should also stop the registration of pesticides with Class I toxicity. Otherwise, sales and use of such insecticides should be done under strict measures.

- The government should intensify monitoring of insecticides activities at all levels as well as ensuring sanctions to defaulters of laws, decrees or decisions.
- Produce official and important pesticide documents in the English and French language as well as making these documents accessible to the public.
- Set up agro-health centres for pesticides and related chemicals. Such centres to be charged with training on pesticides (use, safety measures, pesticide application management, recognizing and managing intoxications).
- Introduce a course (Management of Pesticide Intoxications) in health/agriculture institutes/departments to teach students on pesticide basic classification, merits and demerits, use/misuse, safety measures/first aid, sign/symptoms of poisoning/intoxications, management of intoxications, international/national legislation).
- Promote research and use of biofertilizers and biopesticides especially those made from local bioresources.

9. Conclusions

Agriculture is an important sector in Cameroon with significant contribution to the GDP. Biotic constraints (especially insects and pathogens0 play a major role in reducing the productivity of farmers and therefore most tend to rely on use of agrochemicals such as synthetic insecticides and fungicides. Improper use of these chemicals is not only increasing cost of production but also greatly affecting the health of farmers and consumers. Although there are several conventions and laws in the country to promote best practices in the use of pesticides, the implementation is weak and this needs to be strengthened by continuous monitoring, sensitization and training of stakeholders including health practictioners.

Acknowledgements

Institute of Agricultural Research for Development for help in providing some useful journal articles.

Conflict of interest

No conflict of interest.

Author details

Nambangia Justin Okolle[1,2*], Ekwa Yawa Monono[3,4], Amungwa Ivan Tabikam[5], Mambo Stephania Kinge[3] and Magwell Pierre Fils Rodrique[6,7]

1 Institute of Agricultural Research for Development (IRAD), Kumba, Cameroon

2 Department of Agronomic and Applied Molecular Sciences, University of Buea, Cameroon

3 Institute of Agricultural Research for Development (IRAD), Ekona, Cameroon

4 Department of Plant Science, University of Buea, Cameroon

5 Institute of Agricultural Research for Development (IRAD-CERECOMA), Kribi, Cameroon

6 Laboratory of Biochemistry, University of Douala, Cameroon

7 Institute of Agricultural Research for Development (IRAD), Nko'olong, Cameroon

*Address all correspondence to: okollejustin@yahoo.com

References

[1] ZEF, FARA, IRAD. Innovation for Sustainable Agricultural Growth in Cameroon. 2017

[2] WFP. Cameroon: Comprehensive Food Security and Vulnerability Analysis (CFSVA). 2017. Available from: http://vam.wfp.org/default.aspx%0AFor

[3] Abia WA, Shum CE, Fomboh R. Agriculture in Cameroon: proposed strategies to sustain productivity. International Journal for Research in Agricultural Research. 2016;**2**(2):1-13

[4] Yengoh GT. Determinants of yield differences in small-scale food crop farming systems in Cameroon. Agriculture & Food Security. 2012; **1**(19):1-17

[5] Galani JHY, Houbraken M, Wumbei A, Djeugap FJ, Fotio D, Spanoghe P. Evaluation of 99 pesticide residues in major agricultural products from the western highlands zone of Cameroon using QuEChERS method extraction and LC-MS/MS and GC-ECD analyses. Food. 2018;**7**:184. DOI: 10.3390/foods7110184

[6] Raymond M, Michel NN, Martijn THG, Luc D, Salomon N, Mike R, et al. Pesticides use in cocoa sector in Cameroon: characterization of supply source, nature of actives ingredients, fashion and reasons for their utilization. International Journal of Biological and Chemical Sciences. 2014; **9**(3):1976-1989

[7] CREPD Report. Production durable et formulation de deux plantes aromatiques insecticides alternatives aux pesticides chimiques dangereux, project's report. 2018

[8] Food and Agriculture Organization of the United Nations. FAOSTAT statistical database. Rome: FAO; 2021

[9] Oberson N, Agwanda C, Bateman M, Dougoud J, Wood A. CABI Report. Étude sur la protection des cultures dans les pays où le programme 'Centres d'innovations vertes pour le secteur agro-alimentaire' est actif. Rapport national pour le ProCISA au Cameroun. 2018

[10] Pouokam GB, Album WL, Ndikontar AS, Sidatt MEH. A pilot study in Cameroon to understand safe uses of pesticides in agriculture, risk factors for farmers' exposure and management of accidental cases. Toxics. 2017;**5**:30. DOI: 10.3390/toxics5040030

[11] Tamgno BR, Tinkeu LSN. Diversity of stored grain insect pests in the Logone valley, from northern Cameroon to western Chad Republic in Central Africa. Journal of Agricultural Science and Technology A. 2013;**3**(9):724-731

[12] Tamgno BR, Tinkeu LSN. Application of the flour of four leguminous crops for the control of Tenebrionidae beetle (*Tribolium castaneum*—Herbst). African Journal of Food, Agriculture, Nutrition and Development. 2014;**14**(1):8474-8487

[13] Anon. Atelier de formation sur la gestion, la reglementation et l'evaluation des risqué de pesticides en Afrique Centrale. Cpac Info Pesticide, No.0010, Avril-Juin; 2010. pp. 4-8

[14] MINADER [Ministère de l'Agriculture et du Développement Rural]. Liste des pesticides homologués au Cameroun au18 Avril 2019. Liste réservée au grand public. Belgium. 2019. 212 p

[15] Tetang TJ, Foka G. Utilisation des pesticides dans la zone agricole du Moungo-évaluation de l'impact sur l'environnement, la santé des populations et solutions envisageables: cas de la localité de Njombé dans l'arrondissement de Njombé-Penja. In:

African Front for the Protection of Nature and Man. Pesticide Bull. 2008;**39** (2):12-16

[16] Pimental D. Amount of pesticide reaching target pests: Environmental impact and ethics. Journal of Agricultural and Environmental Ethics. 1995;**8**:17-29

[17] Lo C-C. Effect of pesticides on soil microbial community. Journal of Environmental Science and Health, Part B. 2010;**45**(5):348-359

[18] Rugalema GP, Mnyone LL. Malpractices in the use of agro-veterinary pesticides in plague endemic foci in Tanzania: Potential risk for development of insecticide resistance in flea vectors. Journal of Entomology and Zoology Studies. 2020;**8**(4):1665-1671

[19] Ao D, Ak E, Asuming-Brempong S. Insecticide use pattern on tomatoes produced at Yonso community in the Sekyere West District of Ashanti Region, Ghana. Ghana Journal of Agricultural Scienc. 2009;**42**:55-63

[20] Krishi Shiksha-homepage [Internet]. 2021. Available from: http://ecoursesonline.iasri.res.in [Accessed: August 16, 2021]

[21] Tandi TE, Wook CJ, Shendeh TT, Eko EA, Afoh CO. Small-scale tomato cultivators' perception on pesticides usage and practices in Buea Cameroon. Health. 2014;**6**:2945-2958. DOI: 10.4236/health.2014.621333

[22] Christos AD, Spyridon DK. Farmers' training on pesticide use is associated with elevated safety behavior. Toxics. 2017;**5**(3):19. DOI: 10.3390/toxics5030019

[23] Asongwe GA, Yerima BPK, Tening AS. Vegetable production and the livelihood of farmers in Bamenda Municipality, Cameroon. International Journal of Current Microbiology and Applied Sciences. 2014;**3**(12):682-700

[24] Tarla DN, Manu IN, Tamedjouong ZT, Kamga A, Fontem DA. Plight of pesticide applicators in Cameroon: case of tomato (*Lycopersicon esculentum* Mill.) farmers in Foumbot. Journal of Agriculture and Environmental Sciences. 2015;**4**(2): 87-98. ISSN: 2334-2404 (print), 2334-2412 (online)

[25] Sonchieu J, Nantia EA, Cheche TN, Ngassoum BM. Health risk among pesticide sellers in Bamenda (Cameroon) and peripheral areas. Environmental Science and Pollution Research. 2018;**25**:9454-9460

[26] Blandine TS, Jean S, Samelle EAE. Handling conditions and pesticide-related signs/symptoms among agrochemical sellers in Bafoussam municipality, West-Cameroon. Austin Environmental Sciences. 2020;**5**(2): 1048

[27] Abhilash PC, Singh N. Pesticide use and application: An Indian scenario. Journal of Hazardous Materials. 2009; **165**:1-12

[28] Balasha AM, Nsele MK. Pesticide use practices by Chinese cabbage growers in suburban environment of Lubumbashi (DR Congo): Main pests, costs and risks. Journal of Applied Agricultural Economics and Policy Analysis. 2019;**2**(1):56-64

[29] Amuoh CN. A case study of health risk estimate for pesticide-users of fruits and vegetable farmers in Cameroon [thesis]. Belgium: Ghent University; 2011

[30] Okolle JN, Afari-Sefa V, Bidogeza JC, Tata PI, Ngome FA. An evaluation of smallholder farmers' knowledge, perceptions, choices and gender perspectives in vegetable pests and diseases control practices in humid

tropics of Cameroon. International Journal of Pest Management. 2016. DOI: 10.1080/09670874.2016.1184772

[31] Kenko BDN, Fai PAB, Ngameni TN, Mbida M. Environmental and human health assessment in relation to pesticide use by local farmers and the Cameroon Development Corporation (CDC), Fako Division, South-West Cameroon. European Scientific Journal. 2017;**13**(21):454-473. DOI: 10.19044/esj.2017.v13n21p454

[32] Okolle JN, Ngosong C, Nanganoa LT, Dopgima LL. Alternatives to synthetic pesticides for the management of the banana borer weevil (*Cosmopolites sordidus*) (Coleoptera: Curculioniidae). CAB Reviews. 2020;**15**(026)

[33] Chowański S, Kudlewska M, Marciniak P, Rosiński G. Synthetic insecticides—is there an alternative? Polish Journal of Environmental Studies. 2014;**23**(2):291-302

[34] Endersby NM, Morgan WC. Alternatives to synthetic chemical insecticides for use in crucifer crops. Biological Agriculture and Horticulture. 1991;**8**:33-52

[35] Tanyi CB, Nkongho RN, Okolle JN, Tening AS, Ngosong C. Effect of intercropping beans with maize and botanical extract on fall armyworm (*Spodoptera frugiperda*) infestation. Hindawi International Journal of Agronomy. 2020;**2020**:4618190, 7 pages. DOI: 10.1155/2020/4618190

[36] Tondje PR, Roberts DP, Bon MC, Widmer T, Samuels GJ, Ismaiel A, et al. Isolation and identification of mycoparasitic isolates of *Trichoderma asperellum* with potential for suppression of black pod disease of cacao in Cameroon. Biological Control. 2007;**43**:202-212

[37] Mbarga JB, Ten Hoopen GM, Begoude AD, Tondje PR, Kuate J, Ambang Z, et al. Efficacy of *Trichoderma asperellum* oil formulations on the control of cocoa black pod disease (*Phytophthora megakarya*). Communications in Agricultural and Applied Biological Sciences. 2012;**77**(3):65-73

[38] Ambang Z, Bekolo N, Petga E, Ngoh Dooh JP, Asanga A. Effect of crude extracts of *Thevetia peruviana* seed on development of leaf spot disease of groundnut (*Arachis hypogaea* L.) caused by Cercospora sp. In: African Crops Science Conference Proceeding. Vol. 8. 2007. pp. 797-800

[39] Oben EO, Mcconchie R, Phan-thien K, Ntonifor NN. Essential oil composition of different fractions of *Piper guineense* Schumach. et Thonn from Cameroon using gas chromatography-mass spectrometry and their insecticidal effect on *Sitophilus oryzae* (L.). African Journal of Biotechnology. 2015;**14**(36):2662-2671. DOI: 10.5897/AJB2015.14785

[40] Okolle JN, Lenyu I, Mumah K, Christopher N, Nanganoa LT, Doungous O, et al. Role of combined ash and terbuphos as preplanting corm treatment to manage corm borer weevils (*Cosmopolites sordidus*) on plantains and stimulate growth. Journal of Experimental Agriculture International. 2018;**22**(3): 1-12. DOI: 10.9734/JEAI/2018/39564

[41] Akob CA, Ewete FK. The efficacy of ashes of four locally used plant materials against *Sitophilus zeamais* (Coleoptera: Curculionidae) in Cameroon. International Journal of Tropical Insect Science. 2007;**27**(1):21-26. DOI: 10.1017/S1742758407699615

[42] Kouninki H, Ngamo LS, Hance T, Ngassoum M. Potential use of essential oils from local Cameroonian plants for the control of red flour weevil *Tribolium castaneum* (Herbst.) (Coleoptera: Tenebrionidae). African Journal of Food Agriculture Nutrition and Development. 2007;**7**(5):1-15

[43] Langsi D, Nukenine E, Fokunang C, Suh C, Goudoungou W. Potentials of essential oils of *Chenopodium ambrosioides* L. and *Cupressus sempervirens* L. Motschulsky. Journal of Entomology and Zoology Studies. 2017;**5**(2):309-313

[44] Akono PN, Michel P, Dongmo J, Kouotou S, Kekeunou S, Magne GT, et al. Larvicidal activity of essential oils from pericarps of ripe citrus fruits cultivated in Cameroon on pyrethroids sensitive and resistant strains of *Anopheles gambiae* Giles, 1902. Journal of Entomology and Zoology Studies. 2015;**3**(4):334-339

[45] Tofel KH, Kosma P, Stähler M, Adler C, Nukenine EN. Industrial crops & products insecticidal products from *Azadirachta indica* and *Plectranthus glandulosus* growing in Cameroon for the protection of stored cowpea and maize against their major insect pests. Industrial Crops and Products. 2017;**110**:58-64

[46] Fotso TG, Nukenine EN, Tchameni R, Goudoungou W, Kosini D, Tigamba V, et al. Use of Cameroonian *Hemizygia welwitschii* Rolfe-Ashby (Lamiaceae) leaf powder against *Callosobruchus maculatus* and *Sitophilus zeamais*. Journal of Entomology and Zoology Studies. 2018;**6**(4):1261-1269

[47] Okolle JN, Oumarou PM, Almeck AD, Ntam F, Ngane B, Suh C, et al. Status of agricultural innovations, innovation platforms, and innovations investment. In: 2015 PARI Project Country Report. Accra, Ghana: Republic of Cameroon. Yaounde, Cameroon: Forum for Agricultural Research in Africa (FARA); 2016. p. 43

[48] Oben EOBI, NTONIFOR NN. Comparative insecticidal effects of powders from four plants. Journal Of International Academic Research For Multidisciplinary. 2015;**3**(6):408-418

[49] Kosini D, Nukenine EN. Bioactivity of novel botanical insecticide from *Gnidia kaussiana* (Thymeleaceae) against *Callosobruchus maculatus* (Coleoptera: Chrysomelidae) in stored *Vigna subterranea* (Fabaceae) grains. Journal of Insect Science. 2017;**17**(1):1-7

[50] Ayiki EN, Tamgno BR, Madou C, Vatsou J, Ngassoum MB, Simon L, et al. Toxicity of the powdery formulation of *Hyptis spicigera* (Lamiaceae), *Azadirachta indica* (Meliaceae) and *Vepris heterophylla* (Rutaceae) Single and in combination on *Tribolium castaneum* Herbst. (Coleoptera: Tenebrionidae). International Journal of Biological and Chemical Sciences. 2019;**13**(3):1613-1620

[51] Tinkeu LN, Tamgno BR, Gandebe M. Bioactivity of flours of seeds of leguminous crops *Pisum sativum*, *Phaseolus vulgaris* and *Glycine max* used as botanical insecticides against *Sitophilus oryzae* Linnaeus (Coleoptera: Curculionidae) on sorghum grains. International Journal of Biological and Chemical Sciences. 2016; **10**(3):919-927

[52] Sadou I, Ali M, Noé W, Eware S, Cyrille BW. Evaluation of the efficiency of aqueous extact of neem fruits on insect pest of rice in rice agroecosystem of Maga in the Far North Region of Cameroon. International Journal of Environment, Agriculture and Biotechnology. 2018;**3**(5):1891-1905

[53] Vandi FN, Raoul BB, Ngakou A, Nukenine EN. Potential of botanicals to control cowpea flower thrips (*Megalurothrips sjostedti*) and production in the Guinea-savannah and Sudano-Sahelian zones of Cameroon. Journal of Entomology and Zoology Studies. 2020; **8**(3):1164-1172

[54] Nguemkam EN. Effet larvicide des extraits de trois plantes de la famille des Cupressaceae sur des larves d'Anopheles gambiae, vecteur principal du paludisme au Cameroun [Mémoire de Master]. Faculté des Sciences,

département de Biochimie. Yaounde, Cameroon: Université de Yaoundé I; 2013. p. 70

[55] Bigoga J, Manga L, Titanji V, Coetzee M, Leke R. Malaria vectors and transmission dynamics in coastal south-western Cameroon. Malaria Journal. 2007;**6**:5

[56] Norris RF, Caswell-Chen EP, Kogan M. Concepts in Integrated Pest Management. Upper Saddle River, New Jersey: Pearson Education Inc; 2003. p. 575

Chapter 2

Use of Insecticides in Nepal, Its Impact and Alternatives of Insecticides for Nepalese Farmers

Sushil Nyaupane

Abstract

Nepal is an agrarian country whose population is primarily dependent on agriculture but the contribution to national Gross Domestic Product (GDP) is low as expected. There are many constraints to agricultural crop production and the farmers are facing those problems in their day-to-day lives. Deployment of insecticides and others to mitigate various insects and pests is one of them. Although abundant with locally available plant resources for pest management, farmers, especially in commercial pocket areas, are primarily dependent on conventional pesticides and those chemicals have detrimental effects on human health, including various flora, fauna, and environment. Although the Nepal government has formulated an act and worked on that basis, there is plenty of room to work on. Since farmer knowledge and behavior have a positive impact on reducing the use of conventional insecticides and work on alternative measures for pest management, these sorts of programs should be prioritized by the Government of Nepal and its allied agricultural organizations.

Keywords: agrarian, GDP, insecticides, health, act, farmer

1. Introduction

Nepal is an agrarian country and 60.4% of its population is dependent on agriculture and it contributes to 26.8% of national GDP [1, 2]. Commercialization of agriculture is needed to accelerate the economic growth in the country, which is largely subsistence type. Since Nepal has entered World Trade Organization (WTO) as a member country in 2004, it is necessary to exploit the globalized trade for the nation [3]. Most of the people who are engaged in agriculture are rural dwellings and they are the prime driver of the agriculture of the country, Nepal. However, the commercialization of agriculture demands high-value inputs, which are often associated with higher use of improved, and hybrid cultivars, machinery, fertilizers, pesticides, etc.

Pesticides are those chemical substances that are used to control pests of an agricultural and urban setting. These substances include fungicides, insecticides, rodenticides, herbicides, molluscicides, nematicides, miticides, avicides, etc. Insecticides are used for a very long time to deter, minimize, and manage insect pests in an agricultural field, forest land, and in human settlements. In agricultural crop production only, insects and other pests cause around 35% yield decline [4].

The role of insecticides to reduce the insect pests attack on various crops, damage to the health of humans and livestock is crucial. Due to the advantage of the rapid action of these chemicals over target organisms, these are widely being used all over the world. Nepal could not be an exception regarding the use of chemical insecticides. Insecticides encompass a broad range of chemicals that are toxic not only to insects but also to other organisms. These chemicals often lead to pesticide resistance, the resurgence of insect pests, and the decline of beneficial organisms, along with the detrimental impact on human health and the environment [5]. Unscientific use of pesticides is of major concern to the farmers of the developing countries and Nepal could not be the exception, which further exacerbates the situation.

Phytophagous insects only do the damage to grown crops, on average of 35–40%. Sometimes, it exceeds more than that based on the severity of the pest [6]. Commercial growers mainly depend on various insecticides to get rid of the various insect pests. But, the exact amount of import of these insecticides, their use, and the effect on human health and the environment is of major concern to Nepalese agriculture [7].

2. Material and methods

A rigorous and thorough study was done to collect and synthesize information on the topic of the review. Different research papers, review articles, reports, governmental websites, and their publications were studied and screened for data compilations. Gathered data were coded in the MS-Excel and subsequent tabulation and column graphs were generated.

3. Results and discussion

3.1 History of pesticide use in Nepal

The use of insecticides started in Nepal in early 1950s with intention of control of malaria, especially to eradicate the disease transmitted by mosquitoes for the Gandaki Hydropower Project [8]. First introduced chemicals to Nepal were Paris green, gramaxone, nicotine sulfates, Dichloro-diphenyl –trichloroethane (DDT), and these all were brought from the USA. These chemicals were followed by other organochlorines, organophosphates, carbamates, and synthetic pyrethroids [8, 9]. In the agricultural field, pesticides were started to use in the early sixties. This is the era of the green revolution where farmers were instructed to get maximum yield from a crop by using higher inputs such as improved seeds, chemical fertilizers, pesticides, etc. Until that period, farmers were unaware of the chemicals and insecticides to manage the various insect pests of agricultural crops. At the time, farmers have a preference over broad-spectrum pesticides due to the effective work to knock down the pests [10]. Nepal does not produce any insecticides till now but imported primarily from six countries, that is, India, China, Malaysia, Singapore, Italy, and Japan [3]. Till now, 54 types of insecticides were introduced to Nepal with 14 bio-pesticides, which are depicted in **Tables 1** and **2**. Organochlorines and some other highly toxic chemical pesticides were banned in Nepal, which are shown in **Table 3**. Insecticides were registered in 1787 commercial names by Plant Quarantine and Pesticide Management Center (PQPMC) under the Department of Agriculture, Nepal. In Nepal, there are altogether 16,110 retailers, 5 pesticide formulators, 37 pesticide applicators, and 286 pesticide importers [11]. Traders of pesticides are mainly concentrated in the commercial agricultural areas such as in plain regions, in the valley, and in and around the major cities of the country. Still, pesticide business has not penetrated the mid-hills, hills, and larger rural areas of the country.

S. No.	Insecticide chemical group (in use)	Common names
1	Organophosphate	Acephate, Azamethiphos, Chlorantraniliprole, Chlorpyrifos, Dimethoate, Ethion, Malathion, Phenthoate, Profenofos, Quinalphos, Temephos
2	Carbamates	Propoxur, Thiodicarb
3	Synthetic pyrethroids	Cypermethrin, Permthrin, Alphacypermethrin, Alphamethrin, Bifenthrin, Beta-cyfluthrin, Cyfluthrin, Etofenprox, Fenvalerate, Flumethrin, Lambda Cyhalothrin
4	Nicotinoid	Acetamiprid, Dinotefuran, Imidacloprid, Nitenpyram, Thiacloprid, Thiamethoxam
5	Avermectin	Abamectin, Emamectin benzoate
6	Methyl	Amitrazz
7	Organic thiophosphate	Azamethiphos
8	Nereistoxin analogue	Cartap hydrochloride
9	Halogenated pyrroles	Chlorfenapyr
10	Thioureas	Diafenthiuron
11	Benzoylurea	Diflubenzuron
12	Pyrazole	Fipronil
13	Pyridine compound	Flonicamid, Pymetrozin
14	Diamide	Flubendiamide
15	Isoxazoline	Fluralaner
16	Oxadiazine	Indoxacarb
17	Spinosyns	Spinosad
18	Tetronic acid	Spriomesifen
19	Tetramic acid	Spriotetreamat
20	Insect growth regulator	Novaluron, Lufenuron, Cyromazine, Chlorfluazuron, Buprofezin
21	Dazomet	—

Table 1.
Registered pesticides in Nepal till 14 July, 2020.

In average, consumption of pesticide inactive ingredient is very low, that is, 0.396 kg/ha compared to other countries such as India (0. 481 kg/ha), China (2.0–2.5 kg/ha), Japan (10.8 kg/ha), Europe (1.9 kg/ha) and USA (1.5 kg/ha) [12]. But, in highly commercial agricultural areas have much higher use of pesticides than the national average.

3.2 Trend of insecticide use in Nepal

Since insecticides are imported highly from foreign countries based on higher demand, farmers are using those chemicals in their fields injudiciously. Comparatively use of insecticides and other pesticides used in Nepal are lower than in developed countries, but the real problem is in the commercial pocket areas where growers are using exceedingly higher than they needed. There is a wider perception to the farmers that they have got the only chemical measures to control insect pests. Lack of awareness and knowledge of farmers, lack of alternatives of insect pests' management other than chemicals, lack of governmental regulation

S. No.	Common name	Origin
1	*Azadirachtin*	Neem based
2	*Bacillus amyloliquefaciens D 203*	Bacteria
3	*Bacillus subtilis*	Bacteria
4	*Bacillus thuringiensis*	Bacteria
5	*Pseudomonas fluroscens*	Bacteria
6	*Beauveria bassiana*	Fungus
7	*Metarhizium anisopliae*	Fungus
8	*Verticillium lecanii*	Fungus
9	*Trichoderma viridae*	Fungus
10	*Trichoderma harzianum*	Fungus
11	*Paecilomyces lilacinus*	Fungus
12	*Paecilomyces Spp* (Nematicide)	Fungus
13	*Heterohabditis indica*	Nematode
14	Nuclear polyhedrosis virus	Virus

Table 2.
List of bio-pesticides registered in Nepal.

S. No.	Banned pesticides	Decision year	S. No.	Banned pesticides	Decision year
1	DDT	2001	13	Monocrotophus	2006
2	BHC	2001	14	Methyl Parathion	2006
3	Aldrin	2001	15	Endosulphan	2012
4	Dieldrin	2001	16	Phorate	2015
5	Endrin	2001	17	Carbofuran	2019
6	Heptachlor	2001	18	Dichlorvos	2019
7	Chlordane	2001	19	Triazophos	2019
8	Mirex	2001	20	Carbaryl	2019
9	Phosphamidon	2001	21	Benomyl	2019
10	Organo Murcuric Fungicides	2001	22	Carbosulphan	2019
11	Lindane	2001	23	Dicofol	2019
12	Toxapheone	2001	24	Aluminium Phosphide 56%	2019

Table 3.
Banned pesticides in Nepal.

and monitoring policies and actions for pesticide use are some of the reasons for improper and excessive use of insecticides in Nepal [13]. Insecticide use is reported much higher in vegetables compared to cereal crops and others. Since the vegetable growers are commercial, they tend to use insecticides more often. One study reported that more than 85% of insecticides imported were used in vegetable crops to deter various insect pests and oftentimes farmers are using insecticides even the insects are not at a damaging level. It is reported that a higher concentration of insecticides residues, that is, Cypermethrin than the permissible limit was detected in tomato and brinjal. The same study also showed that the concentration of Deltamethrin was higher in cowpea and was followed by cauliflower, tomato, and brinjal [14]. The residues of carbamate and organophosphate group of insecticides were observed in the vegetables sampled from the leading vegetable market of Nepal located in the heart of the capital city, Kathmandu. Tomato and cowpea were having higher residues of insecticides and these were grown in the commercial pocket of vegetables of Nepal, that is, Sarlahi and Kavre districts. The same study has revealed that 21.38% of tomato samples and 18.75% of cowpea samples were of sub-standard quality among the samples which were tested positive in pesticide residue analysis using the reagent kit method were [15]. The trend of insecticide use is increasing in Nepal by 10–20% per year and this signifies the prevailing crisis of Nepalese agriculture not only in terms of economic losses but also of associated detrimental effects [16].

It is reported that 25% of farmers of plain regions, 9% of mid-hills, and 7% of mountains use pesticides in their fields, and their usage in these ecological zones of Nepal is depicted in **Figure 1** [17]. It is also reported that insecticides application is significantly higher in cotton and tea plantation in Nepal and it is worthwhile to mention that, compared to the cereal crops, use of insecticides and other pesticides is significantly higher in vegetables and other commercial/cash crops, as shown in **Figure 2** [11]. In Kavrepalanchok district, near to the capital city, farmers were using insecticides 1–3 times whereas the same farmers were using 2–15 times in vegetables such as cabbage, potato, tomato, bitter gourd, cucumber, etc. It is even comparable to the share of pesticides in the production of various crops. Wheat has no pesticide application whereas, pesticide application in bitter gourd accounts for an 8.41% share in crop production [18]. Farmers have reported the use of a cocktail spray of insecticides. Some farmers have also malpractice of dipping green vegetables in insecticide solutions such as malathion, mancozeb, etc. for a shiny and fresh look to fetch a good price in the market [12]. Farmers are very unaware and they hardly care for the waiting period to pick their harvest before they take it to the market. And, these products are purchased by the consumer and immediately taken for their food requirement and this makes the case more worsen [15].

Farmers of Nepal are very unaware of pesticide risk and it is the case of the area where people are engaged in conventional agriculture. In one survey conducted in Gaidahawa Rural Municipality of Rupandehi district, about 73% of the vegetable farmers have the practice of reusing the leftover pesticides. In the farmers' field, researchers have reported that farmers have left the pesticide containers and packets in the open field, without thinking about the risk those containers possess [19]. Among the various pesticides reported in the area, chlorpyrifos was with higher concentration, that is, 177 μg/kg from the soil samples collected from three different depths of soil, that is, 0–5 cm, 15–20 cm, and 35–40 cm. DDT although banned in Nepal from 2001, its residues were found at all depths of the soil, which shows its persistent nature in the environment [19, 20]. The DDT mean concentration at 35–40 cm soil depth from the above-mentioned research area was found higher than 10 μg/kg, which is more than the threshold value for the safety of various soil organisms. Other insecticides such as Profenofos and imidacloprid were also found in the soil samples abundantly at different soil samples and found to be toxic to different soil organisms [19].

3.3 Impact of insecticide use

Insecticides can be used in a variety of forms, including liquid, concentrated, powder, dust, particle, aerosol, and fog, to control various insect pests of various crops. Those chemicals sprayed in a crop's field will move and transfer to the

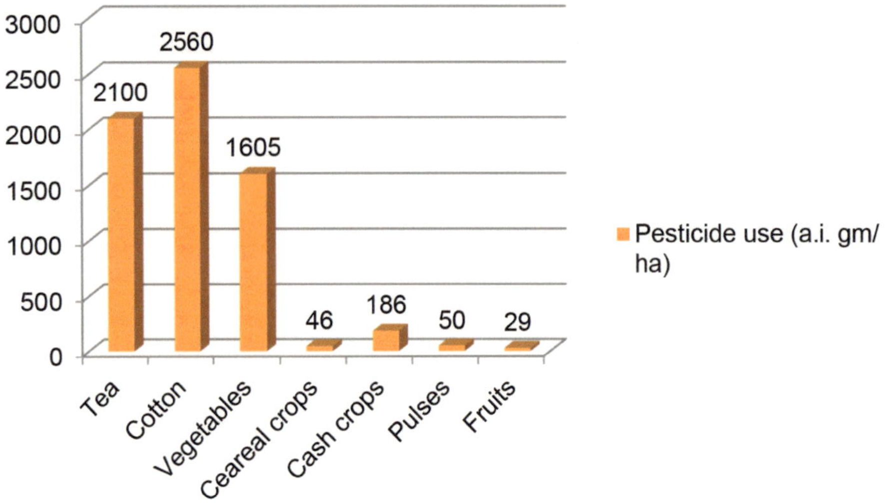

Figure 1.
Crop wise pesticide use (a. i. gm/ha) in Nepal (Source: PQPMC, 2021).

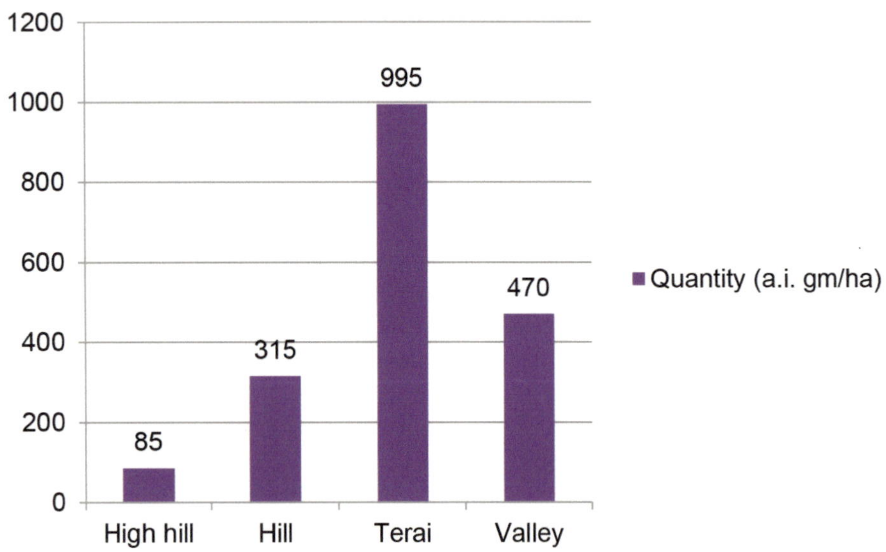

Figure 2.
Ecological scenario of pesticide use in Nepal (Source: PQPMC, 2021).

environment via water, wind, and absorption. It can be transferred to long distances and in various forms. A large part of the most commonly used insecticides do not reach their target insect and may be affecting non-target organisms or polluting the environment. Non-target organisms include not only other insects, but also vertebrates such as wildlife, humans, and domestic animals. Insecticides can enter non-target habitats or ecosystems and affect non-target organisms [20]. Since food is a basic need and the practices of insecticide use do have a greater impact on human health. The most contaminated insecticides group, that is, carbamate and organophosphates are neurotoxic and are acetylcholinesterase inhibitors. These insecticides belong to the toxicity categories I and II. These are categorized under the most dangerous insecticides to the non-target organisms including humans and the environment [21]. These chemical insecticides may have contaminate on the environment such as soil, water (surface and ground), various flora and fauna, etc.

Since the import of pesticides including insecticides is increasing every year. The import of pesticides in the year 2013/14 was 454 tons but now, in the year 2019/20, import has been increased to 681 tons as shown in **Figure 3** [11]. The residues of those chemicals on the soil and water are accumulating every year. One research has highlighted the moderate risk of cancer to the public where the soil is contaminated with organochlorine residues such as DDT and endosulfan [22]. This signifies not only the impending to the human health but also to the rich flora and fauna of the country itself. This sort of unsustainable practices in agriculture could be the cause of the loss of rich fauna which includes 17,097 species [23]. Various biotas inhabiting the soil such as bacteria, fungi, nematodes, earthworms, soil-inhabiting insects, and other arthropods with the presence of other organisms help to maintain the quality of soil and provide major ecosystem services for maintaining soil health and ultimately the quality of food production. The malpractices of insecticides along with other hazardous pesticides could have a detrimental effect on those organisms and ultimately deteriorate the quality and quantity of food production [19]. Another research conducted at Biratnagar of Nepal reported the presence of DDT and endosulfan in soil. The research also suggested that the use of DDT is still ongoing in the region but endosulfan residues were of past use [22].

These insecticides exposure to humans causes detrimental health defects such as hormonal imbalance, immune suppression, lower intelligence, reproductive anomaly, damage on kidney, liver, neural regions, and cancer. Farmworkers who have also exposure to insecticides get the symptoms of headache, drowsiness, dizziness, skin irritation, muscular twitching, respiratory discomfort, etc. [24, 25].

It is reported that the estimated health cost of the pesticide user individual who has got exposure to pesticides is Nepalese Rupee (NPR) 287. Of the total household expenditure, pesticide-induced health costs take 0.2% of annual household expenditure and 10.32% of annual health care expenditure [26].

More than the optimal concentration of insecticides also has unprecedented results human health and their expenditure on health care. One unit increase in insecticide concentration, that is, by 1 ml/L of water, would cause increased

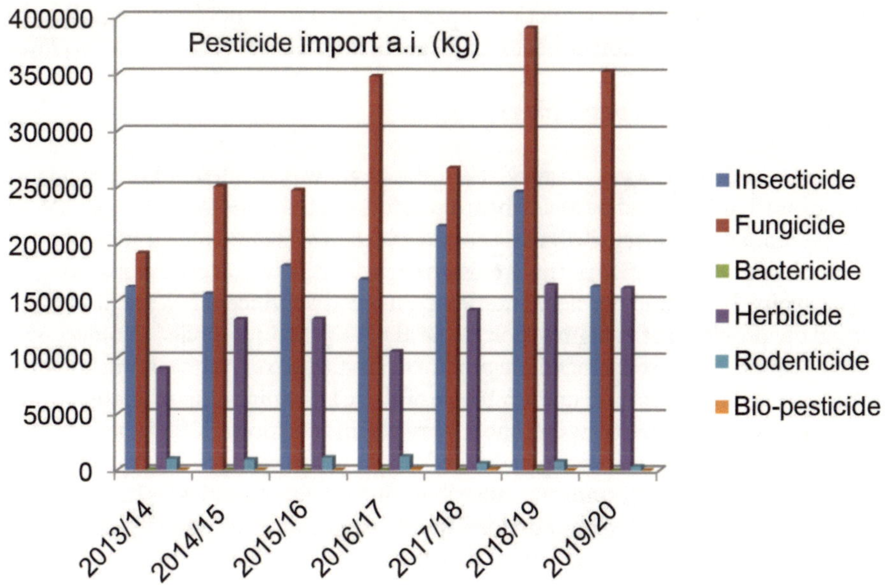

Figure 3.
Scenario of yearly pesticide import into Nepal (Source: PQPMC, 2021).

sickness cases by 6.8% and health costs by nearly NPR 30. Similarly, more hours of insecticide or any other pesticides application would bring unintended results to the health of the farmers and their expenditure [26].

It is also upsetting to mention the intentional or suicidal attempts of pesticide poisoning are common in Nepal. Most of the time, insecticides; mainly organophosphate are used by suicidal attempters. The most commonly used insecticides for self-pesticide poisoning were methyl parathion, dichlorvos, aluminum phosphide, and zinc phosphide [27].

3.4 Lesson learned from other countries

It is speculated that the insecticide reduction will cause a decline in the yield of the crops. But, it is not the case of the countries which are following a reduction in pesticide use because of their focus on the ecology of pests and agro-ecosystem. In that scenario, their production has been affected as speculated. Sweden has reduced pesticide use by 68% and public health poisonings by 77%. Their cutoff to the pesticides did not cause increased crop losses by the various pest species including insects. Indonesia also has reduced pesticide use by 65% and on the contrary, their production of rice has increased by 12%. India is also practicing the same and reducing the use significantly over the past years. But, Nepal is doing the opposite [25]. We are quite increasing the pesticide use for the sake of higher production, but, we are not aware of the fact that we are using unwarranted pesticides. Farmers, the ones who are not trained with the Integrated Pest Management (IPM) practices, are spraying the chemical pesticides more often than the ones who are trained. It is found that the trained farmers are spraying the pesticides 2.7 more times than the optimal whereas; the ones who are not trained are spraying 4.4 times of control [27]. This suggests the need of organizing community-based IPM training and environmental awareness programs about harmful effects of pesticides and sharing the know-how of insect pest management other than chemicals. It is also reported that Nepalese farmers are willing to pay higher prices (53–79%) than the current pesticide costs to mitigate the detrimental effect on their health and environment, and this clearly shows that they are willing to adopt alternative measures of pest management. But, the IPM programs of Nepal do have a contribution to the reduction of pesticide use but do not have a significant contribution to the reduction of health damages associated with the pesticides [25].

3.5 Regulation policy of pesticides in Nepal

For the first time in Nepal's history, the pesticide act was enacted in 1991, regulations were approved in 1993, and pesticide board was formulated in 1994 [18, 28]. Currently, Pesticides Management Act, 2019 was enacted which provisioned registration of bio-pesticides and also included the provision of facilitating warehouses for storing the date expire, band, and obsoleted pesticides in seven provinces of Nepal. It also included the provision of bringing back the pesticides which are spoiled, banned, or obsolete pesticides. It also included the provincial pesticide committee. Punishment was also provisioned in the act and upon defiance of these laws minimum of 25 thousand Nepalese Rupee (NPR) penalty, one-month prison, and maximum 200 thousand NRS penalty, and one-year prison was provisioned. Overall, the pesticide act regulates the manufacture, import, sale, transport, distribution, and use of pesticides in the country. This enabled the registration of pesticides, monitoring and inspection of pesticides, registration of importers and traders, and banning of highly toxic pesticides to minimize the exposure to humans, livestock, and other associated environmental components [29]. But, there is a great scope for proper inaction of law so that the widespread misuse of chemical pesticides in the country either by the importers, traders, and applicators could be minimized greatly. Since Nepal shares an open border with India, there are unintended pesticide imports to the country and many of them are more toxic, banned, and unregistered. Tracking the trade with India is oftentimes difficult since a porous border gives the opportunity to the persons who are involved in illegal trades.

Nepal is also a signatory country for WHO and follows the rules, regulations, and treaties proposed by them. Recently as directed by WHO, the country has banned 1a and 1b types of extremely hazardous pesticides. As a responsible member, Nepal has signed international treaties like the Basal convention, Stockholm convention, and Rotterdam convention, which have aimed to minimize the use of persistent and toxic pesticides [3].

3.6 Alternatives of conventional insecticides to Nepalese farmers

Since Nepali farmers do not have much more information and knowledge about the methods of pest management other than chemicals. But, the Nepal Government and Department of Agriculture have started to prioritize the IPM program. Integrated Pest Management (IPM) is a pest control strategy that aims to combine various techniques of pest management such as mechanical, physical, cultural, biological, and chemical to minimize the risks possessed by the pest in a given ecosystem [30]. IPM always considers the use of chemicals as a last resort and before using chemicals, it seeks out all the possible alternatives for insect pest management.

Since 1999, the Nepalese government has used the Farmer Field School approach to strengthen farmers for cultivating healthy crops with decisions based on an understanding of the field agroecosystem with having eyes on beneficial organisms such as predators and parasites of insect pests. A Farmer Field School, also known as a school without walls, is a school that teaches basic agroecology and crop management skills. A group of farmers gathers in one of their own fields to observe, discuss, record, and analyze real-world field problems from crop planting to harvest. This field school is based on the concept of "learning by doing" rather than "seeing is believing". The FFS was specially designed for farmers to learn and adopt IPM practices to their diverse and ever-changing ecological conditions [31]. Several crop season-long FFS have been organized in Nepal in recent years to provide knowledge and know-how on IPM to vegetable farmers in the hope of reducing their use of pesticides [32].

IPM farmer's field schools in the country have positive impacts on the farmers for using a lesser amount of pesticides. This was evident in the Bhaktapur district of the country, which is also well known for commercial vegetable production, and seasonal and off-seasonal vegetables are produced here. As reported, farmers were using a significantly higher amount of pesticides where mean active ingredient (a.i.) of fungicides and insecticides were 2373 and 1963 g respectively and on average use of pesticide use was 2011 g a.i./ha. Among the used pesticides to cruciferous vegetables, the share of insecticides was more, that is, 76% which was followed by fungicides (19%) and unknown were 5%. The participants of IPM farmer's field school had reduced significantly lower amounts of pesticides compared to non-participants. It was reported the 36% lesser amount of pesticides due to the effect of participation of IPM farmer's field school [32]. In another report, pesticide application by the farmers was decreased by 40% upon participation in farmer's field school [33]. This obviously shows the importance of these programs organized by governmental institutions.

Bio-pesticide could be a viable alternatives for Nepalese farmers since it will not be toxic to humans, other organisms, and the environment at large. There are altogether 14 registered bio-pesticides in Nepal which are effective to manage various insect pests and in some instances, other pests too of various crops. In Nepal, the use of bio-pesticides started commercially roughly after 2000. The share of bio-pesticides in the year 2019/20 is 0.005% of the total quantity of pesticides imported and used. This shows the predominantly higher use of conventional pesticides compared to commercial bio-pesticides. But, the use of locally available plant resources for pest control is a long practiced tradition of the farmers of Nepal. Many plants possess pesticide properties and these are all available all around the country. Three hundred and twenty four species of botanicals are found in Nepal only and among them, 23 species have special importance to the farming community of Nepal. The most common plants used as pesticides are as follows: Neem (*Azadirachta indica*), Garlic (*Allium sativum*), Bojho (*Acorus calamus*), Mint (*Mentha arvensis*), Turmeric (*Curcuma domestica*), Ginger (*Zingiber officinalis*), Marigold (*Tagetus patula*), Tobacco (*Nicotiana tabacum*), Drum-stick (*Moringa oleifera*), Basil (*Ocimum sanctum*), Onion (*Allium sepa*), Sugar apple (*Annonaa squamosa*), Sweet flag (*Acorus calamus*), *Artemesia vulgaris*, Winged prickly ash (*Zanthoxylum armatum*), China berry (*Melia azedarach*), *Urtica dioica*, Malabar nut (*Justicia adhatoda*), Marsh pepper (*Polygonum hydropiper*), *Euphorbia royaleana*, *Jatropha curcus*, Lantana (*Lantana camara*) and *Vitex nigundo* [34]. Botanical pesticides are easily made with these plant materials with pungency, bitterness, sourness, and repellent and antifeedant properties that make insects unhappy or cause death due to toxicity [3]. These botanicals are being used as a pesticide for a very long time. But, commercial production of these botanicals is missing in Nepal and it offers great scope for Nepalese entrepreneurs. Nepal possesses tremendous scope of developing these plants parts as botanical pesticides.

Although Nepal shares larger scope of isolation of different micro-organisms from the soil of Nepal, it offers only the formulation of two funguses, that is, *Metarhizium anisopliae* and *Trichoderma viridae,* only two isolated till now [3]. In the new pesticide act, the government of Nepal has made it easier to register the bio-pesticides compared to conventional pesticides, and obviously that would have positive impacts in the days to come. Nepal Government also have prioritized and started to give emphasis on organic agriculture since the 10th five-year plan [35], the scope of commercializing the bio-pesticides is certainly the need of the country.

4. Conclusion

Nepal, an agrarian country located in Southeast Asia is going to face unprecedented changes in human health, environment, and ecosystems due to more use of insecticides to deter insect pests in the farmer's field. Large amounts of insecticides are imported from foreign countries. These chemicals certainly have negative impacts on the farming community and the environment at large. The situation seems even worse because of a lack of knowledge and skills related to the safety aspects of the farming community about the use of insecticides and its negative effects not only to the consumers but on them too. Many researches have confirmed the presence of undesirable residues of insecticides in vegetables, fruits, and other agricultural commodities. Incidences of human diseases such as immune dysfunction, kidney failure, cancer, etc. are also increasing in the country which somehow has a direct or indirect relation to the more use of insecticides in the field. Because farmer knowledge and behavior can reduce the ecological risk of pesticides, programs such as IPM training and farmer's field school (FFS), etc. could be determined to change the status quo. Prioritizing the botanicals by the Nepal government and its respective agricultural agencies to the area where there is no practice of using conventional pesticides has special significance to protect the health of humans, various flora and fauna, and the environment.

Acknowledgements

The author wishes to appreciate the contribution of all the individuals and organizations who are constantly working on pesticides, their residues, effects, and mitigation in Nepal, and who has helped the author directly and indirectly in preparing this manuscript.

Author details

Sushil Nyaupane
Faculty of Agriculture, Far Western University (FWU), Tikapur, Kailali, Nepal

*Address all correspondence to: sunyaupane@gmail.com

References

[1] AICC. Agriculture Diary (in Nepali). Hariharbhawan, Lalitpur: Agriculture Information and Communication Center; 2021

[2] MOF. Information on Current Economic Condition of Nepal. Singhdurbar, Kathmandu: Ministry of Finance; 2021

[3] GC Y. Use of chemical pesticides in Nepal—The prospects for organic agriculture. Environmental Sciences and Ecology: Current Research. 2020;**1**:1002

[4] Palikhe RB. Challenges and options of pesticide use: In the context of Nepal. Landschaftsökologie und Umweltforschung. 2002;**38**:130-141

[5] Matsukawa M, Ito K, Kawakita K, Tanaka T. Current status of pesticide use among rice farmers in Cambodia. Applied Entomology and Zoology. 2016 Nov;**51**(4):571-579

[6] Pavela R. History, presence and perspective of using plant extracts as commercial botanical insecticides and farm products for protection against insects—A review. Plant Protection Science. 2016 Sep 26;**52**(4):229-241

[7] Rijal JP, Regmi R, Ghimire R, Puri KD, Gyawaly S, Poudel S. Farmers' knowledge on pesticide safety and pest management practices: A case study of vegetable growers in Chitwan, Nepal. Agriculture. 2018 Jan;**8**(1):16

[8] Dahal L. Study on Pesticide Pollution in Nepal. National Conservation Strategy Implementation Project, National Planning Commission, HMG Nepal, in collaboration with IUCN-The World Conservation Union. Kathmandu, Nepal; 1995

[9] Giri YP, Thapa RB, Shrestha SM, Pradhan SB, Maharjan R, Sporleder M, et al. Pesticide use pattern and awareness of pesticides users with special reference to potato growers in Nepal. International Journal of Development Research. 2014;**4**(11):2297-2302

[10] Neupane FP. Agricultural entomology in Nepal. Review of Agricultural Entomology. 1995; **83**(12):1291-1304

[11] PPQMC. List of Registered Pesticides and Consumption Data. Plant Quarantine and Pesticide Management Centre (PQPMC), Ministry of Agriculture and Livestock Development, Government of Nepal. Lalitpur, Nepal; 2021

[12] Sharma DR. Status of Chemical Pesticides Use and Their Regulation in Nepal. Hariharbhawan, Lalitpur: Plant Quarantine and Pesticides Management Centre; 2019

[13] GC Y, Palikhe BR. From the field to dining table: Pesticides residues. Journal of Agriculture and Environment. 2021;**22**(1):1

[14] Sharma DR. Use of pesticides and its residue on vegetable crops in Nepal. Journal of Agriculture and Environment. 2015;**16**:33-42

[15] Ghimire P. Status of Pesticide Residue in Vegetable and Fruit Samples Collected from Kalimati Wholesale Market of Kathmandu, Nepal. Singhdurbar, Kathmandu: Ministry of Agriculture and Livestock Development; 2020

[16] Diwakar J, Prasai T, Pant SR, Jayana BL. Study on major pesticides and fertilizers used in Nepal. The Scientific World Journal. 2008; **6**(6):76-80

[17] CBS. National Sample Census of Agriculture, Nepal, 2001/02: Highlights.

Kathmandu, Nepal: Central Bureau of Statistics; 2003

[18] Shrestha PL, Neupane FP. Socio-economic contexts on pesticide use in Nepal. Landschaftsökologie und Umweltforschung. 2002;**38**:205-223

[19] Bhandari G, Atreya K, Vašíčková J, Yang X, Geissen V. Ecological risk assessment of pesticide residues in soils from vegetable production areas: A case study in S-Nepal. Science of the Total Environment. 2021;**21**:147921

[20] Boul HL. DDT residues in the environment—A review with a New Zealand perspective. New Zealand Journal of Agricultural Research. 1995;**38**(2):257-277

[21] Gill HK, Garg H. Pesticide: Environmental impacts and management strategies. Pesticides - Toxic Aspects. 2014;**8**:187

[22] Van den Bosch H, Chaowen L, Pham Van Hoi TH, Van den Brink PJ, Yunliang P, Groenwold JG, et al. Environmental Risks of Pesticide Use in Intensive Vegetable Farming Systems in Peri-Urban Hanoi (Dong Anh) and Chengdu (Pengzhou). Vol. 1285. Wageningen, The Netherlands: Alterra; 2006. p. 166

[23] Yadav IC, Devi NL, Li J, Zhang G, Shakya PR. Occurrence, profile and spatial distribution of organochlorines pesticides in soil of Nepal: Implication for source apportionment and health risk assessment. Science of the Total Environment. 2016;**573**:1598-1606

[24] MoFE. Nepal's Sixth National Report to the Convention on Biological Diversity. Kathmandu: Ministry of Forests and Environment; 2018. p. 137

[25] Atreya K, Johnsen FH, Sitaula BK. Health and environmental costs of pesticide use in vegetable farming in Nepal. Environment, Development and Sustainability. 2012;**14**(4):477-493

[26] Atreya K, Kumar Sitaula B, Overgaard H, Man Bajracharya R, Sharma S. Knowledge, attitude and practices of pesticide use and acetylcholinesterase depression among farm workers in Nepal. International Journal of Environmental Health Research. 2012;**22**(5):401-415

[27] Jha RK, Regmi AP. Productivity of Pesticides in Vegetable Farming in Nepal. SANDEE, Kathmandu, Nepal; 2009

[28] Gupta SK, Joshi MP. Pesticide poisoning cases attending five major hospitals of Nepal. Journal of Nepal Medical Association. 2002;**41**:447-456

[29] NPPO. Pesticides Management Act. Government of Nepal. Kathmandu, Nepal; 2019

[30] Romeh AA. Integrated pest management for sustainable agriculture. In: Sustainability of Agricultural Environment in Egypt: Part II. Cham: Springer; 2018. pp. 215-234

[31] Kafle L, Yubak Dhoj GC, Yang JT, Bhattarai S, Tiwari S, Katuwal M. Integrated pest management in Nepal. In: The 5th International Conference on Clinical Plant Science. NPUST, Pingtung, Taiwan; 2014

[32] Jha RK. An assessment of farm-level use of biopesticides in Nepal: A case study based on IPM farmers' field schools of Bhaktapur District. In: Third Annual Meeting of Plant Protection Society of Nepal. Plant Protection Society, Kathmandu, Nepal; 2008

[33] GC Y. Status of plant protection activities in Nepal. In: Conference on Capacity Building in Use of the International Phytosanitary Portal and APPPC Website for Information Exchange. IPP & APPPC. Malaysia; 2011. pp. 4-9

[34] Budhathoki P, Gnawali P, Baral D, Gyawali A. Pesticidal potential of

ethnobotanically important plants in Nepal—A review. Acta Scientifica Malaysia. 2020;**4**(2):69-74

[35] NPC. Tenth five year plan, National Planning Commission. Kathmandu, Nepal; 2003

Chapter 3

Chemical Pesticides and Food Safety

Onyeka Kingsley Nwosu and Ayibapreye John

Abstract

Pesticides are usually applied to protect crops against insects and other pests. These pesticides of synthetic origin are potentially toxic to humans and can have both acute and chronic health effects, depending on the quantity and ways in which a person is exposed. They play significant roles in food production as they protect or increase yields due to less attack by insect pests. This is particularly important in countries that face food security challenges. The general population—who are not in the area where pesticides are used—is exposed to significantly lower levels of chemical insecticide residues through food and water. Chemical pesticides are among the leading causes of death by self-poisoning, in particular in low- and middle-income countries. Adverse effects from these synthetic pesticides occur only above a certain safe level of exposure. When people come into contact with large quantities of it in food, it may cause acute poisoning or long-term health effects, including cancer and adverse effects on reproduction. Production, distribution, and use of pesticides require strict regulation and control. Regular monitoring of residues in food and the environment is also required. Best among all is the promotion and adoption of bio-insecticides as a better alternative to chemical insecticides.

Keywords: pesticides, food safety, public health, bioaccumulation, poisoning, chronic condition

1. Introduction

Actions for a noticeable increase in crop yields and food production over the last century have involved the use of pesticides and agrochemicals [1]. These chemical pesticides are used to provide protection to crops against weeds, fungi, insects, and other pests. Consequently, these chemical pesticides are potentially lethal to human and can exert both acute and chronic health effects, depending on the amount and the route of exposure.

World Health Organization reported that there are more than 1000 pesticides used around the world to ensure food is not damaged or destroyed by pests and each of these pesticides has different properties and toxicological effects. The cheaper and older and most likely the off-patent chemical pesticides, such as lindane and dichlorodiphenyltrichloroethane (DDT) tend to remain for years in soil and water. Due to this, some of these chemicals have been banned by countries that signed the Stockholm Convention of 2001—an international treaty that aims to eliminate or restrict the production and use of persistent organic pollutants.

The Stockholm Convention on the production and use of persistent organic pollutants mandates that each Party shall Prohibit, restrict and/or take the (i) legal and administrative measures necessary to eliminate the production and use of + chemicals as listed in the treaty; and (ii) it imports and export of the persistent chemicals as listed in the treaty and (b) it is the production and use of the chemicals as listed in the treaty. It also emphasized that each Party shall take measures to ensure that any chemical listed in the treaty is imported only for

i. the purpose of environmentally sound disposal

ii. a use or purpose which is permitted for that Party under Annex A or Annex B of the treaty

Notwithstanding, the high increasing human population and the need for farmers to guarantee good value for farming has further expressed the need for enhanced agricultural yield towards achieving increased food production. This need is provoked by the intensive damage to agro-products caused by pest attacks and diseases triggered by viruses, fungi, and bacteria. These pest attacks and diseases are also seriously affecting crop yield. It is based on this provocation that the increasing use of chemical pesticides has to be the case. However, [2] report identified that agrochemical residues did spread in the environment and food causing significant contamination of terrestrial ecosystems and poisoning human foods.

Alternatives to the intensive use of crop protection chemicals achieved through a science-based process that promotes efficient food production, enhances food safety, and guarantees environmental protection, are thus the necessary direction in reducing or eliminating the increasing use of chemical pesticides in agriculture, thus ensuring food safety.

2. The use of chemical pesticides in food production

Chemical pesticides in agriculture usually referred to as agricultural chemicals cover a wide range of compounds including insecticides, fungicides, herbicides, rodenticides, nematicides, and others [3]. After the banning of some of these chemicals for use in agriculture by most technologically advanced countries in the 1960s, organophosphates insecticides, carbamates, pyrethroids, herbicides, and fungicides were introduced between 1970 and 1980 [3]. These chemicals are said to have contributed immensely to agricultural pest control and agricultural output.

The benefits of these chemical pesticides cannot be overemphasized as the consequences of their effects lead to the advantages anticipated from their use. In re-emphasizing the benefits of these chemical pesticides in food production, it is important to note that without crop protection, including pesticides, more than half of the world's crops would be lost to insects, diseases, and weeds.

It is of importance to highlight that in the absence of pesticides, food production would be on the decrease, and increased cultivated farm areas would be necessary to produce the same amount of food, consequently impacting the wildlife habitat. The recurrent cultivation of the farm would be increase soil loss due to erosion, too. The other effects will include the decrease in agricultural production, rise in food prices, competitiveness for farmers in global markets would be less, and decrease in exports would drop, leading to many job losses.

Regardless of the benefits of pesticides, they can be harmful or hazardous to both humans and the environment. Innumerable chemicals are environmentally stable, toxic, and disposed to bioaccumulation. In some cases, pesticides can persevere in the environment and remain there for years. Contamination of the environment or increased occupational use can expose the general population to pesticides residues, including physical and biological degradation products present in the air, water, and food.

3. Global application and management of chemical pesticides and its regulation

There is an ever-increasing global population and there has to be food to match the statistics. The United Nations Population Division estimates that there will be a tremendous increase on Earth emanating from developing countries by the year 2050 (9.7 billion people on Earth—around 30% more people than in 2017).

The Food and Agriculture Organization (FAO) estimated that, in countries of dwindling economy, Population growth keeps pace with the required increases in food production. This availability of food is seen to be increasing by 80%, and these increases are anticipated to emanate from rise in produces and the frequency crops are grown on the same land per year. This new production of food projected at 20% is likely to come from an extension of farming land [4]. Pesticide usage is almost inevitable in agriculture to maintain high yields and profits.

Pesticides can prevent large crop losses and will therefore continue to play a role in agriculture. However, the effects on humans and the environment of exposure to pesticides are a continuing concern.

Between the years 1970s and 1990s, most governments encouraged the usage of pesticides. This is evident in the amendment of several policies resulting in lessening in input subsidies as well as less monitoring by the government. This further led to more inflow occurring from the informal channels causing enhanced usage of pesticides and, leading to an increased import value by 261% from 2000 to 2010 [5].

In continents like Africa, nearly 59% of the population makes their living from farming, this is because the economy is highly dependent on agriculture [6]. Despite that, the African continent influences 2–4% of the international market for chemical pesticides which also accounts for the lowermost rate of their usage in the world [6]. Owing to the growing population, the food demand was projected in 2005 to enhance at a rapid rate in the next three decades [7]. This projection after the last decade remains valid considering the increased demand for pesticides, herbicides, and fungicides.

In Africa, the regulatory mechanism for pesticides is inadequate resulting in the import of pesticides that are banned. The farmers also lack awareness which causes poor pesticide practices and the usage of those pesticides which fall under the WHO risk classification system. Also, the registration of pesticides in West Africa is a multi-national process called Comité Sahélien des Pesticides (CSP) [8]. The African market is unregulated and does not comply with the code of conduct laid out by the Food and Agriculture Organization due to which most of the pesticides used are untested leading to the enhanced risks.

According to reports showing a limited capacity of CSP in Niger, 44% of pesticide dealers are unlicensed. Also, the registered chemicals account for only 8%, while 38% of pesticides have incomplete labels and 6% are unlabeled [9]. In the same report, 27% of the tested pesticides did not stipulate the active ingredients and 30% was tagged to be of poor quality. There are myriads of issues associated with pesticides usage in Africa.

In Southeast Asia, the use of pesticides in agriculture is increasing rapidly. An annual increase in the import of pesticides is reported as 61% for Cambodia, 55% for Laos, and 10% for Vietnam [10]. In the past 50 years, China has become the major pesticide manufacturing country, and these pesticides are chiefly used for rice production [11]. The use of pesticides in China has increased from 0.76 million tonnes in 1991 to 1.8 million tonnes in 2011 [11]. In terms of use, Japan is also one of the largest pesticide users in the world and has the biggest pesticide market in Asia [12].

In a survey involving seven European countries including Germany, the Netherlands, Sweden, the UK, Denmark, Finland, and Latvia on the usage of pesticides in the urban or non-agricultural amenities, it was observed that the infiltrate on of arable lands in Europe has occurred swiftly due to improved application of insecticides. This ultimately has resulted in the loss of biodiversity and heterogeneity of the arable lands and other landscapes [13, 14].

The European Union has developed definite imperative regulation regarding pesticides usage. These include:

- Directive 2009/128/EC approved by European Parliament and Council in the year 2009: This directive is employed to manage methods and procedures to sustainably apply chemical pesticides;

- A regulation for sustainability and maintenance of Products of Plant Protection on the Market. This in the year 2009 was proposed in the European Parliament and Council and termed Regulation (EC) No. 1107/2009, and

- A regulation for monitoring the MRLs of chemical pesticides in food and its products, including plant-derived feeds and animals. This regulation was proposed by European Parliament and Council in the year 2005, and was termed Regulation (EC) No. 396/2005

4. Effect of chemical pesticides on safety of food: research studies implicating effect on human health

As population size increases in the world, the industrialization of agriculture and the escalation of animal production to meet the growing demand for food creates both opportunities and challenges for food safety. These challenges put more responsibilities on food producers and handlers to guarantee food safety. Food safety is the prevention of food contamination before its being released to the consumer. Access to sufficient amounts of safe food is key to sustaining life and promoting good health. Unsafe food containing harmful contaminants including chemical substances causes diseases ranging from diarrhea to cancers. Chemical contaminations can lead to acute poisoning or long-term diseases, such as cancer. An estimated 600 million (almost 1 in 10 people) in the world fall ill after eating contaminated food inclusive of chemical contaminants and 420,000 die every year, resulting in the loss of 33 million healthy life years [15].

Of most health concern are industrial chemicals and environmental pollutants which can accumulate in the environment without the exception of food and further accumulates in the human body when ingested. Some of these chemicals are very toxic and can cause reproductive and developmental problems, damage to the immune system, interfere with hormones. They are also made up of heavy metals such as lead, cadmium, and mercury which can cause neurological and kidney damage [15]. The harmful effects on human health linked with pesticide usage are considered by numerous factors, such as the chemical class in which those compounds belong, dosage, time, and exposure route. Insecticides accumulated in food can be lethal to humans at high and/or even lower doses [16]. Several health effects can result from a prolonged exposure including the development of diseases such as cancer and neurodegenerative diseases, reproductive and developmental changes, and respiratory effects.

According to a study, an estimated 35% of all cases of cancer in the U.S. population originate from the diet, and the chemical pesticides present in foods are responsible [17]. Estrogenicity assays made by [18] show that pesticides of organochlorine origin usually act as endocrine disruption via more than one mode of mechanism, including agonist or antagonist effects of different receptors. Pre-emergent pesticides such as chloro-s-triazize which is popularly used in the world, have been generally considered as pesticides of low toxic potential for humans; nevertheless, there are many controversies on this issue. According to the Environmental Protection Agency (EPA), atrazine, for instance, was categorized as a chemical agent undoubtedly oncogenic to humans, even though the basis for this inference was only demonstrated in other animals [19]. This was also reported by the Development for Environmental Assessment Center of the United States, and Monographs of the International Agency for Research on Cancer (IARC). Since atrazine induces mammary tumors in female Sprague-Dawley rats, the EPA Office of Pesticide Program (OPP) through its Peer Review Committee resolved after its deliberations that atrazine be categorized in the Group of "Possibly Carcinogenic to Humans". Nevertheless, EPA has considered this chemical compound as most-likely non-carcinogenic to humans [2].

In certain studies, human exposure to high doses of atrazine can cause loss of body weight. Nevertheless, several epidemiological studies done with workers usually exposed to triazine indicate that these compounds show no potentials of been carcinogenic to the workers [20]. Furthermore, via analyses of different studies, it was observed that, though the chloro-s-triazine interferes in the endocrine responses of different species of mammals, their impending impact on humans seems to be primarily related to reproduction and development and not with human carcinogenesis [21]. An extensive list of epidemiological studies with the atrazine has described that the carcinogenic potential of this compound to humans is not conclusive [22], although there is a relationship between the high risk of prostate cancer and exposure to the insecticide [23].

The study by [24] evaluated the genotoxic and mutagenic effects of low concentrations of terbuthylazine, considered to be safe and, consequently accorded to possibly occur in occupational and residential exposures (ADI—Acceptable Daily Intake, REL—Residential Exposure Level, OEL—Occupational Exposure Level, and 1/100 and 1/16 LD50—Lethal Dose 50%—oral, rat), in human lymphocytes, with and without the use of metabolic activation (S9 fraction), using the FSH cytome assay and pan-centromeric DNA probes to evaluate the content of micronuclei and other chromatinic instabilities. The study showed that, treating terbuthylazine in the absence of metabolic activation indicated a dose-response escalation in the frequency of micronuclei of the lymphocytes exposed. The concentration of 0.0008 μg/mL (REL) tested was the basis of the significant data obtained. The hybridization of the micronuclei with the centromeric probe (C+) significantly occurred due to the concentrations ADI (0.00058 μg/mL), REL (0.0008 μg/mL) and OEL (0.008 μg/mL) of terbuthylazine. This was regardless of the presence or absence of S9, and nuclear buds containing centromeric signals, only in the presence of S9. Considering these outcomes, it was proposed that terbuthylazine presents a predominant aneugenic potential for the genetic material of human lymphocytes.

The chloro-s-triazine insecticide, which constrains the photosynthesis of weeds, by reaching photosystem II and impedes the effect of certain pests on crops has also being a serious food safety concern. It is a chemical used for a variety of crops, such as maize, sugarcane, olive, and pineapple. Since the banishment of atrazine in European countries in 2006, chloro-s-triazines like terbuthylazine were recommended as its substitute since it is suspect of causing diseases in humans, such as non-Hodgkin lymphoma and lung cancer. A study showing the effects of persistent exposure (14 days) to low concentrations of terbuthylazine (0.58 ng/ml and 8 ng/ml) in human lymphocytes, using the comet assay and the comet-FISH assay (with the c-Myc and TP 53 genes) was carried out [24]. Treatment with the compound induced the migration of fragments of DNA in a significant manner, only for the highest concentration treated. The results indicated an impairment in the structural integrity of c-Myc and TP 53, as a result of the prolonged exposure of human lymphocytes to terbuthylazine. For the fact many copies of TP53 were affected by the compound, it indicates the ability of terbuthylazine to interfere in the control of the cell cycle negatively. Nevertheless, it was concluded that a more comprehensive evaluation of the risk of cancer associated with the exposure to terbuthylazine, be evaluated for the impact of these insecticides on other housekeeping genes and markers.

Concerning insecticides, a study by [25] assessed the genotoxic potentials using the FISH and comet assay, and the oxidative damages, by the TBARS lipid peroxidation, of different concentrations of glyphosate in human lymphocytes. These concentrations of glyphosate are similar to those observed in residential and occupational exposures and related to LC50. At concentration of 580 μg/mL, results from the comet assay indicated a stimulation with significant increase in the tail length, while at concentration of 92.8 μg/mL an increase in the tail intensity was noticed, both concerning the control test. However, the addition of the S9 fraction increased the tail length significantly, for all the concentrations tested. In furthering the experiment, an increase in the frequency of micronuclei, nuclear buds, and nucleoplasmic bridges were identified when the lymphocytes were exposed to the three highest concentrations without S9. It was the consequence of the addition of a metabolic activation system that only promoted a significant increase of the nuclear instabilities for the highest concentration tested. It was clearly shown that the values of TBARS significantly increased with the increase of the concentrations tested, regardless of the presence or absence of the S9 fraction. Because dose-dependent effects for all the assays used were not observed, the authors concluded that these concentrations of glyphosate are not relevant for human exposure, since they did not present a significant risk for human health.

According to a study by [26], paraquat, the second most widely used insecticide in the world, selectively accrued in human lungs by causing oxidative injury and fibrosis, causing several individuals to mortality. Chronic exposure to this insecticide is also linked with kidney failure, Parkinson's disease and hepatic lesions [27]. In the study by [26], they assessed the paraquat toxicity on BEAS-2B normal cells (human bronchial epithelial cells), which showed its dose-dependency resulting to death of lung cells exposed, damage of the mitochondria, oxidative stress, as well as production of pro-fibrogenic growth factors, cytokines, and transformation of myofibroblasts. In the study, the authors also demonstrated that polyphenolic phytoalexin naturally produced by several plants, resveratrol, to control bacteria and fungi, inhibited the production of reactive oxygen species, fibrotic reactions, and inflammations when induced by paraquat. This is as a result of the activation of the Nrf2 signaling (Nuclear Factor Erythroid-2), revealing a novel molecular mechanism for the intervention against oxidative damages as well as pulmonary fibrosis which resulted from the action of toxic chemical compound.

The study on the influence of a complex mixture of pesticides in workers exposed to them occupationally was carried out using the comet assay technique and standardly established cytogenetic methods (chromosome aberrations and micronucleus assay). This study indicated that DNA migration significantly increased (P < 0.001). This suggests that over exposure to or ingestion of the pesticide may affect damages in the genome of somatic cells and, therefore, would pose a potential risk to human health [26].

5. Exposure to foods contaminated with chemical pesticides: overview of the public health concerns

Pesticide use has been closely associated with human poisonings and their related illnesses and has been long seen as a severe public health problem. The potential toxicity in pesticides has caused both acute and chronic health effects, depending on the quantity and ways in which the person is exposed (**Figure 1**).

The tenacious and pervasive nature of several pesticides used in agriculture and other carbon-based pollutants has posed chaos to mankind as a result of their high toxicity and potentials to bio-accumulate [28]. These chemical pesticides are identified to impede the usual effectiveness of reproductive and endocrine systems in living organisms [29]. Several pesticides such as dichlorodiphenyltrichloroethane (DDT), chlordane, aldrin, dieldrin, endrin, mirex, heptachlor, and hexachlorobenzene influence lethal effects on the health of human and the environment [28]. In the year 1990, a task force of the World Health Organization (WHO) estimated that about one million unintentional pesticide poisonings occur naturally, leading to approximately 20,000 deaths. There are also an estimated 385 million cases of unintentional acute pesticide poisoning UAPP occur manually worldwide including 11,000 fatalities. This estimation depends on the quality and validity of data as well as the estimation procedure [30].

In most regions of the world, the condition is even worse. Approximately 80% of the pesticides produced per annum in the world are used in developed countries [12], but less than half of all pesticide-induced deaths occur in these countries [31]. Increased proportion of pesticide poisonings and mortality occur in developing countries where there are insufficient occupational safety standards and regulations in its use on foods; insufficient enforcement; poor labelling of pesticides; illiteracy; and deficient knowledge of pesticide [31]. Moreover, usual pesticide residue levels in food are often higher in developing countries than in the developed countries. For example, a study in Egypt reported that most of assayed milk samples, when tested for fifteen

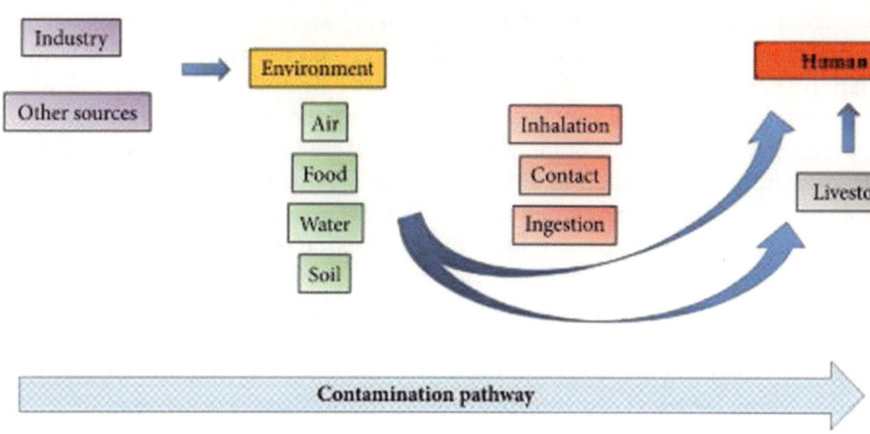

Figure 1.
Sources of chemical pesticide contamination in human foods [28].

different pesticides, contained residue levels between 60% and 80% [32]. By way of contrast, 50% of the milk samples analyzed in a US milk study had pesticide residues, all in trace quantities well below EPA and FDA regulatory limits [33].

Detectable levels of chemical pesticide residues are seen in about 35% of the foods purchased by consumers possess [34, 35]. Between 1 and 3% of these foods possess chemical pesticide residue levels that are beyond the permissible tolerance level [34]. Considering the analytical methods used in the developing countries, the residue levels may even be higher due to the reasons that they may detect just about one-third of the chemical pesticides in use. The rate of contamination is undoubtedly higher for fruits and vegetables because these foods receive the highest dosage of pesticides. One USDA study has shown that some pesticide residue remains in fruits and vegetables even after they have been washed, peeled, or cored [36]. Consequently, there are many justifiable reasons why 97% of the public is concerned about pesticide residues in its food [31].

All over the world, apart from exposure via contaminated food, pesticide exposure at the highest levels are found in farm workers, pesticide applicators, and people who live adjacent or very close to heavily treated agricultural land. Due to the fact that farmers and farm workers directly handle 70–80% of the pesticides they use, they are at the greatest risk of exposure [31, 37]. The epidemiological evidence suggests a significantly higher rate of cancer incidence among farmers and farmworkers in the US and Europe than among non-farm workers in some areas [34, 38]. In these high-risk populations, there is strong evidence for associations between lymphomas and soft-tissue sarcomas as well as between lung cancer and exposure to organochlorine insecticides [27].

Consequently, both the acute and chronic health effects of pesticides warrant attention and concern especially as it was used in farm food production and its storage. While the acute toxicity of most pesticides is well documented [39], there is no sound information on chronic human illnesses such as cancer. Though based on animal studies, the International Agency for Research on Cancer found "sufficient" evidence of carcinogenicity in eighteen pesticides and "limited" evidence in an additional sixteen pesticides [28]. However, some studies found no significant difference in non-Hodgkin's lymphoma mortality between farmers and non-farmers. In addition, [31] estimates that fewer than 1% of the human cancer cases in the US are attributable to pesticide exposure via food or otherwise. With the increasing number of cancer cases annually, [31, 40] assessment indicates that chemical pesticides causes less than 12,000 cases of cancer per year.

Studies with proven confirmation have also suggested that many severe and chronic conditions are linked with the use chemical pesticide [41]. For example, in an animal studies, dibromochloropropane (DBCP), the proscribed pesticide used for plant pathogen control, was found to cause testicular dysfunction [42] and was linked to infertility in human workers who had been exposed to the chemical [43]. Also, a large body of evidence obtained from animal studies suggests that pesticides can produce immune dysfunction [44]. In a study of women who had chronically ingested groundwater contaminated with low levels (mean of 16.6 ppb) of aldicarb, [44] reported evidence of significantly reduced immune response, although these women did not exhibit any overt health problems.

There is also growing evidence of sterility in humans and various other animals, particularly in males, due to various chemicals and pesticides they ingest through contaminated food and in the environment [45]. Sperm counts in Europe have reduced by about 50% and continue to decrease an additional 2% per year. In the study of [46], young male river otters in the lower Columbia River and male alligators in Florida's Lake Apopka have smaller reproductive organs than males in unpolluted regions of their respective habitats.

Even though it is habitually challenging to evaluate the influence of individual chemical pesticides, the serious health issues associated with organophosphorus related pesticides which have basically substituted the proscribed organochlorines are of specific interest [39]. The malady Organophosphate Induced Delayed Polyneuropathy (OPIDP) is well studied, reported, documented and is manifested by irreversible neurological defects. The deterioration of memory, moods, and the capacity for abstract thought has been observed in some cases [47], while other cases indicate that persistent neurotoxic effects may result even after the termination of an acute organophosphorus poisoning incident [39].

Chronic conditions such as OPIDP constitute an important public health issue because of their potential cost to society. For example, the effect of pesticides on children has become a growing concern [48]. Children can be exposed to pesticides daily through the foods they eat [31]. Considering the increased understanding of the distinctive biological differences between children and adults, it has shown noticeably that the current chemical pesticide acceptability level and the system of regulation, as it concerns children, is sternly lacking. Majority of the regulations are based on adult acceptability level and tolerances. Biologically, it is known that children's metabolic rates are higher than adults, and their capability to stimulate, detoxify, and excrete compounds that are xenobiotic in nature is dissimilar from that of adults. Also, considering of their slighter bodily size, children are exposed to increased levels of chemical pesticides per unit of body weight. Indication of this is seen in a study of [49] which reported that 50% of England and Wales pesticide poisonings involved children of or under the age of ten [49]. In general, the realization that children's sensitivities to toxins are much different than those of adults has provided the impetus for the movement towards setting specific pesticide regulations especially the level of residues in food with children in mind [36].

6. Conclusion

Chemical pesticides are often applied to control and manage weeds, and insect pests in the agricultural practices. Water, soil and air serve as dynamic medium for the movement of chemical pesticides from a point to another. Among several types of chemical pesticides, organochlorine and its related pesticides are the utmost risky ones as a result to their slow rate of decomposition, long half-life and greater stability. In the upper trophic levels of the food chain, these pesticides can move and accumulate. In any ecosystem, contamination by chemical pesticide is stern problem due to the harm it causes to all associated organisms. Hence, to control pesticide usage, novel methodologies and techniques are needed in curtailing the effect of widespread use of pesticides on the ecosystem including food production (farm to storage) and efforts should be made to provide awareness among the public to minimize the application of harmful pesticides. The better alternative remains in the use of microbial and plant-based bio-pesticides in control of field and storage pests as part of the integrated pest management (IPM). Also, the adoption of plant-incorporated protectants (PIPs) as seen in plants genetically modified to resist pests should be encouraged over chemical pesticides.

Author details

Onyeka Kingsley Nwosu[1,2*] and Ayibapreye John[2]

1 Department of Biochemistry, School of Life Sciences, Federal University of Technology, Minna, Niger State, Nigeria

2 National Biosafety Management Agency, Abuja, Nigeria

*Address all correspondence to: nwosuonyeka6@gmail.com

References

[1] Alexandratos N, Bruinsma J. World Agriculture Towards 2030/2050. Global Perspective Studies Team, FAO Agricultural Development Economics Division. ESA Working Paper No. 12-03. 2012

[2] European Environmental Agency (EEA). Late Lessons from Early Warnings: Science, Precaution, Innovation. European Environmental Agency, Report No 1/2013. Copenhagen: EEA; 2013

[3] Aktar W, Sengupta D, Chowdhury A. Impact of pesticides use in agriculture: Their benefits and hazards. Interdisciplinary Toxicology. 2009;**2**(1): 1-12

[4] FAO. The Future of Food and Agriculture: Trends and Challenges. Rome: FAO Publications; 2017

[5] De Bon H, Huat J, Parrot L, Sinzogan A, Martin T, Malezieux E, et al. Pesticide risks from fruit and vegetable pest management by small farmers in Sub-Saharan Africa. A review. Agronomy and Sustainable Development. 2014;**34**(4):723-736

[6] Abate T, van Huis A, Ampofo JK. Pest management strategies in traditional agriculture: An African perspective. Annual Review on Entomology. 2000;**45**:631-659. DOI: 10.1146/annurev.ento.45.1.631

[7] Snyder J, Smart J, Goeb J, Tschirley D. Pesticide use in Sub-Saharan Africa: Estimates, projections, and implications in the context of food system transformation. In: IIAM Research Report No. 8E. 2015. DOI: 10.22004/ag.econ.230980

[8] Jepson PC, Guzy M, Blaustein K, Sow M, Sarr M, Mineau P, et al. Measuring pesticide ecological and health risks in West African agriculture to establish an enabling environment for sustainable intensification. Philosophical Transactions of the Royal Society, London B: Biological Science. 2014;**369**(1639):20130491. DOI: 10.1098/rstb.2013.0491

[9] Van der Valk H. Strategic Program for Pesticide Management in the Sahel: A Vision for 2003-2007. Burkina-Faso: Sahelian Pesticide Committee (CSP); 2003

[10] Schreinemachers P, Afari-Sefa V, Heng CH, Dung PTM, Praneetvatakul S, Srinivasan R. Safe and sustainable crop protection in Southeast Asia: Status, challenges and policy options. Environmental Science Policy. 2015; **54**:357-366

[11] Lai W. Pesticide use and health outcomes: Evidence from agricultural water pollution in China. Journal of Environmental Economics and Management. 2017;**86**:93-120

[12] Sharma A, Kumar V, Shahzad B, et al. Worldwide pesticide usage and its impact on ecosystem. SN Applied Sciences. 2019:1446

[13] Attwood S, Maron M, House A, Zammit C. Do arthropod assemblages display globally consistent responses to intensified agricultural land use and management? Global Ecology and Biogeography. 2008;**17**(5):585-559

[14] Hendrickx F, Maelfait JP, Van Wingerden W, Schweiger O, Speelmans M, Aviron S, et al. How landscape structure, land-use intensity, and habitat diversity affect components of total arthropod diversity in agricultural landscapes. Journal of Applied Ecology. 2007;**44**(2):340-351

[15] World Health Organization (WHO). Food Safety. 2020. Available from: https://www.who.int/news-room/fact-sheets/details/food-safety

[16] Marin-Morales MA, Ventura-Camargo B, Hoshina MM. Toxicity of herbicides: Impact on aquatic and soil biota and human health. In: Herbicides—Current Research and Case Studies in Use, Chapter 16. London, UK: IntechOpen; 2013. pp. 400-443

[17] Zhang FF, Cudhea F, Shan Z, Michaud DS, Imamura F, Eom H, et al. Preventable cancer burden associated with poor diet in the United States. JNCI Cancer Spectrum. 2019;**3**(2):pkz034

[18] Hernandez LG, Van Steeg H, Luijten M, Van Benthem J. Mechanisms of non-genotoxic carcinogens and importance of a weight of evidence approach. Mutagenesis Research. 2009;**682**(2-3):94-109

[19] EPA. EPA Releases Draft Biological Evaluations for Atrazine, Simazine and Propazine. 2020. Available from: https://www.epa.gov/pesticides/epa-releases-draft-biological-evaluations-atrazine-simazine-and-propazine

[20] Klementova S, Keltnerova L. In: Price A, Kelton J, Sarunaite L, editors. Triazine Herbicides in the Environment, Herbicides, Physiology of Action, and Safety. London, UK: IntechOpen; 2015. DOI: 10.5772/60858. Available from: https://www.intechopen.com/chapters/48620

[21] Sanderson JT, Seinen W, Giesy JP, van den Berg M. 2-Chloro-s-triazine herbicides induce aromatase (CYP19) activity in H295R human adrenocortical carcinoma cells: A novel mechanism for estrogenicity? Toxicological Science. 2000;**54**(1):121-127

[22] Gammon DW, Aldous CN, Carr WC, Sanborn JR, Pfeifer KF. A risk assessment of atrazine use in California: Human health and ecological aspects. Pest Management Science. 2005;**61**(4):331-355

[23] Pardo LA, Beane Freeman LE, Lerro CC, et al. Pesticide exposure and risk of aggressive prostate cancer among private pesticide applicators. Environmental Health. 2020;**19**:30. DOI: 10.1186/s12940-020-00583-0

[24] Mladinic M, Perkovic P, Zeljezic D. Characterization of chromatin instabilities induced by glyphosate, terbuthylazine, and carbofuran using cytome fish assay. Toxicology Letters. 2009;**189**(2):130-137

[25] Mladinic M, Berend S, Vrdoljak A, Kopjar N, Radic B, Zeljezic D. Evaluation of genome damage and its relation to oxidative stress induced by glyphosate in human lymphocytes *in vitro*. Environmental Molecular Mutagen. 2009;**50**(9):800-807

[26] He X, Wang L, Szklarz G, Bi Y, Ma Q. Resveratrol inhibits paraquat-induced oxidative stress and fibrogenic response by activating the nuclear factor erythroid 2-related factor 2 pathways. The Journal of Pharmacology and Experimental Therapeutics. 2012;**342**(1):81-90

[27] Sarbawal A, Kumar K, Singh RP. Hazardous effects of chemical pesticides on human health- cancer and other associated disorders. Environmental Toxicology and Pharmacology. 2018;**63**:103-114

[28] Thompson LA, Darwish WS. Environmental chemical contaminants in food: Review of a global problem. Journal of Toxicology. 2019; Article ID: 2345283

[29] Vos JG, Dybing E, Greim HA, Ladefoged O, Lambre C, Tarazona JV, et al. Health effects of endocrine-disrupting chemicals on wildlife, with special reference to the European situation. Critical Review of Toxicology. 2000;**30**(1):71-133. DOI: 10.1080/10408440091159176

[30] Boedeker W, Watts M, Marquez E. The global distribution of acute unintentional pesticide poisoning: Estimations based on a systematic review. BMC Public Health. 2020;**20**(1875):1-19

[31] Pimentel D. Environmental and economic costs of the application of pesticides primarily in the United States. Environment, Development, and Sustainability. 2005;**7**:229-252

[32] Dogheim SM, Mohamed E-Z, Gad Alla SA, El-Saied S, Emel SY, Mohsen AM, et al. Monitoring of pesticide residues in human milk, soil, water, and food samples collected from Kafr El-Zayat Governorate. Journal of AOAC International. 1996;**79**(1):111-116

[33] Trotter WJ, Dickerson R. Pesticide residues in composited milk collected through the U.S pasteurized milk network. Journal of AOAC International. 1993;**76**(6):1220-1225

[34] FDA. Pesticide program. Residue monitoring 1993. Food and Drug Administration. Journal of AOAC International. 1994;**77**(5):161A-185A

[35] World Health Organization. Pesticide Residues in Food. Toxicological Evaluations. Berlin, Germany: A Joint FAO/WHO Meeting on Pesticide Residues; 18-27 September 2018

[36] Wiles R, Campbell C. Washed, Peeled-Contaminated: Pesticide Residues in Ready-To-Eat Fruits and Vegetables. Washington, D.C.: Environmental Working Group; 1994

[37] Jallow MF, Awadh DG, Albaho MS, Devi VY, Thomas BM. Pesticide knowledge and safety practices among farm workers in Kuwait: Results of a survey. International Journal of Environmental Research and Public Health. 2017;**14**(4):340

[38] Alavanja MC, Bonner MR. Occupational pesticide exposures and cancer risk: A review. Journal of Toxicology and Environmental Health. 2012;**15**(4):238-263

[39] Ecobichon DJ. Pesticides use in developing countries. Toxicology. 2001;**160**(1-3):27-33

[40] Bassil KL, Vakil C, Sanborn M, Cole DC, Kaur JS, Kerr KJ. Cancer health effects of pesticides: A systematic review. Canadian Family Physician. 2007;**53**(10):1704-1711

[41] Abarikwu SO. Causes and risk factors for male-factor infertility in Nigeria: A review. African Journal of Reproductive Health. 2013;**17**(4): 150-166

[42] Yoshida S, Yamada H, Sugawara I, Takeda K. Effect of dibromochloropropane (DBCP) on the hormone receptors of the male rat reproductive system. Bioscience, Biotechnology and Biochemistry. 1998;**62**(3):479-483

[43] Corsini E, Liesivuori J, Vergieva T, Van Loveren H, Colosio C. Effects of pesticide exposure on the human immune system. Human and Experimental Toxicology. 2008; **27**(9):671-680

[44] Fiore MC, Anderson HA, Hong R. Chronic exposure to aldicarb-contaminated groundwater and human immune function. Environmental Research. 1986;**41**:633-645

[45] World Health Organization. Dioxins and their Effects on Human Health. 2016. Available from: https://www.who.int/news-room/fact-sheets/detail/dioxins-and-their-effects-on-human-health

[46] Guillette LJ, Crain DA, Gunderson MP, Kools SAE, Milnes MR, Orlando EF, et al. Alligators and endocrine-disrupting contaminants: A current perspective. American Zoology. 2000;**40**:438-452

[47] Sarkar SR, Banerjee S. Gut microbiota in neurodegenerative disorders. Journal of Neuroimmunology. 2019;**328**:98-104

[48] Pimentel D, Culliney TW, Bashore T. Public health risks associated with pesticides and natural toxins in foods. 2018. Available from: ipmworld.umn.edu

[49] Casey P, Vale JA. Deaths from pesticide poisoning in England and Wales: 1945-1989. Human and Experimental Toxicology. 1994; **13**(2):95-101

Chapter 4

Insecticide's Disappearance after Field Treatment and during Processing into Byproducts

Alberto Angioni and Nicola Arru

Abstract

Insecticide's disappearance after field treatments could be ascribed to different factors such as sunlight photodegradation, dilution effect due to fruit growth, co-distillation during fruit respiration and evaporation. Moreover, the epicuticular waxes could speed or slow down the degradation rate, and the cultivation in an open field or greenhouses could affect the residues dramatically. After harvest, the processing techniques to produce byproducts deeply influence insecticide residues. For example, fruit drying, winemaking, the industrial processing of tomatoes to produce purée, triple-concentrated paste, fine pulp, diced, olive processing to obtain table olive and olive oil, and other industrial applications on fruits affect residues and their half-life time. The scope of this chapter is to highlight the major factors responsible for the disappearance of insecticides after treatment. Moreover, the chapter intends to review the influence of the industrial processes on insecticide behaviour when the raw material is transformed into its byproducts.

Keywords: insecticide disappearance, photodegradation, fruit growth, food processing

1. Introduction

Insecticides are plant protection products, namely pesticides, intended to prevent, destroy, repel, or mitigate pests, regulated in the EU by Regulation (EC.) No 1107/2009 and Regulation (EC.) No 396/2005 [1, 2]. Their toxicological and legal limits on food (MRL), human and environmental risk assessment are defined, set and continuously revised.

Before an active substance can be used within a plant protection product, the European Commission must approve it. Active substances undergo an intensive evaluation process before a decision can be made on approval.

Insects represent the most abundant group of animals on the Earth; the number of species is more than any other group with an estimated number of living species of insects of 30 million and is the major competitors with humans for agriculture resources [3].

Insecticides act in different modes, disrupting the nervous system, impacting the development of the exoskeleton, or have a repelling activity. In addition, insecticides can be applied in various formulations such as sprays, dust, gels and baits, showing a different level of risk to non-target insects, people, pets and the environment [4].

In natural environments, insects live in a complex relationship with mutual limitations and constrictions, whereas in defined agroecosystems, the natural regulations are limited, and insect pest outbreaks can occur [5].

The factors influencing insecticide disappearance after treatment and during processing are primary to evaluate the possible dangerous effect of insecticides on human health and environmental safety.

2. Materials and methods

Photodegradation studies have been carried out under sunlight in field and in model systems; moreover, trials on model systems under artificial lights have been widely used. Sunlight experiments in the field have studied the disappearance rate of insecticides comparing open field treatments versus greenhouse treatments, and the degradation compounds in the greenhouse have been investigated simulating with petry dishes the effect of the covering glasses [6]. Other model systems built up trials with catalysator to investigate photocatalytic degradation with TiO_2 or oxidating compounds such as H_2O_2 and O_3 [7–14], and the different effects of soil thin layers or epicuticular waxes extracted from different commodities [15–17]. Model systems in laboratories trials employed different light sources, and the most common are represented by the xenon arc lamp and mercury lamps [12, 15, 18]. Trials were performed in water or methanol solutions in Pyrex (adsorb length < 290 nm) and quartz glass filter photochemical reactors.

Drying experiments have been conducted following traditional processes under sunlight exposure or the industrial process in an oven with controlled temperature and humidity [19–28].

The effect of fruit growing was assessed in field and greenhouses trials, correlating fruit growing with insecticide disappearance rate [29].

The effect of the technological process has been studied at industrial to avoid laboratory mystification of the results. Vinification, olive oil production and tomatoes processing have been carried out in real conditions allowing a real risk assessment [30–52].

Analytical measures of the insecticides and their metabolites have been made mainly using liquid chromatographic apparatus with UV–VIS or mass spectrometry detectors (MS).

3. Factors affecting insecticide's depletion

3.1 Photodegradation

Insecticides released in the environment are actively photodegraded by sunlight radiation, representing one of the most destructive pathways [5]; this is because most pesticides absorb UV–VIS bands at medium-short wavelengths. However, the sunlight reaching the Earth is mainly composed of UV-A and variable amounts of UV-B with only small amounts of short wavelengths of UV radiation; therefore, the effect on pesticides should be only of limited importance [7].

Photodegradation of insecticides could lead to non-toxic metabolites or more toxic compounds, such as systemic insecticides that are converted into more water-soluble compounds. For example, 95% of the nicotinoid imidacloprid is degraded into olefine, and 4-5-dihydroxy when applied as a drench or granule. These two metabolites are active against insect pests such as aphids. Thiamethoxam is converted to clothianidin, whereas acetamiprid is converted into five active metabolites [53].

The possible toxic effect of insecticides metabolites has led to the development of environmental monitoring studies to evaluate the presence of the parent compound and its metabolite in water and soil.

The mechanisms involved in the photodegradation process could be direct photodegradation, photosensitized degradation, photocatalytic degradation and degradation mediated by hydroxyl radicals [54].

When dealing with food and the environment, UV radiation comes from the sunlight, and direct photodegradation represents the primary mechanism, while other mechanisms are involved when remediation studies occur [7].

Plant surfaces, especially leaf surfaces, are the first reaction environment for a pesticide molecule after application, and spray drift would indirectly present a similar situation. Photolysis on soil surfaces becomes essential when a pesticide is directly applied to soil or not significantly intercepted by plants, providing that the leaf cover does not shade the ground from sunlight. Because the foliar interception of pesticides depends on plant species and usually increases with their growth stage [55], the importance of soil photolysis is considered to be lessened when plants become mature. Spray drift after pesticide application or wash off from plants by rain is the indirect route by which a pesticide reaches the soil [5].

Photodegradation of insecticides in typical environmental conditions did not lead to individual responses (homolysis, heterolysis, or photoionization). Therefore, trials in model systems have been carried out to define the effect of direct irradiation [56, 57]. These trials investigate photodegradation in water solutions in the presence of catalysts or OH donators.

Photodegradation experiments showed a higher rate in aqueous solution than in dry film; therefore, the results of model systems should be correct accordingly when dealing with fruit and vegetables in the field.

Amidinohydrazone insecticides led to three products in an aqueous medium within 10 hours and were only slightly affected by pH [9]; on the other hand, carbaryl and propoxur photodegradation increased with the pH, and the main photochemical products were derived from the ester bond cleavage [54].

Carbamate insecticide photodegradation has been widely studied, and different UV irradiation lengths and mediums have been used in model systems to reproduce the natural condition. Apolar solvent (hexane, cyclohexane, isopropanol) was used to simulate plant surface, and water at different pH was used to evaluate the influence of acidity; in addition, additional studies added O_3 or other oxidants (H_2O_2, TiO_2) to verify the enhanced oxidant effect of UV irradiation at selected wavelengths [7, 9–14]. Dry film photodegradation trials in the presence of epicuticular waxes extracted from fruits and vegetable commodities or soil with humic substances led to different results in the half-life time of insecticides [12, 13].

The epicuticular waxes from different vegetables and fruits did not show univocal results, decreasing or enhancing photodegradation (**Table 1**) [15].

The photochemical degradation of carbaryl and carbofuran under sunlight and UV (λ 290 nm) exposure in the natural waters of Northern Greece has been investigated. The major photoproducts observed were carbofuran–phenol from carbofuran and naphthol from carbaryl [54].

The analysis of the degradation pathways of aldicarb and carbaryl in water showed as major photodegradation products aldicarb sulfoxide and 1-naphthol, respectively. In addition, the study highlighted the influence of the UV source on the rate of photodegradation [28].

Organophosphorus insecticides in water solutions showed first-grade rate constants with an observed half-life in the range of hours [58, 59].

Neonicotinoid pesticides have shown a high degree rate of photodegradation on tomato leaves after treatment in water solutions [60, 61].

		$K_{obs} \times 10^{-5}$ s^{-1}		T ½ (min)		
Compound	K^w/K^b	$Wax^a \pm SD$	Blank ± SD	Wax^a	Blank	r^2
Methiocarb	1.0	2.5 (0.06)	2.5 (0.08)	436	438	0.9976
Methiocarb sulfone	0.9	0.35 (0.08)	0.39 (0.11)	3351	2964	0.9955
Fenthion	2.9	5.6 (0.4)	1.9 (0.5)	204	593	0.9889
Aminocarb	0.3	5.2 (0.2)	19.2 (1.8)	222	60	0.9921
Pirimicarb	3.0	19.4 (1.2)	6.4 (1.5)	59	181	0.9959

[a] *70ug cm*$^{-2}$
[b] *blank*
[w] *waxes*

Table 1.
Photolysis rates of some insecticides in the presence of epicuticular waxes of Persica laevis *[15].*

3.2 Fruit and vegetable growing and shape

Insecticide amount on fruit and vegetable is related to the moment of treatment, the shape and cultivar. Fruits and vegetable sizes increase during development depending on the species and the agricultural practice adopted, modifying the rate surface/weight (s/w) and influencing the residue amount profoundly. The maximum residue limit (MRL) is the highest level of pesticide residue acceptable in food or feed when pesticides are applied following Good Agricultural Practice (GAP) [62] and is expressed as mg/Kg or mg/L of active ingredients (a.is.); therefore, the rate surface/weight (s/w) represents the discriminant leading to the residue amount.

The first treatment is made after the fruit set; when no other factors are involved in the residue decrease, the amount of the insecticides in the harvested commodities at commercial ripening is reduced by the growth factor and the residues are lower when the treatment is done much in advance of the harvest.

Therefore, when insecticides are applied to fruit and vegetable before harvest when the development is concluded, the growth dilution could not more influence the residue amount.

On the other hand, the shape and final size of the commodities have great importance. For example, tomatoes have different dimensions and shapes depending on the cultivar ranging from 12 to 200 g from cherry to beef heart. The surface/weight ratio for 1 Kg is notably higher for cherry; therefore, the same treatment applied on the cultivar would lead to entirely different results with higher residue expressed in mg/Kg in the cherry tomatoes. The cultivar (CV) Koreniki has small olives (1–2 g), whereas Yacouti has big olives (5–6 g), accounting for a lower s/w ratio and minor final residues (**Table 2**).

In iceberg and romana lettuce CV, the edible parts have different shapes: an open calyx in romana and ball-shaped in the iceberg; similar consideration can be made for artichoke when comparing meda or masedu cv (calyx shape) to spinoso sardo and romanesco cv (close shape). The calyx shape allows the deposition of treatment solution among the fruit leaf, with prolonged contact and final higher residue concentration, whereas the ball and close shape let the solution slide down, allowing only a short contact (**Figure 1**). Moreover, the outer leaves in both cases are removed before eating [29].

3.3 Drying

Drying is one of the oldest and most common preservation methods; water removal minimizes many moisture-driven deterioration reactions. However, drying fruit to obtain raisins, prunes and apricots could increase insecticide levels in the

Commodities	Cultivar	s/w ratio	
Olive	Yacouti	0.7	
	Koroneiki	1.3	
Tomatoes	Shiren	1.8	
	Caramba	0.8	
		Weight (diluting factor/day)	
Peaches	Flavorcrest	45 (1.07/1)	135 (3.21/15)

Table 2.
Surface/weight ratio of different cultivars and after development.

Figure 1.
Different shapes of lettuce and artichoke cultivars.

final product due to the concentration factor (4, 3 and 6, respectively) [19]. In addition, sunlight or oven drying processes lead to different results [20].

In raisins, insecticides decrease has been ascribed to the temperature of oven drying and lesser effect to sunlight, correlating the decrease to the degradation of the pesticides [21]. However, trials in model systems evaluating the process of disappearance during oven drying showed that insecticide co-distillation with water could be as important as heat degradation [22, 23].

Apricots, plums and prunes showed lower residues in the dried product; however, results were not the same for the different pesticides; moreover, the effect of sunlight in apricot was more efficacious (**Table 3**) [24–26].

Chilli pepper subjected to oven drying showed a concentration factor of almost 5, and the insecticides tested had different behaviour some were more concentrated in the final product while other decreased [27]. A similar experiment on orange slices showed a general decrease in the pesticide investigated even if the mechanism was not explained [28].

3.4 Technological processes

3.4.1 Wine, beer and byproducts

The fate of insecticides residues on grapes during winemaking has been widely studied [30–37]. Depending on the technology adopted, two categories of wine

Commodities	Insecticides	Fruit weight (g)	mg/kg ± SD	Conc. factor	Reference
Apricots					[25]
Diazinon	Fresh fruit	43.3 ± 3.3	0.50 ± 0.13		
	Dried fruit	6.6 ± 0.6	0.63 ± 0.20	6.56	
Phosalone	Fresh fruit	42.5 ± 2.0	0.48 ± 0.14		
	Dried fruit	7.6 ± 1.1	1.56 ± 0.37	5.59	
Prunes					[24]
Phosalone	Fresh fruit	38.7 ± 1.3	0.21 ± 0.06		
	Dried fruit	12.5 ± 1.2	0.62 ± 0.16	3.09	

Table 3.
Insecticides level after drying process.

are obtained, white wines and red wines. The former are wines produced in the absence of skins, whereas red wines are produced with maceration in the presence of the skins. During fermentation, two main waste products are generated, lees and grapes. These fractions adsorb insecticides due to the affinities for the solid residues during alcoholic fermentation and, therefore, sequestered them from the must, which results free from residues or with reduced amounts with respect to the grapes. Moreover, active degradation by the yeast could be encountered, and these results were also confirmed during beer preparation [31, 32]. In the second stage, the malolactic fermentation by bacteria can actively decrease pesticide levels, which would encounter an added decrease with the clarification step using fining agents such as activated carbon, bentonite, polyvinylpolypyrrolidone (PVPP), gelatin, egg albumin, isinglass-fish glue, and casein [31, 38–41].

The spirit drink industry can use grapes and lees to produce alcohol, and distilled beverage spirits and insecticides could concentrate in the distillate of a factor between one (cake) and six hundred (lees) (Garoglio 1973). When wines are used to produce distillate, the concentration factor would be 10 times.

Literature data showed that only small amounts of fenthion (2%) and quinalphos (1%), and other organophosphate insecticides passed during the distillation process from artificially contaminated lees [42–44], indicating that insecticides hardly migrate to the distilled spirits.

3.4.2 Olive oil

The transfer to virgin olive oil is related to the active ingredients' octanol/water partition coefficient during the production step. Since no MRL is set in olive oil, the values for olive are adapted to the oil relating to the partition coefficient. In 2015, EU differentiated between fat-soluble and fat insoluble compounds, setting processing factors of 5 and 1, respectively [45]. On the other hand, highly polar insecticides showed negligible transfer rates in the oil being concentrated in the aqueous phase [46].

Insecticides with Kow > 0 increase their concentration in the oil while decreasing their polarity. For example, organophosphorus insecticides concentrate 7 times in the oil following the concentration factor from olive to oil (7 kg olives for 1 l of oil) [47]. Although, on the other hand, triazoles and neonicotinoids displayed different behaviour, so that an increase of hydrophobicity did not cause such an increase of pesticide transfer efficiency for these two classes, water addition during the extraction step caused a decrease of the insecticides with lower Kow [46–48].

3.4.3 Tomatoes processing

Europe is the most important producer of tomatoes, and integrated pest management strategies (IPM) are applied widely. For example, insecticides applied in the field could be transferred and concentrate into processing products such as purée, triple concentrated paste, fine pulp, and diced tomatoes [49].

Washing and peeling led to a decrease in insecticide residues, even if washing affected only the pesticide adsorbed on the dust adhering to the surface [50, 51].

Different batches of tomatoes from different fields subjected to different agriculture procedures and pesticide treatments are processed jointly during the industrial process. Therefore, a dilution effect would occur during the various production steps. The main effect related to the decrease of insecticides during tomatoes processing are represented by peeling and the dilution effect [52].

4. Conclusions

The disappearance of insecticides after treatment could be related to many different paths. The main effects in the field are related to sunlight photodegradation and the development of the fruits during fruit growing leading to a decrease of the residues below the legal limit and sometimes the analytical detectable levels. Run-off and washing could affect only the residues adsorbed in the adhering dust on fruits and vegetable surfaces not influencing the residues adsorbed in the epicuticular waxes. Drying causes a reduction in weight, theoretically the residue could increase giving a value that is a function of the concentration factor, however the results showed a decrease in insecticide residues in foodstuffs. The different processing steps affect insecticides residues involving partition, microbiological and chemical degradation, and adsorption on the waste such as lees, marc, vegetation water and pomace.

Acknowledgements

The chapter has been written with the active help of researchers from the Italian Society of FOO Chemistry (ItaChemFood).

Conflict of interest

The authors declare no conflict of interest.

Author details

Alberto Angioni* and Nicola Arru
Department of Life and Environmental Science, University of Cagliari, Italy

*Address all correspondence to: aangioni@unica.it

References

[1] Regulation (EC.) No 1107/2009 of the European parliament and of the council of 21 October 2009Official Journal of the European Union;**L 309**: 1-50

[2] Regulation (EC.) No 396/2005 of the European Parliament and of the Council of 23 February 2005 on maximum residue levels of pesticides in or on food and feed of plant and animal origin and amending Council Directive 91/414/ EECText with EEA relevance. Available from: http://data.europa.eu/eli/reg/2005/396/oj

[3] Sharma S, Kooner R, Arora R. Insect Pests and Crop Losses. In: Arora R, Sandhu S, editors. Breeding Insect Resistant Crops for Sustainable Agriculture. Singapore: Springer Nature Singapore Pte Ltd.; 2017. DOI: 10.1007/978-981-10-6056-4_2

[4] EPA Insecticides. 2021. Available from: https://www.epa.gov/caddis-vol2/insecticides

[5] Gill HK, Garg H, editors. Pesticides: Environmental Impacts and Management Strategies, Pesticides - Toxic Aspects. Rijeka: IntechOpen. DOI: 10.5772/57399 Available from: https://www.intechopen.com/chapters/46083

[6] Garau V, Angioni A, Real AAD, Russo M, Cabras P. Disappearance of azoxystrobin, pyrimethanil, cyprodinil and fludioxinil on tomatoes in a greenhouse. Journal of Agricultural and Food Chemistry. 2002;**50**:1929-1932

[7] Katagi T. Photodegradation of Pesticides on Plant and Soil Surfaces. In: Ware GW, editor. Reviews of Environmental Contamination and Toxicology. Continuation of Residue Reviews. Vol. 182. New York, NY: Springer; 2004. DOI: 10.1007/978-1-4419-9098-3_1

[8] Assalin MR, Ferracini VL, Nascimento Queiroz SC, Jonsson CM, Clemente Z, Silva SRCM. Photocatalytic degradation of an organophosphorus pesticide from agricultural waste by immobilized TiO_2 under solar radiation. Revista Ambiente & Água. 2016;**11**(4). DOI: 10.4136/ambi-agua.1824

[9] Adityachaudhury N, Chowdhury A, Das AK, Bhattacharyya A, Pal S. Transformation of some selected pesticides. Journal of the Indian Chemical Society. 1994;**71**:425-433

[10] Kanan M. A study of the photodegredation of carbaryl: The influence of natural organic matter and the use of silver zeolite Y as a catalyst. Electronic Theses and Dissertations. 2002:211

[11] Benitez FJ, Acero JL, Real FJ, Roldan G, Rodriguez E. Modeling the photodegradation of emerging contaminants in waters by UV radiation and UV/H2O2 system. Journal of Environmental Science and Health, Part A, Toxic/Hazardous Substances and Environmental Engineering. 2013;**48**(1):120-8. doi: 10.1080/10934529.2012.707864

[12] Pirisi F, Cabras P, Garau VL, Melis M, Angioni A, Bullita D. Photodegradation of pesticides. 1. Photolysis rates and half-life of acylanilides and their major metabolites in water. Toxicological and Environmental Chemistry. 1996;**55**:199-214. DOI: 10.1080/02772249609358336

[13] Mansour SA, Tony MA, Tayeb AM. Photocatalytic performance and photodegradation kinetics of Fenton-like process based on haematite nanocrystals for basic dye removal. SN Applied Science. 2019;**1**:265. DOI: 10.1007/s42452-019-0286-x

[14] Tanaka K, Reddy KSN. Photodegradation of phenoxyacetic acid

and carbamate pesticides on TiO_2. Applied Catalysis B: Environmental. 2002;**39**(4):305-310

[15] Pirisi FM, Angioni A, Cabizza M, Cabras P, Falqui CC. Photolysis of pesticides: influence of epicuticular waxes from *Persica laevis* DC on the photodegradation in the solid phase of aminocarb, methiocarb and fenthion. Pest Managment Science. 2001;**57**:522-526. DOI: 10.1002/ps.329

[16] Pirisi FM, Angioni A, Cabizza M, Cabras P, Maccioni E. Influence of epicuticular waxes on the photolysis of pirimicarb in the solid phase. Journal of Agricultural and Food Chemistry. 1998;**46**:762-765

[17] Graebing P, Frank M, Chib JS. Soil photolysis of herbicides in a moisture- and temperature-controlled environment. Journal of Agricultural and Food Chemistry. 2003;**51**: 4331-4337

[18] de Bertrand N, Barceló D. Photodegradation of the carbamate pesticides aldicarb, carbaryl and carbofuran in water. Analytica Chimica Acta. 1991;**254**:1-2 235-244

[19] Zhao X, Kong W, Wei J, Yang M. Gas chromatography with flame photometric detection of 31 organophosphorus pesticide residues in *Alpinia oxyphylla* dried fruits. Food Chemistry. 2014;**162**:270-276

[20] Alasalvar C, Shahidi F. Composition, phytochemicals, and beneficial health effects of dried fruits: An overview. In: Alasalvar C, Shahidi F, editors. Dried fruits: Phytochemicals and health effects. Oxford: Wiley-Blackwell; 2013. pp. 1-18

[21] Ayşe O, Şeyda K, Ali C. Effetc of drying process on pesticide residues in grapes. GIDA/The Journal of Food. 2017;**42**:204-209. DOI: 10.15237/gida.GD16098

[22] Cabras P, Angioni A. Pesticide residues in grapes, wine, and their processing products. Journal of Agricultural and Food Chemistry. 2000;**48**(4):967-973. DOI: 10.1021/jf990727a

[23] Athanasopoulos PE, Pappas C, Kyriakidis NV, Thanos A. Degradation of methamidophos on soultanina grapes on the vines and during refrigerated storage. Food Chemistry. 2005;**91**(2): 235-240

[24] Cabras P, Angioni A, Garau VL, Pirisi FM, Brandolini V, Cabitza F, et al. Pesticide residues in prune processing. Journal of Agricultural and Food Chemistry. 1998;**46**:3772-3774

[25] Cabras P, Angioni A, Garau VL, Melis M, Pirisi FM, Cabitza F, et al. Pesticide Residues on Field-Sprayed Apricots and in Apricot Drying Processes. Journal of Agricultural and Food Chemistry. 1998;**46**:2306-2308

[26] Cabras P, Angioni A, Garau VL, Minelli EV, Cabitza F, Cubeddu M. Pesticide residues in plums from field treatment to the drying process. Italian Journal of Food Science. 1998;**10**(1): 81-85

[27] Noh HH, Kim DK, Lee EY, Chang MI, Im MH, Lee YD. K KS. Effects of oven drying on pesticide residues in field-grown chili peppers. Journal of Korean Society for Applied Biological Chemistry. 2015;**58**:97-104. DOI: 10.1007/s13765-015-0016-z

[28] Acoglu B, Omeroglu PY. The effect of drying processes on pesticide residues in orange (Citrus sinensis). Drying Technology. An International Journal. 2021;**39**(13). DOI: 10.1080/07373937.2021.1946078

[29] Cabras P. Part II. Anthropogenic contaminants Pesticides: toxicology and residues in food. In D'Mello JPF, editor Food safety: contaminants and toxins.

2003; CABI Pubblishing, UK. 91-124. 10.1079/9780851996073.0000

[30] Flori P, Frabboni B, Cesari A. Pesticide decay models in the winemaking process and wine storage. Italian Journal of Food Science. 2001;**12**(1):279-289

[31] Cabras P, Angioni A. Pesticide Residues in Grapes, Wine, and Their Processing Products. Journal of Agricultural and Food Chemistry. 2000;**48**(4):967-973

[32] Navarro S, Pérez G, Navarro G, Mena L, Vela N. Variability in the fermentation rate and color of young lager beer as influenced by insecticide and herbicide residues. Food Chemistry. 2007a;**105**:1495-1503. DOI: 10.1016/j.foodchem.2007.05.035

[33] Cabras P, Angioni A, Garau VL, Pirisi FM, Brandolini V. Gas Chromatographic Determination of Azoxystrobin, Fluazinam, Kresoxim-Methyl, Mepanipyrim, and Tetraconazole in Grapes, Must, and Wine. Journal of AOAC International. 1998;**81**(6):1185-1189

[34] Navarro S, Barba A, Oliva J, Navarro G, Pardo F. Evolution of residual levels of six pesticides during elaboration of red wines. Effect of winemaking procedures in their disappearance. Journal of Agricultural and Food Chemistry. 1999;**47**:264-270. DOI: 10.1021/jf980801

[35] Dumitriu (Gabur) GD, Teodosiu C, Valeriu V. Management of Pesticides From Vineyard to Wines: Focus on Wine Safety and Pesticides Removal By Emerging Technologies. 2021. DOI: 10.5772/intechopen.98991

[36] Cus F, Cesnik HB, Bolta SV, Gregorcic A. Pesticide residues in grapes and during vinification process. Food Control. 2010;**21**:1512-1518. DOI: 10.1016/j.foodcont.2010.04.024

[37] Navarro S, Barba A, Oliva J, Navarro G, Pardo F. Evolution of residual levels of six pesticides during elaboration of red wines. Effect of winemaking procedures in their disappearance. Journal of Agricultural and Food Chemistry. 1999;**47**:264-270. DOI: 10.1021/jf980801

[38] Doulia DS, Anagnos EK, Liapis KS, Klimentzos DA. Effect of clarification process on the removal of pesticide residues in red wine and comparison with white wine. Journal of Environmental Science and Health. Part. B. 2018;**53**(8):534-545. DOI: 10.1080/03601234.2018.1462937

[39] Marangon M, Vincenzi S, Curioni A. Wine fining with plant proteins. Molecules. 2019;**24**(11):2186. DOI: 10.3390/molecules24112186

[40] Sen K, Cabaroglu T, Yilmaz H. The influence of fining agents on the removal of some pesticides from white wine of Vitis vinifera L. cv. Emir. Food and Chemical Toxicology. 2012;**50**(11): 3990-3995. DOI: 10.1016/j.fct.2012.08.016

[41] Oliva J, Payá P, Cámara MA, Barba A. Removal of pesticides from white wine by the use of fining agents and filtration. Communications in Agricultural and Applied Biological Sciences. 2007;**72**(2):171-180 PMID: 18399438

[42] He NX, Bayen S. An overview of chemical contaminants and other undesirable chemicals in alcoholic beverages and strategies for analysis. Comprehensive Reviews in Food Science and Food Safety. 2020:3916-3950. DOI: 10.1111/1541-4337.12649

[43] Shin JA, Cho H, Seo DW, Jeong HG, Kim SC, Lee JH, et al. Approach Study for Mass Balance of Pesticide Residues in Distillers' Stillage along with Distillate and Absence Verification of Pesticides in Distilled Spirits from Pilot-Scale of

Distillation Column. Molecules. 2019;**24**(14):2572. DOI: 10.3390/molecules24142572

[44] Cabras P, Angioni A, Garau VL, Minelli EV, Melis M, Pirisi FM. Pesticides in the distilled spirits of wine and its byproducts. Journal of Agricultural and Food Chemistry. 1997;**45**:2248-2251. DOI: 10.1021/jf960457l

[45] European Commission. Commission Implementing Regulation (EU) 2015/595 of 15 April 2015 concerning a coordinated multiannual control programme of the Union for 2016, 2017 and 2018 to ensure compliance with maximum residue levels of pesticides and to assess the consumer exposure to pesticide residues in and on food of plant and animal origin. The Official Journal of theEuropean Union. 2015;**L99**:7-20

[46] López-Blanco R, Moreno-González D, Nortes-Méndez R, García-Reyes JF, Molina-Díaz A, Gilbert-López B. Experimental and theoretical determination of pesticide processing factors to model their behavior during virgin olive oil production. Food Chemistry. 2018;**239**:9-16

[47] Cabras P, Angioni A, Melis M, Minelli EV, Pirisi FM. Simplified multiresidue method for the determination of organophosphorus insecticides in olive oil. Journal of Chromatography A. 1997;**761**(1-2):327-331

[48] Amvrazi EG, Alabanis TA. Multiclass pesticide determination in olives and their processing factors in olive oil: Comparison of different olive oil extraction systems. Journal of Agricultural and Food Chemistry. 2008;**56**:5700-5709

[49] Silva-Rodríguez A, Acedo-Valenzuela MI, Diez NMM, de la Peña AM, Galeano-Díaz T. Multiresidue method for the control of pesticide residues in tomatoes and derived products. Analytical Methods. 2012;**4**:2543-2549

[50] Kwon H, Kim TK, Hong SM, Se EK, Cho NJ, Kyung KS. Effect of household processing on pesticide residues in field-sprayed tomatoes. Food Science and Biotechnology. 2015;**24**:1-6

[51] Al-Taher F, Chen Y, Wylie P, Cappozzo J. Reduction of pesticide residues in tomatoes and other produce. Journal of Food Protection. 2013;**76**:510-515

[52] Corrias F, Atzei A, Lai C, Dedola F, Ibba E, Zedda G, et al. Effects of industrial processing on pesticide multiresidues transfer from raw tomatoes to processed products. Food. 2020;**9**:1497. DOI: 10.3390/foods9101497

[53] McKenzie R, Madronich S, Ultraviolet surface, In Encyclopedia of Atmospheric Sciences 1st edition, Holton JR edt., London, UK: Academic Press; 2003. pp 2474

[54] Burrows HD, Canle M, Steenken S. Reaction pathways and mechanisms of photodegradation of pesticides. Journal of Photochemistry and Photobiology B: Biology. 2002;**67**(2):71-108

[55] Cloyd RA. Pesticide Metabolites. Manhattan, KS: Kansas State University; 2012

[56] Linders J, Mensink H, Stephenson G, Wauchope D, Racke K. Pure and Applied Chemistry. 2000;**72**:2199-2218

[57] Bustosa N, Cruz-Alcaldeb A, Iriela A, Cirellia AF, Sans C. Sunlight and UVC-254 irradiation induced photodegradation of organophosphorus pesticide dichlorvos in aqueous matrices. The Science of the Total

Environment. 2019;**649**:592-600. DOI: 10.1016/j.scitotenv.2018.08. 2540048-9697

[58] Hebert JVR, Hoonhout C, Miller GC. Use of stable tracer studies to evaluate pesticide photolysis at elevated temperatures. Journal of Agricultural and Food Chemistry. 2000;**48**(5):1916-1921. DOI: 10.1021/jf990699w

[59] Doong RA, Chang WH. Photoassisted titanium dioxide mediated degradation of organophosphorous pesticides by hydrogen oxidation of herbicides using natural sunlight. Journal of Photochemistry and Photobiology A. 1997;**107**:239-244

[60] Kole RK, Banerjee H, Bhattacharyya A, Chowdhury A, Adityachaudhury N. Photo transformation of some pesticides. Journal of the Indian Chemical Society. 1999;**76**:595-600

[61] Scholz K, Reinhard F. Photolysis of imidacloprid (NTN 33893) on the leaf surface of tomato plants. Pesticide Science. 1999;**55**:633-675

[62] Codex Alimentarius, International Food Standards. Maximum residue limits (MRLs) and risk management reccommendations (RMRs) for residues of veterinary drugs in foods CX/MRL 2-2018. FAO

Chapter 5

An Overview of the Biochemical and Histopathological Effects of Insecticides

Haci Ahmet Deveci, Gökhan Nur, Ayla Deveci, Inan Kaya, Müge Mavioglu Kaya, Abdulsamed Kükürt, Volkan Gelen, Ömer Faruk Başer and Mahmut Karapehlivan

Abstract

The number of studies on insecticides, which constitute an important class of pesticides, is increasing day by day. These chemicals used in the fight against pests in the field of agriculture; It is also used to fight mosquitoes and insects in homes, gardens and rural areas. Excessive use of insecticides has adversely affected many beneficial organisms besides target organisms. As a result of the negative effects of insecticides on non-target organisms, the normal balance of nature has been disturbed and this has led to the decline of some living species in the world. In many studies from the past to the present, it has been shown that these insecticides have negative effects on the environment, human and animal health. Some of these chemicals, which have many benefits in the fight against agricultural pests, have been banned due to their harmful effects on non-target organisms as a result of excessive use. Today, however, these chemicals are still used intensively against agricultural pests, threatening non-target organisms and human health. For this reason, in this book chapter we have prepared, the negative effects of insecticides on living things are examined by reviewing biochemical and histopathological studies.

Keywords: Pesticide, insecticide, oxidative stress parameters, histopathology changes

1. Introduction

The rapid and uncontrolled increase in the world population brings along the need for nutrition and food. Today, the failure to provide agricultural products at the level to meet the increase in world population causes problems of nutritional and food needs. In order to meet the nutritional and food needs, it is important to produce agricultural products with low cost and high quality and efficiency. In the production of agricultural products, losses of up to 65% may occur in products due to some pests and diseases. For this reason, producers use different methods to increase the yield in agricultural areas and to prolong the durability of foodstuffs. One of these methods is the chemical control method, which is carried out under the name of agricultural control, which increases product productivity in the agricultural field. In fact, the basis of this method is the use of pesticides. In the global pesticides market, herbicides rank first among pesticides with a share of 47%. This is followed by insecticides (29%), fungicides (19%) and other pesticides (5%). In Turkey, insecticides (47%) take the first place in the use of pesticides, which is followed by herbicides (24%), fungicides (16%) and other pesticides (13%) [1, 2].

Pesticides, which are chemical and biological substances used in the fight against pests, are widely used for protective purposes against internal and external parasites in veterinary medicine and agricultural control. Pesticides are substances that are frequently used to obtain more products in the control of insects that damage agricultural products and various vectors that are the cause of disease [3, 4].

Pesticides are widely used in the field of agriculture to control pests. In addition to agricultural purposes, pesticides are also utilized in the fight against mosquitoes and weeds in houses, gardens and rural areas. However, pesticides which remain the same in the soil for a long-time cause water, soil and air pollution and ruin the ecological balance. The most important harmful effects of pesticides are that they enter the body through the food chain and cause acute and chronic poisoning in humans and animals. In addition, it has been reported that pesticide degradation products cause damage to biological systems as a result of accumulation in tissues and organs over time [5–7].

Pesticides are classified in different ways according to their formulation forms, the chemical structure of their active substances and the pest group they are used in. The most widely used classification is the classification made based on the pest group they are used in. Insecticide (insecticide), fungicide (fungicide), herbicide (weed killer), acaricide (spider killer), bactericide (bactericide), rodenticide (rodenticide), nematocide (nematode killer), aphicide (aphid killer), and algicide (algae killer) can be given as examples to the pesticides in this classification. Among them, herbicides, insecticides, and fungicides are the most widely used and studied in the world [8, 9].

Knowing the types of insecticides, their chemical structures, and their harmful effects on the environment and living things will guide the studies that can prevent the damage of these substances. This study, presented in line with this information, it is aimed to explain the biochemical and histopathological effects of insecticides on the organism by considering them with current articles.

2. Classification of insecticides

Insects are one of the factors that threaten our health in the environments we live in. In addition to its psychological effects such as disgust in humans, it leads to the spreading of diseases such as plague, jaundice, and typhoid. Apart from these,

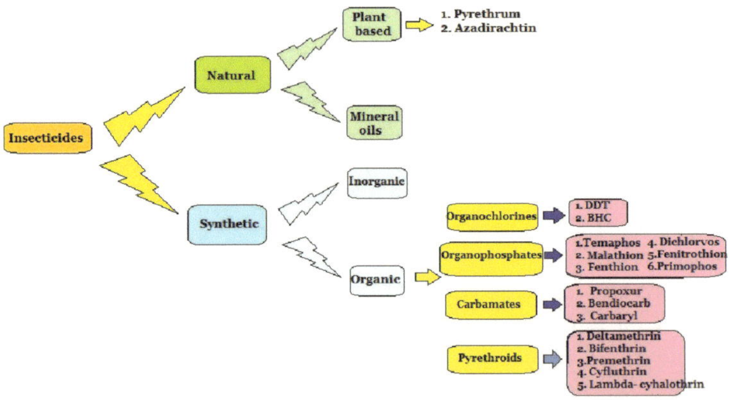

Figure 1.
Classification of insecticides [10].

insect bites can cause complaints such as itching, skin peeling, and pain in humans. Pesticides used to kill insects in many living areas such as agriculture, livestock farming, houses and workplaces are called as 'insecticide'. Insecticides are the second most widely used pesticide type in the world after herbicides [1, 9]. Insecticides are generally classified as in **Figure 1** [10].

3. Bioaccumulation of insecticides and their transfer in the food chain

As from the middle of the twentieth century, human health and the natural environment have begun to be adversely affected upon the excessive use of pesticides, including insecticides. These substances, which are used against insects that

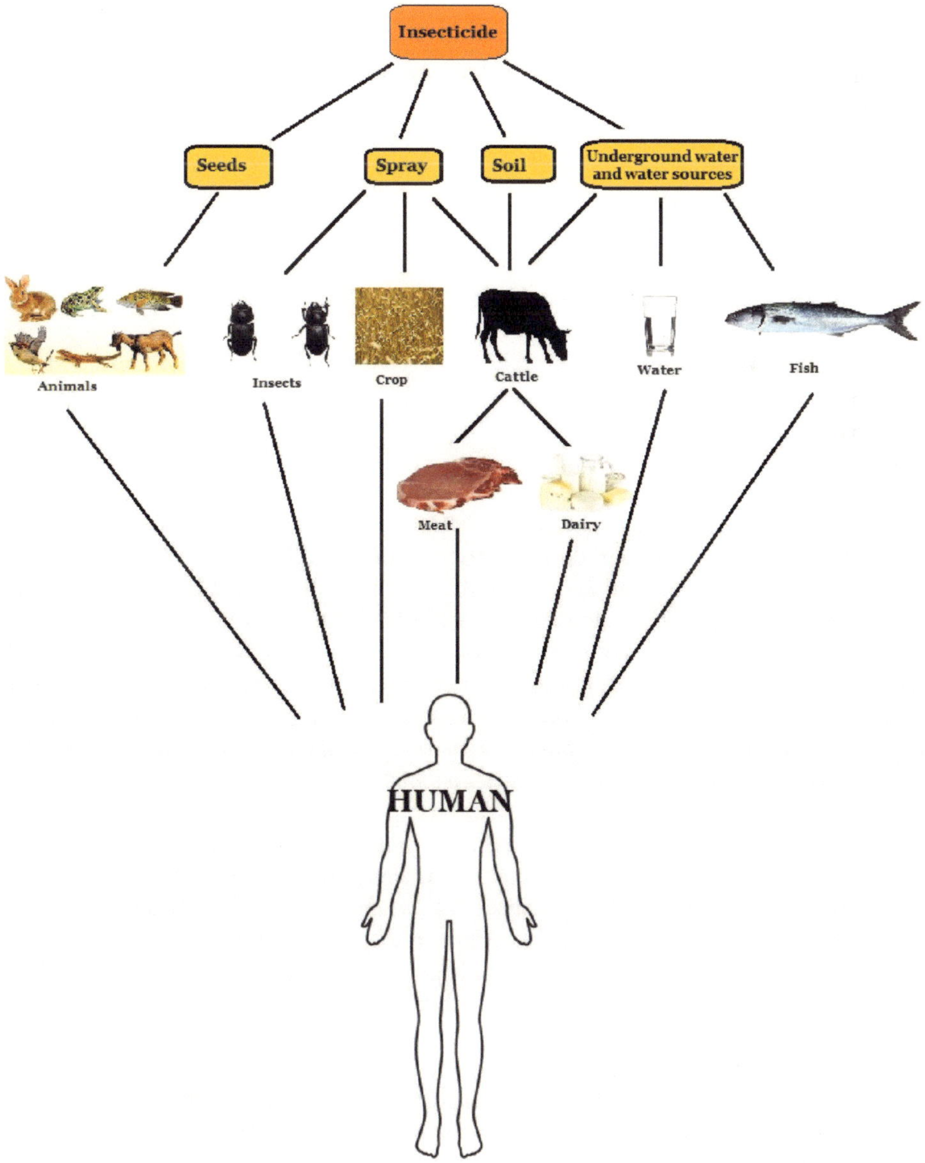

Figure 2.
Bioaccumulation of insecticide residues in the food chain [9].

harm agricultural products in the beginning, have indeed brought about significant increases in the quality and quantity of agricultural products. Insecticides, which serve to protect agricultural products by destroying unwanted insects, have provided great benefits for both the environment and public health by also controlling pathogenic vectors. However, as a result of excessive use of these substances to obtain better quality and more products later on, many unexpected harmful effects have been experienced. These substances remain intact for a long time in water, soil, fruits and vegetables and cause environmental pollution. Since insecticides applied to agricultural products can remain intact for a long time, they can reach the food chain by mixing with soil and water, and then passing through plants and animals [11–13]. These pollutants enter through the skin, gills, and digestive systems of aquatic organisms and begin to accumulate in the organs of these living creatures. These pollutants entering the body are metabolized and excreted with the help of body fluids. If they cannot be excreted from the body, they accumulate in tissues with high fat content and may remain in these tissues for many years [13, 14]. The accumulation increases exponentially when small aquatic organisms carrying insecticide residues are eaten by larger aquatic organisms and plants with insecticide residues are eaten by animals. Thus, the residue level in humans, who are in the last link of the food chain, becomes much higher and concentrated [11, 12]. The bioaccumulation of pesticide residues in the food chain is shown in **Figure 2** [9].

4. Action mechanisms of insecticides

The nervous system of insects is developed and it has characteristics similar to the nervous system of mammals. Therefore, insecticides do not have species-specific selective effects, and all mammals, including humans, are extremely sensitive to the toxic effects of insecticides. Selective action between insects and mammals is usually the result of differences in detoxification mechanisms or differential interactions in their target structures. Insecticides cause more acute poisoning in non-target organisms compared to the other pesticides [8]. The effects of insecticides may vary depending on features such as their chemical formulations, amount and duration of application, temperature and pH of the environment etc. [13–15]. The continuous use of the same insecticide species in agricultural practices causes insect species to lose their sensitivity and become resistant to these insecticides over time. As a result, the need for continuous renewal has emerged due to the decreasing effectiveness of organophosphate, carbamate, organochlorine, and pyrethroid insecticides, and alternative new insecticides such as neonicotinoid insecticides have been developed [16, 17].

The majority of insecticides used today are neurotoxic substances and act by poisoning the nervous systems of target organisms [18, 19]. The action mechanism of organophosphate and carbamate insecticides is based on the inhibition of the acetylcholinesterase enzyme (AChE). Organophosphate insecticides or their active metabolites covalently bind to the hydroxyl group of serine in the active site of AChE with phosphate radicals and cause inhibition of the enzyme. Detoxification of organophosphates includes hydrolysis reaction catalyzed by A-esterases such as paraoxonase (PON) and stoichiometric binding reactions to B-esterases such as acetylcholinesterase, butyrylcholinesterase, and carboxylesterase in plasma [19–22]. Carbamate insecticides also involve "carbamylation" of the enzyme. The cholinergic syndrome that develops in acute poisonings caused by this group of insecticides is short-lived, since the enzyme's reactivation time is short after carbamylation [23]. Organochlorine insecticides are effective on the central nervous system and

Insecticide Class	Example	Mechanism	Effect
Organophosphate	Diazinon Chlorpyrifos Mipafox	Inhibition of AChE	Cholinergic syndrome Peripheral axonopathy
Carbamate	Aldicarb Carbaryl	Inhibition of AChE	Cholinergic syndrome
Pyrethroid	Deltamethrin Allethrin	Prolonged opening of sodium Channels	Hyperexcitability
Organochlorine	DDT Lindane	Prolonged opening of sodium channel Inhibition of GABA- and voltage-dependent chloride channels	Hyperexcitability, tremors Seizures

Table 1.
Mechanisms and effects of neurotoxic insecticides in mammals [18].

their action mechanism may vary according to the structure of the insecticide. DDT (Dichlorodiphenyltrichloroethane) changes the permeability of sodium-potassium channels in the nerve membrane and causes excessive nerve stimulation by causing slow closure of sodium channels. In cyclodiene and lindane exposure, neurotransmitter release from synapses is affected. These compounds antagonize GABA (γ-aminobutyric acid) and cause depolarization and overstimulation in the postsynaptic membrane [23, 24]. Pyrethroid insecticides, on the other hand, change the properties of voltage-dependent sodium channels and cause prolonged opening of the channel. In this way, excessive excitation occurs in the central nervous system [18, 23]. **Table 1** shows the action mechanisms of insecticides [18]. In recent years, it has been reported that insecticides, in addition to these effects, increase the production of reactive oxygen species (ROS), thus causing an increase in oxidant molecules and a decrease in antioxidant molecules in the organism [5–7, 17, 19, 25, 26]. ROS formation rate and elimination work in balance. If it breaks down in favor of ROS, oxidative stress occurs. Due to oxidative stress, peroxidative damage to macromolecules and membranes of cells occurs in organisms. Moreover, their metabolic activities in cell components are impaired. Known to tissue and organ pathologies occur in the presence of oxidative stress in the organism [27–34].

5. Biochemical effects of insecticides

Pesticides are metabolized in the liver by cytochrome P450 enzyme systems, passing through the human body through the skin, respiration, and digestion. Pesticides stimulate lipid peroxidation in hepatic microsomes and cause a decrease in cytochrome P450 enzymes, glucose 6-phosphatase and pyrophosphatase activities [35]. Detoxification of organophosphate pesticides, including organophosphate insecticides, is provided by A-esterases such as paraoxonase (PON) and B-esterases such as acetylcholinesterase, butyrylcholinesterase, and carboxylesterase in plasma [19, 20].

In a study conducted on agricultural laborers who were exposed to pesticides for a long time, it has been observed that protein levels significantly reduced and aspartate aminotransferase (AST), alanine aminotransferase (ALT), alkaline phosphatase (ALP), and lactate dehydrogenase (LDH) activities significantly increased in these people, compared to people who were not directly exposed to pesticides [36]. On the other hand, in a study conducted on pesticide sales people in the GAP Region, it was observed that AST and ALT activities increased, while ALP and LDH activities decreased in those who worked at pesticide sales locations for a long time [37].

In an experimental study in which the organophosphate insecticide diazinon was applied to rats for 4 weeks, it was reported that this insecticide caused significant changes in hematological and biochemical parameters. Accordingly, it was reported that serum biochemical parameters AST, ALT, ALP, LDH, creatine kinase (CK) activities, and urea, uric acid, and creatinine values of rats, to which diazinon was administered, significantly increased, compared to the control group [38]. It was reported that DDT administration to rats increased serum AST and ALT levels,

Insecticide Used	Experimental Model	Oxidative Stress Parameters	References
Chlorpyriphos	Fish	Increase in MDA, NO	[48]
Chlorpyriphos	Mice	Increase in TOC Decrease in TAC, PON	[19]
Chlorpyriphos	Rat	Increase in lipid peroxidation Decrease in TAC, GSH, GPx, SOD, CAT	[49]
Cypermethrin	Fish	Increase in MDA, SOD, CAT Decrease in GSH, GPx	[50]
Cypermethrin	Mice	Increase in lipid peroxidation (TBARS) Decrease in GST, SOD	[51]
Cypermethrin	Rat	Increase in MDA, CAT Decrease in GSH, GPx, SOD	[52]
Deltamethrin	Fish	Increase in MDA Decrease in CAT, SOD, GSH, GPx	[43]
Deltamethrin	Mice	Increase in MDA	[53]
Deltamethrin	Rat	Increase in MDA Decrease in GSH	[54]
Diazinon	Fish	Increase in MDA, CAT Decrease in SOD, TAC	[55]
Diazinon	Mice	Increase in MDA Decrease in GSH, SOD	[56]
Diazinon	Rat	Increase in MDA, NO Decrease in GSH, GPx, SOD, CAT, TAC	[38]
Dichlorvos	Fish	Increase in lipid peroxidation (TBARS) No significant changes in GST	[57]
Dichlorvos	Mice	Increase in MDA Decrease in CAT, SOD, GPx	[58]
Dichlorvos	Rat	Increase in MDA, NO Decrease in PON	[7]
Malathion	Fish	Increase in LPO, CAT, SOD, GST, GPx Decrease in GSH, TAO	[59]
Malathion	Mice	Increase in MDA Decrease in GSH, GPx, SOD, CAT	[60]
Malathion	Rat	Increase in MDA Decrease in GSH	[61]

Table 2.
The effects of insecticides on oxidative stress parameters in experimental animal models.

stimulated inflammation, and suppressed the immune system [39]. In another study conducted on rats with the carbamate insecticide carbofuran, it was observed that this insecticide increased cholesterol level and AST, ALT, LDH activities and decreased high-density lipoprotein (HDL) level and AChE activity in rat serum after 24-hour treatment [40]. In a study conducted on fish with the organochlorine insecticide lindane, it was reported that glucose increased and total protein decreased at low doses and increased at high doses [41]. In another study conducted on fish, it was stated that the pyrethroid insecticide deltamethrin increased the biochemical parameters cholesterol and glucose values and AST, ALT, ALP activities and decreased the total protein and albumin values [42].

Oxidative stress parameters are among the most important biochemical parameters affected by pesticides. Most environmental pollutants, including pesticides, have the ability to induce oxidative stress in almost all organisms, especially fish. Some studies with pesticide-treated fish revealed that pesticide treatment caused oxidative stress by increasing reactive oxygen species (ROS) in the cells and tissues of fish [43–45]. Oxidative stress negatively affects life of living creatures by causing genotoxic effects, lipid peroxidation and enzyme inhibitions. Lipid peroxidation, which occurs as a result of the toxic effects of pesticides, is an important indicator of oxidative stress and can be demonstrated by measuring malondialdehyde (MDA) levels [16, 46].

As in the other higher organisms, fish have important defense mechanisms to cope with oxidative stress. This defense mechanism, generally called as antioxidant, plays an important role in the survival of fish and in their adaptation to chemical stress. Antioxidant defense systems are composed of enzymatic components such as Paraoxonase (PON), superoxide dismutase (SOD), catalase (CAT), glutathione-S-transferase (GST), glutathione peroxidase (GPx), glutathione reductase (GR) and non-enzymatic components such as glutathione (GSH). SOD and CAT are important antioxidant enzymes that form the first defense mechanism against pesticides. GSH is also an important non-enzymatic antioxidant molecule that protects cells against the harmful effects of oxidative stress [16, 19, 46, 47].

Insecticides that contaminate aquatic systems not only cause toxic effects on fish but also adversely affect living creatures at higher trophic levels through the food chain and cause many negative situations in humans and animals. Insecticides cause significant changes on oxidative stress parameters in humans and animals as well as in fish. **Table 2** shows the effects of insecticides on oxidative stress parameters in experimental animal models.

6. Histopathological effects of insecticides

Pesticides have been widely used from past to present so that food production in the world is not affected by external factors. The most common use in the world is seen in the United States, and almost 15 billion dollars are spent annually for the pesticides [62, 63] and the most common of them is the herbicide glyphosate [62, 64]. Besides their use for sectoral beneficial results, they cause many metabolic disorders and lead to even death, especially due to their intake and absorption into the living body in various ways. OP compounds [65] are responsible for half of the deaths by inhibiting the acetylcholinesterase enzyme (AChE) in the central and autonomic nervous systems, lungs and neuromuscular junctions [62, 66, 67]. AChE inhibition increases cholinergic activity in both the central and peripheral nervous systems. Loss of consciousness, diarrhea, bronchospasm, paralysis and vomiting are the most typical symptoms of poisoning [62, 68], and death can occur as a result of respiratory failure [69, 70]. The toxicity of the substances taken into the organism, its chemical structure, the human resources involved in the poisoning event and the quality of the institution providing medical support affect the mortality rate [71]. Almost half of the patients who are affected by pesticides and apply to the hospital are intubated due to their symptoms and receive ventilation support. Approximately 23–50% of patients in this condition die [72, 73]. The way to diagnose a significant part of the diseases and to obtain sufficient information about clinicopathological parameters is performed by assessing the samples taken from the organism in diseases or suspicious cases. In addition to determining the morphological characteristics of the tissues taken in this process, the role of scientists and especially the contributions of histopathologists in this field are undeniably important in evaluating and interpreting from different scientific dimensions also by using the latest developments in which science has evolved [74]. Histopathological changes are associated with complex biochemical and physiological responses to any stressor. Although histopathological parameters are not highly specific and do not provide quantitative information, they are popular biomarkers for environmental pollution [75]. The histopathological studies, one of the most promising areas for assessing animal health and response to different chemical species, include various studies that show generally cellular differences between control and pesticide-exposed animals [76]. Histopathological markers are considered very important in terms of showing the health status of the organism, together with other branches of science that provide data [77].

As a result of the intake, absorption and participation of pesticides in the systemic circulation and their effects on a cellular basis, the formation of biochemical and histopathological changes in tissue integrity and the emergence of negative symptoms are provided [7, 78–82]. The effects of environmental pollutants on fish tissues can also be determined by histopathological methods. Gills are especially important biomarkers as they are the first organ to encounter pollutants in the environment [13, 83]. In addition, the liver and kidneys are also target organs for the examination of histopathological and biochemical parameters [84]. In *Oncorhynchus mykiss*, as a result of application of clothianidin at different doses for 21 days, the histopathological state caused by this application in muscle, gill, brain and kidney tissues was examined and necrosis ranging from mild to severe in muscle tissue, atrophy and edema in myocytes, hyaline degeneration in muscle fibers and dissolution in connective tissue between myotomes were determined. In the gill tissue, primary and secondary lamella edema, secondary lamella fusion and hyperplasia, primary lamella hyperplasia, secondary lamella lifting, vasodilation, primary lamella thinning and secondary lamella shortening, and secondary lamella peculiar malformations were reported. Pericellular edema and necrosis, Purkinje cell degeneration, cell infiltration, congestion, gliosis, vascular dilatation and dystrophic changes were detected in the brain tissue. In kidney tissue, glomerular atrophy, decrease in hematopoietic tissue cells, tubular degeneration, and an increase in the number and spread of melanomacrophage centers depending on the increasing dose of clothianidin were observed [26].

The LC50 value (50% mortality) of malathion at the end of 96 hours in *Orthrias angorae* exposed to malathion administration was determined to be 3.0237 mg L^{-1}, and it was reported that the frequency of micronucleus formation in erythrocytes increased due to the increasing dose [85]. In a study with rainbow trout (*Oncorhynchus mykiss*) juveniles, the acute effects of maneb and carbaryl were examined and it was reported that edema and lamellar fusion, epithelial swelling and necrosis were observed in the gill lamellae of the fish [86]. In a study investigating the histopathological effects on the gill and kidney tissues of *Cyprinus carpio* as a result of acute application of deltametrine, necrosis, spills, aneurysm, hemorrhages, edema, and hyperplasia were reported in the gills of fish [87]. Different doses of chlorpyrifos-based termifos pesticide were applied to *Clarias gariepinus*

(African catfish) fish for 5, 10 and 15 days. An increase in white blood cell counts and a decrease in erythrocyte counts and hematocrit levels were observed in fish [88]. In a study in which clothianidin, a neoniconioid insecticide, was applied in rainbow trout for 7 and 21 days, it was reported that clothianidin caused hepatocellular degeneration, focal necrosis areas, sinusoidal dilatation and congestion, fibrous and vacuole formation, mild steatosis and pycnosis in the liver tissue, depending on the increasing dose [17].

In a study in which dichlorvos was administered in rats in a subacute manner [7], enlargement of Bowman's capsule, inflammatory cell infiltration, vascular occlusion, glomerular atrophy, and tubular degeneration areas were demonstrated in kidney tissue obtained from the substance-administered group. On the other hand, glomerular lobulation, tubular degeneration, separation in the basal lamina and inflammatory cell infiltration were observed in the group in which dichlorvos and vitamin E were administered. No significant decrease was observed in the severity and frequency of histological changes compared to the dichlorvos administered group (**Figure 3a–f**).

Carbon tetrachloride (CCI4) is a fumigant used to kill insects in cereals and in a study examining the effects of green tea (*Camellia sinensis*) and parsley (*Petroselinum crispum*) diets against carbon tetrachloride hepatoxicity in albino mice, liver degeneration, cellular infiltration, sinusoidal bleeding focuses, congestion and necrotic areas were observed in the CCI4-administered group. No significant decrease in lesion severity and frequency was observed in the histopathological evaluation obtained from the groups using parsley and green tea separately with CCI4 [89]. It was demonstrated that histopathological changes occurred in tissues at doses of 0.1 and 0.05 mg kg^{-1} in mice exposed to deltamethrin. Degenerative and vascular changes in the liver, polymorphonuclear cell infiltration and focal necrosis

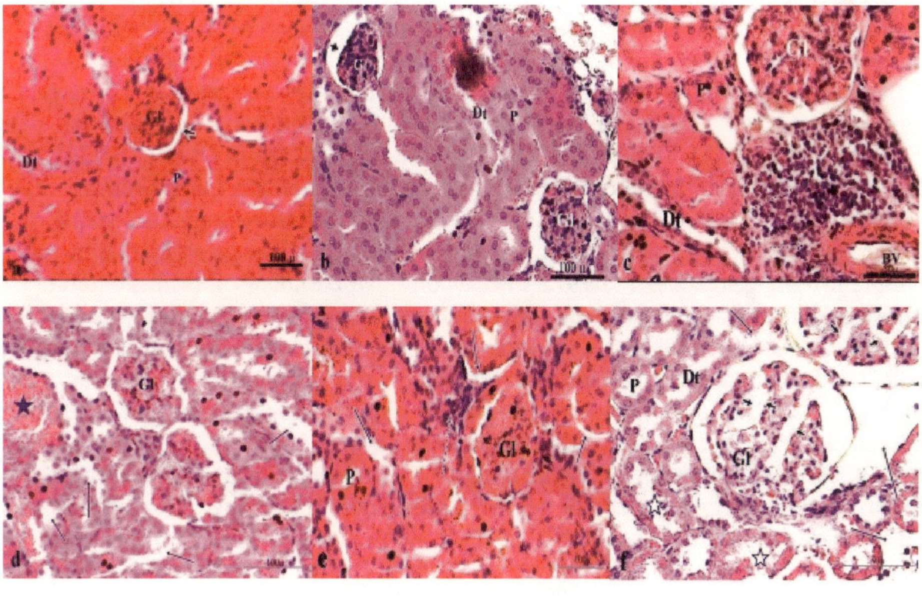

Figure 3.
The effect of dichlorvos (5 mg/kg), corn oil (5 ml/kg) and vit-E (120 mg/kg) either separately or 1 h ago dichlorvos histology of renal section by using hematoxylin and eosin staining (H&E): (a) vit-E group; Gl: Glomeruli, P: Proximal tubule, Dt: Distal tubule, bowman capsule (arrow head), (b, c, d) Dichlorvos group; enlargement in bowman capsule and glomerular atrophy (arrow head), inflammatory cell infiltration (split arrows), BV: Blood vessel, vascular occlusion (asterisks), tubular degeneration (arrows), (e, f) Dichlorvos+vit-E; Gl: Separation in the basal lamina (arrows), tubular degeneration (asterisks), glomerular lobulation (arrow head), magnification: X 400, x 600, (bar: 50 μ, 100 μ) [7].

in hepatocytes were detected. Tubule degeneration and polymorphonuclear cell infiltration in the kidneys, and polymorphonuclear cell infiltration in the peribronchial and perivascular areas of the lungs were reported. Spermatogenic cell degeneration, tubule degeneration, and hyalinization in the seminiferous tubules were demonstrated in the testicles [90]. As a result of malathion application on *Channa punctatus*, the 96-hour LC50 value was determined to be 8.0 mg L^{-1}, sinusoidal dilatation and congestion were observed in the liver, and hypertrophy and pyknotic nuclei were detected in the hepatocytes. In the kidney, histopathological separation of the renal tubular epithelial layer from the basal membrane, vacuolization in the cytoplasm, renal tubule degeneration and necrosis, nuclear pyknosis and hypertrophy were reported in parallel with biochemically high creatinine, urea and BUN values. As the duration of exposure to malathion increased, damage to tissues increased in terms of severity and frequency [91].

7. Conclusion

Objective evaluation of biochemical markers is more practical than histological changes, and the formation of pathological damage in tissue occurs after the reflection of pesticide effects on biochemical parameters [92]. However, the results from histological evaluations alone do not necessarily indicate a direct effect of pesticides. In addition to the histopathological evaluation, examining other results and determining the source of the effects in the organism in this way will be accepted as a more accurate scientific approach. In response to the increasing human population on the planet we live in, the ever- increasing use of pesticides in order for agricultural production to meet this, and their presence at certain doses in food and the risk of mixing with the aquatic ecosystem is an important public health problem that will adversely affect the health of living creatures. Even exposure to pesticides in the above- mentioned amounts that are allowed to be taken into the body daily will cause accumulation in the body over a long period of time and, thus the changes primarily reflected in biochemical parameters, and, in the longer term, histopathological changes will occur due to increased tissue destruction. The oxidant/antioxidant balance in the organism may change in favor of oxidants as a result of external factors such as irregularity in physiological reactions or pesticide accumulation that may occur in the body. Thus, the changes occurring in cellular basis primarily provide outputs as a result of biochemical assessments, and when the damage reaches a textural dimension, histopathological results emerge following the changes in tissue integrity. The necessity of use of pesticides is an undeniable fact, considering the objectives it aims, but it is also known that there will be a decrease in pesticide use as a result of obtaining plant breeds, especially by making use of the developments in the field of biotechnology, that are more resistant to foreign factors such as insects, fungi, algae, weeds, bacteria, nematodes and rodents. In addition, more use of the biological control option against the factors that reduce the plant yield at a higher level will also produce beneficial results in terms of human health.

Author details

Haci Ahmet Deveci[1*], Gökhan Nur[2], Ayla Deveci[3], Inan Kaya[4], Müge Mavioglu Kaya[5], Abdulsamed Kükürt[6], Volkan Gelen[7], Ömer Faruk Başer[8] and Mahmut Karapehlivan[8]

1 Faculty of Health Sciences, Department of Nutrition and Dietetics, Gaziantep University, Gaziantep, Turkey

2 Faculty of Engineering and Natural Sciences, Department of Biomedical Engineering, Iskenderun Technical University, Hatay, Turkey

3 Vocational School of Technical Sciences, Kilis 7 Aralık University, Kilis, Turkey

4 Faculty of Science and Letters, Department of Biology, Kafkas University, Kars, Turkey

5 Faculty of Science and Letters, Department of Molecular Biology and Genetics, Kafkas University, Kars, Turkey

6 Faculty of Veterinary Medicine, Department of Biochemistry, Kafkas University, Kars, Turkey

7 Faculty of Veterinary Medicine, Department of Physiology, Kafkas University, Kars, Turkey

8 Faculty of Medicine, Department of Biochemistry, Kafkas University, Kars, Turkey

*Address all correspondence to: h.ahmet_deveci@hotmail.com

References

[1] Akdoğan A, Divrikli U, Elçi L. Importance of pesticides and their effects on ecosystem. Academic Food. 2012; 1:125-132.

[2] Kekillioğlu A, Bıçak Z. Investigation of pesticides and their effects on bees. EJONS International Journal on Mathematic, Engineering and Natural Sciences. 2020; 4(13):26-44.

[3] Abdollahi M, Mostafalou S, Pournourmohammadi S, Shadnia S. Oxidative stress and cholinesterase inhibition in saliva and plasma of rats following subchronic exposure to malathion. Comp. Biochem. Physiol. Part C. 2004; 137: 29-34.

[4] Ekebaş S, Çakır S, Ertuğrul O, Kence A. The detection of mutagenic activity of some chemicals (azamethypos, dichlorvos, methyl parathion, aflatoxin b1) by the smart test in drosophila melanogaster. Turk. J. Vet. Anim. Sci. 2000; 24(6):563-569.

[5] Deveci HA, Karapehlivan M, Kaya I, Kükürt A, Alpay M. Protective effect of caffeic acid phenethyl ester against acute chlorpyrifos-ethyl poisoning. Journal of Ankara University Faculty of Veterinary Medicine. 2015; 62: 255-260.

[6] Nur G, Husunet MT, Güler I, Deveci A, Koç E, Nur O, Kilicle PA. The effect of caffeic acid phenethyl ester (CAPE) on hepatic histopathology and oxidative stress in rats treated with malathion. Medicine Science. 2018; 7(3): 604-609.

[7] Nur G, Deveci HA, Koç E. Preservation of vitamin-E against nephrotoxic effect induced by subacute Dichlorvos application. Fresenius Environmental Bulletin. 2021; 30(7): 8651-8659.

[8] Daş YK, Aksoy A. Pesticides. Turkiye Klinikleri Veterinary Sciences- Pharmacology and Toxicology - Special Topics. 2016; 2(2): 1-17.

[9] Chawla P, Kaushika R, Swarajb VJS, Kumarb N. Organophosphorus pesticides residues in food and their colorimetric detection, Environmental Nanotechnology, Monitoring & Management, 2018; 10:292-307.

[10] Hassaan MA, El Nemr A. Pesticides pollution: Classifications, human health impact, extractionand treatment techniques. Egyptian Journal of Aquatic Research. 2020; 46: 207-220.

[11] Ceron JJ, Panizo CG, Montes A. Toxicological effects in rabbits induced by endosulfan, lindane and methylparathion representing agricaltural by products contamination. Bull. Environ. Contam. Toxicol., 1995; 54: 258-265.

[12] Madej K, Kalenik TK, Piekoszewski W. Sample preparation and determination of pesticides in fat-containing foods. Food Chemistry. 2018; 269: 527-541.

[13] Kayhan FE. Cycle of Insecticides in Nature and Effects on Aquatic Environment. Selçuk University Faculty of Science Journal of Science. 2020; 46(2): 29-40.

[14] Buah-Kwofie A, Humphries MS, Pillay L. Bioaccumulation and risk assessment of organochlorine pesticides in fish from a global biodiversity hotspot: Simangaliso Wetland Park, South Africa. Science of The Total Environment 2018; 621: 273-281.

[15] Kapsi M, Tsoutsi C, Paschalidou A, Albanis T. Environmental monitoring and risk assessment of pesticide residues in surface waters of the Louros River (N.W. Greece). Science of The Total Environment. 2019; 650: 2188-2198.

[16] Fırat Ö, Aytekin T. Effect of neonicotinoid insecticide Thiamethoxam on oxidative stress parameters in Oreochromis niloticus. Journal of Balikesir University Institute of Science and Technology., 2018; 20(2): 224-234.

[17] Dogan D, Deveci HA, Nur G. Manifestation of oxidative stress and liver injury in clothionidin exposed Oncorhynchus mykiss. Toxicology Research, 2021a; 10 (3): 501-510. DOI: 10.1093/toxres/tfab027.

[18] Costa LG, Giordano G, Guizzetti M, Vitalone A. Frontiers in Bioscience 2008; 13: 1240-1249.

[19] Deveci HA, Karapehlivan M. Chlorpyrifos-induced parkinsonian model in mice: Behavior, histopathology and biochemistry. Pesticide Biochemistry and Physiolology. 2018; 144: 36-41.

[20] Eyer P, Szinicz L, Thiermann H, Worek F, Zilker T. Testing of antidotes for organophosphorus compounds: Experimental procedures and clinical reality. Toxicology. 2007; 233(1-3): 108-119.

[21] Karalliedde LD, Edwards P, Marrs TC. Variables influencing the toxic response to organophosphates in humans. Food Chem Toxicol, 2003; 41:1-13

[22] Demirdöğen BC. Organophosphate pesticide poisoning and the role of serum Paraoxonase 1 (PON1) enzyme in organophosphate metabolism. Turkish Hygiene Experimental Bio. Journal., 2010; 67(2): 97-112.

[23] Ozkaya G, Çeliker A, Koçer-Giray B. Insecticide poisoning and evaluation of the situation in Turkey. Turkish Hygiene Experimental Bio. Journal., 2013; 70(2): 75-102.

[24] Evangelista de Duffard AM, Duffard R. Behavioral toxicology, risk assessment, and chlorinated hydrocarbons. Environ Health Perspect, 1996; 104(2): 353-360.

[25] Nur G, Deveci HA, Ersan Y, Merhan O, Nazli M, Nur O. Protective role of caffeic acid phenethyl ester against tetramethrine-induced toxicity in mice. Medicine Science. 2016; 5(4): 972-978.

[26] Dogan D, Nur G, Deveci HA. Tissue-specific toxicity of clothianidin on rainbow trout (*Oncorhynchus mykiss*). Drug and Chemical Toxicology. 2021b; 1:1-11. DOI: 10.1080/01480545.2021.1892128.

[27] Kükürt A, Kuru M, Karapehlivan M. Nitrik oksit, nitrik oksit sentaz ve dişi üreme sistemindeki rolleri. In: Evereklioğlu C, editor. Sağlık Bilimleri Alanında Akademik Çalışmalar-II, Gece Kitaplığı; 2020, p. 113-23.

[28] Kükürt A. Doğal bir antioksidan olarak propolis tedavisinin koruyucu etkileri. In: Evereklioğlu C, editor. Sağlık Bilimlerinde Teori ve Araştırmalar II, Gece Kitaplığı; 2020, p. 501-15.

[29] Kükürt A, Kuru M, Faruk Başer Ö, Karapehlivan M. Kisspeptin: Role in Female Infertility. In: Marsh C, editor. Reproductive Hormones, IntechOpen; 2020. https://doi.org/10.5772/intechopen.94925.

[30] Başer ÖF, Kükürt A, Karapehlivan M. Oksidatif stresin azaltılmasında anjiyotensin dönüştürücü enzimin rolü. In: Evereklioğlu C, editor. Sağlık Bilimlerinde Teori ve Araştırmalar II, Gece Kitaplığı; 2020, p. 243-53.

[31] Kuru M, Kükürt A, Oral H, Öğün M. Clinical Use of Progesterone and Its Relation to Oxidative Stress in Ruminants. In: Drevenšek G, editor. Sex Hormones in Neurodegenerative Processes and Diseases, InTechOpen; 2018, p. 303-27. https://doi.org/10.5772/intechopen.73311.

[32] Gelen V, Kükürt A, Şengül E, Başer ÖF, Karapehlivan M. Can Polyphenols be Used as Anti-Inflammatory Agents against Covid-19 (SARS-CoV-2)-Induced Inflammation? [Online First], IntechOpen, DOI: 10.5772/intechopen.98684. Available from: https://www.intechopen.com/online-first/77746

[33] Gelen V, Kükürt A, Şengül E. Role of the Renin-Angiotensin-Aldosterone System in Various Disease Processes: An Overview [Online First], IntechOpen, DOI: 10.5772/intechopen.97354. Available from: https://www.intechopen.com/online-first/76612

[34] Kükürt A, Gelen V, Başer OF, Deveci HA, Karapehlivan M. Thiols: Role in Oxidative Stress-Related Disorders [Online First], IntechOpen, DOI: 10.5772/intechopen.96682. Available from: https://www.intechopen.com/online-first/75841

[35] Kurutaş EB, Kılınç M. Effect of pesticides on biological systems. Archive Source Review Journal, 2003; 12:215-228

[36] Çömelekoğlu Ü, Mazmancı B, Arpacı A. Examination of liver functions in agricultural workers exposed to the chronic effects of pesticides. Turk J Biol, 2000; 24: 461-466.

[37] Mutlu F, Şimşek Z, Cici M. Investigation of cholinesterase inhibitors and liver enzyme levels in pesticide sales workers in the GAP Region. Harran Univ Veterinary Faculty Journal, 2017; 6 (1): 57-62.

[38] Abdel-Daim MM, Abushouk AI, Alkhalf MI, Toraih EA, Fawzy MS, Ijaz H, Aleya L, Bungau SG. Antagonistic effects of Spirulina platensis on diazinon-induced hemato-biochemical alterations and oxidative stress in rats. Environmental Science and Pollution Research. 2018; 25:27463-27470.

[39] Karamese M, Aydin H, Gelen V, Sengul E, Karamese SA. The antiinflammatory, anti-oxidant and protective effects of a probiotic mixture on organ toxicity in a rat model. Future Microbiol. 2020;15:401-12. https://doi.org/10.2217/fmb-2020-0005.

[40] Jaiswal SK, Gupta VK, Siddiqi NJ, Pandey RS, Sharma Bechan. Hepatoprotective effect of *Citrus limon* fruit extract against Carbofuran induced toxicity in Wistar Rats. Chinese Journal of Biology. 2015; 1-10.doi.org/10.1155/2015/686071.

[41] Saravanan M, Prabhu Kumar K, Ramesh M. Haematological and biochemical responses of freshwater teleost fish Cyprinus carpio (Actinopterygii: Cypriniformes) during acute and chronic sublethal exposure to lindane. Pest Biochem Physiol. 2011; 100:206-211.

[42] El-Sayed YS, Saad TT. Subacute intoxication of a deltamethrinbased preparation (Butox® 5% EC) in monosex Nile tilapia, Oreochromis niloticus L. Bas. Clinic Pharmacol Toxicol. 2007; 102:293-299.

[43] Yonar, ME, Sakin F. Ameliorative effect of lycopene on antioxidant status in Cyprinus carpio during pyrethroid deltamethrin exposure. Pesticide Biochemistry and Physiology.2011; 99: 226-231.

[44] Deveci HA, Unal S, Karapehlivan M, Ayata MK, Gaffaroğlu M, Kaya I, Yılmaz M. Effects of glyphosate (herbicide) on serum paraoxonase activity, high density lipoprotein, total antioxidant and oxidant levels in Kars Creek Transcaucasıan Barbs (Capoeta capoeta [Guldenstaedt, 1773]). Fresenius Environmental Bulletin. 2017; 26(5), 3514-3518.

[45] Nur G, Deveci HA. Histopathological and biochemical

responses to the oxidative stress induced by glyphosate-based herbicides in the rainbow trout (Oncorhynchus mykiss). Journal of Cellular Neuroscience and Oxidative Stress. 2018; 10(1), 656-665.

[46] Toroser D, Orr WC, Sohal RS. Carbonylation of mitochondrial proteins in Drosophila melanogaster during aging, Biochemical and Biophysical Research Communications.2007; 363: 418-424.

[47] Zirong X. Shijun B. Effects of waterborne Cd exposure on glutathione metabolism in Nile tilapia (Oreochromis niloticus) liver. Ecotoxicology and Environmental Safety.2007; 67: 89-94.

[48] Deveci HA, Kükürt A, Nur G, Kaya I. Effect of Chlorpyrifos-ethyl application on sialic acid, malondialdehyde and nitric oxide levels in Cyprinus carpio (L. 1758). Kafkas University, Journal of Science Institute. 2016; 9 (2): 46-51.

[49] Kopjar N, Žunec S, Mendaš G, Micek V, Kašuba V, Mikolić A. Evaluation of chlorpyrifos toxicity through a 28-day study: cholinesterase activity, oxidative stress responses, parent compound/metabolite levels, and primary DNA damage in blood and brain tissue of adult male Wistar rats. Chem Biol Interact. 2018; 279:51-63.

[50] Yonar ME. Protective Effect of Lycopene on oxidative stress and antioxidant status in Cyprinus carpio during Cypermethrin exposure. Environmental Toxicology. 2013; 28(11):609-616. doi: 10.1002/tox.20757

[51] Al-Shaikh TM. Protective antioxidant effect of garlic against cypermethrin induced lung toxicity in adult male mice. Biochemical and Histopathological studies Life Science Journal. 2012; 9(4): 4017-4025.

[52] Raina R, Verma PK, Pankaj NK, Kant V, Prawez S. Protective role of L-ascorbic acid against cypermethrin-induced oxidative stress and lipid peroxidation in Wistar rats. Toxicological & Environmental Chemistry. 2010; 92(5): 947-953.

[53] Bagherpour H, Malekshah AK, Amiri FT, Azadbakht M. Protective effect of green tea extract on the deltamethrin-induced toxicity in mice testis: An experimental study. International Journal of Reproductive BioMedicine. 2019; 17: 337-348. doi.org/10.18502/ijrm.v17i5.4601.

[54] Shivanoor SM, David M. Reversal of deltamethrin-induced oxidative damagein rat neural tissues byturmeric-diet: Fouriertransform-infrared and biochemical investigation. Zoology the Journal of Basic & Applied Zoology. 2016; 77:56-68.

[55] Ali M, Mirvaghefi A, Asadi F. Effects of vitamin E, selenium and vitamin C on various biomarkers following oxidative stress caused by diazinon exposure in rainbow trout. Ege J Fish Aqua Sci. 2015; 32(3): 151-158. DOI: 10.12714/egejfas.2015.32.3.05.

[56] El-Shenawy NS, El-Salmy F, Al-Eisa RA, El-Ahmary B. Amelioratory effect of vitamin E on organophosphorus insecticide diazinon-induced oxidative stress in mice liver. Pesticide Biochemistry and Physiology. 2010; 96:101-107. doi:10.1016/j.pestbp.2009.09.008.

[57] Varó I, Navarro JC, Nunes B, Guilhermino L. Effects of dichlorvos aquaculture treatments on selected biomarkers of gilthead sea bream (*Sparus aurata* L.) fingerlings. Aquaculture. 2007; 266:87-96

[58] Chaudhary B, Bist R. Protective manifestation of bacoside A and bromelain in terms of cholinesterases, gamma-amino butyric acid, serotonin level and stress proteins in the brain of dichlorvosintoxicated mice. Cell Stress Chaperones. 2017; 22(3):371-376.

[59] Ibrahim AT. Biochemical and histopa-thological response of Oreochromis niloticus to malathion hepatotoxicity. J. Royal. Sci. 2019; 1(1): 10-15.

[60] Ali RI, Ibrahim MA. Malathion induced testicular toxicity and oxidative damage in male mice: the protective effect of curcumin. Egypt J Forensic Sci. 2018; 8(70):1-13 doi.org/10.1186/s41935-018-0099-x. Protective Effect of Lycopene on Oxidative Stress and Antioxidant Status in *Cyprinus carpio* during Cypermethrin Exposure

[61] Mohammadzadeh L, Hosseinzadeh H, Abnous K, Razavi BM. Neuroprotective potential of crocin against malathion-induced motor deficit and neurochemical alterations in rats. Environ Sci Pollut Res. 2018; 25(5): 4904-4914.

[62] Hulse EJ, Smith SH, Wallace WA, Dorward DA, Simpson AJ, Drummond G, Clutton RE, Eddlestonet M. Development of a histopathology scoring system for the pulmonary complications of organophosphorus insecticide poisoning in a pig model. PLoS ONE. 2020; 15(10): e0240563.

[63] A look at fertilizer and pesticide use in the US New York, USA: gro-intelligence.com; 2018. Available from: https://gro-intelligence.com/insights/articles/a-look-at-fertilizer-and-pesticide-use-in-the-us.

[64] Atwood D, Paisley-Jones C. Pesticides Industry Sales and Usage 2008-2012 Market Estimates. Washington, USA: Office of Pesticide Programs US Environmental Protection Agency, 2017.

[65] Gunnell D, Eddleston M, Phillips MR, Konradsen F. The global distribution of fatal pesticide self-poisoning: Systematic review. Bmc Public Health. 2007; 7:357-373.

[66] Patočka J, Kuča K, Jun D. Acetylcholinesterase and butyrylcholinesterase–important enzymes of human body. Acta Medica (Hradec Kralove). 2004; 47(4):215-228.

[67] Fryer AD, Lein PJ, Howard AS, Yost BL, Beckles RA, Jett DA. Mechanisms of organophosphate insecticide-induced airway hyperreactivity. American Journal of Physiology-Lung Cellular and Molecular Physiology. 2004; 286(5):L963-L9L9.

[68] Namba T. Poisoning due to organophosphate insecticides-acute and chronic manifestations. American Journal of Medicine. 1971; 50(4):475-492.

[69] Chang C, Chuang M, Lee C, Hsieh M. Outcome of Organophosphate Poisoning with Acute Respiratory Failure: A 10-Year Study. AJRCCM. 2020; 201: A5354.

[70] Banday TH, Tathineni B, Desai MS, Naik V. Predictors of Morbidity and Mortality in Organophosphorus Poisoning: A Case Study in Rural Hospital in Karnataka, India. North American journal of Medical Sciences. 2015; 7(6):259-265.

[71] Eddleston M. Novel clinical toxicology and pharmacology of organophosphorus insecticide self-poisoning. Annual review of pharmacology and toxicology. 2019; 59:341-360.

[72] Alahakoon C, Dassanayake TL, Gawarammana IB, Weerasinghe VS, Buckley NA. Differences between organophosphates in respiratory failure and lethality with poisoning post the 2011 bans in Sri Lanka. Clinical Toxicology (Phila). 2019; 466-470.

[73] Muley A, Shah C, Lakhani J, Bapna M, Mehta J. To identify morbidity and mortality predictors in acute organophosphate poisoning. Indian

journal of critical care medicine: peer-reviewed, official publication of Indian Society of Critical Care Medicine. 2014; 18(5):297-300.

[74] Doğan Ö. Standardization in the process of histopathological diagnosis. Aegean Pathology Journal. 2005; 2:8-28.

[75] Hinton DE, Lauren DJ. Liver structural alterations accompanying chronic toxicity: potential biomarkers of exposure. In: McCartney JF, Shugart LR (eds): Biomarkers of environmental contamination (pp 15-57). Boca Raton, FL: Lewis Publ., 1990.

[76] Niimi AJ. Review of biochemical methods and other indicators to assess fish health in aquatic ecosystems containing toxic chemicals. J Great Lakes Res. 1990; 16:529-541.

[77] Van der Oost R, Beyer J, Vermeulen NP. Fish bioaccumulation and biomarkers in environmental risk assessment: a review. Environ Toxicol Pharmacol. 2003; 13:57-149.

[78] Clasen B, Loro VL, Murussi CR, Tiecher TL, Moraes B, Zanella R. Bioaccumulation and oxidative stress caused by pesticides in Cyprinus carpio reared in a rice-fish system. Science of The Total Environment. 2018; 626(1):737-743.

[79] Gaaied S, Oliveira M, Le Bihanic F, Cachot J, Banni M. Gene expression patterns and related enzymatic activities of detoxification and oxidative stress systems in zebrafish larvae exposed to the 2,4-Dichlorophenoxyacetic acid herbicide. Chemosphere. 2019; 224:289-297.

[80] Volschenka CM, Ikenak Y, Yohannes YB, Nakayama SM, Ishizuka M, Smith E, Van Vuren JMJ, Greenfield R. Baseline bio-accumulation concentrations and resulting oxidative stress in Synodontis zambezensis after an acute laboratory exposure to 4,4′- DDT. Pesticide Biochemistry and Physiology. 2019; 156:44-55.

[81] Kaya I., Yılmaz M., Koç E., Deveci H.A., Ersan Y., Karapehlivan M. Investigation of the serum total antioxidant, oxidant and sialic acid levels of *Cyprinus carpio* (l. 1758) treated with tebuconazole (fungicide). Journal of FisheriesSciences com. 2014; 8(3): 214-219.

[82] Kaya, I., Onen O., Kaya M.M, Kükürt A., Deveci H.A, Karapehlivan M. The effects of organophosphorus pesticides on some biochemical markers in Teleosts. In: New Trends in Health Sciences. Duran N. (ed.), Livre de Lyon. Lyon. France. 2021; p. 213-227.

[83] Akbulut C, Kaymak G, Esmer HE, Yön ND, Kayhan FE. Mechanisms of oxidative stress induced by heavy metals and pesticides in fish. Ege J Fish Aqua Sci. 2014; 31(3):155-160.

[84] Akbulut C, Özturk B, Uzun A, Yon ND. Tribenuron methyl exposure inhibits oogenesis in zebrafish Danio rerio (Hamilton, 1822). Indian Journal of Fisheries. 2017; 3:127-131.

[85] Gül S, Nur G, Kaya Ö.T, Kamber U, Gürgegin B. Detection of micronuclei in peripheral erythrocytes of Orthrias angorae (Steindachner, 1897) exposed to malathion. Fresenius Environmental Bulletion. 2007; 16(5):472-476.

[86] Boran H, Altinok I, Capkin E. Histopathological changes induced by maneb and carbaryl on some tissues of rainbow trout, *Oncorhynchus mykiss*. Tissue and Cell. 2010; 42(3):158-164.

[87] Cengiz EI. Gill and kidney histopathology in the freshwater fish Cyprinus carpio after acute exposure to deltamethrin. Environmental Toxicology and Pharmacology. 2006; 22(2):200-204.

[88] Nwani CD, Ugwu DO, Okeke OC, Onyishi GC, Ekeh FN, Atama C, Eneje LO. Toxicity of the chlorpyrifos-based pesticide Termifos: effects on behaviour and biochemical and haematological parameters of African catfish Clarias gariepinus. African Journal of Aquatic Science. 2013; 38(3):255-262.

[89] Guven A, Nur G, Deveci HA. Effect of Green tea (Camellia sinensis L.) and Parsley (Petroselinum crispum) Diets Against Carbon Tetrachloride Hepatotoxicity in Albino Mice. Fresenius Environmental Bulletin. 2019; 28(9):6521-6527.

[90] Tewari A, Banga HS, Gill JPS. Sublethal chronic effects of oral dietary exposure to deltamethrin in Swiss albino mice. Toxicology and Industrial Health. 2018; 34(6):1-10.

[91] Bharti S, Razool F. Analysis of the biochemical and histopathological impact of a mild dose of commercial malathion on Channa punctatus (Bloch) fish. Toxicology Reports. 2021; 8:443-455.

[92] Ewald G. Chronic measures of toxicant-induced effects on fish. Ann Zool Fennici. 1995; 32:311-316.

Chapter 6

Review on the Impact of Insecticides Utilization in Crop Ecosystem: Their Prosperity and Threats

S.A. Dwivedi, V.K. Sonawane and T.R. Pandit

Abstract

Pesticide covers a wide range of compounds including insecticides, fungicides, herbicides, rodenticides, molluscicides, nematicides, plant growth regulators, and others. Among them, organochlorine insecticides, used successfully in controlling a number of diseases, such as malaria and typhus, were banned or restricted after the 1960s in most of the technologically advanced countries. The introduction of other synthetic insecticides—organophosphate insecticides in the 1960s, carbamates in the 1970s, and pyrethroids in the 1980s, and the introduction of herbicides and fungicides in the 1970s–1980s contributed greatly to pest control and agricultural output. Ideally, a pesticide must be lethal to the targeted pests, but not to nontarget species, including man. Unfortunately, this is not the case, so the controversy of the use and abuse of pesticides has surfaced. The rampant use of these chemicals, under the adage, "if little is good, a lot more will be better" has played with humans and other life forms. The known ecological impacts of insecticides on terrestrial and aquatic ecosystems are reviewed in this chapter. Awareness of the impacts that insecticides are having in our world may help to introduce the management practices that aim at reducing and mitigating those impacts.

Keywords: pest control, insecticides, organophosphate, organochlorine, environment, human health

1. Introduction

Insecticides (natural or synthetic) are used in agriculture to control pests, weeds, and diseases in plant species. Herbicides, insecticides, fungicides, rodenticides, nematicides, and others are examples of pesticides. Pesticides became an important tool for plant protection and crop yield enhancement during the agricultural development process. A pest infestation accounts for approximately 45% of annual food production: therefore, effective pest management through the use of a diverse range of pesticides is required to combat pests and increase crop production [1]. However, in the latter half of the nineteenth century, rapid growth in the global economy, including both the industrial and agricultural sectors, resulted in a steady increase in the production and use of agricultural-based chemicals, which frequently have disastrous effects on the environment. Inadequate use of pesticides and some other persistent organic pollutants in agricultural soils has wreaked havoc on future consequences. Due to their bioaccumulation properties and high toxicity, the persistent and ubiquitous nature of various agricultural-based pesticides and other organic pollutants has caused destruction to humanity. These pesticides are known to impair the functioning of living organisms' metabolic and reproductive systems [2]. Certain pesticides, such as DDT, aldrin, hexachlorobenzene, dieldrin, mirex, endrin, chlordane, and heptachlor, have negative effects on human health and the environment [3].

Currently, approximately 2 million tonnes of pesticides are used globally, with 47.5% being herbicides, 29.5% being insecticides, 17.5% being fungicides, and 5.5% being other pesticides [4]. China, the United States, Argentina, Thailand, Brazil, Italy, France, Canada, Japan, and India are the top 10 pesticide-consuming countries in the world. Furthermore, it is predicted that global pesticide usage will rise to 3.5 million tonnes by 2020 [5].

They include ovicides and larvicidal, which are used to kill insect eggs and larvae. Insecticides are used extensively in agriculture, medicine, and industry, and by consumers. Insecticides are credited with significantly increasing agricultural productivity in the twentieth century [6]. Almost all insecticides have the potential to drastically change ecosystems; many are harmful to humans and also to animals and some become focused as they move up the food chain.

2. Classification of insecticides

2.1 Organochlorine

Organochlorines (chlorinated hydrocarbons) having organic compounds are attached to five or more chlorine atoms. It is the first group of pesticides that are synthesized and used in public health and agriculture. It is used against the management of a wide range of insects and having long-term residual effects in the atmosphere. The mode of action of these insecticides is disrupting the nervous system of the insects which caused paralysis and convulsions, which leads to death (**Figure 1**).

2.2 Organophosphates

The group "organophosphate" pesticides are considered to be one of the broad-spectrum pesticides, which manage various numbers of pests because of their multiple functions. The organophosphate pesticides are biodegradable, cause very low environmental pollution, and are slow pest resistance [8].

2.3 Carbamates

The carbamates insecticides are similar to the organophosphates insecticides. Carbamates are derived from carbamic acid where organophosphates are the derivatives of phosphoric acid. Mode of action of carbamate pesticides and organophosphate pesticides are similar, that is, affecting the nervous system, that is, the transmission of nerve signals that caused the death [9].

2.4 Synthetic pyrethroid

A synthetic pyrethroid is a group of pesticides that are organic pesticide group that can be synthesized from natural pyrethrins.

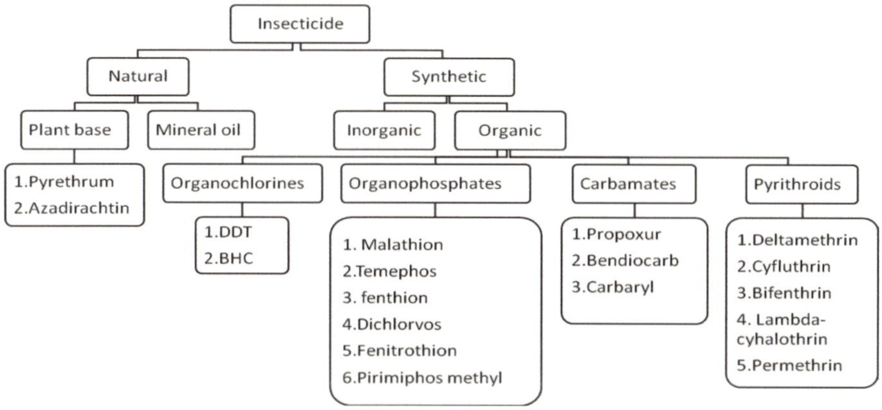

Figure 1.
Classification of insecticides (Source: [7]).

3. Pesticides production in India

India is the fourth largest producer of agrochemicals followed by the USA, Japan, and China. Worldwise utilization of agrochemical regarding pest management India has the ninth rank (**Figure 2**). In India, Maharashtra ranks first in the consumption of pesticides regarding crop protection followed by Uttar Pradesh and West Bengal having 13,243, 11,557, and 3630 MT consumption of pesticide, respectively [10]. In India, maximum utilization of pesticides takes place regarding the management of pest of cotton (45%) followed by rice (22%) and vegetable (9%) as globally comparing the highest in fruit and vegetable (26%) followed by cereals (18%) (**Figure 3**).

3.1 Benefits of pesticides

Farmers have made significant growth in food production by using pesticides over the last 60 years. They did this primarily to prevent or reduce agricultural

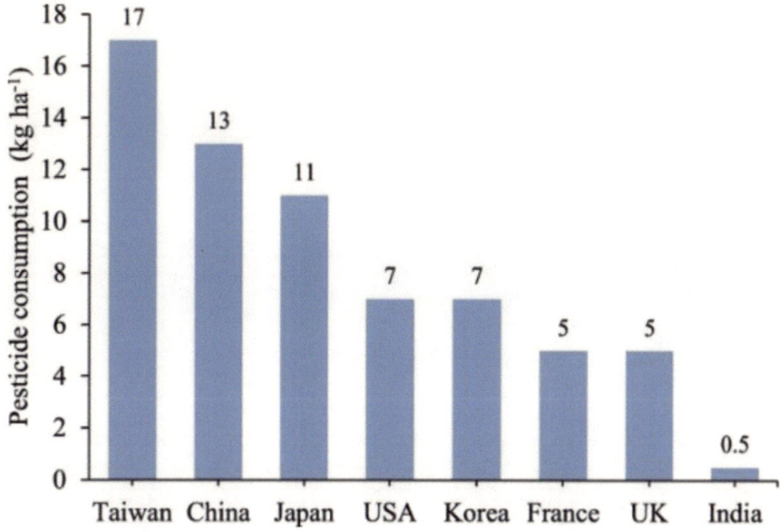

Figure 2.
Pesticide consumption in different countries of the world (Source: Indian Agrochemical Industry Report, 2016).

losses caused by pest activity, which resulted in increased yield and greater availability of food at a reasonable price throughout the year. In most countries, agricultural productivity has increased dramatically as a result of the use of pesticides (**Figure 4**). For example, wheat yields in the United Kingdom [12] corn yields in the USA [13], and total yields in Russia and other countries were enhanced enormously [14–16]. It has long been assumed that diets rich in fresh fruits and vegetables far compensate the risks of eating crops with very low pesticide residues [17]. Better nutrition and less drudgery both improve the quality of life and the length of life [18]. Improved medical care and drug treatments, as well as hygiene, having all played a significant role in extending lives, but the importance of nutritious, safe, and affordable food as a health promoter that increases life expectancy should not be underestimated [19, 20]. Controlling a wide range of vectors of human and livestock disease, thereby reducing the number of infected people and deaths, as well as preventing the spreading of international disease, is one of the most obvious benefits of widespread pesticide use. The most effective way to combat vectors is to kill them. According to the World Health Organization, life will be unacceptably dangerous for a large proportion of humanity if chemical control methods are not available. Pesticides are essential in the destruction of many living things, which

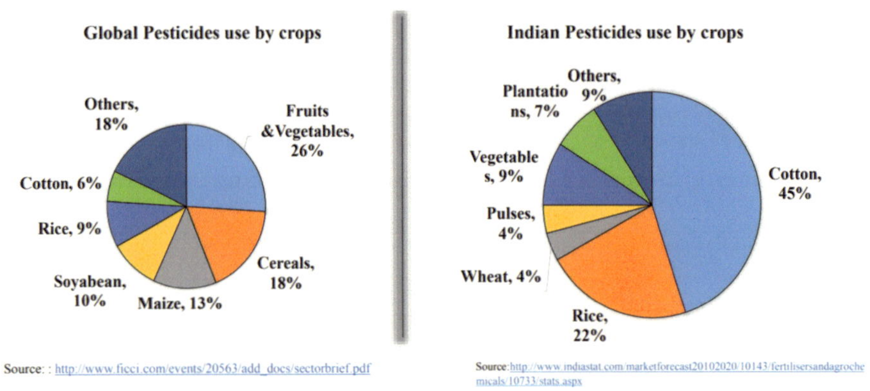

Figure 3.
Crop wise consumption of pesticides.

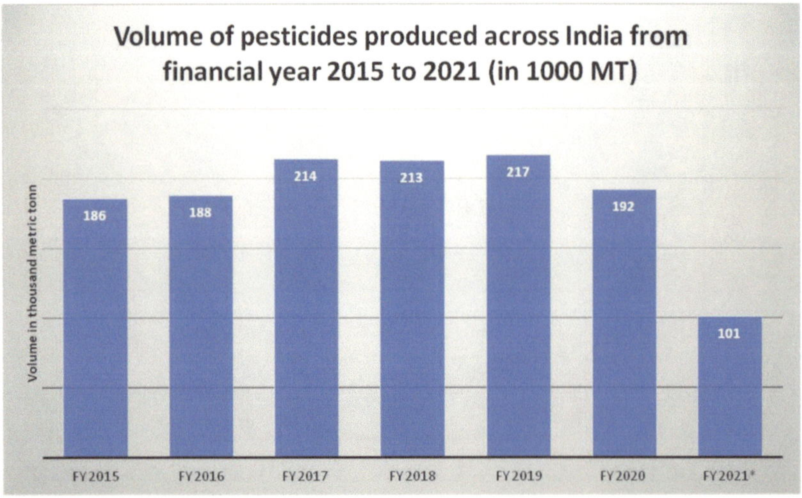

Figure 4.
Pesticides production in India (Source: [11]).

has a negative effect on human activities, infrastructure, and everyday activities. Insecticides have been used to kill unwanted organisms in many particular sectors of human action, such as the prevention of accelerated corrosion of metal structures, the maintenance of turf on sports pitches, cricket grounds, and golf courses, and assisting in the facilitation of a hugely popular pastime that provides fresh air and sunshine for thousands of people in domestic and decorative gardening.

Pesticides can provide a variety of benefits, but many of these advantages go unobserved by the general public. The most important and simplest to calculate advantages are also the financial advantages for farmers derived from the protection of farm outcome and quality as well as the reduction of other expensive inputs such as labor and fuel. Estimates of global pest failures for eight crops in some regions revealed that pest-induced losses exceeded 50% of attainable agricultural production [21]. Pests destroyed 15% of crops, disease pathogens and weeds each accounted for 13%, and post-harvest pest infestations accounted for 10%. Agricultural production would decline and food costs would rise steeply if pesticides were not used. Farmers would be less competitive in the international markets for major commodities if they produced less and charged more. Preventing or decreasing agriculture sector losses to pests through the use of pesticides improves yields and thus ensures consistent supplies of agricultural produce at reasonably priced prices for consumers. It also enhances the quality of the food in terms of esthetic appeal, which is important to customers.

Insecticides are also commonly used in a number of other situations, many of which the general public is unaware of. In the same manner that pests in agriculture and public health cause adverse effects such as losses, contamination, and damage, those organisms have a negative effect on social activities, infrastructure, and everyday materials when left unchecked. Pesticides play a significant, but often unseen, role in mitigating this negative effect. Thus, the advantages of pesticides can accumulate to a range of different recipients, not just farmers or buyers, but also to society as a whole.

Other advantages include the preservation of esthetic quality, the protection of human health from disease-carrying organisms, the eradication of nuisance that causes diseases, and the preservation of other organisms, such as endangered species. Insecticides are often used in ways which we often take it for granted in our businesses and homes. For example, the controlled use of insecticides in processing, manufacturing, and packaging facilities protects raw commodities and packaged grocery products from insect contamination. Pesticides are often used in supermarkets to control rodents and insects drawn to food and food waste.

According to Davis et al. [22], nearly all families (97.8%) used pesticides at least once per year, and two-thirds used insecticides more than five times each year. The home was the most common location for family pesticide use, with 80% of families using pesticides at least once per year. This was followed by the use of herbicides to control yard weeds (57% of families) and insecticides to control fleas and ticks on pets (50% of families). Pesticides were also used by a significant number of families in their gardens or orchards (33%). It is obvious that proper pesticide use improves our quality of life, protects our property, and promotes a healthier environment. These nonmonetary advantages of pesticide use are difficult to quantify. Policymakers have long struggled with how to assign monetary values to things like esthetic value, the survival of certain endangered species, and peacefulness. In practice, such nonmarket advantages are rarely recognized as key by policymakers as positive effects that can be quantified in the marketplace, and thus, they are largely unnoticed. The innovation of a pesticide use profile is usually the first step in calculating the beneficial effects of each pesticide. The lack of an insecticide use database is a major barrier to determining accurate estimates of the impact of changes in insecticide availability (**Table 1**).

Sr. No.	The primary advantages	The secondary advantages
1.	Controlling disease vectors and nuisances organism	Global benefits
	Human lives saved Human disturbance reduced Animal suffering reduced Increased livestock quality	Less pressure on uncropped land Fewer pest introductions else where International tourism revenue
2.	Controlling pests and plant disease vectors	National benefits
	Improved crop/livestock quality Reduced fuel use for weeding Reduced soil disturbance Invasive species controlled	National agricultural economy Increased export revenues Reduced soil erosion/moisture loss
3.	Prevent or control of organisms that harm other human activities and structure	Community benefits
	Tree/bush/leaf hazards prevented Recreational turf protected Wooden structures protected	Nutrition and health improved Food safety/security Life expectancy increased Reduced maintenance costs

Source: Cooper and Dobson [23].

Table 1.
The complexity of the effects, and primary and secondary benefits of pesticides.

3.2 Impacts of pesticides

They are potentially harmful to humans, animals, other living organisms, and the environment if used incorrectly. It is estimated that about 5000–20,000 people died and about 500,000 to 1 million people get poisoned every year by pesticides [24, 25] at least half of the intoxicated and 75% of those who die due to pesticide are agricultural workers. The rest is being poisoned due to the eating of contaminated food. For control of aphid population on mustard crop, imidacloprid 17.8 SL and thiamethoxam 25 WP were recorded as the most effective. In the biopesticide point of view, *M. anisopoliae* 1.15 WP was recorded as more effective than *B. bassiana* 1.15 WP [26].

3.3 Impact on human health

Pesticides might enter the human body through the inward breath of dirty air, residue, and fume that contain pesticides; through oral openness by burning through sullied food and water; and through dermal openness by direct contact with pesticides [27]. Pesticides are showered onto food, particularly foods grown from the ground, and they emit into soils and groundwater, which can wind up in drinking water and pesticide splash can float and contaminate the air. Poisonousness of synthetic substances, length, and greatness of openness decides the level of spiteful effect on human well-being [28]. Pesticide float from rural fields, openness to pesticides during the application and deliberate or inadvertent harming for the most part, prompts the intense sickness in people [29, 30]. Studies build up a connection between pesticides openness and the frequencies of human persistent infections influencing anxious, regenerative, renal, cardiovascular, and respiratory frameworks [31]. The credits of pesticides remember upgraded financial potential for terms of expanded production of food and fiber and the board of vector-borne infections and afterward, their charges have brought about genuine well-being suggestions to man and climate. There is presently overpowering proof that a portion of these synthetic substances does represent an expected danger to people and other living things and undesirable incidental effects to the climate [32]. Exact insights on the well-being impacts of pesticides are not accessible. Notwithstanding, it is assessed that all around the world, consistently somewhere in the range of 1 and 41 million individuals experience the ill effects of openness to pesticides assessed that at least 300,000 individuals kick the bucket from pesticide harming every year, with close to 100% of them from low and center pay nations. In 2008, the World Bank put the quantity of passing at 355,000. Notwithstanding, FAO (2005) suggesting to ongoing information from Sri Lanka demonstrated that 300,000 passing each year might happen in the Asia-Pacific locale alone because of pesticide harming. The study of disease transmission of pesticide openness worldwide is not completely perceived and more often than not under-analyzed as indicated by the Dish American Wellbeing Association, a global general wellbeing organization situated in Washington, D.C. "Pesticide harming cases are under-revealed by 50% to 80% area wide," detailed the PAHO in 2011, introducing to the Americas (**Table 2**).

3.4 Impacts on environment

At the point when pesticides are showered in farming yield, they might discover their direction through the air and ultimately end up in different portions of the climate, for example, in soil or water. Pesticides that are applied straightforwardly to the dirt might be washed off and reach close surface water bodies through surface spillover or may permeate through the dirt to bring down soil layers and groundwater [34]. The impacts of pesticides on the natural framework might go from minor deviation on the typical working of the environment to the deficiency of species variety. At some point, utilization of pesticides might cause long-haul remaining impacts while in any case intense deadly impacts. For instance, most organochlorine pesticides are persevering in the climate for a long time, thus bringing about tainting of groundwater, surface water, food items, air, and soil (**Table 3**).

Sr. No.	Active ingredient	Brand name	Manufactory company	Signs and symptoms
1.	Acephate (organophosphate)	Orthene	Kalyani Industries Ltd., India	Headache, excessive salivation and tearing, muscle twitching, nausea, diarrhea. Respiratory depression, seizures, loss of consciousness. Pinpoint pupils.
2.	Aldicarb (*N*-methyl carbamate)	Temik	Yangzhou Xinhua Chemical Industries, China	Malaise, muscle weakness, dizziness, sweating. Headache, salivation, nausea, vomiting, abdominal pain, diarrhea. Nervous system depression, pulmonary edema in serious cases.
3.	Carbaryl (*N*-methyl carbamate)	Sevin	Yangzhou Xinhua Chemical Industries, China	Malaise, muscle weakness, dizziness, sweating. Headache, salivation, nausea, vomiting, abdominal pain, diarrhea. Nervous system depression, pulmonary edema in serious cases.

Sr. No.	Active ingredient	Brand name	Manufactory company	Signs and symptoms
4.	Chlorpyrifos (organophosphate)	Dursban	Dow Chemical Company, USA	Headache, excessive salivation and tearing, muscle twitching, nausea, diarrhea. Respiratory depression, seizures, loss of consciousness. Pinpoint pupils.
5.	Endosulfan (organochlorine)	Thiodan	Hindustan Insecticides Ltd. (HIL),India	Itching, burning, tingling of skin. Headache, dizziness, nausea, vomiting, lack of coordination, tremor, mental confusion. Seizures, respiratory depression, coma.
6.	Malathion (organophosphate)	Cythion	Shri Ram Agro Chemicals, India	Headache, excessive salivation and tearing, muscle twitching, nausea, diarrhea. Respiratory depression, seizures, loss of consciousness. Pinpoint pupils.
7.	Methyl Parathion (organophosphate)	Penncap-M	European Chemicals Agency, Europe	Headache, excessive salivation and tearing, muscle twitching, nausea, diarrhea. Respiratory depression, seizures, loss of consciousness. Pinpoint pupils.
8.	Phosmet (organophosphate)	Imidan	Abhayam Cropsafe Private Limited, India	Headache, excessive salivation and tearing, muscle twitching, nausea, diarrhea. Respiratory depression, seizures, loss of consciousness. Pinpoint pupils.
9.	Pyrethrins (Plant origin)		Green Heaven India (A Herbal Manufacturing Unit) India	Irritating to skin and upper respiratory tract. Contact dermatitis and allergic reactions—asthma.
10.	Pyrethroids (synthetic pyrethrin)	Cypermethrin, permethrin	Sumitomo Chemical India	Abnormal facial sensation, dizziness, salivation, headache, fatigue, vomiting, diarrhea. Irritability to sounds or touch. Seizures, numbness.

Source: Lorenz [33].

Table 2.
Signs and symptoms of acute exposure for several insecticide-active ingredients.

4. Impacts on nontarget organism

Most insect sprays are once applied to kill nuisance; it might form likewise unfavorably non-objective life forms such as worm, normal hunters, and pollinator [52].

Sr. No.	Diseases	Reference
1	Diabetes (type 2 diabetes)	Son et al. [35]
2	Birth defects	Winchester et al. [36]; Mesnage et al. [37]
3	Reproductive disorders	Petrelli and Mantovani [38]; Greenlee et al. [39]
4	Cancer (childhood and adult brain cancer; renal cell cancer; lymphocytic leukemia (CLL); prostate cancer)	Lee et al. [40]; Shim et al. [41]; Heck et al. [42]; Xu et al. [43]; Band et al. [44]; Cocco et al. [45]
5	Reproductive disorders	Petrelli and Mantovani [38]; Greenlee et al. [39]
6	Neuro degenerative diseases including Parkinson's disease, Alzheimer disease	Elbaz et al. [46]; Hayden et al. [47]; Tanner et al. [48]
7	Respiratory diseases (asthma, chronic obstructive pulmonary disease [COPD])	Chakraborty et al. [49]; Hoppin et al. [50]
8	Hormonal imbalances including infertility and breast pain	Xavier et al. [51]

Table 3.
Chronic disease to human due to the pesticides application.

Pesticide applications can cause a decrease in the population of the worm. For instance, carbamate bug sprays are exceptionally poisonous to nightcrawlers and a few organophosphates have been displayed to lessen worm populaces [53]. Disgracefully, regular hunters, for example, parasitoids and hunters (fundamental for controlling pest populace level), are generally vulnerable to insect sprays and are seriously influenced [54].

Pollinators such as honey bees, organic product flies, a few scarabs, and birds can be utilized as bio-pointers of biological system measures from various perspectives as their exercises are influenced by natural pressure brought about by pesticides application and living space adjustments [55]. Utilization of pesticides may likewise cause direct loss of creepy-crawly pollinators and indirect calamity to crops as a result of the absence of satisfactory populaces of pollinators [56]. *M. anisopliae*, *B. bassiana*, imidacloprid, and thiamethoxam are noted as less toxicity to beneficial insects while managing mustard aphid [57].

5. Impacts on soil micro-flora

The lost and reused utilization of pesticides disturbs this dirt growth. Soil properties and soil miniature vegetation get influenced because of pesticides that may go through an assortment of exploitation, transport, and adsorption/desorption measures [58]. The tarnished pesticides connect with the dirt and its native microorganisms, along these lines changing its microbial variety, biochemical responses, and enzymatic movement [58, 59]. Any adjustment in the microbial variety and soil biomass, in the long run, prompts the unsettling influence in the soil environment and loss of soil fruitfulness. Pesticide application may likewise restrain or kill certain gatherings of microorganisms and dwarf different gatherings by delivering them from the opposition [58]. They may likewise antagonistically influence the dirt essential biochemical responses including nitrogen obsession, nitrification, and ammonification by initiating/deactivating explicit soil microorganisms and additional chemicals [58, 59].

6. Impacts on water and air ecosystem

There are diverse ways by which pesticides can get into water such as unplanned spillage, mechanical profluent, surface runoff and transport from pesticide-treated soils, washing of shower gear after splash activity, float into lakes, lakes, streams and waterway water, ethereal showers to control water-repressing bugs [60]. Pesticides move from fields to different water sources by overflow or in waste actuated by rainstorm or water system [59]. The instability or semi-unpredictability nature of the pesticide compounds likewise establishes a significant danger of barometrical contamination of huge urban communities [61–63].

7. Conclusion

Pesticides act as the backbone of farmers as well as people all around the world by enhancing crop production and providing innumerable benefits to society indirectly. Due to indiscriminate application of pesticide, it causes harmful impact on human health and environment pollution. However, farmers are unable to completely eliminate the hazards caused by the application of synthetic insecticides but by need-base utilization try to reduce it. Exposure to pesticides and hence the harmful consequences and undesirable effects of this exposure can be minimized by several means such as applied integrated pest management techniques. Through organic farming, the highly nutritive production of better, safe, and eco-friendly pesticide formulations can reduce the harmful effects.

8. Future scope

Now in the coming days, the chemical formulation can be utilized by a combination of biological and botanical products that give a positive impact on the sustainable reduction of pest population. This combination not only promises environmental sustainability, but also has diverse applications in controlling urban pests and invasive species. Pesticides have also posed a serious threat to the biological integrity of marine and aquatic ecosystems. It is the need of time to integrate the studies of different disciplines including toxicology, environmental chemistry, population biology, community ecology, conservation bioagents, and landscape ecology to understand the direct and indirect effects of pesticides on the environment.

Author details

S.A. Dwivedi*, V.K. Sonawane and T.R. Pandit
Department of Entomology, School of Agriculture, Lovely Professional University, Phagwara, Punjab, India

*Address all correspondence to: dwivedi.sunilkumar46@gmail.com

References

[1] Abhilash PC, Singh N. Pesticide use and application: An Indian scenario. Journal of Hazardous Materials. 2009;**165**(1-3):1-12

[2] Vos JG, Dybing E, Greim HA, Ladefoged O, Lambre C, Tarazona JV, et al. Health effects of endocrine-disrupting chemicals on wildlife, with special reference to the European situation. Critical Reviews in Toxicology. 2000;**30**(1):71-133

[3] UNEP. Stockholm Convention on Persistent Organic Pollutants (POPs). Geneva, Switzerland: United Nations Environment Programme; 2001

[4] De A, Bose R, Kumar A, Mozumdar S, editors. Worldwide pesticide use. In: Targeted Delivery of Pesticides Using Biodegradable. New Delhi, India: Springer; 2014. pp. 5-6

[5] Zhang W. Global pesticide use: Profile, trend, cost/benefit and more. Proceedings of the International Academy of Ecology and Environmental Sciences. 2018;**8**(1):1-27

[6] Mansour R, Grissa-Lebdi K, Suma P, Mazzeo G, Russo A. Key scale insects (Hemiptera: Coccoidea) of high economic importance in a Mediterranean area: host plants, bio-ecological characteristics, natural enemies and pest management strategies – A review. Plant Protection Science. 2017;**53**(1):1-14

[7] Yadav IC, Devi NL. Pesticides classification and its impact on human and environment. Environmental Science and Engineering. 2017;**6**:140-158

[8] Martin H. Pesticides Manual. London, UK: British Crop Protection Council; 1968

[9] Drum C. Soil Chemistry of Pesticides. USA: PPG Industries, Inc.; 1980

[10] Anonymous. Statistical Database. India: Directorate of Plant Protection, Quarantine and Storage; 2021

[11] Statista. Volume of pesticides produced across India from financial year 2015 to 2021. Chemicals and Resources; 2021. Available from: https://www.statista.com/statistics/726938/india-pesticides-production-volume/

[12] Austin RB. Yield of wheat in the United Kingdom: Recent advances and prospects. Crop Science. 1999;**39**(6):1604-1610

[13] Kucharik CJ, Ramankutty N. Trends and variability in US corn yields over the 20th century. Earth Interactions. 2005;**9**:1-29

[14] Chenkin AF. Economic effects of plant protection in the Russian Federated Republic. In: Papers at Sessions, VIII International Congress of Plant Protection. USSR: Moscow; 1975

[15] Keiserukhsky MG, Kashirsky OP. Economics of plant protection in the USSR. In: VIII International Congress of Plant Protection, Papers at Sessions. Moscow: USSR; 1975

[16] Zakharenko VA. Economics and prospects of pesticide application as related to the intensification of farming. In: Papers at Sessions, VIII International Congress of Plant Protection. Moscow: USSR; 1975

[17] Brown I. UK Pesticides Residue Committee Report. 2004

[18] WHO. The World Health Organization. Geneva: WHO; 2004

[19] Atreya N. Chemophobia – Pesticide residues in food. Outlooks on Pest Management. 2006;**17**:242

[20] Thobaben M. Dietary counseling using dietary guidelines for Americans 2005 and anatomy of MyPyramid. Home Health Care Management and Practice. 2007;**19**(2):140-142

[21] Oerke EC, Dehne HW, Schonbeck F, Weber A. Crop Production and Protection: Estimated Losses in Major Food and Cash Crops. Elsevier. 1994

[22] Davis JR, Brownson RC, Garcia R. Family pesticide use in the home, garden, orchard, and yard. Archives of Environmental Contamination and Toxicology. 1992;**22**:260-266

[23] Cooper J, Dobson H. The benefits of pesticides to mankind and the environment. Crop Protection. 2007;**26**(9):1337-1348

[24] FAO/WHO. Pesticide Residues in Food—1999 Evaluations. Part II—Toxicological. Joint FAO/WHO Meeting on Pesticide Residues. Geneva: World Health Organization; 2000

[25] Yadav IC, Devi NL, Syed JH, Cheng Z, Li J, Zhang G, et al. Current status of persistent organic pesticides residues in air, water, and soil, and their possible effect on neighboring countries: A comprehensive review of India. Science of the Total Environment. 2015;**511**:123-137

[26] Dwivedi SA, Singh RS. Evaluation of the new molecular insecticides and biopesticides against mustard aphid *Lipaphis erysimi* (Kalt.) on yield parameter in mustard. Journal of Biopesticides. 2019;**12**(2):203-214

[27] Sacramento CA. Department of pesticide regulation "What are the potential health effects of pesticides?" In: Community Guide to Recognizing and Reporting Pesticide Problems; 2008. pp. 27-29

[28] Lorenz ES. Potential health effects of pesticides. Agricultural Communication and Marketing. 2009b:1-8

[29] Dawson AH, Eddleston M, Senarathna L, Mohamed F, Gawarammana I, Bowe SJ, et al. Acute human lethal toxicity of agricultural pesticides: A prospective cohort study. PLoS Medicine. 2010;**7**:1-10

[30] Lee SJ, Mehler L, Beckman J, DieboltBrown B, Prado J, Lackovic M, et al. Acute pesticide illnesses associated with off target pesticide drift from agricultural applications: 11 States, 1998-2006. Environmental Health Perspectives. 2011;**119**:11-62

[31] Mostafalou S, Abdollahi M. Concerns of environmental persistence of pesticides and human chronic diseases. Clinical and Experimental Pharmacology. 2012;**S5**:e002

[32] Aktar W, Sengupta D, Chowdhury A. Impact of pesticides use in agriculture: Their benefits and hazards. Interdisplainary Toxicology. 2009;**2**(1):1-12. DOI: 10.2478/v10102-009-0001-7

[33] Lorenz ES. The Pennsylvania State University. 2009a

[34] Harrison SA. The Fate of Pesticides in the Environment. PA, USA: Agrochemical Fact Sheet # 8; 1990

[35] Son HK, Kim SA, Kang JH, Chang YS, Park SK, Lee SK, et al. Strong associations between low dose organochlorine pesticides and type 2 diabetes in Korea. Environment International. 2010;**36**:410-414

[36] Winchester PD, Huskins J, Ying J. Agrichemicals in surface water and birth defects in the United States. Acta Paediatrica. 2009;**98**:664-669

[37] Mesnage A, Esnouf S, Jégou P, Deniau G, Palacin S. Understanding the redox-induced polymer grafting

process: A dual surface-solution analysis. Chemistry of Materials. 2010;**22**(23):6229-6239

[38] Petrelli G, Mantovani A. Environmental risk factors and male fertility and reproduction. Contraception. 2002;**65**:297-300

[39] Greenlee AR, Arbuckle TE, Chyou PH. Risk factors for female infertility in an agricultural region. Epidemiology. 2003;**14**:429-436

[40] Lee WJ, Colt JS, Heineman EF, McComb R, Weisenburger DD, Lijinsky W, et al. Agricultural pesticide use and risk of glioma in Nebraska, United States. Occupational and Environmental Medicine. 2005;**62**:786-792

[41] Shim YK, Mlynarek SP, van Wijngaarden E. Parental exposure to pesticides and childhood brain cancer: US Atlantic coast childhood brain cancer study. Environmental Health Perspectives. 2009;**117**:1002-1006

[42] Heck JE, Charbotel B, Moore LE, Karami S, Zaridze DG, Matveev V, et al. Occupation and renal cell cancer in Central and Eastern Europe. Occupational and Environmental Medicine. 2010;**67**(1):47-53

[43] Xu X, Dailey AB, Talbott EO, Ilacqua VA, Kearney G, Asal NR. Associations of serum concentrations of organochlorine pesticides with breast cancer and prostate cancer in US adults. Environmental Health Perspectives. 2010;**118**:60-66

[44] Band PR, Abanto Z, Bert J, Lang B, Fang R, Gallagher RP, et al. Prostate cancer risk and exposure to pesticides in British Columbia farmers. Prostate. 2011;**71**:168-183

[45] Cocco P, Satta G, Dubois S, Pili C, Pilleri M, Zucca M, et al. Lymphoma risk and occupational exposure to pesticides: Results of the Epilymph study. Occupational and Environmental Medicine. 2013;**70**:91-98

[46] Elbaz A, Clavel J, Rathouz PJ, Moisan F, Galanaud JP, Delemotte B, et al. Professional exposure to pesticides and Parkinson disease. Annals of Neurology. 2009;**66**:494-504

[47] Hayden KM, Norton MC, Darcey D, Ostbye T, Zandi PP, Breitner JCS, et al. Occupational exposure to pesticides increases the risk of incident AD the cache county study. Neurology. 2010;**74**:1524-1530

[48] Tanner CM, Kamel F, Ross GW, Hoppin JA, Goldman SM, Korell M, et al. Rotenone, paraquat, and Parkinson's disease. Environmental Health Perspectives. 2011;**119**:866-872

[49] Chakraborty S, Mukherjee S, Roy S, Siddique S, Lahiri T, Ray MR. Chronic exposures to cholinesterase inhibiting pesticides adversely affect respiratory health of agricultural workers in India. Journal of Occupational Health. 2009;**51**:488-497

[50] Hoppin JA, Umbach DM, London SJ, Henneberger PK, Kullman GJ, Coble J, et al. Pesticide use and adult-onset asthma among male farmers in the Agricultural Health Study. The European Respiratory Journal. 2009;**34**:1296-1303. DOI: 10.1183/09031936.00005509

[51] Xavier R, Rekha K, Bairy KL. Health perspective of pesticide exposure and dietary management. Malaysian Journal of Nutrition. 2004;**10**:39-51

[52] Ware GW. Effect of pesticides on non-target organisms. Residue Reviews. 1980;**76**:173-201

[53] Edwards CA. The environmental impact of insecticides. In: Delucchi V, editor. Integrated Pest Management, Protection Integàee Quo vadis? An International Perspective. Geneva,

Switzerland: Parasitis 86; 1987. pp. 309-329

[54] Vickerman GP. Farm scale evaluation of the long-term effects of different pesticide regimes on the arthropod fauna of winter wheat. In: Fields Methods for the Study of Environmental Effects of Pesticides. Symposium; 1988. pp. 127-135

[55] Kevan PG. Pollinators as bio-indicators of the state of the environments: Species, activity and diversity. Agriculture, Ecosystems and Environment. 1999;**74**:373-393

[56] Fishel FM, Ferrell JA. Managing Pesticide Drift. Gainesville, FL, USA: Agronomy Department, University of Florida; 2013. Available from: http://edis.ifas.ufl.edu/pi232 [Accessed: 14 October 2013]

[57] Dwivedi SA, Singh RS. Synthetic insecticides and bio-pesticide affect natural enemies of aphid (*Lipaphis erysimi* Kalt) in mustard. Indian Journal of Agricultural Research. 2020;(10):1-9. DOI: 10.18805/IJARe.A-5613

[58] Hussain S, Siddique T, Saleem M, Arshad M, Khalid A. Impact of pesticides on soil microbial diversity, enzymes, and biochemical reactions. Advances in Agronomy. 2009;**102**:159-200

[59] MunozLeoz B, RuizRomera E, Antiguedad I, Garbisu C. Tebuconazole application decreases soil microbial biomass and activity. Soil Biology and Biochemistry. 2011;**43**:2176-2183

[60] Singh B, Mandal K. Environmental impact of pesticides belonging to newer chemistry. In: Dhawan AK, Singh B, Brar Bhullar M, Arora R, editors. Integrated Pest Management. Jodhpur, India: Scientific Publishers; 2013. pp. 152-190

[61] Trajkovska S, Mbaye M, Seye MG, Aaron JJ, Chevreuil M, Blanchoud H. Toxicological study of pesticides in air and precipitations of Paris by means of a bioluminescence method. Analytical and Bioanalytical Chemistry. 2009;**394**:1099-1106

[62] IUPAC. Glossary of Terms Relating to Pesticides (PDF). IUPAC. p. 2123 [Retrieved: January 28, 2014]. Pure Applied Chemistry 78(11):2075-2154. DOI: 10.1351/pac200678112075

[63] Larson SJ, Capel PD, Majewski M. Pesticides in Surface Waters: Distribution, Trends, and Governing Factors (No. 3). Boca Raton CRC Press; 2010

Section 2

Microbiological Control

Chapter 7

Role of Microbial Biopesticides as an Alternative to Insecticides in Integrated Pest Management of Cotton Pests

Lawrence N. Malinga and Mark D. Laing

Abstract

Cotton (*Gossypium hirsutum* L.) is the most produced natural fibre worldwide, and it contributes significantly to the economy of almost 80 cotton-producing countries. Given the high pest infestation, huge amounts of insecticides have been used in cotton production. However, this has resulted in the development of resistance from primary cotton pests and contamination of the environment. Furthermore, the reduction of beneficial insects and outbreaks of secondary pests have been observed. Many arthropod pests are associated with cotton, most of which belong to the orders Lepidoptera, Thysanoptera, and Hemiptera. Biocontrol agents play a critical role in preventing pests in most cotton-growing areas globally. Biological control of cotton pests forms part of integrated pest management as most of these pests have developed resistance against synthetic pesticides. This chapter focuses on the effects of some of the biopesticides, on cotton insect pests. It examines the control of cotton pests using microbial-based products *Bacillus thuringiensis*, *Beauveria bassiana*, *Helicoverpa armigera* nucleopolyhedrovirus and *Metarhizium rileyi*. Furthermore, the chapter summarizes the application of microbial biopesticides as well as the advantages and disadvantages of using these biocontrol agents in agriculture.

Keywords: Cotton, Insecticides, Microbial biopesticides, *Bacillus thuringiensis*, *Beauveria bassiana*, *Metarhizium rileyi*, Nucleopolyhedrovirus

1. Introduction

Pests and diseases are estimated to cause 60% losses in cotton production throughout the world [1]. A successful control strategy requires integrated pest management (IPM) that prevents or suppresses damaging populations of insect pests by applying the comprehensive and coordinated integration of multiple control tactics, including chemical, cultural and biological methodologies.

Synthetic insecticides are mainly used on cotton to control insect pests rapidly [2], and farmers opt for insecticides as the first line of defense [3]. Since the development of synthetic insecticides after World War II, they have been extensively used in agriculture due to their efficiency in pest control and yield increment of many crops [4]. Cotton has been reported to receive more chemical control than most other arable crops [5]. Cotton uses up to 60% of all commercialized agrochemicals globally [6]. Various insect pests and beneficial insects coexist in a cotton ecosystem; however, insecticides have reduced the impact of beneficial insects [7]. As one of the management tools for pests, synthetic insecticides can be used as part of integrated pest management to promote sustainable pest control methods [8]. When synthetic insecticides such as organophosphate (1960s), carbamates (1970s), and pyrethroids (1980s) were introduced, they had an impact on agricultural pest control and resulted in high yields [9].

Although chemical control remains a key method to control targeted pests, a controversy has surfaced regarding the use and abuse of pesticides [9]. The diversity of pests found on cotton requires serious control, mostly with pesticides, which subsequently has a negative impact on natural enemies and the environment [10]. The continuous use of synthetic chemicals to protect crops may also result in resistance to insecticides in pest populations [3]. Combining chemical and biological controls is important for integrated pest management; however, this has not been entirely explored due to, among others, the insufficient information on the insecticide tolerance or resistance of natural enemies [11]. The development of integrated pest management strategies is required to reduce insecticide use and maximize the impact of natural enemies.

Biological control includes introducing a natural enemy or living organisms [12], and cultural control focuses on manipulating the environment to reduce the pest's populations [13]. Pest management has evolved to include integrated pest management that focuses on biological control strategies, including biopesticides. It has been widely reported that chemical pesticides have a negative impact on the environment; therefore, efforts have been made to minimize their use in controlling insect pests. Biopesticides are commonly used to manage agricultural pests through specific biological effects [14] compared to wider control of synthetic pesticides. They contain organisms or substances derived from natural resources in nature and have inhibitory effects on insect pests.

Biopesticides are cheaper, take less time to develop [15], and are naturally less toxic to humans and the environment [16] compared with synthetic pesticides. They are mainly categorized into biochemical, plant, and microbial pesticides [17–19]. Biochemical pesticides include plant extracts, pheromones, plant and insect growth regulators that control pests by non-hazardous mechanisms [20]. Plant pesticides, also known as plant-incorporated protectants, include genetically modified crops using protein from the bacterium *B. thuringiensis* [15]. Microbial pesticides consist of viruses, fungi, and bacteria [21]. Biopesticides form only around 5% of the global pesticides [22], while microbial pesticides account for over 75% worldwide [23]. This chapter provides an overview of microbial-based products *B. thuringiensis*, *B. bassiana*, *H. armigera* nucleopolyhedrovirus, *M. rileyi* and their application to control cotton pests. The chapter further explores the constraints and opportunities for the use of these biopesticides.

2. *Bacillus thuringiensis*

Bacillus thuringiensis (Bacillaceae) is a spore-forming gram-positive bacterium that produces poisonous insecticidal crystal proteins used on more than 3 000 different insects [24, 25]. The bacterium commonly lives in soil, water, plants and dead insects [26]. It was first isolated by Shigetane Ishiwatari in 1901 and first used commercially in the 1920s [27]. *B. thuringiensis* accounts for 95% of the biopesticide market worldwide [28]. The bacterium plays a significant role in biological control because it is the most widely used microbial control agent [19]. Different strains

Control	Findings
Larvicidal activity of *B. thuringiensis* strains against *B. tabaci* [43]	The second instar larvae of *B. tabaci* exhibited mortalities of up to 69%.
Interaction of *B. thuringiensis* and *B. bassiana* for biological control of *B. tabaci* [44]	Higher concentrations of *B. thuringiensis* and *B. bassiana* had above 90% mortality of *B. tabaci* nymphs
Efficacy of *B. thuringiensis* spray applications for the control of *E. biplaga* [45]	*B. thuringiensis* spray provided between 77 and 88% control of *E. biplaga* after ten days
Effects of *B. thuringiensis* on *A. argillacea* and *A. gossypii* of cotton [42]	Dipel® had good control on *A. argillacea,* selective for *A. gossypii,* and caused an increase in cotton yield
Evaluation of *B. thuringiensis* strain when applied to *B. tabaci* nymphs [46]	*B. thuringiensis* strain had 88–92% mortality of the third and fourth instar of *B. tabaci* nymphs
Efficacy of biopesticides and chemical insecticide to control *H. armigera* [47]	*B. thuringiensis* showed the highest mortality rate of *H. armigera* larvae in the shortest period
Efficacy of *B. thuringiensis* against *H. armigera* under laboratory and field conditions [48]	*B. thuringiensis* showed 95–100% and 76% *H. armigera* mortality under laboratory and field conditions, respectively
Influences of *B. thuringiensis* cotton on *A. gossypii* [49]	*B. thuringiensis* cotton efficiently prevented *A. gossypii* resurgence in response to insecticide use
Effects of *B. thuringiensis* on larva and adult of *B. tabaci* [50]	*B. thuringiensis* showed latent effects on the reproductive potential of *B. tabaci*
Evaluation *B. thuringiensis* for control of *Heliothis* spp. on cotton [51]	Dipel® exhibited higher mortality of *Heliothis* spp. larvae

Table 1.
Summary of some studies on the control of cotton pests using Bacillus thuringiensis.

of *B. thuringiensis* have been produced with different spectrums of activity [29]. Although there are massive spectrums with different cry toxin genes, kurstaki and aizawaï are the only two *B. thuringiensis* subspecies developed into products used to control lepidopteran pests [30]. *B. thuringiensis* commonly attacks larval stages of different insects rather than adults or other stages [31, 32]. As a target-specific pathogen, *B. thuringiensis* only attacks the target insects [33] without disturbing non-target insects and natural enemies [32, 34]. *B. thuringiensis* does not kill the target pest on contact but through disruption of the midgut tissue of the insect [31]. Therefore, it is difficult for the pathogen to attack those insects that feed inside the plant [32]. *B. thuringiensis* toxins have shown well-documented toxicity against various insects, including Lepidoptera, Diptera, Hemiptera, Coleoptera, and nematodes [35–40]. In cotton, *B. thuringiensis* has been widely reported as a biopesticide to control various insect pests [27, 41, 42]. **Table 1** provides an overview of some studies conducted to control some cotton pests using *B. thuringiensis*.

3. *Beauveria bassiana*

Beauveria bassiana (Ascomycota: Cordycipitaceae) is a fungus that grows naturally in soils. It is one of the commercial alternatives to chemical insecticides [52]. Its strains have been used as the active ingredient in several biopesticides to control a diversity of agricultural pests [53]. The genus *Beauveria* contains at least 49 species, of which approximately 22 are considered pathogenic [54]. Notwithstanding its importance as a biological control agent, *B. bassiana* is also an organism used to examine fungal growth and development, such as host-pathogen interactions [55, 56]. Its strains can be developed as host-specific, considering their broad-spectrum as

Control	Findings
The activity of protease and the virulence of *B. bassiana* isolates against *T. urticae* [66]	The isolate of *B. bassiana* caused 15 to 70% mortality of *T. urticae*
Pathogenicity of *B. bassiana* isolates against *H. armigera* larvae [67]	Of 22 *B. bassiana* isolates, four exhibited ˃80% larval mortality
Assessment of the effects of exposure of *H. armigera* larvae to *B. bassiana* [68]	Pre-adult duration of *H. armigera* was extended, and longevity and fecundity were decreased
Effect of isolates of *B. bassiana* against different life stages of *B. tabaci* on cotton [65]	*B. bassiana* isolate had the highest eggs (65.30%) and nymphs (88.82%) mortality
Effect of *B. bassiana* on cotton growth and control of cotton bollworm [54]	*B. bassiana* significantly reduced boll damage, increased plant dry biomass and seed cotton yield
Infection of *H. armigera* by endophytic *B. bassiana* colonizing tomato plants [69]	*B. bassiana* has potential as an effective strategy to control *H. armigera*
Susceptibility of different stages of *T. urticae* to *B. bassiana* in the laboratory [55]	*B. bassiana* gave 90% mortality of *T. urticae*
Effect of *B. bassiana* against *A. gossypii* on cotton [52]	Plants inoculated with *B. bassiana* had significantly lower numbers of *A. gossypii*
Control of *H. armigera* (Hubner) with *B. bassiana* [70]	The highest dose of *B. bassiana* gave 76.7% mortality on the fourth instar larvae of *H. armigera*
Effect of *B. bassiana* on the control of *T. urticae* [58]	*B. bassiana* had 81.8% control of *T. urticae*
Biological control of *T. urticae* [71]	Two strains of *B. bassiana* caused 80% mortality of *T. urticae* in the laboratory and one strain-controlled *T. urticae* in the field

Table 2.
Summary of some studies on the control of cotton pests using Beauveria bassiana.

an insect pathogen [57]. *B. bassiana* has good control by coming into contact with the insect pests [58]. *B. bassiana* attacks its host by penetrating the exoskeleton or cuticle [59], producing a toxin that prevents the immune response of the host [52]. Even though *B. bassiana* based biopesticides may reduce the application of chemical pesticides; their effectiveness requires enhanced formulation or combining them with other pesticides [60]. *B. bassiana* is a promising pathogen against a variety of cotton pests, including spider mites [61], stainers [62], thrips [63], whiteflies [64, 65], aphids and bollworms [52, 54]. Some research on the efficacy of *B. bassiana* on cotton pests are documented in **Table 2**.

4. *Metarhizium rileyi*

Metarhizium rileyi (Farlow) Kepler S.A. Rehner & Humber (Ascomycota: Clavicipitaceae), formerly known as *Nomuraea rileyi* (Farlow) Samson, is a potential agent for microbial control of insect pests that can cause considerable agricultural productivity loss [72]. *M. rileyi* was firstly described as *Botrytis rileyi* in 1883, then as *Spicaria rileyi* (Charles 1936) and later moved to the genus *Nomuraea* (Kish, Samson, and Allen 1974). In 2014, Kepler, Humber, Bischoff, and Rehner transferred the fungi to the genus *Metarhizium*. It is an entomopathogenic fungus commonly known to infect and cause mortality in insects, particularly the lepidopterans [73–75]. The spores of this fungus penetrate the body of the host through the cuticle or by ingestion when the larvae are feeding. This fungus grows inside the larvae and

Control	Findings
The potential of *M. rileyi* as a biological control agent of *B. tabaci* [86]	*M. rileyi* isolate had a high mortality rate and control efficiency against *B. tabaci*
Field evaluation *N. rileyi* against *H. armigera* [87]	*N. rileyi* significantly reduced *H. armigera* (74.58%) larval population
Effect of *N. rileyi* on *H. armigera* cellular immune responses [85]	*N. rileyi* suppressed the cellular immune response of *H. armigera*
The occurrence of an entomopathogenic fungus on *H. armigera* larvae [84]	The natural occurrence of *N. rileyi* caused 33% of the total mortality of *H. armigera* larvae
The effective dose of *N. rileyi* against *H. armigera* [83]	*N. rileyi* was effective against the developmental stages of *H. armigera*
Bio-efficacy of *N. rileyi* against *H. armigera* [88]	*N. rileyi* revealed 30–83% mortality against different instars of *H. armigera*
The efficiency of *N. rileyi* against *B. tabaci* [89]	The percentage of infested plants with *B. tabaci* significantly decreased after treatments with *N. rileyi* under the field conditions
Comparison of *N. rileyi* with *B. bassiana* and *I. fumosorosea* against *H. armigera* in the laboratory [90]	*N. rileyi* performed the best with a mortality rate of 87 ± 1.4% against *H. armigera*.
Pathogenicity of *N. rileyi* against *H. armigera* larvae [91]	*N. rileyi* showed 73 to 87% mortality of *H. armigera* larvae within eight days
Application of *N. rileyi* for the control of *H. armigera* [92]	*N. rileyi* showed an average of 95% mortality in fourth instar and fifth instar larvae of *H. armigera*
Effects of *N. rileyi* in a field population of *H. armigera* [93]	*N. rileyi* showed higher rates of fungal infection (37%) in *H. armigera* found on pigeon pea

Table 3.
Summary of some studies on the control of cotton pests by using Metarhizium rileyi.

reproduces, resulting in internal tissue destruction. The fungus is host-specific and eco-friendly, making it significant in integrated pest management [76]. However, *M. rileyi* has been rarely developed and commercialized [77]. As a result, the host range of *M. rileyi* has been reported to be only around 60 species compared to fungi such as *B. bassiana* [74]. Under favourable environmental conditions, caterpillars from the Noctuidae family are mostly attacked by this pathogen [78].

This fungus is a biological control agent for about 30 species of orders Lepidoptera [79], although two species of order Coleoptera are also found to be susceptible [80]. As a well-known entomopathogenic fungus used in the biological control of pests, limitations such as the long pathogenic process and its application are limited [81]. On the contrary, Jaronski and Mascarin [82] have claimed that *M. rileyi* can be easily produced than other fungi. *M. rileyi* has been broadly studied, mainly on its efficacy against cotton bollworm *H. armigera* [83–85]. **Table 3** presents some research work on the control of cotton pests using *M. rileyi*.

5. Nucleopolyhedrovirus

Baculoviruses belong to the family Baculoviridae, which consists of four genera, including Alphabaculovirus [94]. Viruses from this family have been recorded since 1911, and their natural hosts include almost 700 insect species, mainly belonging to the orders Lepidoptera, Hymenoptera, and Diptera [95]. Baculoviruses are insect-specific [96, 97] and are usually limited to one or a few insect species [98, 99].

Because of their specificity, these viruses can form part of the resistance management strategy [100], demonstrating genetic variations among species [98].

Several members of baculoviruses that display promising results have been successfully developed into commercial biopesticides to control agricultural and forest insect pests worldwide [101]. However, the application of these pesticides has a limited acceptance due to marketing, slow speed of kill, and difficulties with registration and mass production [102]. The production relies mainly on baculoviruses infection and transmission in vulnerable hosts as well as harvesting and purification [103]. Although viruses can be an alternative to synthetic insecticides, they depend on integrating other management strategies [104]. Baculoviruses are part of integrated pest management programmes to control pests in field crops [102]. Despite the regular use of baculoviruses as biopesticides, biological insecticides based on the bacterium *B. thuringiensis* remain the most used biopesticides [94].

Nucleopolyhedrovirus (NPV) is a naturally occurring pathogen that belongs to the group of Alphabaculovirus, and it is a lepidopteran-specific virus [105]. The virus reproduces in the host cells, causing nuclear polyhedrosis disease, and the outbreak of the virus may assist in controlling the host population [106]. The nucleopolyhedrovirus has the potential to control the target insects without harming the environment, pest predators, and parasitoids [107]. *Helicoverpa armigera* nucleopolyhedrovirus (HearNPV) is specifically developed to control *H. armigera,* and the formulations are commercially available throughout the world [108]. The first commercialization of HearNPV was done in China in 1993 [106]. It is reported to have significant potential as a biopesticide in the field [102, 109]. Nucleopolyhedrovirus can be used in conjunction with other insecticides to control

Control	Findings
Assessment of NPV and spinosad against *H. armigera* in a controlled environment [113]	The highest concentrations of NPV had the highest mortality of 95%
Pathogenicity of HearNPV against *H. armigera* [114]	HearNPV had 90–100% mortality effects of newly hatched and second instars larvae
Evaluation of different HearNPV concentrations on neonate, 3rd, and 5th instars larvae of *H. armigera* [115]	The highest dose of HearNPV showed 92% mortality within 14 days
The ability of HearNPV to kill each *H. zea* instar, and a second infestation [108]	HearNPV was successful in controlling early instars of *H. zea* in 5 days
The efficiency of production of HearNPV in *H. armigera* [116]	HearNPV exhibited 80–93% of virus-induced mortality in individualized *H. armigera* larvae
Insecticidal efficacy of HearNPV on *H. armigera* [117]	Larval mortality of *H. armigera* ranged from 97.9–100% at ten days post-application of HearNPV
Efficacy of HearNPV as a control in the cell transfection analysis [118]	HearNPV caused paralysis, weight loss, and suppressed growth and feeding of *H. armigera* larvae
Bio-efficacy of NPV against *H. armigera* [119]	NPV significantly reduced both larval population and boll damage
Field efficacy of (HaNPV) isolates and insecticide control against *H. armigera* on cotton [120]	HaNPV isolates significantly reduced *H. armigera* larvae and recorded the highest yield of over 2 000 kg ha^{-1}
Evaluation of HearNPV for control of *H. armigera* in citrus [109]	HearNPV had a 100% reduction of *H. armigera* larval infestation within 7–16 days

Table 4.
Summary of some studies on the control of cotton pests using nucleopolyhedrovirus.

H. armigera [110, 111]. It is recommended that the application of HearNPV must commence when cotton starts flowering, and the pests are observed in the field [108]. However, the interaction between HearNPV and host insects remains poorly understood [112]. Bolldex™ is one of the commercial labels currently registered as a HearNPV to control *H. armigera* [102]. Below (**Table 4**) is a summary of some studies on the efficacy of the nucleopolyhedrovirus against cotton pests.

6. Application of microbial biopesticides

Majority of biopesticides that show a reduction of pest populations under controlled environments have not succeeded under field conditions [121]. This is due to that, application methods of biopesticides have not been effectively explored. Most of the equipment used to apply biopesticides were developed for synthetic pesticides and are not suitable for biorational agents [122]. The use of application

Microbial	Product name	Target insect	Manufacturer
Bacillus thuringiensis (kurstaki)	Delfin	*Helicoverpa armigera*	Certis
Bacillus thuringiensis (kurstaki)	Dipel	African (American) bollworm	Valent BioSciences
Bacillus thuringiensis (kurstaki)	Javelin	*Helicoverpa armigera* cotton cutworm	Certis
Bacillus thuringiensis (kurstaki)	Biobit	Cotton bollworm	Valent BioSciences
Bacillus thuringiensis (kurstaki)	Condor	Cotton bollworm	Certis
Bacillus thuringiensis (kurstaki)	Crymax	Cotton bollworm	Certis
Bacillus thuringiensis (aizawai)	Florbac	American bollworm	Valent BioSciences
Bacillus thuringiensis (aizawai)	XenTari	Fall armyworm	Valent BioSciences
Beauveria bassiana	Eco-Bb/ Bb-Protec	Whitefly, red spider mite	Andermatt PHP
Beauveria bassiana	BotaniGard	Whiteflies, spider mites, leafhoppers, aphids, thrips	Lam International Corporation
Beauveria bassiana	Mycotrol	Mealybugs, leafhoppers, aphids, thrips, whiteflies	Lam International Corporation
Beauveria bassiana	Broadband	Stink bugs, red spider mites, thrips, whiteflies	BASF
Beauveria bassiana	Naturalis-L	Thrips, whiteflies, red spider mites	Fargro
Metarhizium rileyi	Nomu-Protec	*Helicoverpa armigera*	Andermatt PHP
Nuclear polyhedrosis virus	Heli-cide	*Helicoverpa armigera*	Pest Control India
Nuclear polyhedrosis virus	Bolldex	*Helicoverpa armigera*	Andermatt PHP
Helicoverpa armigera Nucleopolyhedrovirus	Helicovex	*Helicoverpa* species	Andermatt Biocontrol
Helicoverpa armigera Nucleopolyhedrovirus	ViVus	*Helicoverpa armigera*	AgBiTech

Table 5.
Some of the commercially available biopesticides used to control cotton pests.

equipment designed for uniform application of biopesticides such as air-assisted spraying is essential [123]. The design of methods for biopesticide application also relies on the material used and the shape of the crop canopy [124]. Therefore, thorough coverage of all the surfaces reached by a pest is required for effective control. Over and above the correct equipment, precise microbial inoculants are key for a successful biocontrol programme. Microbial biopesticides can be applied in the field as a powder or in a liquid form through seed treatment, root dip, soil or foliar application [125]. Biopesticides must be applied as per the instructions provided in order to apply the correct dosage and the amount of water. As the persistence of biopesticides is an important factor in their efficacy, the timing of application plays a crucial role in pest control. These biocontrol agents tend to be less effective when applied during hotter day times and high rainfall.

Therefore, applications may be were administered late afternoon due to the UV sensitivity of the biological agents [25]. Alternatively, ultraviolet (UV) absorbents or protectants are necessary to combat this degradation and protect the microbes from sunlight. The UV absorbents or protectants dissolves in the insect stomach and release the virus that kills the pest [126]. However, more commercial UV-resistant biopesticides need to be improved to be readily accepted by farmers. It is also important to carefully select a biopesticide specific for the pests that have to be controlled. Furthermore, the level of toxin in the selected biopesticide as well as the feeding behaviour of the target pest is essential to determine the efficacy of the product [127]. Some of the common trade names for commercially available microbial biopesticides are listed in **Table 5**, and many small manufacturers distribute similar biopesticides using different trade names.

7. Challenges and opportunities for the use of biopesticides

Over-reliance on chemical control results in changes in the status of cotton pests and environmental pollution [7]. There are still challenges to sustain the environment for cotton production [128]. Much research has focused on advancing pest control, and biological control agents are an important criterion for sustainable agriculture [129, 130]. Biopesticides or biological pesticides are an eco-friendly alternative to chemical pesticides [131]. They can play a significant role in the integrated pest management of many insect pests [132]. They are obtained from the environment to control agricultural diseases and insects [15]. They are only about 5% of the total crop protection market; however, they are expected to surpass synthetic pesticides by 2050 [133]. The production of biopesticides is sometimes highly labour intensive and difficult to produce at levels that are economically viable and profitable [134]. Enhancement of biopesticides has been explored by improving different compounds to sustain their efficacy as well as the shelf life [135, 136]. The development of non-toxic and effective biopesticides requires a holistic approach, which will turn most of the research results into profitable business products. Although this section provides generalities, each biopesticide needs to be individually assessed to determine its impacts on pest control, humans, the environment, and other factors associated with the adoption by farmers. The adoption of biopesticides by farmers relies on their efficacy, increased yield, lower prices, and an efficient supply [137]. They have been unreliable and very costly due to their limited market share [138]. However, Sharma et al. [107] reported that bacterial biopesticides are the most widely used and less expensive than other control measures. Biopesticides benefit the farmers due to target specificity, the ability to manage the pest rather than eradicate, and conservation of environmental balance [131]. The very high specificity of the products might be a disadvantage when a complex pest species needs to be controlled. Baculovirus-based insecticides have been considered safe on non-target organisms and can be used as part of integrated pest management to ease the risks of synthetic insecticides [99]. However, baculoviruses are reported to act slowly in killing the targeted pests [60], which has led to the development of faster killing products through genetic modifications [94, 102]. Baculoviruses are also reported to be less effective due to their high susceptibility to ultraviolet radiation, and this requires the reapplication of the virus over time [139, 140]. This effectively increases input costs that farmers may incur. The activity of nucleopolyhedrovirus has been found to decrease significantly over time after applying the virus on the plant leaves [116]. When exposed to direct sunlight, nucleopolyhedrovirus has been reported to be inactivated within a day or two [141]. *B. thuringiensis* has a vast spectrum of insecticidal activity compared to other bacteria, and it is safe for the environment and humans [142]. *B. thuringiensis* does not affect non-target organisms, except for some closely related insects to the target pests [143]. The application of *B. thuringiensis* as a biopesticide is potent and biodegradable than synthetic insecticides [144]. However, the bacterium is effective when the present part of the plant that the target insect feeds on and when larvae are still early instars [144].

The efficiency of entomopathogens mainly relies on their ability to infect the target insect and their persistence [145]. Microbial insecticides have low persistence in the environment, and they require accurate application because many of these pathogens are insect-specific [33]. Namasivayam and Vidyasankar [130] recorded that various formulations of *M. rileyi* are persistent under different temperatures. They further recommended that using bio gel formulation of *M. rileyi* might play a role in controlling pests under field conditions. However, Edelstein et al. [146] reported that this pathogen is extremely sensitive to nutritional and environmental conditions, affecting the virulence of the asexual reproductive spore of fungi and stability in storage [147]. Further research is required to stabilize *M. rileyi* in storage and determine the insecticidal activity of formulated conidia [148]. The persistence of *B. bassiana* under field conditions is negatively affected due to ultraviolet light, extreme temperatures and rain [58]. Sandhu et al. [149] have reported that this pathogen can live longer at lower temperatures and relative humidity. Bouslama et al. [150] demonstrated that some formulations of *B. thuringiensis* could be persistent after rain wash compared to treatment with an unformulated bacterium. Biopesticides that degrade rapidly in the environment may have a short field persistence resulting in numerous product applications [60]. The major constraints of biopesticides are limited to, among others, environmental conditions such as solar ultraviolet radiation, temperature, humidity and their ability on spreading on the surface [145, 151]. Since biopesticides often contain living material, the products have reduced shelf life. Temperature, moisture or humidity also plays a role in the shelf life of the biopesticides [152]. Due to their practical limitations, such as rapidly washing away in rain and degradation by the sunlight, biopesticides may not be as effective as synthetic pesticides. The impact of rain on the persistence of entomopathogenic fungi is less when the conidia are in direct contact with the cuticle of leaves and larvae [153]. Under natural conditions, biopesticides often cause natural mortalities of insect populations [149]. Inglis et al. [154] noted that the influence of rain has a minimal effect on *B. bassiana* persistence; however, high rains washed away significant quantities of *B. bassiana* from leaves. *B. thuringiensis* is reported to persist for few days after application due to weather, UV light, chemical environment and the presence of proteinases [144]. Like the other biopesticides, most spores are washed off into the soil.

8. Conclusion

All cotton pests have the potential to cause enormous damage to the crop if left uncontrolled. Structurally integrated pest management is essential to control the existing or new infestation of pests. Although the use of biocontrol agents on cotton does not eliminate pest populations, their application is crucial to suppress the infestations. Therefore, it is essential to acquire and study pest-related information to make appropriate decisions regarding which control methods to implement. The advantage of using biopesticides rather than complete reliance on synthetic pesticides is that these biocontrol agents are cheaper, target-specific, effective in very small quantities, reduce pesticide resistance, environmental and human friendly. Biocontrol agents must not be regarded as a substitute for synthetic insecticides; therefore, to realize the advantage of using biocontrol agents, integration with other crop protection strategies in the IPM programme is necessary.

Acknowledgements

The authors would like to acknowledge the University of KwaZulu-Natal for funding the publication and the South African Sugarcane Research Institute for the support of the first author during the preparation of the book chapter.

Conflict of interest

The authors have declared that no conflict of interests exists.

Author details

Lawrence N. Malinga[1*] and Mark D. Laing[2]

1 South African Sugarcane Research Institute, South Africa

2 University of KwaZulu-Natal, Pietermaritzburg, South Africa

*Address all correspondence to: lawrence.malinga@sugar.org.za

References

[1] UIA. Pests and diseases of cotton. The Encyclopedia of World Problems. 2019. http://encyclopedia.uia.org/en/problem/135349

[2] Asif MU, Muhammad R, Akbar W, Tofique M. Relative efficacy of some insecticides against the sucking insect pest complex of cotton. The Nucleus. 2016;53 (2), 140-146. www.thenucleuspak.org.pk.

[3] Kone PWE, Didi GJR, Ochou GEC, Kouakou M, Bini KKN, Mamadou D, Ochou OG. Susceptibility of cotton leafhopper *Jacobiella facialis* (Hemiptera: Cicadellidae) to principal chemical families: Implications for cotton pest management in Côte D'ivoire. Journal of Experimental Biology and Agricultural Sciences. 2018;6(5):774-781. DOI: 10.18006/2018.6(5).774.781.

[4] National Research Council. The Future Role of Pesticides in US Agriculture. National Academies Press. 2000. DOI: 10.17226/9598.

[5] Matthews G. Integrated pest management practice. In: Thomas, B, Murphy, DJ, Murray, BG, (eds), Encyclopedia of Applied Plant Sciences. Academic Press. 2003;1(1). https://www.sciencedirect.com/topics/biochemistry-genetics-and-molecular-biology/gossypium.

[6] Yadav S, Dutta S. Evaluation of organophosphorus pesticide residue in cotton of Tijara Tehsil, Alwar, Rajasthan. Nature Environment and Pollution Technology. 2019;18(4): 1455-1458.

[7] El-Wakeil N, Abdallah A. Cotton pests and the actual strategies for their management control. In: Giuliano, B., Vinci, E. J., (eds.), Cotton: Cultivation, Varieties and Uses. Nova Science Publishers. 2012;1(1):1-59

[8] Chamuene A, Araújo TA, Lopes MC, Ramos Pereira R, Berger PG, Picanço MC. Investigating the natural mortality of *Aphis gossypii* (Hemiptera: Aphididae) on cotton crops in tropical regions using ecological life tables. Environmental Entomology. 2020;49(1):66-72. DOI: 10.1093/ee/nvz148.

[9] Aktar W, Sengupta D, Chowdhury A. Impact of pesticides use in agriculture: their benefits and hazards. Interdisciplinary Toxicology. 2009;2(1):1-12. DOI: 10.2478/v10102-009-0001-7.

[10] Machado AVA, Potin DM, Torres JB, Silva Torres CSA. Selective insecticides secure natural enemies action in cotton pest management. Ecotoxicology and Environmental Safety. 2019;184:109669. DOI: 10.1016/j.ecoenv.2019.109669.

[11] Rodrigues ARS, Spindola AF, Torres JB, Siqueira HAA, Colares F. Response of different populations of seven lady beetle species to lambda-cyhalothrin with record of resistance. Ecotoxicology and Environmental Safety. 2013;96:53-60. DOI: 10.1016/j.ecoenv.2013.06.014.

[12] le Hesran S, Ras E, Wajnberg E, Beukeboom LW. Next generation biological control – an introduction. Entomologia Experimentalis et Applicata. 2019;167(7):eea.12805. DOI: 10.1111/eea.12805

[13] de Franca SM, Oliveira MAC, de Oliveir JV. The use of behavioral manipulation techniques on synthetic insecticides optimization. In: Trdan, S, (ed), Insecticides - Development of Safer and More Effective Technologies. IntechOpen. 2013. DOI: 10.5772/53354

[14] Dimetry NZ. Different plant families as bioresource for pesticides. In: Singh, D, (ed), Advances in Plant

Biopesticides. Springer India. 2014;1-20. DOI: 10.1007/978-81-322-2006-0_1

[15] Liu X, Cao A, Yan D, Ouyang C, Wang Q, Li Y. Overview of mechanisms and uses of biopesticides. International Journal of Pest Management. 2019;61(1):65-72. DOI: 10.1080/09670874.2019.1664789

[16] Leahy J, Mendelsohn M, Kough J, Jones R, Berckes N. Biopesticide oversight and registration at the U.S. Environmental Protection Agency. ACS Symposium Series. 2014;1172:3-18. DOI: 10.1021/bk-2014-1172.ch001

[17] Ojha PK, Kumari R, Pandey NK. Laboratory evaluation of certain bio-pesticides against the larvae of *Helicoverpa armigera* Hubner. Legume Research. 2018;41(5):784-787. DOI: 10.18805/lr.v0i0.7852

[18] Fathipour Y, Sedaratian A, Bagheri A, Talaei-Hassanlouei R. Increased food utilization indices and decreased proteolytic activity in *Helicoverpa armigera* larvae fed sublethal *Bacillus thuringiensis*-treated diet. Physiological Entomology. 2019;44(3-4):178-186. DOI: 10.1111/phen.12288

[19] Nuruzzaman M, Liu Y, Rahman MM, Dharmarajan R, Duan L, Uddin AFMJ, Naidu R. Nanobiopesticides: Composition and preparation methods. In: Koul O, (ed) Nano-Biopesticides Today and Future Perspectives. Elsevier. 2019;69-131. DOI: 10.1016/B978-0-12-815829-6.00004-8

[20] Sarwar ZM, Ijaz M, Sabri MA, Yousaf H, Mohsan M. Effects of selected synthetic insecticides on the total and differential populations of circulating haemocytes in adults of the red cotton stainer bug *Dysdercus koenigii* (Fabricius) (Hemiptera: Pyrrhocoridae). Environmental Science and Pollution Research. 2018;25(17):17033-17037. DOI: 10.1007/s11356-018-1898-1

[21] Marrone PG. Pesticidal natural products – status and future potential. Pest Management Science. 2019;75(9):5433. DOI: 10.1002/ps.5433

[22] Seiber JN, Coats J, Duke SO, Gross AD. Biopesticides: State of the art and future opportunities. Journal of Agricultural and Food Chemistry. 2014;62(48):11613-11619. DOI: 10.1021/jf504252n

[23] Leppla NC, Johnson MW, Merritt JL, Zalom FG. Applications and trends in commercial biological control for arthropod pests of tomato. In: Wakil W, Brust GE, Perring TM (eds), Sustainable Management of Arthropod Pests of Tomato. Elsevier Inc. 2018;283-303. DOI: 10.1016/B978-0-12-802441-6.00013-9

[24] Wu Y, Lei CF, Yi D, Liu PM, Gao MY. Novel *Bacillus thuringiensis* δ-endotoxin active against *Locusta migratoria manilensis*. Applied and Environmental Microbiology. 2011;77(10):3227-3233. DOI: 10.1128/AEM.02462-10

[25] Zhang L, Zhang X, Zhang Y, Wu S, Gelbiĉ I, Xu L, Guan X. A new formulation of *Bacillus thuringiensis*: UV protection and sustained release mosquito larvae studies. Scientific Reports. 2016;6(1):1-8. DOI: 10.1038/srep39425

[26] Argôlo-Filho RC, Loguercio LL. *Bacillus thuringiensis* is an environmental pathogen and host-specificity has developed as an adaptation to human-generated ecological niches. Insects. 2014;5:62-91; DOI: 10.3390/insects5010062

[27] Gorashf NE, Tripatki M, Kalia V, Gujar GT. Identification and characterization of the Sudanese *Bacillus thuringiensis* and related

bacterial strains for their efficacy against *Helicoverpa armigera* and *Tribolium castaneum*. Indian Journal of Experimental Biology. 2014;52(6): 637-649

[28] Devi PSV, Duraimurugan P, Chandrika KSVP. *Bacillus thuringiensis*-based nanopesticides for crop protection. In: Koul O, (ed), Nano-Biopesticides Today and Future Perspectives. Elsevier. 2019;249-260. DOI: 10.1016/B978-0-12-815829-6.00010-3

[29] Thorne L, Garduno F, Thompson T, Decker D, Zounes M, Wild M, Walfield AM, Pollock TJ. Structural similarity between the Lepidoptera- and Diptera-specific insecticidal endotoxin genes of *Bacillus thuringiensis* subsp. "kurstaki" and "israelensis." Journal of Bacteriology. 1986;166(3):801-811. DOI: 10.1128/jb.166.3.801-811.1986

[30] Sanchis V, Bourguet D. *Bacillus thuringiensis*: applications in agriculture and insect resistance management: A review. Agronomy for Sustainable Development. 2008; 28:11-20. DOI: 10.1051/agro:2007054

[31] Raymond B, Johnston PR, Nielsen-LeRoux C, Lereclus D, Crickmore N. *Bacillus thuringiensis*: An impotent pathogen? Trends in Microbiology. 2010;18(5):189-194. DOI: 10.1016/j.tim.2010.02.006

[32] Ring D. *Bacillus thuringiensis* as a biopesticide applied to plants. The LSU AgCenter and the LSU College of Agriculture. 2017. https://www.lsuagcenter.com/articles/page1508333080784

[33] Bravo A, Likitvivatanavong S, Gill SS, Soberón M. *Bacillus thuringiensis*: A story of a successful bioinsecticide. In: Jindra, M, Kanost, MR, (eds), Insect Biochemistry and Molecular Biology. 2011;41(7):423-431. DOI: 10.1016/j.ibmb.2011.02.006

[34] Pujiastuti Y, Irmawati Arsi A, Sulistiyani DP. Effects of *Bacillus thuringiensis*-based bio-insecticides on the presence of *Aphis gossypii* and Coccinellid predators on intercropping cultivation. IOP Conference Series: Earth and Environmental Science. 2019;347(1):012056. DOI: 10.1088/1755-1315/347/1/012056

[35] Dulmage HT. Insecticidal activity of isolated of *Bacillus thuringiensis* and their potential for pest control. Microbial Control of Pests and Plant Diseases. 1981;1970-1980. https://agris.fao.org/agris-search/search.do?recordID=US201302855632

[36] Brousseau R, Masson L. *Bacillus thuringiensis* insecticidal crystal toxins: Gene structure and mode of action. Biotechnology Advances. 1988;6(4): 697-724. DOI: 10.1016/0734-9750(88)91920-9

[37] Acosta N, Dicklow MB. A strain of *Bacillus thuringiensis* for the control of plant-parasitic nematodes. Biocontrol Science and Technology. 1993;3(1):41-46. DOI: 10.1080/09583159309355257

[38] Wei JZ, Hale K, Carta L, Platzer E, Wong C, Fang SC, Aroian RV. *Bacillus thuringiensis* crystal proteins that target nematodes. Proceedings of the National Academy of Sciences of the United States of America. 2003;100(5):2760-2765. DOI: 10.1073/pnas.0538072100

[39] Torres-Quintero MC, Gómez I, Pacheco S, Sánchez J, Flores H, Osuna J, Mendoza G, Soberón M, Bravo A. Engineering *Bacillus thuringiensis* Cyt1Aa toxin specificity from dipteran to lepidopteran toxicity. Scientific Reports. 2018;8(1):4989. DOI: 10.1038/s41598-018-22740-9

[40] Fernández-Chapa D, Ramírez-Villalobos J, Galán-Wong L. Toxic potential of *Bacillus thuringiensis*: An overview. In: Jia Y, (ed) Protecting Rice Grains in the Post-Genomic Era.

IntechOpen. 2019. DOI: 10.5772/intechopen.85756

[41] Togbé CE, Zannou E, Gbèhounou G, Kossou D, van Huis A. Field evaluation of the synergistic effects of neem oil with *Beauveria bassiana* (Hypocreales: Clavicipitaceae) and *Bacillus thuringiensis* var. kurstaki (Bacillales: Bacillaceae). International Journal of Tropical Insect Science. 2014;34(4):248-259. DOI: 10.1017/S1742758414000447

[42] De Bortoli SA, Vacari AM, do Carmo Fernandes M, De Bortoli CP, De Bortoli SLP, Ramalho DG. Effects of planting distances and *Bacillus thuringiensis* Berliner on *Alabama argillacea* (Hubner), *Aphis gossypii* Glover and natural enemies of cotton. Comunicata Scientiae. 20156(2): 202-212

[43] Cabra ET, Fernandez JAH. Larvicidal activity of *Bacillus thuringiensis* Colombian native strains against *Bemisia tabaci* (Hemiptera: Aleyrodidae). Journal of Plant Protection Research. 2019;59(4):503-511. DOI: 10.24425/jppr.2019.131259

[44] Somoza-Vargas CE, Hernández-Velázquez VM, Peña-Chora G, Torres-García G, Huerta-De La Peña A, Ortega-Martínez LD, Salazar-Magallón JA. Interaction of *Beauveria bassiana* strain HPI-019/14 and *Bacillus thuringiensis* strain GP139 for the biological control of *Bemisia tabaci* in strawberry. Bulletin of Insectology. 2018;71(2):201-209

[45] Fourie D, van den Berg J, du Plessis H. Efficacy of *Bacillus thuringiensis* sprays and cotton cultivars expressing cry proteins in the control of *Earias biplaga* (Walker) (Lepidoptera: Noctuidae). African Entomology. 2017;25(2):335-340. DOI: 10.4001/003.025.0335

[46] Salazar-Magallon JA, Hernandez-Velazquez VM, Alvear-Garcia A, Arenas-Sosa I, Peña-Chora G. Evaluation of industrial by-products for the production of *Bacillus thuringiensis* strain GP139 and the pathogenicity when applied *to Bemisia tabaci* nymphs. Bulletin of Insectology. 2015;68(1):103-109

[47] David D, Simon S, Kumar A. Efficacy of biopesticides and chemical insecticide to control *Helicoverpa armigera* (Hubner) (Lepidoptera: Noctuidae) fed on chickpea (*Cicer arietinum* lin.) in in vitro condition. Pestology. 2013;37(8):32-36

[48] Shanker S, Boora KS, Chaudhary K. Bioefficacy of native isolate of *Bacillus thuringiensis* S6 against *Helicoverpa armigera* under field conditions in cotton. Annals of Biology. 2009;25(1):1-6

[49] Wu K, Guo Y. Influences of *Bacillus thuringiensis* Berliner cotton planting on population dynamics of the cotton aphid, *Aphis gossypii* Glover, in Northern China. Environmental Entomology. 2003;32(2):312-318. DOI: 10.1603/0046-225X-32.2.312

[50] Al-Shayji Y, Shaheen N, Saleem M, Ibrahim M. The efficacy of some *Bacillus thuringiensis* formulations against the whitefly *Bemisia tabaci* (homoptera: Aleyrodidae). Kuwait Journal of Science and Engineering. 1998;25(1):222-229

[51] Patti JH, Carner GR. *Bacillus thuringiensis* investigations for control of *Heliothis* spp. on cotton. Journal of Economic Entomology. 1974;67(3):415-418. DOI: 10.1093/jee/67.3.415

[52] Lopez DC, Zhu-Salzman K, Ek-Ramos MJ, Sword GA. The entomopathogenic fungal endophytes *Purpureocillium lilacinum* (formerly *Paecilomyces lilacinus*) and *Beauveria bassiana* negatively affect cotton aphid reproduction under both greenhouse and field conditions. PLOS ONE.

2014;9(8):e103891. DOI: 10.1371/journal.pone.0103891

[53] Zanwar PR, Deosarkar DB, Yadav GA, Shelke LT. Bioefficacy of entomopathogen, *B. bassiana* 1.15 per cent wp (larvocel) against cotton bollworm. Pestology. 2010;34(9):35-38

[54] Lopez DC, Sword GA. The endophytic fungal entomopathogens *Beauveria bassiana* and *Purpureocillium lilacinum* enhance the growth of cultivated cotton (*Gossypium hirsutum*) and negatively affect survival of the cotton bollworm (*Helicoverpa zea*). Biological Control. 2015;89:53-60. DOI: 10.1016/j.biocontrol.2015.03.010

[55] Bugeme DM, Knapp M, Boga HI, Ekesi S, Maniania NK. Susceptibility of developmental stages of *Tetranychus urticae* (Acari: Tetranychidae) to infection by *Beauveria bassiana* and *Metarhizium anisopliae* (Hypocreales: Clavicipitaceae). International Journal of Tropical Insect Science. 2014;34(3):190-196. DOI: 10.1017/S1742758414000381

[56] Arthurs SP, Bruck DJ. Microbial control of nursery ornamental and land-scape plant pests. In Lacey LA [ed.], Microbial Control of Pests. Elsevier. 2017;355-365. London, United Kingdom

[57] Uma Devi K, Padmavathi J, Uma Maheswara Rao C, Khan AAP, Mohan MC. A study of host specificity in the entomopathogenic fungus *Beauveria bassiana* (Hypocreales, Clavicipitaceae). Biocontrol Science and Technology. 2008;18(10):975-989. DOI: 10.1080/09583150802450451

[58] Gatarayiha MC. Biological control of the two-spotted spider mite, *Tetranychus urticae* Koch (Acari: Tetranychidae). Thesis. 2009. Discipline of Plant Pathology, School of Agricultural Sciences and Agribusiness, Faculty of Science and Agriculture, University of KwaZulu-Natal, Pietermaritzburg

[59] Mousumi DM, Sabu A. Agro-processing residues for the production of fungal bio-control agents. In: Zakaria ZA, Aguilar CN, Kusumaningtyas RD, Binod P (eds), Valorization of Agro-Industrial Residues. Non-Biological Approaches. 2020;2:110.

[60] Islam MT, Omar DB. Combined effect of *Beauveria bassiana* with neem on virulence of insect in case of two application approaches. Journal of Animal and Plant Sciences. 2012;22(1):77-82

[61] Seyed-Talebi FS, Kheradmand K, Talaei-Hassanloui R, Talebi-Jahromi K. Sublethal effects of *Beauveria bassiana* on life table parameters of two-spotted spider mite, *Tetranychus urticae* (Acari: Tetranychidae). Biocontrol Science and Technology. 2012;22(3):293-303. DOI: 10.1080/09583157.2012.655709

[62] Moorthi PV, Balasubramanian C, Avery PB, Banu AN, Kubendran T. Efficacy of fungal entomopathogens against red cotton stainer, *Dysdercus cingulatus* Fabricius (Hemiptera: Pyrrhocoridae). Journal of Biopesticides. 2012.5(2):140-143

[63] Abe M, Ikegami T. Susceptibility of five species of thrips to different strains of the entomopathogenic fungus, *Beauveria bassiana*. Applied Entomology and Zoology. 2005;40(4):667-674. DOI: 10.1303/aez.2005.667

[64] Lacey L. Microbial Control of Insect and Mite Pests (1st ed.). Academic Press. 2016

[65] Zafar J, Freed S, Khan BA, Farooq M. Effectiveness of *Beauveria bassiana* against cotton whitefly, *Bemisia tabaci* (gennadius) (aleyrodidae: Homoptera) on different host plants. Pakistan Journal of Zoology. 2016;48(1): 91-99

[66] Elhakim E, Mohamed O, Elazouni I. Virulence and proteolytic activity of entomopathogenic fungi against the two-spotted spider mite, *Tetranychus urticae* Koch (Acari: Tetranychidae). Egyptian Journal of Biological Pest Control. 2020;30(1):1-8. DOI: 10.1186/s41938-020-00227-y

[67] Tahir M, Wakil W, Ali A, Sahi ST. Pathogenicity of *Beauveria bassiana* and *Metarhizium anisopliae* isolates against larvae of the polyphagous pest *Helicoverpa armigera*. Entomologia Generalis. 2019;38(3):225-242. DOI: 10.1127/entomologia/2019/0460

[68] Kalvnadi E, Mirmoayedi A, Alizadeh M, Pourian HR. Sub-lethal concentrations of the entomopathogenic fungus, *Beauveria bassiana* increase fitness costs of *Helicoverpa armigera* (Lepidoptera: Noctuidae) offspring. Journal of Invertebrate Pathology. 2018;158:32-42. DOI: 10.1016/j.jip.2018.08.012

[69] Qayyum MA, Wakil W, Arif MJ, Sahi ST, Dunlap CA. Infection of *Helicoverpa armigera* by endophytic *Beauveria bassiana* colonizing tomato plants. Biological Control. 2015;90:200-207. DOI: 10.1016/j.biocontrol.2015.04.005

[70] Prasad A, Syed N, Purohit S. *Beauveria bassiana* (Balsamo) Vuillemin: A successful biopesticide against key pest *Helicoverpa armigera* (Hubner). International Journal of Pharma and Bio Sciences. 2010;6: 1-7

[71] Armes NJ, Jadhav DR, DeSouza KR. A survey of insecticide resistance in *Helicoverpa armigera* in the Indian subcontinent. Bulletin of Entomological Research. 1996;86(5):499-514. DOI: 10.1017/s0007485300039298

[72] Binneck E, Lastra CCL, Sosa-Gómez DR. Genome sequence of *Metarhizium rileyi*, a microbial control agent for Lepidoptera. Microbiology Resource Announcements. 2019;8(36):e00897-e00819. DOI: 10.1128/mra.00897-19

[73] Kogan M, Gerling D, Maddox JV. Enhancement of biological control in annual agricultural environments. In: Bellows TS, Fisher TW, (eds), Handbook of Biological Control. Academic Press. 1999;789-818. DOI: 10.1016/B978-012257305-7/50077-1

[74] Fronza E, Specht A, Heinzen H, de Barros NM. *Metarhizium* (*Nomuraea*) *rileyi* as biological control agent. Biocontrol Science and Technology. 2017;27(11):1243-1264. DOI: 10.1080/09583157.2017.1391175

[75] Loureiro E de S, Tosta RA de S, Dias PM, Pessoa LGA, Oliveira Neto FM de, Devoz GLR, Muchalak F. Performance of *Metarhizium rileyi* applied on *Helicoverpa armigera* (Hubner) (Lepidoptera: Noctuidae). Journal of Neotropical Agriculture. 2020;7(1):60-65. DOI: 10.32404/rean.v7i1.4208

[76] Sinha KK, Choudhary AK, Kumari P. Entomopathogenic fungi. In: Omkar, Ecofriendly Pest Management for Food Security. Elsevier. 2016;475-505. DOI: 10.1016/B978-0-12-803265-7.00015-4

[77] Jaronski ST. Mass production of entomopathogenic fungi: State of the art. In: Morales-Ramos JA, Guadalupe Rojas M, Shapiro-Ilan DI, (eds), Mass Production of Beneficial Organisms: Invertebrates and Entomopathogens. Academic Press. 2013;357-413. DOI: 10.1016/B978-0-12-391453-8.00011-X

[78] Dias PM, Loureiro ESL, Pessoa LGA, Oliveira Neto FM, Tosta RAS, Teodoro PE. Interactions between fungal-infected *Helicoverpa armigera* and the predator *Chrysoperla externa*. Insects. 2019;10(309):1-11

[79] Loureiro ES, Tosta RAS, Dias PM, Pessoa LGA, Oliveira Neto FM, Devoz GLR, Muchalak F. Performance of *Metarhizium rileyi* applied on *Helicoverpa armigera* (Hubner) (Lepidoptera: Noctuidae). Revista de Agricultura Neotropical. 2020;7(1):60-65

[80] Ingle YV, Wadaskar RM, Burgoni EB. Host spectrum and relative virulence of entomogenous fungi, *Nomuraea rileyi* and *Metarhizium anisopliae*. Current Biotica. 2016;9(4):348-355

[81] Liu S, Xu Z, Wang X, Zhao L, Wang G, Li X, Zhang L. Pathogenicity and in vivo development of *Metarhizium rileyi* against *Spodoptera litura* (Lepidoptera: Noctuidae) larvae. Journal of Economic Entomology. 2019;112(4):1598-1603. DOI: 10.1093/jee/toz098

[82] Jaronski ST, Mascarin GM. Mass production of fungal entomopathogens. In: Lacey LA, (ed) Microbial Control of Insect and Mite Pests: From Theory to Practice. Academic Press. 2017;141-155. DOI: 10.1016/B978-0-12-803527-6.00009-3

[83] Barad AH, Jethva DM, Balas TK, Chandravadiya TK. Impact of different doses of *Nomuraea rileyi* (Farlow) Samson against *Helicoverpa armigera* (Hubner) on pigeonpea. Pestology. 2015;39(12):58-60

[84] Costa VHDD, Soares MA, Rodríguez FAD, Zanuncio JC, Silva IMD, Valicente FH. *Nomuraea rileyi* (Hypocreales: Clavicipitaceae) in *Helicoverpa armigera* (Lepidoptera: Noctuidae) larvae in Brazil. Florida Entomologist. 2015;98(2):796-798. DOI: 10.1653/024.098.0263

[85] Zhong K, Liu Z, Wang J, Liu X. The entomopathogenic fungus *Nomuraea rileyi* impairs cellular immunity of its host *Helicoverpa armigera*. Archives of Insect Biochemistry and Physiology. 2017;96(1):e21402. DOI: 10.1002/ARCH.21402

[86] Espinosa DJL, Silva IHS da, Duarte RT, Gonçalves KC, Polanczyk RA. Potential of entomopathogenic fungi as biological control agents of whitefly (*Bemisia tabaci* biotype B) (Genn.) (Hemiptera: Aleyrodidae). Journal of Experimental Agriculture International. 2019;1-8

[87] Sharmila T, Manjula K. Field evaluation of oil formulations of *Nomuraea rileyi* (Farlow) Samson against *Spodopfera litura* and *Helicoverpa armigera* in groundnut. Pestology. 2017;41(10):21-26. DOI: 10.15740/has/ijpp/8.1/142-147

[88] Ingle YV, Wadaskar RM, Gathe AG. Bio-efficacy of *Nomuraea rileyi* against *Helicoverpa armigera* (Hubner). Indian Journal of Ecology. 2015;42(1):54-58

[89] Matter MM, Sabbour MM. Differential efficacies of *Nomuraea rileyi* and *Isaria fumosorosea* on some serious pests and the pests' efficient predator prevailing in tomato fields in Egypt. Journal of Plant Protection Research. 2013;53(2):103-109. DOI: 10.2478/jppr-2013-0015

[90] Hatting JL. Comparison of three entomopathogenic fungi against the bollworm, *Helicoverpa armigera* (Hübner) (Lepidoptera: Noctuidae), employing topical vs per os inoculation techniques. African Entomology. 2012;20(1):91-100. DOI: 10.4001/003.020.0112

[91] Padanad MS, Krishnaraj PU. Pathogenicity of native entomopathogenic fungus *Nomuraea rileyi* against *Spodoptera litura*. Plant Health Progress. 2009;10(1):11. DOI: 10.1094/php-2009-0807-01-rs

[92] Tang L, Hou RF. Potential application of the entomopathogenic

fungus, *Nomuraea rileyi*, for control of the corn earworm, *Helicoverpa armigera*. Entomologia Experimentalis et Applicata. 1998;88(1):25-30. DOI: 10.1046/j.1570-7458.1998.00342.x

[93] Gopalakrishnan C, Narayanan K. Epizootiology of *Nomuraea rileyi* (Farlow) Samson in field populations of *Helicoverpa* (=*Heliothis*) *armigera* (Hubner) in relation to three host plants. Journal of Biological Control. 1989;3(1):50-52. DOI: 10.18311/JBC/1989/4213

[94] Moscardi F, de Souza ML, de Castro MEB, Moscardi ML, Szewczyk B. Baculovirus pesticides: Present state and future perspectives. In: Iqbal A, Farah A, John P, (eds), Microbes and Microbial Technology: Agricultural and Environmental Applications. Springer New York. 2011;415-445. DOI: 10.1007/978-1-4419-7931-5_16

[95] Eroglu GB, Inan C, Nalcacioglu R, Demirbag Z. Genome sequence analysis of a *Helicoverpa armigera* single nucleopolyhedrovirus (HearNPV-TR) isolated from *Heliothis peltigera* in Turkey. PLOS ONE. 2020;15(6):e0234635. DOI: 10.1371/journal.pone.0234635

[96] Nai YS, Huang YF, Chen TH, Chiu KP, Wang CH. Determination of nucleopolyhedrovirus' taxonomic position. In: Shields VDC, (ed) Biological Control of Pest and Vector Insects. IntechOpen. 2017. DOI: 10.5772/66634

[97] Rohrmann GF. Baculoviruses as insecticides: Four examples. In: Rohrmann GF, (ed), Baculovirus Molecular Biology (4th ed.). National Center for Biotechnology Information (US). 2019. https://www.ncbi.nlm.nih.gov/books/NBK543459/

[98] Cory JS, Myers JH. The ecology and evolution of insect baculoviruses. In: Futuyma DJ, (ed), Annual Review of Ecology, Evolution, and Systematics. Annual Reviews Inc. 2003;34:239-272. DOI: 10.1146/annurev.ecolsys.34.011802.132402

[99] Haase S, Sciocco-Cap A, Romanowski V. Baculovirus insecticides in Latin America: Historical overview, current status and future perspectives. Viruses. 2015;7(5):2230-2267. DOI: 10.3390/v7052230

[100] Jehle JA, Lange M, Wang H, Hu Z, Wang Y, Hauschild R. Molecular identification and phylogenetic analysis of baculoviruses from Lepidoptera. Virology. 2006;346(1):180-193. DOI: 10.1016/j.virol.2005.10.032

[101] Grzywacz D. Basic and applied research: Baculovirus. In: Lacey LA, Microbial Control of Insect and Mite Pests: From Theory to Practice. Elsevier Inc. 2017;27-46. DOI: 10.1016/B978-0-12-803527-6.00003-2

[102] Knox C, Moore SD, Luke GA, Hill MP. Baculovirus-based strategies for the management of insect pests: a focus on development and application in South Africa. Biocontrol Science and Technology. 2015;25(1):1-20. DOI: 10.1080/09583157.2014.949222

[103] Sokolenko S, George S, Wagner A, Tuladhar A, Andrich JMS, Aucoin MG. Co-expression vs. co-infection using baculovirus expression vectors in insect cell culture: Benefits and drawbacks. Biotechnology Advances. 2012;30(3):766-781. DOI: 10.1016/j.biotechadv.2012.01.009

[104] Endersby NM, Morgan WC. Alternatives to synthetic chemical insecticides for use in crucifer crops. Biological Agriculture and Horticulture. 1991;8:33-52. http://www.eap.mcgill.ca/MagRack/BAH/BAH 8.htm

[105] Sosa-Gómez DR. Microbial control of soybean pest insects and mites. In: Lacey L (ed), Microbial Control of

Insect and Mite Pests: From Theory to Practice (1st Ed.) Elsevier, 2017199-208. DOI: 10.1016/B978-0-12-803527-6. 00013-5

[106] Sun X. History and current status of development and use of viral insecticides in China. Viruses. 2015;7(1):306-319. MDPI AG. DOI: 10.3390/v7010306

[107] Sharma A, Sandhi RK Reddy GV. A review of interactions between insect biological control agents and semiochemicals. Insects. 2019;10(12):439. DOI: 10.3390/insects10120439

[108] Black J. Horizontal transmission of *Helicoverpa armigera* Nucleopolyhedrovirus (HearNPV) in soybean fields infested with *Helicoverpa zea* (Boddie). In Theses and Dissertations. 2017 https://scholarworks.uark.edu/etd/2533

[109] Moore SD, Pittaway T, Bouwer G, Fourie JG. Evaluation of *Helicoverpa armigera* nucleopolyhedrovirus (HearNPV) for control of *Helicoverpa armigera* (Lepidoptera: Noctuidae) on citrus in South Africa. Biocontrol Science and Technology. 2004;14(3):239-250. DOI: 10.1080/09583150310001655666

[110] Reddy GVP, Manjunatha M. Laboratory and field studies on the integrated pest management of *Helicoverpa armigera* (Hubner) in cotton, based on pheromone trap catch threshold level. Journal of Applied Entomology. 2000;124(5-6):213-221. DOI: 10.1046/j.1439-0418.2000.00466.x

[111] Mtambanengwe KTE. Genetic characterization of a range of geographically distinct *Helicoverpa armigera* nucleopolyhedrovirus (HearNPV) isolates and evaluation of biological activity against South African populations of the African bollworm, *Helicoverpa armigera* (Hubner). Thesis. 2019Faculty of Science, Biochemistry, Microbiology and Biotechnology, Rhodes University

[112] Xing L, Yuan C, Wang M, Lin Z, Shen B, Hu Z, Zou Z. Dynamics of the interaction between cotton bollworm *Helicoverpa armigera* and nucleopolyhedrovirus as revealed by integrated transcriptomic and proteomic analyses. Molecular and Cellular Proteomics. 2017;16(6):1009-1028. DOI: 10.1074/mcp.M116.062547

[113] Nawaz A, Ali H, Sufyan M, Gogi MD, Arif MJ, Ranjha MH, Arshid M, Waseem M, Mustafa T, Qasim M, Khan KA, Ghramh HA. Comparative bio-efficacy of nuclear polyhedrosis virus (NPV) and spinosad against American bollworm, *Helicoverpa armigera* (Hubner). Revista Brasileira de Entomologia. 2019;63(4):277-282. DOI: 10.1016/j.rbe.2019.09.001

[114] Ginting S, Santoso T, Munara Y, Anwar R, Sudirman LI. Pathogenicity of *Helicoverpa armigera* nucleopolyhedro virus (HearNPV) and *Lecanicillium* sp. against *Helicoverpa armigera*. Journal of Biopesticides. 2018;11(2):128-135

[115] Eroglu GB, Demir I, Demirbag ZB. A novel alphabaculovirus isolated from the cotton bollworm, *Helicoverpa armigera* (Hubner) (Lepidoptera: Noctuidae): characterization and pathogenicity. Biologia. 2018; 73:545-551

[116] Arrizubieta M, Simón O, Torres-Vila LM, Figueiredo E, Mendiola J, Mexia A, Caballero P, Williams T. Insecticidal efficacy and persistence of a co-occluded binary mixture of *Helicoverpa armigera* nucleopolyhedrovirus (HearNPV) variants in protected and field-grown tomato crops on the Iberian Peninsula. Pest Management Science. 2016;72(4): 660-670. DOI: 10.1002/ps.4035

[117] Arrizubieta M, Simó O, Williams T, Caballero P. Determinant factors in the production of a co-occluded binary mixture of *Helicoverpa armigera* alphabaculovirus (HearNPV) genotypes with desirable insecticidal characteristics. PLOS ONE. 2016;11(10):e0164486. DOI: 10.1371/journal.pone.0164486

[118] Yu H, Meng J, Xu J, Liu TX, Wang D. A Novel neurotoxin gene ar1b recombination enhances the efficiency of *Helicoverpa armigera* nucleopolyhedrovirus as a pesticide by inhibiting the host larvae ability to feed and grow. PLOS ONE. 2015;10(8):e0135279. DOI: 10.1371/journal.pone.0135279

[119] Pugalenthi P, Dhanasekaran S, Elumali K, Krishnappa K. Bio-efficacy of nuclear polyhedrosis virus (NPV) tested against American bollworm *Helicoverpa armigera* (hub.) (Lepidoptera: Noctuidae) and protection of cotton boll damage. International Journal of Renewable and Environmental Sciences. 2013;1(3):22-26

[120] Jeyarani S, Sathiah N, Karuppuchamy P. Field efficacy of *Helicoverpa armigera* nucleopolyhedrovirus isolates against *H. armigera* (Hübner) (Lepidoptera: Noctuidae) on cotton and chickpea. Plant Protection Science. 2010;46(3): 116-122. DOI: 10.17221/30/2009-PPS

[121] Gan-Mor S, Matthews GA. Recent developments in sprayers for application of biopesticides—An overview. Biosystems Engineering. 2003;84(2):119-125.

[122] Steinke WE, Giles DK. Delivery systems for biorational agents. Biorational Pest Control Agents 1995;6:80-94. DOI: 10.1021/bk-1995-0595.ch006.

[123] Gan-Mor S, Palevsky E, Matthews GA. Equipment for Application of Biopesticides. In: Sahayaraj, K (Ed), Basic and Applied Aspects of Biopesticides. Springer 2014; 333-343.

[124] Behle R, Compton D, Kenar J, Shapiro-Ilan D. Improving formulations for biopesticides: enhanced UV protection for beneficial microbes. In: Zollinger R, Lindsay A, Goldschmidt E, (Eds) Pesticide Formulations and Delivery Systems, 30th Volume: Regulations and Innovation. ASTM International. 2011

[125] Jambhulkar PP, Sharma P, Yadav R. Delivery systems for introduction of microbial inoculants in the field. In: Singh DP, Singh HB, Prabha R, (eds), Microbial inoculants in sustainable agricultural productivity. Springer, New Delhi. 2016;199-218.

[126] Wilson, K., Grzywacz, D., Curcic, I. et al. A novel formulation technology for baculoviruses protects biopesticide from degradation by ultraviolet radiation. Sci Rep 2020;10:13301. DOI: 10.1038/s41598-020-70293-7

[127] Hall FR, Chapple AC, Taylor RA, Downer RA. Modeling the dose acquisition process of *Bacillus thuringiensis*: influence of feeding pattern on survival. Biorational Pest Control Agents, ACS Symposium Series. 1995; 5(595)68-78. DOI: 10.1021/bk-1995-0595.ch005

[128] Partzsch L, Zander M, Robinson H. Cotton certification in Sub-Saharan Africa: Promotion of environmental sustainability or greenwashing? Global Environmental Change 201957, 101924. DOI: 10.1016/J.GLOENVCHA.2019.05.008

[129] Bale JS, van Lenteren JC, Bigler F. Biological control and sustainable food production. Philosophical Transactions of the Royal Society B: Biological Sciences. 2008;363(1492):761-776. DOI: 10.1098/rstb.2007.2182

[130] Namasivayam SKR, Vidyasankar A. Biocompatible formulation of potential fungal biopesticide *Nomuraea rileyi* (f.) Samson for the improved post treatment persistence and biocontrol potential. Nature Environment and Pollution Technology. 2014;13(4):835-838

[131] Gupta S, Dikshit AK. Biopesticides: An ecofriendly approach for pest control. Journal of Biopesticides. 2010;3(1):186-188

[132] Usta C. Microorganisms in biological pest control — A review: Bacterial toxin application and effect of environmental factors. In: Silva-Opps, M, (ed), Current Progress in Biological Research. IntechOpen. 2013;287-317. DOI: 10.5772/55786

[133] Damalas CA, Koutroubas SD. Current status and recent developments in biopesticide use. Agriculture. 2018;8(1):13. DOI: 10.3390/agriculture8010013

[134] Claus JD, Gioria VV, Micheloud GA, Visnovsky G. Production of insecticidal baculoviruses in insect cell cultures: Potential and limitations. In: Soloneski, S., (ed), Insecticides - basic and other applications. IntechOpen. 2012;127-152. DOI: 10.5772/29551

[135] Leggett M, Leland J, Kellar K, Epp B. Formulation of microbial biocontrol agents-an industrial perspective. Canadian Journal of Plant Pathology. 2011;33(2):101-107. DOI: 10.1080/07060661.2011.563050

[136] Ravensberg WJ. A roadmap to the successful development and commercialization of microbial pest control products for control of arthropods. Springer Netherlands. 2011. DOI: 10.1007/978-94-007-0437-4

[137] Shukla N, Negi A, Singh A, Kabadwa BC, Sharma R, Kumar J. Present status and future prospects of bio-agents in agriculture. International Journal of Current Microbiology and Applied Sciences. 2019;8(4):2138-2153. DOI: 10.20546/ijcmas.2019.804.251

[138] Siegwart M, Graillot B, Lopez CB, Besse S, Bardin M, Nicot PC, Lopez-Ferber M. Resistance to bio-insecticides or how to enhance their sustainability: A review. Frontiers in Plant Science. 2015;6:1-19. DOI: 10.3389/fpls.2015.00381

[139] Arthurs SP, Lacey LA. Field evaluation of commercial formulations of the codling moth granulovirus: persistence of activity and success of seasonal applications against natural infestations of codling moth in Pacific Northwest apple orchards. Biological Control. 2004;31(3): 388-397. DOI: 10.1016/j.biocontrol.2004.05.004

[140] Jeyarani S, Sathiah N, Karuppuchamy P. An in vitro method for increasing UV-tolerance in a strain of *Helicoverpa armigera* (Lepidoptera: Noctuidae) nucleopolyhedrovirus. Biocontrol Science and Technology. 2013;23(3):305-316. DOI: 10.1080/09583157.2012.757296

[141] Kranthi KR, Kranthi S, Mohan KS. Potential of mango leaf extract containing flavonoids as UV protectant against inactivation of *Heliothis armigera* (Hubner) NPV. Journal of Entomological Research. 2001;25(2):155-160. DOI: 10.5958/0974-4576.2020.00033.x

[142] Ibrahim MA, Griko N, Junker M, Bulla LA. *Bacillus thuringiensis*: a genomics and proteomics perspective. Bioengineered bugs. 2010;1(1):31-50. DOI: 10.4161/bbug.1.1.10519

[143] Betz FS, Hammond BG, Fuchs RL. Safety and Advantages of *Bacillus thuringiensis*-protected plants to control insect pests. Regulatory Toxicology and

Pharmacology. 2000;32(2):156-173. DOI: 10.1006/rtph.2000.1426

[144] Sanahuja G, Banakar R, Twyman RM, Capell T, Christou P. *Bacillus thuringiensis*: a century of research, development and commercial applications. Plant Biotechnology Journal. 2011;9(3):283-300. DOI: 10.1111/j.1467-7652.2011.00595.x

[145] Patil S D, Jadhav RS, Raut SA. Chemical science review and letters effect of UVC rays on aqua suspension formulations of *Nomuraea rileyi* (Farlow) Samson. Chemical Science Review and Letters. 2017;6(21):159-165

[146] Edelstein JD, Lecuona RE, Trumper EV. Selection of culture media and *in vitro* assessment of temperature-dependent development of *Nomuraea rileyi*. Neotrop. Entomol. 2004;33: 737-742. DOI: 10.1590/S1519-566X2004000600011

[147] Caro L, Villamizar L, Espinel C, Cotes AM. Efecto del medio de cultivo en la virulencia de *Nomuraea rileyi* sobre *Spodoptera frugiperda* (Lepidoptera: Noctuidae). Revista Colombiana de Entomologia. 2005;31:79-88

[148] Grijalba EP, Espinel C, Cuartas PE, Chaparro ML, Villamizar LF. *Metarhizium rileyi* biopesticide to control *Spodoptera frugiperda*: stability and insecticidal activity under glasshouse conditions. Fungal biology. 2018;122(11):1069-1076

[149] Sandhu SS, Rajak RC, Agarwal GP. Studies on prolonged storage of *Beauveria bassiana* conidia: Effects of temperature and relative humidity on conidial viability and virulence against chickpea borer, *Helicoverpa armigera*. Biocontrol Science and Technology. 1993;3(1):47-53. DOI: 10.1080/09583159309355258

[150] Bouslama T, Chaieb I, Rhouma A, Laarif A. Evaluation of a *Bacillus thuringiensis* isolate-based formulation against the pod borer, *Helicoverpa armigera* Hübner (Lepidoptera: Noctuidae). Egyptian Journal of Biological Pest Control. 2020;30(1):16. DOI: 10.1186/s41938-020-00218-z

[151] Saranraj P, Jayaparakash A. Agrobeneficial entomopathogenic fungi – *Beauveria bassiana*: A review. Asian Journal of Multidisciplinary Research. 2017;3(2):1051-1087. DOI: 10.22192/iajmr.2017.3.2.4

[152] Hong TD, Ellis RH, Moore D. Development of a model to predict the effect of temperature and moisture on fungal spore longevity. Annals of Botany. 1997;79(2):121-128. DOI: 10.1006/anbo.1996.0316

[153] Inglis GD, Johnson DL, Goettel MS. Effects of simulated rain on the persistence of *Beauveria bassiana* conidia on leaves of alfalfa and wheat. Biocontrol Science and Technology. 1995;5(3):365-370. DOI: 10.1080/09583159550039819

[154] Inglis GD, Ivie TJ, Duke GM, Goettel MS. Influence of rain and conidial formulation on persistence of *Beauveria bassiana* on potato leaves and Colorado potato beetle larvae. Biological Control. 2000;18(1):55-64. DOI: 10.1006/bcon.1999.0806

Section 3

Biological Control versus Insecticides

Chapter 8

Biological Control of Agricultural Insect Pests

Mrinalini Kumari, Atul Srivastava, Shyam Babu Sah and Subhashini

Abstract

Pests are highly responsible for heavy crop losses and reduced food supplies, poorer quality of agricultural products, economic hardship for growers and processor. Generally, chemical control methods are practiced for their control which is neither always economical nor effective and may have associated unwanted health, safety and environmental risks. However, to meet the challenge of feeding to the ever increasing human population, an efficient, economical and environment friendly disease control methods are requisites. In this regard, biological control may be an effective means of reducing or mitigating the pests and pest effects through the use of natural enemies. Biological control is an environmentally sound which involves the use of beneficial microorganism to control plant pathogens and diseases they cause. Therefore, in this chapter we will provide a comprehensive account of this environmental friendly approach for effectively management of plant diseases. This chapter will also accentuate the development of biological control agents for practical applications and the underlying mechanism. The contents in the chapter will be beneficial and advantageous to all those working in academia or industry related to crop protection.

Keywords: bioinsecticides, biological control, crop protection, insecticides, pests

1. Introduction

1.1 Definition

Since the beginning of agriculture, farmers had to compete with the harmful organisms called "Pests". These organisms are not only responsible for potential loss of revenue due to heavy crop damage and reduced food supplies but also significantly damage the machinery, equipment and property as well. They are prevented, destroyed, repelled or mitigated using different types of pesticides such as insecticides, herbicides, rodenticides and fungicides. Such chemical control methods are neither always economical nor effective and are generally associated with unwanted health, safety and environmental risks.

Therefore, in the recent decades, elevated awareness on the impacts of pesticide use on the human health and environment has fostered to decrease the reliance on chemical controls. Furthermore, the development of pesticide resistance is also indicative for the need of change for pests management with emphasis on the human health and environment. In this regard, biologically based technologies

Biocontrol agents	Examples
Predators	Ladybugs, Dragonflies, Lacewings, Pirate Bugs, Rove and Ground Beetles, Aphid midge, Centipedes
Parasitoids	Ichneumonid wasps, Braconid wasps, Chalcid wasps, Tachinid flies
Nematodes	*Heterorhabditidae* spp., *Mermithidae* spp., *Rhabditidae* spp., *Steinernematidae* spp.
Bacteria	*Bacillus thuringiensis, Bacillus popillae*
Viruses	Cytoplasmic polyhedrosis (CPV), Granulosis (GV), Entomopox viruses (EPN)
Fungi	*Metarhizium anisopliae, Beauveria bassiana, Trichoderma viride*

Source: Homes et al. [2].

Table 1.
Different types of biological control agents.

could be more convectional to solve the urgent needs in pest management. Biological control or biocontrol is particularly the use of animals, fungi, or other microbes to feed upon, parasitize or interfere with a targeted pest species. When the chemical pesticides were not appropriate for controlling the specific pest, the use of biological control came as a practical solution to the pest problem. The biological control method is innovative and sustainable way to control pests. This method leaves no chemical residues and has no harmful impact on the humans or other organisms. If the method is successfully implemented following introduction, it may provide a permanent control with favaourable cost–benefit ratio. A commonly accepted definition of biological control is:

The use of living organisms to suppress the population of a specific pest organism, making it less abundant or less damaging than it would otherwise be [1].

The organisms used to feed on, parasitize, or otherwise interfere with targeted pests are called as the *biocontrol agents*. **Table 1** below summarizes some of the different types of biological control agents.

Though the biological agents controls the pest populations with the use of natural predators and minimizes their impact on economic and environmental practices, however, this control method can offer a few distinct advantages as well as disadvantages.

1.2 Advantageous and disadvantages of biological control over chemical/other control methods

Entomologists consider biological control as an option when the widespread and repetitive usage of chemicals develops resistance in insects. It is a great deal of concern in the recent years.

Extensive application of chemicals is the primary reason for undesirable ecological side effects. Agrochemicals cause many environmental adverse consequences. They may contaminate the groundwater; some may enter the food-chains as a consequence producing a threat to the health of human and other organisms. Besides, the spraying of pesticides is also unsafe for the user. Frequent use of chemical causes inefficacy and is an important part of agricultural enterprise cost [3, 4].

Besides the known adverse consequences of pesticides use, it is not possible to uproot chemicals from the pest management. However, pest mortality caused by chemicals should be added/included to natural causes of death rather than considering the death as substitutive. And here the importance of natural enemies becomes more comprehensible, as the eradication of one pest species could lead to enormous increase in other pest numbers [4].

Therefore, the increasing concern of adversities associated with pesticide use is resulting in a more environmental friendly and sustainable agriculture. Several acceptable regulations are imposed with chemical restrictions or bans, especially in the developed nations. And here natural control method is gaining more attention for pest management at the better prices in market. Such factors hopefully create a favorable condition for the widespread of biological control methods [5].

1.2.1 Advantages

Biological control has several advantages as a pest control method, especially when it is compared with insecticides. The most pivot benefit is that such control methods are environmental friendly and do not add any pollutant into the environment. Kok [6] stated that biological control method may be applied whenever required as it does not pollute the environment.

Another most important advantage of this method is its selectivity i.e., specific pest target strategy. Unlike the chemical/other control method, this method controls the pest meant to target and do not harm the other species or plant. Therefore, danger of damage to non target plant species is restricted. Weeden and Shelton [7] have confirmed that this natural control method do not arise any new complication, like conventional pesticides. However, the side effects may not be totally excluded, though it rarely appears [8]. When discussing the balance of agricultural ecosystems, selectivity plays a vital role because a great damage to non target species can lead to the restriction of natural enemies' populations. Once the natural enemies are introduced into the environment, they reduce the target pest population and sustain their own population as well. Therefore, after initial introduction very little effort is required to keep the system in balance. How successfully a Biological control agent (BCA) can be deployed in an agricultural ecosystem, so as not to damage non target pests, depends on appropriate host specificity tests which determine the potential host range [6].

Another interesting advantage of biological control method is the ability to self-perpetuate. According to Kok (1999), biological control agents will increase in their number and spread. They are self-propagating, dispersing, and self-perpetuating too. This becomes important in relation to the economic feasibility of biological control [9].

An important advantage of biological control method is that the pest is unable or very slow to develop resistance [7]. It is generally not possible for a target pest to develop mechanisms of defense when attacked by a natural enemy [10]. Some examples of defense mechanisms that could develop by pests are escape behavior and repellent chemicals. However, as Van Emden [8] states that "*we know of no cases where previously successful biological control has failed because of selection for resistance*".

The adopted and established biological control methods in a specific area can be kept in a place for longer time than the chemical or other control methods which require repetitive application [6]. Hence, biological control methods are also cost effective as they are needed to be applied only once. The effectiveness of biological control methods is higher than the other control methods and is based on self-perpetuation and self-propagation as mentioned earlier. A small number of biocontrol agents can grow to very high densities and provide continuous control of a pest over a large area. When the cost of deployment of biological control agent is considered in contrast to pesticide applications, biological control is generally less expensive than the chemical control [11]. The financial benefit of biological control is greatest in cases when there is no other option. Another interesting point regarding the cost efficiency of this method is that the yield benefit of biological control is probably less than yield achieved by agrochemicals, but the primary cost of biological control agent is generally lower than chemical pesticides [9].

1.2.2 Disadvantages

There are several cases where we also find the breakdown of successfully implemented biological control programmes. The story of the Cane toad introduction in Australia and *Harmonia axyridis* introduced across continental Europe as a biological control agent is the best example. In 1935 canes toads were introduced in Australia as a biological control method against the Greyback cane beetle damaging the sugarcane crops. However, the management went horribly wrong because the life history and ecology of cane toads were not fully considered before its introduction and the cane toads today have become an invasive pest in Australia. Similarly, *H. axyridis* (Coleoptera: Coccinellidae), the harlequin ladybird, was from Sible Hedingham, Essex, England, in 2004 [12] and introduced in the areas of continental Europe as a biological control agent against aphids. The individuals dispersed and became invasive from Europe to Britain due to their excellent dispersal abilities as well as some anthropogenic activities also [13, 14].

Therefore, there are also serious disadvantages that restrict the popularity and use of biological control agents to the growers promoting the chemical use. The most important disadvantage is the probability of revenue stability. Reichelderfer [9] well mentioned that the biological control agents are highly proned to environmental conditions rather than the chemical control. As a consequence the pest population is highly fluctuated. And this is a challenge for the growers in relation to the product quality, to the crop yield and obviously to the price of product on market. Moreover, if the annual harvest of a crop is not stable, it will affect grower's income stability.

Another important disadvantage is the incompatibility with conventional pesticides by the growers. Growers are characterized by lack of patience and chemical control is one of the quick fix to any pest population [6]. Hence, the growers prefer the potent pesticides rather than the biological control method which is a slow process and requires lot of time and patience producing long term effect. Van Emden [8] states the limitation of biological control over subsequent use of pesticides, "*where biological control agents are being used against one pest, it is clearly difficult to continue using insecticides against other pests on the same crop or other disease vectors in the same area. This may make the use of biological control impossible*". The slow action of biological control lacks the immediacy of chemical control [6]. This method only reduces the number of pest population; it does not completely wipe out the pest as the chemical control methods. The pests are present in intolerable populations. And the pesticides cannot be used as it will destroy the biological control system. Therefore, shifting to the use of biological control from chemical control is unattractive for growers [8].

Due to the reliability of natural enemies on environmental conditions biological control is often unpredictable. It is well reported in the biological control of whitefly in glasshouses, that a sudden change in weather or a period of extreme hot or cold may lead to a breakdown of the system" [8]. The introduction of natural enemies in a new environment needs to carry out extensive research work to achieve the desirable results against climatic constraints.

Another disadvantage of biological control is that it does not exterminate the pest. As Weeden and Shelton [7] points out that: *the general aim of biological control is to depress the pest population below the Economic Injury Level (EIL): i.e., where the costs of the control measures start to exceed those of the extra revenue.* When this method is used to control the pests in fresh fruits and vegetable, where certain quality standards are demanded by the consumers, the incomplete pest control is not desirable. And damage of product appearance is, therefore, not acceptable by the growers [9].

Selectivity is a major advantage of biological control method, however, it could also be disadvantageous. Since, natural enemies are species specific, the other pests which are not affected could cause damage, so that the benefit of bio control technique could be extremely eliminated. Reichelderfer [9] has stated that *when several insect species of the same general type are potential pests of the crop, the economic efficiency of biological control technique is extremely restricted.*

Though this method of control is cost effective, a lot is expensed for its successful implementation in the environmental system. Butt [3] mentions that the lack of infrastructure which facilitates transfer of new technologies and research knowledge to the growers is a major inhibitor factor to its commercial perspective. It is not easy and sometimes expensive too for implementing biological control in field because it requires high qualified scientific staff [7]. The growers generally choose the easy applications of pesticides. There is also relatively less investment in biological control research in compare with chemical pesticides.

Variability in the production batches is also one of the significant disadvantage. "The variation and changes in behaviour of natural enemies that can be caused by rearing conditions are manifold" [15]. This variation arises due to lack of appropriate rearing procedures and often leads to incompatibility. The application of appropriate rearing procedures and the production of high quality biological control agents ultimately increase the cost production of natural enemies. Due to this cause, quality measures in mass rearing are often not applied by companies and consequently production of good quality natural enemies becomes challenging [15].

Even though the biological control method is environmentally safe as it provides less risk of residues in food chain, there are risks associated with disruption of biological control agents in the natural food chain. Kok [6] has reported that "*biological control is most suited for exotic pest that are not closely related to indigenous beneficial species*". Thus, the natural enemies must be exotic species too. Lenteren [15] has mentioned few negative effects for the import of natural enemies and many countries deal with risk issues concerning a release of a new natural enemy. However, none of the biological control agent alone provides a completely satisfactory solution to crop pest control problem [4].

Table 2 summarizes the advantages and disadvantages of biological control over chemical/other control methods:

The use of biological agents to control pest population has long history. Biological control has been in practice since ancient times, however, they were not scientifically validated. During ancient times, Chinese, observed that ants were effective predators of many citrus pests. They multiplied and increased the populations of ants and took their nests from surrounding habitats and placed them into

Advantages	Disadvantages
Specific to a particular pest	Can sometimes fail in its specificity
Self-sustaining system	a slow process
Cheap after startup	Expensive at startup
Works most of the time	Does not completely destroy a pest
Higher recycling potential	Requires expert supervision
Capability to withstand polluted water	Often unpredictable
Low toxicity level	
Slow development of resistance	

Table 2.
Advantages and disadvantages of biological control over chemical/other control methods.

their orchards. Thus, the use of natural enemies to reduce the impacts of pests is just a modern adaptation of the original ideas from the history. In this chapter we will discuss the theories of biological control and examine their approaches and applications in the modern pest management.

2. Theories of biological control

2.1 Biological control success and its correlation with the geographical area

2.1.1 Island theory-Islands vs. continental areas

67% of biological control successes which occurred on islands later occurred in continental areas.

2.1.2 Tropical areas vs. temperate areas

All the successes in tropical areas were strengthened by the fact that initial biological control successes occurred on tropical islands. Success is the consequence of latitude and is reinforced on a physiological basis that is explained comparing the Heterodynamous insects and Homodynamous insects as below:

1. Heterodynamous insects

    a. are unable to continuously reproduce throughout year.

    b. generally have disconnected generations, i.e., all their stages may not be present at any one time.

    c. require conjunction between natural enemy and pest.

2. Homodynamous insects

    a. are capable to reproduce throughout year.

    b. have synchronized or connected generations and all their stages may be present at any one time.

    c. are easiest species to control with biological method.

2.2 Introduction strategies

2.2.1 Pros and Cons of multiple introductions

1. Pemberton & Willard [16] gave the theory that multiple parasitism was destructive. Natural enemy species may be pulled down by superior natural enemies when in competition.

2. The above theory of Pemberton & Willard was disapproved by H. S. Smith who stated that biotic potential is solely an indicator for the success of a parasitoid and competition between parasitoids for the same host will subsequently result to increased host mortality rather than natural enemy individually.

3. Multiple parasitism could be detrimental under two theoretical situations:

    a. Parasitization of host (with overlapping generations) giving rise to disconnected generations because of the eliminated host stage.

    b. Parasitization of host by incapable parasitoid that may lead to reduced intraspecific competition between individuals of the host species at high densities.

4. Advantages of multiple parasitism.

    a. Enhanced effective biological control may be achieved over complete geographical range of the host.

    b. Increased death of the host with a single natural enemy alone.

    c. Higher chances of introduced natural enemy to utilize other hosts when primary host population is low.

    d. Invasion on all host stages (sequence theory).

Biological control could be controlled by adding and subtracting predators.

Larval ectoparasitoids (e.g., Diglyphus spp.) and larval-pupal endoparasitoids (e.g., Chrysocharis oscinidis, Ganaspidium utilis) may interfere with each other when present in a cropping system. Larval ectoparasitoids which parasitize the leafminers, also indirectly harm the living endoparasitoid larva, already present in the leafminers, thus, leading to the death of the endoparasitoid. However, the endoparasitoids do not parasitize the leafminers with ectoparasitoids because if they do so, the parasitized hosts will not pupate, so the endoparasitoids can complete their life cycles. Therefore, numerous species may be needed to biologically control the leafminers in various crop systems.

2.2.2 Single introductions vs. multiple introductions

1. Turnbull and Chant [17] again gave the theory that single species introduction is best for biological control.

2. To gain a desired level and effective biological control result of 2nd or 3rd natural enemy species are valuable.

3. The lack of predictive theory is a major problem to implement biological control as the outcome of introduced natural enemy, with no prior history of classical biological control efforts, may not be assumed or predicted for new exotic species.

2.2.3 The "Sequence Theory"

1. Howard and Fiske [18] more elaborated the above theories. The authors stated that biological control would be more efficient if only one developmental stage of the pest species is attacked. Biological control should be achieved through a variety of natural enemies attacking several developmental stages of the host, making a sequence of natural enemies for satisfactory control.

2.2.4 Time factor with respect to expected results from introduced biological control agent

1. Curtis P. Clausen (1951) highlighted the time factor theory for introduction of biological control agents. He stated that an effective natural enemy might be expected to show evidence of control at the point of release within a period of 3 host generations or 3 years and concluded the following:

    a. A fully effective natural enemy is always easily and quickly established.

    b. If failure occurs, it indicates that the control will not be fully effective even after establishment is achieved.

    c. Further, the colonization of exotic species, to be used as biological control agent, may be discontinued after 3 years, if it fails to establish.

2.2.5 Hyperparasites elimination prior to parasitoid introduction

Hyperparasitism can significantly decline the ability of parasite to control the host. Therefore, their elimination prior to the parasitoid introduction must be of prime consideration in introduction phase of biological control agent.

Geographic races should be neglected.

3. Approaches to biological control

The use of biological control suppresses the pest populations, making them less damaging than they would be. They play an important role in limiting the densities of potential pest and include natural enemies such as predators, parasitoids, and pathogens. There are three general approaches to biological control; importation, augmentation and conservation of natural enemies (**Figure 1**). These have been accepted as an effective, environmentally non-degrading, technically appropriate, economically viable and socially acceptable method of pest management. Each of these techniques can be used either alone or in combination in a biological control program.

3.1 Importation

Importation of natural enemies, is also called as *classical biological control*. It refers to the planned introduction of an exotic biological control agent for permanent establishment and long-term pest control to an area that is invaded by pest. Its

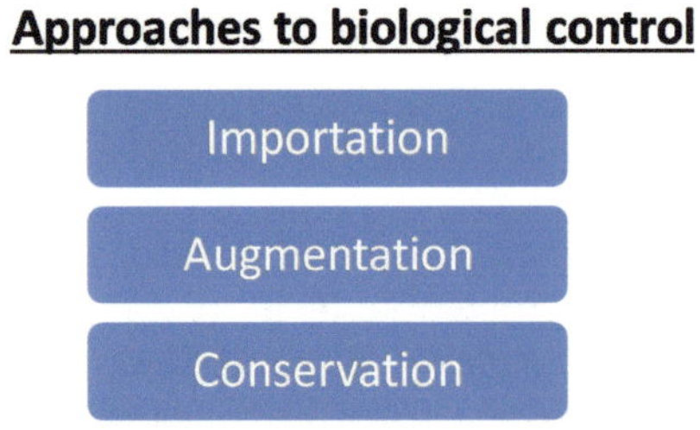

Figure 1.
Three general approaches to biological control.

target is to restore the balance between pest and natural enemy populations in the area invaded by pest without its natural enemies [1].

The import of pests either accidentally, or in some cases, intentionally in any countries where they are not native is continuous. However, due to a lack of natural enemies to suppress their populations, these introduced organisms of exotic origin may become pests, In these cases, importation of natural enemies can be highly effective [19]. Following the identification of the country of origin of the imported pest, a search may be conducted to explore a promising natural enemy. If the natural enemies are identified, their potential impact on the pest in the native country may be evaluated and imported into the new country for further study. Natural enemies are imported into the country under permit by the concerned authorities. The introduced natural enemies are first placed in quarantine for one or more generations to ensure the accidental importation of undesirable species (diseases, hyperparasitoids etc.). Further permits are required from concerned authorities for shipping to different states and field release.

The alfalfa weevil, *Hypera postica* (Gyllenhall) is a native of Europe. The species was introduced and detected in several countries. The first introduction was detected in the US in Utah in 1904. Biological control of this pest is best example of a successful program using importation of natural enemies [20].

3.2 Augmentation

Manipulation of natural enemies for enhancing the effectiveness of biological control is termed as augmentation. This can be adopted by one, or both, of the two general methods as given below:

1. mass production and periodic colonization; or.

2. Genetic enhancement of natural enemies

The first one mass production and periodic colonization is most commonly used. The natural enemies are produced in insectaries, and thereafter, released either inoculatively or inundatively. Augmentation is used where populations of a natural enemy are not present or cannot respond quickly enough to the pest population. This approach, therefore, does not provide permanent solution for the suppression of pests, as the outbreak of pest may occur with importation or conservation methods.

The use of the parasitoid wasp, *Encarsia formosa* Gahan, to suppress populations of the whitefly, *Trialeurodes vaporariorum* is one of the best example of the inoculative release method [21, 22]. The whiteflies are a global pest of vegetable and floriculture crops that is very difficult to manage with pesticides. Immediate release of *E. Formosa* following detection of the first whitefly on the crop effectively prevent the populations from developing to damaging levels. The releases should be made in context of an integrated crop management program taking into account the low tolerance of the parasitoids to pesticides.

3.3 Conservation

When we are to introduce any biological control attempt, conservation of natural enemies is key element for successful effectiveness. The factors which may limit the effectiveness of natural enemy must be identified, further modifying them to enhance the effectiveness.

This approach may be adapted by two ways as follows:

1. reduce the factors which interfere with natural enemies or

2. provide the resources that natural enemies need in their environment

Several factors are responsible for reducing the effectiveness of a natural enemy. Pesticide applications may directly kill natural enemies or have indirect effects through crop reduction in the numbers or availability of hosts. Cultural practices such as tillage or burning of crop debris may be detrimental for natural enemies by killing them or reducing their population by destroying the habitat. In orchards, frequent tillage may generate dust deposits on leaves, killing small predators and parasites and further, increasing certain insect and mite pests.

A study revealing the biological control of California red scale, *Aonidiella aurantii* natural (Maskell), suggests that the control may be achieved through periodic washing of citrus tree foliage that increases the parasitoids efficiency [23]. Some host plant effects such as chemical defenses which are harmful to natural enemies but the pest on the host plant is best adapted to it, also reduces the effectiveness of biological control. There are some pests that are able to sequester toxic components of their host, and use them as defense against their own enemies. In such cases also the effectiveness of biological control is reduced. Some cases like physical characteristics of the host plant such as leaf hairiness, may reduce the ability of the natural enemy to find and attack hosts.

Therefore, conservation ensures that the ecological requirements of the natural enemy are reached out in the cropping environment. To be effective, natural enemies may need access to; alternate hosts, adult food resources, overwintering habitats, constant food supply, and appropriate microclimates [24]. In a study reported by Doutt and Nakata [25] *Anagrus epos* Girault, is the fundamental parasitoid of the grape leafhopper, *Erythroneura elegantula.* A substitute is needed in grape vineyards for overwintering. This host, another leafhopper, overwinters on blackberry foliage in riparian areas, at some distance from the vineyards. Thus, during spring season the occurrence of early colonization by the parasitoids is often observed in the vineyards near to the natural blackberry. This forms the healthier and preferable biological control. Wilson et al. [26] have reported that the French prune trees also harbor another overwintering host. Their plantation in the upwind of the vineyards will effectively conserve *A. epos*.

4. Current applications of biological control

Biological control is an interesting science. This control method is constantly incorporating and introducing new knowledge and techniques. This part will deal with different ways by which efficient biological control may be adapted to meet the current pest management challenges.

4.1 Modern approaches in augmentation of natural enemies

Augmentation generally involves mass-production and periodic colonization of natural enemies. This has imparted to its commercial development. Recently, there are hundreds of commercially available biological control products for pest invertebrates, vertebrates, weeds, and plant pathogens.

The practice of augmentation not only differs for importation and conservation or in making a change in an agro ecosystem to improve its efficacy. Rather, this approach seeks to adapt natural enemies to fit into existing systems.

Inundative release of *Trichogramma* wasps is an excellent example of an augmentative practice is successfully adapted in agricultural systems. These are minute endoparasitoids Their eggs are released on the crops timed to the presence of pest eggs. *Trichogramma* is highly efficient biological control agent and most widely augmented species of natural enemy. Worldwide, over 32 million ha of agricultural crops and forests are treated annually with *Trichogramma* spp. in 19 countries, mostly in China and republics of the former Soviet Union [27].

Developed countries such as China, generally follow a simple, low labour cost, innovative technology for agricultural production and pest management systems. They highly use the *Trichogramma* spp. for the management *Chilo* spp., populations in sugarcane. The natural enemies are inundatively released and are protected from rain and predators inside emergence packets. Their eggs commercially reared in insectaries are wrapped in sections of leaves and slipped by hand over blades of sugarcane. *Trichogramma* is mostly produced in localized areas of China.

Implementation of biological control in western countries have to face socio-economics issues for its implementation [28]. Current in large-scale production agricultural systems, some incentives are there on the efficiency and economy of scale. Large industries have developed around the application of agrichemicals, including application equipment manufacturing, distribution and sales, as well as application services. Therefore, biological control products have to compete strongly with pesticides, they should be as effective as pesticides and they should have the capacity to be applied quickly on a large scale with conventional application equipment. So it is expected that the biological agents must have many characteristics same as pesticides.

In Western country such as Europe, commercial marketing of three products utilizing the European native, *Trichogramma brassicae* Bezdenko, to suppress the European corn borer, *Ostrina nubilalis* Hübner, in corn fields was almost possible following two decades of intensive research [29]. Annual application of these products in approximately 7,000 ha, 150 ha, and 15,000 ha is carried out in Switzerland and Germany, Austria and France respectively. All the three products are manufactured in plastic or paper packets for safeguarding the wasps against weather extremes and predation until their application in the field.

Trichogramma products are mostly manually applied to crop fields. With the exception of Trichocaps, which may be disseminated either by hand or by aircraft using conventional application equipment. Their packets are walnut-shaped cardboard capsules (2 cm. diam.) and contain approximately 500 parasitized Mediterranean flour moth, *Ephestia kuehniella* Zwolfer, eggs [30]. Developing *Trichogramma* inside capsules are induced into an overwintering (diapause) state in the insectaries. These are then stored in refrigerated conditions for nine months without loss of quality. By this system, production of *Trichogramma* product during winter months may be possible. The product may then be distributed to growers when needed in the summer.

When the refrigerated *Trichogramma* is removed from cold storage, it will start its development inside the capsules and begin emergence approximately 100°C. It is required to control this 'reactivation' process for uniform emergence of *Trichogramma*, at different developmental stages, in the fields. The companies only make planning and preparation of the product for its application. The growers are only responsible for applying the product to crop fields.

4.2 Landscape ecology and the conservation of natural enemies

The land disturbance studies and its effects on insect community dynamics as well as the emergence of the discipline of landscape ecology have imparted the way to think about the conservation of natural enemies. Since last 20 years, ecologists have recognized the central role of disturbance in the structuring of ecological communities [21–32]. Among the various ecosystems, the terrestrial ecosystems is highly disturbed one and this ecosystem experiences one disturbance event every several years (e.g. fire in grasslands), however, in agricultural ecosystems multiple events occur in each growing season (plowing, planting, nutrient and pesticide applications, cultivation and harvest) and their outcomes may be anticipated from an ecological point of view [33]. Highly disturbed systems exhibit reduced species diversity and short food chains, resulting to well adaptation of the species (i.e. pests) which have only few natural enemies to reduce their populations. Therefore, the role of additional disturbance events, such as pesticide applications, is needed to be initiated that controls the initial negative symptom, may also precipitate its reoccurrence.

Due to increasing reliance on mechanization and pesticides, diversity in farmlands has rapidly disappeared and the impacts on natural enemies must be studied and understood [34]. Increased habitat fragmentation, isolation and decreased landscape structural complexity destabilizes the biotic interactions which serve to regulate natural ecosystems [35, 36]. Therefore, this current systems of crop production (mechanization and use of pesticides) shape the physical structure of our agricultural landscapes [37].

The goal of an ecological approach to conservation biological control is just to modify the intensity and frequency of disturbance to a point where the natural enemies can effectively function. This requires its occurrence in field, farm and larger landscape-levels. Few modifications of tillage intensity and frequency (reduced tillage or no-tillage) in fields leave behind increased plant residue on the soil surface and have a positive influence on the predators (ground beetles and spiders) as well. Similarly, intercropping may be also modified, changing the microclimate of crop fields will make them more favorable for the parasitoids [38].

When taking at the farm level, the presence and distribution of non-crop habitats can be dangerous for natural enemy survival. *Eriborus terebrans* (Gravenhorst) is a wasp which parasitizes European corn borer larvae. They grow at moderate temperatures and require a source of sugar (nectar of flowering plants or aphid honeydew). But they are unable to met these demands in a conventionally managed corn field. Therefore, wasps seek more sheltered locations in wooded fence rows and woodlots where they find reduced temperatures, higher relative humidity and abundant sources of adult food. Besides, they also parasitize European corn borer larvae in corn field edges near their habitats at two to three times the rate of those in field interiors (up to 40%) [39]. Thus, the current research is creating natural enemy resource habitats and examining the potential of modifying corn production systems to increase natural control of European corn borers. Intercrops, strip crops, as well as modification of grass waterways, shelterbelts, buffer and riparian zones are some of the promising techniques in this regard.

Now at the landscape-level, the physical structure of agricultural production systems can have an impact on the pest and natural enemy diversity and abundance. Ryszkowski et al. [34] has reported in his study in the mosaic landscapes that natural enemies are highly dependent on refuge habitats than are pests and the greater abundance of these refuges in the mosaic landscapes resulted in their higher diversity, abundance and ability to respond to prey numbers. Further studies have also revealed enhanced parasitism of true armyworm, *Pseudaletia unipuncta* (Haworth), in structurally-complex versus simple agricultural landscapes. The parasitism in the complex sites was three times higher than in the simple sites (13.1% versus 3.4%). This differences was attributed to the abundant population of one wasp species, the braconid, *Meterous communis* (Cresson) in the complex sites.

Earlier, conservation was endeavored with introduction of one species at a time, concentrating to fulfill the needs of natural enemy in a particular system. Though it is a useful approach, now it seems possible that basic ecological theory could inform the design and management of landscapes to conserve and enhance the effectiveness of entire communities of natural enemies.

5. Conclusion

Biocontrol is a progressive and environment friendly way to control the pest. It leaves behind no chemical residues that may have a harmful impact on humans or other organisms. Importation, augmentation and conservation of natural enemies form the three basic approaches to biological control. Specified techniques underlying these approaches are developing at constant and are modified to meet the switching requirements of pest management. Modifications and improvements in rearing and application techniques and genetic advancement of natural enemies have increased the effectiveness of biological control agents. Further, application of new ecological theory is transforming the research need at conservation of natural enemies. For its successful implementation with full potentiality, continued refinement and adaptation of approaches and applications are necessary. Additional burden from the consumers and the expanding organic market requirements for biological control, come up with advantageous conditions for future development of the biological control agents in agriculture.

Conflict of interest

The authors declare no conflict of interest.

Author details

Mrinalini Kumari[1*], Atul Srivastava[2], Shyam Babu Sah[1] and Subhashini[3]

1 Mandan Bharti Agriculture College, Saharsa, Bihar, India

2 Institute of Medical Sciences, Banaras Hindu University, Varanasi, UP, India

3 Mahila Maha Vidhyalaya, Banaras Hindu University, Varanasi, UP, India

*Address all correspondence to: m.mrinalini35@gmail.com

References

[1] Eilenberg J, Hajek AE, Lomer C. Suggestions for unifying the terminology in biological control. BioControl. 2001;**46**:387-400

[2] Holmes LD, Mandjiny S, Upadhyay D. Biological control of agriculture insect pests. European Scientific Journal. 2016;**12**:1857-1887

[3] Butt MT, Jackson CW, Magan N. Introduction: Fungal as biological control agents: Progress, problems and potential. In: Butt TM, Jackson CW, Magan N, editors. Fungi as Biocontrol Agents. Progress, Problems and Potential. Oxon/New York: CABI Publishing; 2001. pp. 1-8

[4] Strickland AH. Ecological problems in crop pest control. In: Symposia of the Institute of Biology, editor. Biological Problems Arising from the Control of Pests and Diseases. London: Institute of Biology; 1960. pp. 1-11

[5] Whipps J, Lumsden R. Commercial use of fungi as plant disease biological control agents: Status and prospects. In: Butt MT, Jackson CW, Magan N, editors. Fungi as Biocontrol Agents: Progress, Problems and Potential. Oxon/ New York: CABI Publishing; 2001. pp. 71-153

[6] Kok LT. Biological control for the public. 1999. http://www.ento.vt.edu/~kok/Biological_Control/BC_html.hm.

[7] Weeden CR, Shelton AM. Biological control of pests. 2005. http://www.nysaes.cornell.edu/ent/biocontrol/

[8] Van Emden HF, Service MW. Pest and Vector Control. Cambridge: University Press; 2004

[9] Reichelderfer K. Economic feasibility of biological control of crop pests. In: Beltsville Symposia in Agricultural Research, editor. Biological Control in Crop Production. London/Torondo: Osmun; 1980. pp. 403-417

[10] Holt R, Hochberg M. When is biological control evolutionarily stable? *New York Times*. 2005. http://www.findarticles.com/p/articles/mi_m2120/is_n6_v78/ai_20413010.

[11] Agriculture, Food and Rural Development of Canada. Weed management: Biological control of weeds on the prairies. http://www.gov.mb.ca/agriculture/crops/weeds/weedfacts/fba07s00.html

[12] Majerus M, Strawson V, Roy H. The potential impacts of the arrival of the harlequin ladybird, *Harmonia axyridis* (Pallas) (Coleoptera: Coccinellidae), in Britain. Ecological Entomology. 2006;**31**:207-215

[13] Brown PMJ, Roy HE, Rothery P, Roy DB, Ware RL, Majerus MEN. Harmonia axyridis in Great Britain: Analysis of the spread and distribution of a non-native coccinellid. BioControl. 2008;**53**:55-67

[14] Jeffries DL, Chapman J, Roy HE, Humphries S, Harrington R, Brown PMJ, et al. Characteristics and drivers of high-altitude ladybird flight: Insights from vertical-looking entomological radar. PLoS One. 2013;**8**:e82278

[15] Lenteren JC. Quality Control and Production of Biological Control Agents: Theory and Testing Procedures. Wallingford: CABI Publishing; 2003

[16] Pemberton CE, Willard HF. Interaction of fruit-fly parasites in Hawaii. Journal of Agricultural Research. 1918;**12**:285-295

[17] Turnbull AL, Chant DA. The practice and theory of biological control of insects in Canada. Canadian Journal of Zoology. 1961;**39**:697-753

[18] Howard LO, Fiske WF. *Bull. U.S. Bur.* Entomologiste 1911; no. 91.

[19] Caltigirone LE. Landmark examples in classical biological control. Annual Review of Entomology. 1981;**26**:213-232

[20] Bryan MD, Dysart RJ, Burger TL. *Releases of Introduced Parasites of the Alfalfa weevil in the United States, 1957-88*. USDA APHIS Misc. 1993; Publ. No. 1504. 203 pp.

[21] Hussey NW, Scopes N. Biological Pest Control: The Glasshouse Experience. Ithaca: Cornell University Press; 1985

[22] Parrella ML. Biological pest control in ornamentals: Status and perspectives. SROP/WPRS Bulletin. 1990;**XIII/5**: 161-168

[23] Debach P, Rosen D. Biological Control by Natural Enemies. Cambridge, UK: Cambridge University Press; 1991. p. 440

[24] Rabb RL, Stinner RE, Van den Bosch R. Conservation and augmentation of natural enemies. In: Huffaker CB, Messenger PS, editors. Theory and Practice of Biological Control. New York: Academic Press; 1976. pp. 233-254

[25] Doutt RJ, Nakata J. The Rubus leafhopper and its egg parasitoid: An endemic biotic system useful in grape pest management. Environmental Entomology. 1973;**3**:381-386

[26] Wilson LT, Pickett CH, Flaherty DL, Bates TA. French prune trees: Refuge for grape leafhopper parasite. California Agriculture. 1989;**43**(2):7-8

[27] Li L-Y. Worldwide use of Trichogramma for biological control on different crops: A survey. In: Wajnberg E, Hassan SA, editors. Biological Control with Egg Parasitoids. Wallingford: CAB International; 1994. pp. 37-54

[28] Van Lenteren JC. Implementation of biological control. American Journal of Alternative Agriculture. 1988;**3**:102-109

[29] Bigler F, Bosshart S, Walburger M. Bisherige und neue Entwicklungen bei der biologischen Bekampfung des Maiszunslers, *Ostrinia nubilalis* Hbn., mit *Trichogramma maidis* Pint. et Voeg. in der Schweiz. Landwirtscahft Schweiz Band. 1989;**2**:37-43

[30] Kabiri F, Frandon J, Voegele J, Hawlitzky N, Stengel M. Evolution of a strategy for inundative releases of Trichogramma brassicae Bezd. (Hym Trichogrammatidae) against the European corn borer, *Ostrinia nubilalis* Hbn. (Lep. Pyralidae). In Proceedings, ANPP-Second International Conference on Agricultural Pests, Versailles, 1990.

[31] Pickett STA, White PS. The Ecology of Natural Disturbance and Patch Dynamics. San Diego: Academic Press; 1985

[32] Reice SR. Nonequilibrium determinants of biological community structure. American Scientist. 1994;**82**:424-435

[33] Odum E. Trends expected in stressed ecosystems. Bioscience. 1985;**35**(7):419-421

[34] Ryszkowski LJ, Karg G, Margarit M, Paoletti G, Zlotin R. Above-ground insect biomass in agricultural landscapes of Europe. In: Bunce RGH, Ryszkowski L, Paoletti MG, editors. Landscape Ecology and Agroecosystems. Boca Raton: Lewis Publication; 1993. pp. 71-82

[35] Kruess A, Tscharntke T. Habitat fragmentation, species loss, and biological control. Science. 1994;**264**: 1581-1584

[36] Robinson GR, Holt RD, Gaines MS, Hamburg SP, Johnson ML, Fitch HS, et al. Diverse and contrasting effects of

habitat fragmentation. Science. 1992;**257**:524-526

[37] Forman RTT, Godron M. Landscape Ecology. New York: John Wiley & Sons; 1986

[38] Headley JC, Hoy MA. Benefit/cost analysis of an integrated mite management program for almonds. Journal of Economic Entomology. 1987;**80**:555-559

[39] Landis DA, Haas M. Influence of landscape structure on abundance and within-field distribution of *Ostrinia nubilalis* Hübner (Lepidoptera: Pyralidae) larval parasitoids in Michigan. Environmental Entomology. 1992;**21**:409-416

Chapter 9

Biological Control of Tetranychidae by Considering the Effect of Insecticides

Samira Khodayari and Nayereh Hamedi

Abstract

Spider mites (family Tetranychidae) are important pests of many agricultural, medicinal and ornamental plants worldwide. They possess needle-like chelicerae which pierce plant cells, often feeding on chloroplasts on the under surface of the leaf and cause upper leaf surfaces develop whitish or yellowish stippling. Additionally spider mites produce silk webbing which covers the leaves. In this chapter we present common control methods of these mites including biological control with emphasizing on the prey preference, switching behavior and mutual interference of a biological control agent, *Phytoseius plumifer* (Canestrini and Fanzago). Additionally the side effects of two acaricides, abamectin and fenpyroxymate, on this predator will be discussed.

Keywords: Phytoseiidae, *Tetranychus urticae*, sublethal dose, abamectin, fenpyroxymate

1. Introduction

Spider mites (family Tetranychidae) contains many species that are important pests of agricultural crops. According to Migeon & Dorkeld [1], who provided a database for spider mites of the world, 1300 species had been described until now. Practically all the major food crops and many ornamental plants are subject to attack [2]. Tetranychid mites feed by penetrating the plant tissue with sharp cheliceral stylets and removal of the cell contents. The chloroplasts disappear and the small amount of remaining cellular material coagulates to form an amber mass. The amount of chlorophyll in the leaves may be decreased as much as 60 percent. The mite feeding also causes inhibition of photosynthesis. Small chlorotic spots can be found at feeding sites as the mesophyll tissue collapses due to the destruction of 18–22 cells per minute. Additionally they produce silk webbing which covers the leaves. Continued feeding leads to irregular spots formed by the integration of primary suction spots; finally the leaves turn yellow, gray or bronze. In the case of sever infestation the death of plants occur [3].

A rapid rate of mite development and high reproductive abilities allow spider mites to reach harmful population levels very quickly when the conditions for growth are permissive. A great number of experimental work has been directed toward the control of these mites since they have become resistant to a number of pesticides and their control has become very difficult. Moreover, chemical suppression of mite populations leads to residues on crops, environmental contaminations and toxicity to humans and non-target organisms. For these reasons, research has increasingly been performed to identify alternative methods to chemical control [4].

2. Tetranychidae control Methods

2.1 Chemical control

Prior to world war II, spider mites were minor pests of agricultural crops. This changed rapidly after war, with the extensive use of chemical pesticides, such as DDT [5]. The chemical acaricides used to control Tetranychidae are characterized by a large variety of chemical structures and mode of actions which were reviewed by Attia et al. [6], Knowles [7] and Dekeyser [8]. A pesticide may have both direct and indirect effects on Tetranychidae. Some may kill immediately, while other pesticides take longer to kill. Others may affect mite performance by inhibiting movement and reducing searching ability or lowering oviposition rates. In addition some pesticides (such as carbaryl and DDT) have a stimulatory effect on spider mite reproduction when present in low concentrations. The stimulatory effect on mite reproduction is called hormoligosis. Hormoligosis is an ongoing problem, although it may not be recognized [9]. The chemical control of these mites has become increasingly difficult because of their short life cycle, abundant progeny and arrhenotokous reproduction system. The repeated use of pesticides can lead to the development of resistant population and also can disrupt the natural control of Tetranychidae. Because of its resistance to a large number of chemical compounds, the two-spotted spider mite, *Tetranychus urticae* Koch, is considered most resistant species nearly in all over the world [10].

2.2 Cultural control

Cultural control involve all agronomic practices that are intended to reduce pest population. Cultural practices include changing the time of planting and harvest to avoid or minimize pest damage. It is known that high humidity reduce the reproductive potential of Tetranychidae whose optimal environment is hot and dry air [11]. Proper management of temperature and humidity can be useful to reduce pests' populations in greenhouses. Managing fertilizer applications is another important cultural practice. Large quantities of nitrogen or deficiency of potassium can increase the amount of soluble nitrogen available in the plant so that cause population increase of *T. urticae* [11]. In our previous work on the effect of fertilizer Fosfalim-k application on cucumber and its effect on population growth of *T. urticae* we showed that its application in the recommended dose had a controlling effect [12].

Another example of cultural control is dust management. Dust management is important for control of Tetranychidae, especially in climates that crop irrigation occurs. Whether the dust makes the foliage more suitable for spider mites or interferes with the spider mites predators' performance is in controversy. The elimination of crop residues is another way that can destroy pests and prevent transferring to subsequent crops. Crop rotation and polycropping are other methods that can be used to manage pest population. It is not clear that polycropping is useful in phytophagous mites control but if natural enemies are retained in the crops it could be helpful [9]. In our previous work we showed that the intercropping of sunflower and soybean increased natural enemies compared with monocultures [13].

2.3 Host plant resistance

Host plant resistance along with cultural control, is a component of any pest management program. Resistance of plants to pests enables them to avoid or inhibit host selection, inhibit oviposition and feeding, reduce pest survival and development and tolerate or recover from injury of pests that would cause greater damage to other plants of the same species under similar environmental conditions [14, 15]. Three mechanisms of plant resistance to pests have been categorized by Horber [16]: antixenosis, antibiosis and tolerance. Antixenosis describe the inability of a plant to serve as a host to a pest. The basis of this resistance mechanism can be morphological (e.g. leaf hairs, surface waxes and tissue thickness) or chemical (e.g. repellents or antifeedants). Antibiosis is the mechanism that describe the negative effects of a resistant plant on the biology of a pest which has colonized on the plant (e.g. adverse effects on development, survival and reproduction). Both morphological and chemical characteristics of plants can induce antibiosis. Tolerance is the degree to which a plant can tolerate a pest population that under similar conditions would severely damage a susceptible plant [17]. Resistance against spider mites is known to occur in many crops, including melon, pepper, soybean, cotton, cucumber, bean, eggplant and tomatoes. Resistant cultivars can be discovered by comparing mite populations on different crop varieties grown under the same conditions with equivalent initial mite populations [9]. We discovered the antibiosis mechanism of resistance to *T. urticae* in pepper varieties (unpublished data).

2.4 Biological control

Biological control is the use of natural enemies to manage pests' populations. Natural enemies are very important agents in reducing or regulating populations of pests and include parasitoids, predators and pathogens. A parasitoid is an organism that spends its larval stage in or on another organism, also known as a host. The larval parasitoid feeds only on the host as it develops, eventually killing the host. There are no report of mite's parasitoids. Predators are free living organisms, each of which will consume a number of pests (prey) in their lifespan. More than 65 predators have been recorded for European red mite, *Panonychus ulmi* (Koch), alone. Among the more important of these biological agents are predatory mites and insects, but others include spiders and disease-producing pathogens [3]. Three major methods exist for the use of natural enemies: conservation, classical biological control and augmentation.

Conservation seeks to identify and rectify negative influences of human activities that suppress natural enemies and to enhance agricultural fields as habitats for natural enemies. In conservation, the assumption is that the species of natural enemies already exist locally and have potential to effectively control the pest if given an opportunity to do so [18]. Classical biological control involves importation, evaluation, release and permanent establishment of natural enemies in the environment from the area of origin of a foreign pest. It assume that natural enemies from the area of the pest's origin will be more effective than natural enemies in the pest's new environment [9]. Augmentation involves the mass rearing and release of natural enemies to control target pest. The natural enemies must be capable of being mass reared and must be released at an appropriate time and in sufficient number to be effective. Two approaches are taken in augmentation. Inoculation involves releasing small number of natural enemies early in crop cycle with the expectation that they will reproduce and their offspring will provide pest control for an extended period of time. Inundation involves releasing large number of natural enemies for immediate control of pest when insufficient reproduction of the released natural enemies is likely to occur [18].

We found predatory mites from families Phytoseiidae, Ameroseiidae, Parasitidae, Stigmaeidae, Anystidae and Bdellidae as natural enemies of Tetranychidae during our sampling from Northwestern Iran (2007–2008). Among predator insects, we found *Stethorus gilvifrons* Mulsant (Col.: Coccinellidae), *Oenopia conglobata* (Linnaeus) (Col.: Coccinellidae), *Exochomus quadripustulatus* (Linnaeus) (Col.: Coccinellidae), *Chrysoperla carnea* (Stephens) (Neuroptera: Chrysopidae), *Scolothrips* sp. (Thysanoptera: Thripidae) and *Orius horvathi* Reuter (Het.: Anthocoridae). Among the predatory mites that we found, here we describe, *Phytoseius plumifer* (Acari: Phytoseiidae), which we have been worked on it.

Predaceous mites of the family Phytoseiidae are important natural enemies of several phytophagous mites and other pests on various crops. Phytoseiid mites occur throughout the world. Several authors have considered *Phytoseius plumifer* among the most important predators of phytophagous mites infesting fruit trees [19]. Before using natural enemies in biological control programs, it is essential to evaluate their efficiency and therefore, knowledge of the behavioral attributes of *P. plumifer* is essential for understanding the efficiency of this predator in the biological control of two-spotted spider mite.

2.4.1 *Prey stage preference, switching and mutual interference of* Phytoseius plumifer

Prey stage preference may affect prey–predator population dynamics, if the prey stage affects the development and reproduction of the predator. Prey preference by biological control agents can affect their ability to effectively control target pests too [20]. Preference may vary with the relative abundance of two prey types, in which case if the predator or parasitoid eats or oviposits in disproportionately more of the more abundant type, it is said to display switching behavior. In other words, switching is a behavioral phenomenon whereby a predator alters its preference for the prey species or type as prey relative densities change [21]. Murdoch et al. [22] found that switching could result from several different mechanisms including when (1) the predator develops a search image for the prey type with the highest relative abundance, (2) capture success on a prey type increases with increase in its relative abundance and (3) when the predator's habitat contains sub-habitats that are occupied by different prey types.

Aggregation of predators in space to prey patches causes the prey–predator interaction occur and searching efficiency to decrease with increasing predator density. Inverse density dependence in searching efficiency is known as predator interference or mutual interference. However, it was found that increasing the number of biological control agents released into an environment did not always increase the level of pest control [23]. This occurs when parasites/predators that are searching for a host/prey encounter each other, which can cause one or both to stop searching and possibly leave the area [24].

In our previous work we determined some aspects of the behavioral characteristics of *P. plumifer* on the two-spotted spider mite. We studied the preference of *P. plumifer* for different life stages of the two-spotted spider mite under choice and no-choice conditions. Switching of *P. plumifer* was tested with deutonymphs and larvae of the prey with different ratios too. Also, since the success of a predator in biological control programs is dependent on its behavior under the presence of other con-specific individuals, we investigated the mutual interference of *P. plumifer* in different densities of predator mites [25].

2.4.1.1 Materials and methods

2.4.1.1.1 No-choice experiment

In the feeding tests, we offered a total of 30 prey individuals of egg, larva, protonymph, deutonymph, male and female separately to a 24 h starved unmated female predator on soybean leaf arena and then allowed each predator to feed on the prey individuals for a total of 24 h. At the end of the experiment we estimated the number of prey individuals consumed per predator on each life stage of the prey.

2.4.1.1.2 Choice experiment

In this experiment we exposed total of 30 prey items i.e. equal number (5) of all stages of *T. urticae* (egg, larva, protonymph, deutonymph, male and female) to the predator females.

2.4.1.1.3 Switching

Switching of *P. plumifer* was tested with deutonymphs and larvae of the prey. Deutonymphs (D) and larvae (L) of *T. urticae* were presented in five different ratio treatments: 30 L:70D, 40 L:60D, 50 L:50D, 60 L:40D and 70 L:30D. The total prey number was 30. For evaluating the value of selectivity the following equation were used:

$$C = E_1 / E_2 \qquad (1)$$

where E_1 and E_2 are the proportion of larvae and deutonymphs killed in 50 L:50D ratio, respectively. To find the expected ratio of killed larvae and deutonymphs in no-choice position the obtained data were analyzed by Murdoch [22] formula as follow:

$$Y = C_x / (1 - X + C_x) \qquad (2)$$

where C_x is $C\times$ ratio of stage and X is the ratio of a prey stage on a leaf disc.

2.4.1.1.4 Mutual interference

In this experiment, 160 immature individuals (larvae and protonymphs) of *T. urticae* were placed on each leaf arena. In the next step, female predators at densities of 1, 2, 4, 8 and 16 per leaf arena were allowed to search the prey for 24 h. After this time period, the predators were removed from the arena and the number of eaten preys was counted. Finally, the per capita searching efficiency (a) of the predator at different densities was calculated according to the Nicholson [26] equation as follows:

$$a = (1/PT)\ln\left(N_t / (N_t - N_a)\right) \qquad (3)$$

where N_t is the total number of available prey (160), N_a is the total number of eaten preys, P is the number of predators, and T is the duration of the experiment (set to 1.0 for one day).

The calculated searching efficiency (*a*) was fitted against predator density (both on a logarithmic scale). The points were fitted to a linear regression by the least square method, according to the inductive model given by Hassell and Varley [27] as follows:

$$a = QP^{-m} \text{ or } \log a = \log Q - m \log P \quad (4)$$

where a is the searching efficiency of the predators, Q is the quest constant, and m includes only the component of interference due to behavioral interactions between predators [28].

2.4.1.2 Results

Our results indicated that in our no-choice preference experiments the predation preference of this predator on the different stages of *T. urticae* was as follow: eggs >protonymphs>larvae>males>deutonymphs>females of *T. urticae*. The preferred stage of two-spotted spider mite in choice preference experiments was protonymph. There was no tendency to the adult females of *T. urticae* in our results maybe because of their big size and the feeding rate was zero. Females of the predator killed more larvae than deutonymphs in switching experiments and they preferred larval stage compared to deutonymphs. There was positive switching behavior of predator for larval stage of prey at all ratios except 40%Larva: 60%Deutonymph (**Figure 1**) maybe because of their smaller size.

The values of total predation rate of *P. plumifer* were significantly different at different densities of the predator and the highest and lowest values of this parameter were recorded at 16 and 1 density of this predator, respectively. Furthermore, the per capita predation rate decreased to 1/4 with increasing the predator density from 1 to 16 and consequently the per capita searching efficiency also decreased significantly. According to results of Murdoch et al. [22] mechanisms one and two appear likely for our predator and capture increases on a prey type with increasing in its relative abundance.

The linear relationship between the natural logarithm of the predator density and the natural logarithm of per capita searching efficiency in mutual interference analysis has been demonstrated a negative slope. The negative value of the

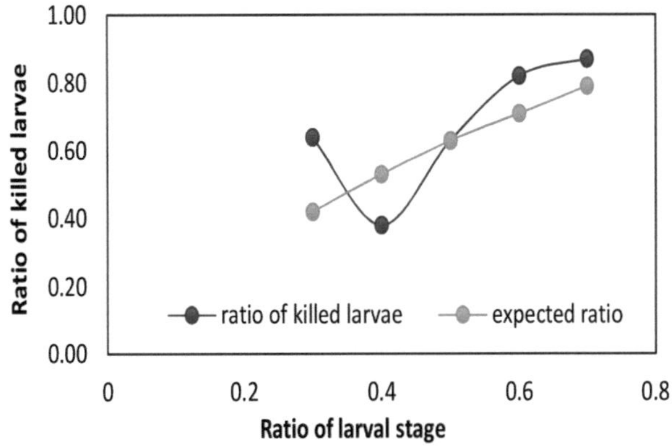

Figure 1.
Switching behavior of Phytoseius plumifer *females to different ratios of larval stage and deutonymph of* Tetranychus urticae.

interference coefficient in the mutual interference analysis showed an inverse relationship between the predator density and per capita searching efficiency and this fact revealed that the searching efficiency of *P. plumifer* significantly decreased with increasing predator density as a result of mutual interference. For most augmentative biological control agents, there is an optimal release rate that produces effective control of a pest species. Increasing the release rate above the optimal rate does not improve the control of pest species and is potentially economically detrimental [29]. In our study although with an increasing number of predators, greater numbers of preys have been consumed but, a doubling in the number of predator employed for *T. urticae* predation did not result in a doubling in the number of mite consumed, because of mutual interference. A significant decrease of the number of prey consumed per predator with an increased predator density suggests that interference among predators also increase at higher predator density. This is probably due to a closed experimental arena with limited predation time and high probability of mutual interference. However, under field conditions, factors such as large searching areas, the effects of other predator species, spatial complexity, and weather may affect the effectiveness of natural enemies [30].

2.4.2 Phytoseius plumifer *performance feeding on corn pollen*

Although phytoseiid mites have been mainly described as predators of mites and small insects, several species can feed and reproduce on pollen as well. The potential of phytoseiids to regulate phytophagous mites at low equilibrium densities has been more attended recently and studies have examined some of the characteristics that contribute to the survival of populations at low prey densities, such as feeding on pollens [31]. Pollen is utilized as an easy food source for phytoseiid mites rearing and also has been recognized as an important factor in the successful biological control of spider mites [32].

McMurtry and Croft [31] categorized the life style of phytoseiids based on feeding habitats and related biological and morphological traits. The life styles are: Type I, specialized predators of *Tetranychus urticae*; Type II, selective predators of tetranychids; Type III, generalist predators that may feed on pollen but perform better on prey; Type IV, specialized pollen feeders-generalist predators. *Phytoseius* species are categorized as Type III predators. Knowledge of the nutritional value of different plant pollens for *P. plumifer* could be important not only for mass rearing of the mite, but also for a better understanding of its population dynamics in the field.

In our previous work we described the effect of corn pollen on the life table parameters of *P. plumifer* at laboratory conditions according to Carey [33] method. We showed that *P. plumifer* can develop and reproduce on corn pollen under laboratory conditions, so the predator can persist in the field when its main prey is scarce or absent. Survival rate was 97% at immature stages and adult females appeared in 10th day and started laying eggs. On day 16 a sharp decline observed in survival curve and all of individuals died until 20th day (**Figure 2**). By comparing with Hamedi et al. [34] results we can conclude that corn pollen as lonely food source increases longevity of immature stages and decreases longevity and fecundity of adults of the predator considerably, although the predator can develop and reproduce successfully.

2.4.3 Side effect of acaricides on phytoseiid mites with an emphasis on Phytoseius plumifer

Use of pesticides cannot be eliminated in a short period of time in perennial crops because phytoseiid mites, as the most important predators of phytophagous mites, might not be able to maintain the spider mite populations below the

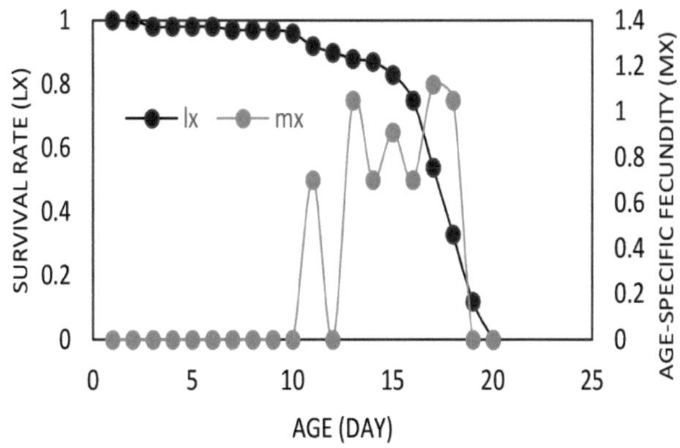

Figure 2.
The age specific survival (l_x) and age-specific fecundity (m_x) (♀/♀) of Phytoseius plumifer *on corn pollen.*

economically acceptable level on their own. Therefore successful utilization of biological control agents could depend on the compatibility of the natural predators with pesticides [35]. Most of the phytoseiid mites that naturally occur on plants, even in the absence of tetranychids, are generalist predators [36] and must be preserved using selective plant protection products [37]. Studying the side-effects of pesticides on natural enemies, including predaceous mites is an important task in pest management program, however, the use of pesticides remains necessary due to inadequate control achieved by natural enemies. The combination of biological and chemical control as an IPM program is only possible when the side-effects of pesticides on natural agents are well known [38].

Any indirect effects, which are referred to as sublethal, latent, or cumulative adverse effects may be associated with inhibiting longevity, fecundity, reproduction (based on the eggs laid by females), development time, mobility, prey consumption, emergence rates, and sex ratio and effects of sublethal concentration on the subsequent generation. In our previous study, the subletal effects of two acaricides abamectin (Vermectin_ 1.8% EC, Giah, Iran) and fenpyroximate (Ortus 5% SC, Giah, Iran) on the predatory mite *P. plumifer* fed on *T. urticae* was assessed in laboratory conditions. The adult predators were exposed to the residues of these acaricides on fig leaves for LC_{50} value determination based on a concentration–response analysis. Then sublethal effects of acaricides on performance of treated females and their offspring of *P. plumifer* were assessed.

2.4.3.1 Materials and methods

The *P. plumifer* individuals were originally collected from unsprayed (for ten years) fig orchards of Iran. The rearing method were explained comprehensively [34, 39]. All laid eggs were transferred daily from rearing arena to new arenas and were reared to adulthood and then used in the bioassay experiments. Pollen grains and *T. urticae* were used as food source in rearing and treatment arena.

2.4.3.1.1 Concentration-response bioassay

Concentration–response bioassay was carried out for acaricides using adult females and males at the first day of emergence. A modification of the leaf-dip technique was used [34, 39]. The sublethal concentrations consisted of LC_{10}, LC_{20} and LC_{30} were evaluated and used for assessment of sublethal effects on biological performance of *P. plumifer* [34, 39, 40].

2.4.3.1.2 Sublethal effects of acaricides on biological performance of treated females

Leaf discs with 3.3 cm diameter were treated with sublethal concentrations ($LC_{10, 20, 30}$) of acaricides and distilled water (as control) and then let to dry. The discs were placed on cotton pads as the same manner as rearing arena [39]. 40 less than 24-h-old unmated females were used in each concentration and stored at 27 ± 1°C, 50% RH and a photoperiod of 16:8 h (L:D). After 72 h treated mites were considered as alive if they were able to move for a distance without losing their balance during the movement and did not turn upside down. The survived females were selected for assessing sublethal effects of acaricides on them. Then each female was exposed to an untreated male from stock colony. Mortality and oviposition were recorded daily until the death of the last female in both treatments and controls. The dead males were replaced with new ones through the experiments.

2.4.3.1.3 Sublethal effects of acaricides on the developmental and biological performance of the offspring from treated females

The eggs laid by the treated and untreated (control) females were collected daily and life-table parameters of both groups were determined and compared to evaluate any possible carry-over activity of acaricides on the offspring. The subsequent generation were checked daily from eggs to dead of the last female. Development time, mortality, oviposition parameters and voracity were recorded daily and life-table parameters were taken until the death of the last female.

2.4.3.1.4 Sublethal effects of acaricides on prey consumption of treated female and the subsequent generation

For assessment of any sublethal effect on prey consumption of treated predators 20 to 30 only protonymphal stage (to decrease the adverse effect of prey webbing on predator) of *T. urticae* were placed on each treated and untreated (control) leaf disc as predator food source. Forty-eight hours after treatment, an unexposed male from the rearing arena was presented to each surviving female. Males that died during the experiments were replaced. The prey consumption of *P. plumifer* females was recorded separately for their pre-oviposition, oviposition and post-oviposition periods, because of the different rates for each one, were observed previously in our experiments [39]. Fresh preys were replaced with consumed ones in treated and untreated arena every 24 hours to maintain a constant daily food supply. Through the experiment adult male and female *P. plumifer* were kept pair. Consumption by the male measured previously as two protonymph per day, which subtracted from the total.

The eggs laid by the treated and untreated females were collected daily and moved to untreated leaf disc for assessment of sublethal effect on prey consumption of *P. plumifer* treated female's offspring from nymph to dead of the last female. Depending on the number of eggs, that laid by exposed females, approximately 10 and 30 replications were carried out for abamectin and fenpyroximate treatments, respectively. After emergence of the adults, males and females were paired and male consumption was subtracted as described previously. Individuals were checked daily and the number of protonymphs of *T. urticae* that had been consumed were counted, recorded and replaced with fresh ones until the death of the last predator. 10, 20, 30 and 20 protonymph stage of *T. urticae* were provided daily for proto- and deutonymphal stages and the pre-oviposition, oviposition and post-oviposition periods of predators, respectively. This was in excess of that required for daily consumption, as observed by our earlier experiments [39].

2.4.3.1.5 Data analysis

Mortality was corrected by using Abbott's Equation [41]. The LC_{50}, other sublethal concentrations and the regression equation were evaluated for the dose mortality line were extracted by using a probit program of SAS. The 95% confidence intervals of LC_{50} obtained from 72 h acute concentration–response curves developed from the responses of adult females and males, for comparing susceptibility of them. Any deviation from the expected sex ratio of 1:1 was determined using a chi-square analysis. For comparing longevity, fecundity, and duration of each stage among different concentrations and the control, analysis of variance (ANOVA) was used. Least Significant Difference (LSD) sequential test was used for comparing the means.

Based on the procedures developed by some authors [33, 42], the following life-table parameters were calculated: gross reproductive rate (*GRR*), net reproductive rate (R_0), intrinsic rate of increase (r_m), finite rate of increase (k), doubling time (D), mean generation time (T), intrinsic rate of birth (b) and intrinsic rate of death (d). Jackknife method was used to generate and compare mean demographic parameter estimates with SE values [43]. For comparing life table parameters among different concentrations and controls analysis of variance (ANOVA) was used. The means were compared using LSD sequential test.

2.4.3.2 Results

Our results of several experiments on side effects of acaricides on predatory mite *P. plumifer* demonstrated that, to evaluate the total effects of acaricides, in spite of effects on treated predator, assessment of all effects on offspring from treated females (subsequent generation) is necessary. Otherwise the real effects of residual exposure on performance of predatory mites would have incomplete end points. Our study proved that abamectin and fenpyroximate had an adverse effect on biological performance of *P. plumifer* females and their offspring [34, 39, 40]. Many other studies showed these effects on phytoseiid mites too [38, 44–47].

2.4.3.2.1 Sublethal effects of acaricides on mortality

Reduction in settlement ratio of phytoseiid mites treated by abamectin reported in our study and several other studies too [34, 36, 44]. Our results along with other studies on predators of *T. urticae* showed that most mortality occurred in 3 days after exposure to abamectin while in the first day there was no effect or a few effects [36, 48–50]. Abamectin was too toxic for *P. plumifer* in our study; it caused 100% mortality in female predators in 0.1 concentration that recommended for *T. urticae* control in the field. Moreover *P. plumifer* males were more susceptible than females to abamectin and fenpyroximate residue.

2.4.3.2.2 Sublethal effects of acaricides on eggs hatch and sex ratio of subsequent generation

The eggs laid by treated females were hatched at least 96.08% in fenpyroximate treatment so this parameter was not affected significantly. The sex ratio of *P. plumifer* was affected by fenpyroximate and the treatment caused a reverse in sex ratio. Sex ratio was 16:8 (female:male) in subsequent generation of untreated females that changed to 10:26 (female:male) in subsequent generation of treated females with LC_{30} of fenpyroximate. Increasing the number of male in comparison with female in subsequent generation of treated female with fenpyroximate can be the other reason of decreasing the predator population after two generations [36, 39]. The sex ratio and egg hatch rate of *P. plumifer* were not significantly affected by abamectin sublethal concentrations.

2.4.3.2.3 Sublethal effects of acaricides on longevity of females and subsequent generation

Our findings revealed that nymphal periods of offspring of exposed females to acaricides (fenpyroximate and abamectin) were shortened significantly. Moreover, the duration of pre-oviposition, oviposition and post-oviposition periods, and female longevity were significantly affected by sublethal concentrations of acaricides in both treated and their subsequent generation [36, 39]. This is in agreement with another research on *Neoseiulus longispinosus* (Evans, 1952) [36]. Our results indicated that longevity of treated females and their offspring were adversely affected by abamectin and fenpyroximate treatments. Reduction in female longevity of *N. longispinosus* after using abamectin, was reported too [36]. We assumed that shortened longevity of both treated females and their offspring may be partially explained by reduced food uptake as a consequence of acaricides effects [40].

2.4.3.2.4 Sublethal effects of acaricides on reproductive performance of females and subsequent generation

Acaricides, abamectin and fenpyroximate caused an overall reduction of *P. plumifer* population by increasing pre-oviposition period, decreasing oviposition period, decreasing fecundity in both treated female and their offspring. The number of eggs laid by treated female was so affected in both abamectin and fenpyroximate treatment. The total laid eggs were 46.57 eggs in control that decreased to 0.57 and 1.08 eggs in LC_{20} and LC_{30} treatment of abamectin and fenpyroximate, respectively. The treated females with LC_{30} of abamectin laid no egg (**Figure 3**).

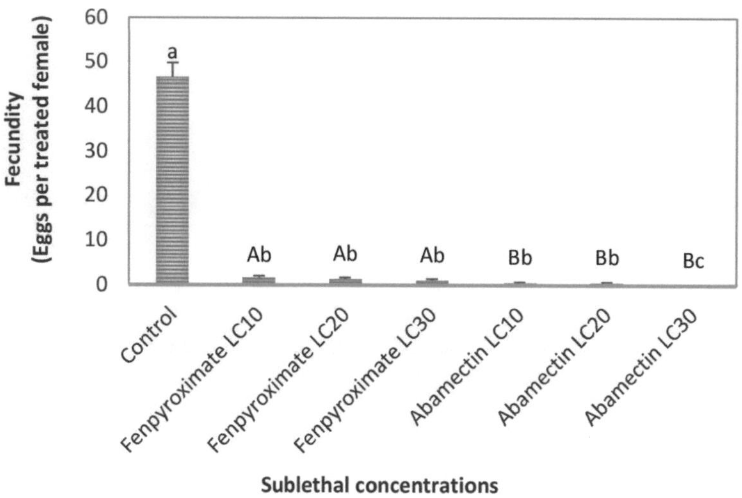

Figure 3.
Effects of sublethal concentrations of acaricides (fenpyroximate and abamectin) on fecundity of Phytoseius plumifer. *Different small letters above each bar indicate a statistically significant difference between concentrations. Different capital letters above each bar indicate a statistically significant difference between acaricides ($P < 0.05$) (LSD).*

2.4.3.2.5 Sublethal effects of acaricides on demographic parameters

The intrinsic rate of increase (r_m) is based on both survivorship and fecundity. So it has been recommended to use for evaluating the total effects of pesticides [51]. Our results along with several other studies have reported that life-table parameters of phytophagous and predatory mites were affected by sublethal concentrations of acaricides [36, 39, 45–47]. In our study, the life-table parameters showed significant differences, in population growth and reproductive performance, between offspring from females treated with sublethal concentrations of acaricides (fenpyroximate and abamectin) and untreated females of *P. plumifer* even in the

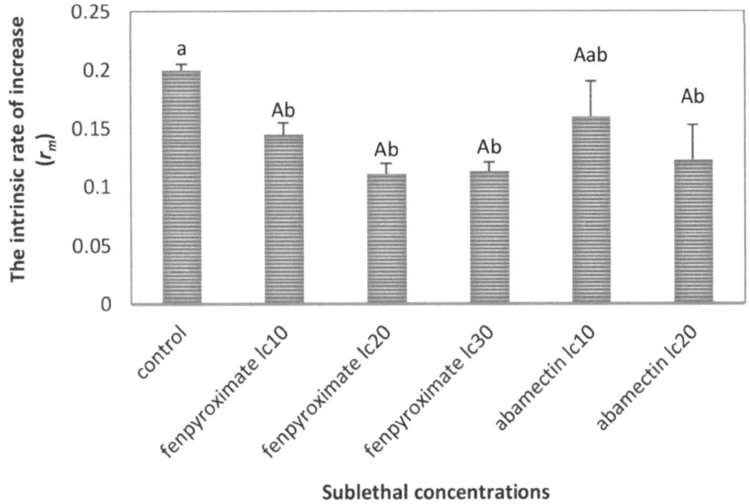

Figure 4.
The intrinsic rate of increase (r_m) of offspring of the treated and untreated females of Phytoseius plumifer. *Different small letters above each bar indicate a statistically significant difference between concentrations. Different capital letters above each bar indicate a statistically significant difference between acaricides ($P < 0.05$) (LSD).*

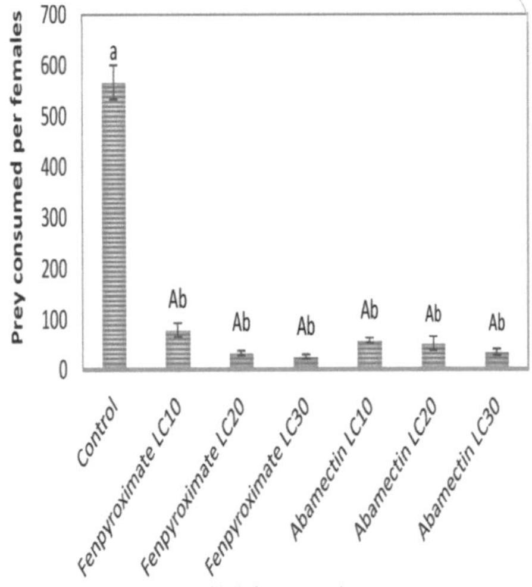

Figure 5.
Total voracity of treated and untreated females of Phytoseius plumifer. *Different small letters above each bar indicate a statistically significant difference between concentrations. Different capital letters above each bar indicate a statistically significant difference between acaricides ($P < 0.05$) (LSD).*

lowest concentration (LC_{10}). The intrinsic rate of increase (r_m), (**Figure 4**) the net reproductive rate (R_0) and the finite rate of increase (λ) of the offspring of treated females with both acaricides were markedly lower compared with the offspring of untreated females. This in turn resulted in a longer doubling time (*DT*). Moreover, in our laboratory observations the decrease in r_m values in sublethal concentrations maybe due to reduction of the mating rate and mobility of the offspring from treated females than untreated ones [36, 39].

2.4.3.2.6 Sublethal effects of acaricides on prey consumption of females and the subsequent generation

Our study revealed that prey consumption of treated females were considerably affected by sublethal concentrations of acaricides (abamectin and fenpyroximate) (**Figure 5**). But these concentrations slightly affected the prey consumption of subsequent generation. Daily prey consumption in the oviposition period was affected more than the other periods in both treated females and their offspring by both of acaricides. Decreasing longevity is another factor that may cause reduction in total prey consumption.

3. Conclusion

The low concentrations of pesticides may be used in combination with biological control agents within an IPM system to reduce the selective pressure and development of resistance in pests, but this study showed that adverse effects of fenpyroximate and abamectin on *P. plumifer* were significant, indicating that this acaricide may not be advisable for combined use with *P. plumifer* in IPM programs for controlling *T. urticae*. Even the low concentrations of acaricides that was suggested could be used in combination with biological control agents [52] had considerable adverse effects on this predator.

Author details

Samira Khodayari[1*] and Nayereh Hamedi[2]

1 Department of Plant Protection, Faculty of Agriculture, University of Maragheh, Maragheh, Iran

2 Department of Plant Protection, Agricultural Research, Education and Extension Organization of Khuzestan, Khuzestan, Iran

*Address all correspondence to: smrkhodayari@gmail.com; khodayari@maragheh.ac.ir

References

[1] Migeon A, Dorkeld F. Spider Mites Web: a comprehensive database for the Tetranychidae. 2021. Available from http://www1.montpellier.inra.fr/CBGP/spmweb (Accessed 17/07/2021)

[2] Pritchard AE, Baker EW. A Revision of the Spider Mite Family Tetranychidae. Vol. 2. San Francisco Pacific Coast Entomological Society; 1955. 472 p.

[3] Jeppson LR, Keifer HH, Baker EW. Mites Injurious to Economic Plants. University of California Press; 1975. 614 p.

[4] Kirisik M, Erler F, Boyaci F, Bayram Y. Evaluation of resistance in 16 eggplant genotypes to the two-spotted spider mite, *Tetranychus urticae* Koch (Acari: Tetranychidae). Phytoparasitica. 2020;49(2):275-285. DOI: org/ 10.1007/s12600-020-00856-x

[5] Stern VM, Smith RF, van den Bosch R, Hagen KS. The integrated control concept. Hilgardia. 1959;29(2): 81-101.

[6] Attia S, Grissa KL, Lognay G, Bitume E, Hance T, Mailleux AC. A review of the major biological approaches to control the worldwide pest *Tetranychus urticae* (Acari: Tetranychidae) with special reference to natural pesticides. Journal of Pest Science. 2013;86:361-386. DOI: 10.1007/s10340-013-0503-0

[7] Knowles CO. Mechanisms of resistance to acaricides. In: Sjut V. (ed) Molecular mechanisms of resistance to agrochemicals. Springer, Berlin, 1997. P. 57-77. DOI: 10.1007/978-3-662-03458-3_3

[8] Dekeyser MA. Acaricide mode of action. Pest Management Science. 2005;61:103-110. DOI: org/10.1002/ps.994

[9] Hoy MA. Agricultural Acarology. Introduction to Integrated Mite Management. CRC Press; 2011. 386 p.

[10] Whalon AE, Mota-Sanchez RM, Hollingworth, RM, Duynslager L. Arthropods Resistant to Pesticides Database (ARPD). 2008. http://www.pesticideresistance.org. Accessed 18 August 2021.

[11] Sabelis MW. The Functional Response of Predatory Mites to the Density of Two-Spotted Spider Mites. In: Metz JAJ, Diekmann O (eds) Dynamics of Structured Populations. 1986. Springer, Berlin; 298-344.

[12] Khodayari S, Abedini F, Renault D. The response of cucumber plants subjected to different salinity or fertilizer concentrations and reproductive success of *Tetranychus urticae* mites on these plants. Experimental & Applied Acarology. 2018;75:41-53. DOI: 10.1007/s10493-018-0246-y

[13] Javanmard A, Amani Machiani M, Ostadi A, Seifi A, Khodayari S. Evaluation of land productivity, competition and insect diversity in different intercropping patterns of sunflower (*Helianthus annuus* L.) and soybean (*Glycine max* L.) under low-input condition. Iran Agricultural Research. 2018;37(2):105-116. DOI: 10.22099/iar.2019.5190

[14] Smith CM. Plant Resistance to Insects. John Wiley and Sons. New York, USA. 1989. 286 p.

[15] Sharma HC, Ortiz R. Host plant resistance to insects: an eco-friendly approach for pest management and environment conservation. Journal of Environmental Biology. 2002;23: 111-135.

[16] Horber E. Types and Classification of Resistance. In: Maxwell FG,

Jennings PR. (eds.). Breeding Plants Resistant to Insects. Wiley, New York. 1980. p. 15-21.

[17] Thomas M, Waage J. Integration of biological control and host plant resistance breeding: a scientific and literature review: Technical Centre for Agricultural and Rural Cooperation. 1996. 99 p.

[18] Van Driesche RG, Bellows TS. Biological Control. New York: Chapman & Hall. 1995. 433 p.

[19] Rasmy AH, Elbanhawy EM. The phytoseiid mite *Phytoseius plumifer* as a predator of the eriophyid mite *Aceria ficus* (Acarina). Entomophaga. 1974;19(4):427-430.

[20] Pandey S, Singh R. Host size induces variation in progeny sex ratio of an aphid parasitoid *Lysiphlebia mirzai*. Entomologia Experimentalis et Applicata. 1999;90:61-67. DOI:10.1046/j.1570-7458.1999.00423.x

[21] Jervis M, Kidd N. Insect Natural Enemies, Practical approaches to their study and evaluation. Chapman & Hall, London. 1996. 491 p.

[22] Murdoch WW, Avery S, Smyth MEB. Switching in predatory fish. Ecology. 1975;56:1094-1105.

[23] Crowder DW. Impact of release rates on the effectiveness of augmentative biological control agents. Journal of Insect Science, 2007;7:1-11. DOI: org/10.1673/031.007.1501

[24] Hassell MP, Varley CG. A new inductive population model for insect parasites and its bearing on biological control. Nature. 1969;223:1133-1136.

[25] Khodayari S, Fathipour Y, Sedaratian A. Prey stage perfence, switching and mutual interference of *Phytoseius plumifer* (Acari: Phytoseiidae) on *Tetranychus urticae* (Acari: Tetranychidae). Systematic and Applied Acarology. 2016;21(3):347-355. DOI: org/10.11158/saa.21.3.9

[26] Nicholson AJ. The balance of animal populations. Journal of Animal Ecolpgy. 1933;2:132-178.

[27] Hassell MPC, Varley G. A new inductive population model for insect parasites and its bearing on biological control. Nature, 1969;223:1133-1136.

[28] Free CA, Beddington JR, Lawton JH. On the inadequacy of simple models of mutual interference for parasitism and predation. Journal of Animal Ecology. 1977;46:543-544.

[29] Stansly PA, Calvo J, Urbaneja A. Release rates for control of *Bemisia tabaci* (Homoptera: Aleyrodidae) biotype "Q" with *Eretmocerus mundus* (Hymenoptera: Aphelinidae) in greenhouse tomato and pepper. Biological Control. 2005;35:124-137. DOI: org/10.1016/j.biocontrol. 2005.07.004

[30] Gitonga LM, Overholt WA, Lohr B, Magambo JK, Mueke JM. Functional response of *Orius albidipennis* (Hemiptera: Anthocoridae) to *Megalurothrips sjostedti* (Thysanoptera: Thripidae). Biological Control. 2002;24:1-6. DOI: 10.1016/S1049-9644(02)00001-4

[31] McMurtry JA, Croft BA. Life-styles of phytoseiid mites and their role in biological control. Annual Review of Entomology, 1997;42:291-321.

[32] Bouras SL, Papadoulis GTH. Influence of selected fruit tree pollen on life history of *Euseius stipulatus* (Acari: Phytoseiidae). Experimental & Applied Acarology, 2005;36:1-14. DOI: 10.1007/s10493-005-2381-5

[33] Carey JR. Applied demography for biologists with special emphasis on insects--- Oxford University Press. 1993. 211 p.

[34] Hamedi N, Fathipour Y, Saber M. Sublethal effects of abamectin on the biological performance of the predatory mite, *Phytoseius plumifer* (Acari: Phytoseiidae). Experimental & Applied Acarology. 2011;53:29-40. DOI: 10.1007/s10493-010-9382-8

[35] El-Wakeil N, Gaafar N, Sallam A, Volkmar C. Side Effects of Insecticides on Natural Enemies and Possibility of Their Integration in Plant Protection Strategies. In: S. Trdan editor. Online book of insecticides. DOI: 10.5772/54199

[36] Ibrahim YB, Yee TS. Influence of sublethal exposure to abamectin on the biological performance of *Neoseiulus longispinosus* (Acari: Phytoseiidae). Journal of Economic Entomology. 2000;93:1085-1089. DOI: 10.1603/0022-0493-93.4.1085

[37] Croft BA. Arthropod biological agents and pesticides. John Wiley and Sons, New York; 1990. 723p. DOI: 10.1017/S000748530005080X

[38] Noii S, Talebi K, Saboori A, Allahyari H, Sabahi Q, Ashouri A. Study on the side-effects of three pesticides on the predatory mite *Phytoseius plumifer* (Canestrini & Fanzago) (Acari: Phytoseiidae) under laboratory conditions. Pesticides and Beneficial Organisms IOBC/wprs Bulletin. 2008;35:146-151

[39] Hamedi N, Fathipour Y, Saber M. Sublethal effects of fenpyroximate on life table parameters of the predatory mite *Phytoseius plumifer*. Biocontrol. 2010;55: 271-278. DOI: 10.1007/s10526-009-9239-4

[40] Hamedi N, Fathipour Y, Saber M, Sheikhi Gargan A. Sublethal effects of two common acaricides on the consumption of *Tetranychus urticae* (Prostigmata: Tetranychidae) by *Phytoseius plumifer* (Mesostigmata: Phytoseiidae). Systematic and Applied Acarology. 2009;14:197-205. DOI: org/10.11158/saa.14.3.4

[41] Abbott WS. A method of computing effectiveness of an insecticide. Journal of Economic Entomology. 1925; 18: 265-267.

[42] Pielou EC. Mathematical ecology. Wiley, New York; 1977. p. 627-628. DOI: 10.1002/bimj.4710200616

[43] Maia AHN, Luiz AJB, Camponhola C. Statistical inference on associated fertility life table parameters using Jackknife technique: computational aspects. Journal of Economic Entomology. 2000;93:511-518. DOI: org/10.1603/0022-0493-93.2.511

[44] Nadimi A, Kamali K, Arbabi M, Abdoli F. Selectivity of three miticides to spider mite predator, *Phytoseius plumifer* (Acari: Phytoseiidae) under laboratory conditions. Agricultural Sciences in China. 2009;8:326-331. DOI: 10.1016/S1671-2927(08)60216-3

[45] Debora B, Lima DB, Melo JWS, Gondim MGC, Guedes RNC, Oliveira JEM. Population-level effects of abamectin, azadirachtin and fenpyroximate on the predatory mite *Neoseiulus baraki*. Experimental and Applied Acarology. 2016;70:165-177. DOI: 10.1007/s10493-016-0074-x

[46] Alinejad M, Kheradmand K, Fathipour Y. Demographic analysis of sublethal effects of propargite on *Amblyseius swirskii* (Acari: Phytoseiidae): Advantages of using age-stage, two sex life table in ecotoxicological studies. Systematic and Applied Acarology. 2020;25(5):906-917. DOI: org/10.11158/saa.25.5.11

[47] Ahmed MM, Abdel-Rahman HR, Abdelwines MA. Application of demographic analysis for assessing effects of pesticides on the predatory mite, *Phytoseiulus persimilis* (Acari: Phytoseiidae). Persian Journal of

Acarology. 2021;10(3):281-298. DOI:. org/10.22073/pja.v10i3.66756

[48] Cote KW, Lewis EE, Schultz PB. Compatibility of acaricide residues with *Phytoseiulus persimilis* and their effects on *Tetranychus urticae*. Hortscience. 2002;37:906-909. DOI: org/10.21273/HORTSCI.37.6.906

[49] Chen T, French JV, Liu T, Graca V. Residual toxicities of pesticides to the predaceous mite *Galendromus helveolus* (Acari: Phytoseiidae) on Texas citrus. Subtropical Plant Science. 2003;55:40-45.

[50] Kim DS, Brooks DJ, Riedl R. Lethal and sublethal effects of abamectin, spinosad, methoxyfenozide and acetamiprid on the predaceous plant bug *Deraeocoris brevis* in the laboratory. Biocontrol. 2006;51:465-484.

[51] Stark JD, Wennergren U. Can population effects of pesticides be predicted from demographic toxicological studies? Journal of Economic Entomology. 1995;88:1089-1096. DOI: org/10.1093/jee/88.5.1089

[52] Dent D. Insect pest management. CABI Publishing, Wallingford, 2th edition. 2000;510 p. DOI: 10.1046/j.1439-0418.2001.0538a.x

Chapter 10

Effect of Insecticides on Natural-Enemies

Mohamed Abdel-Raheem

Abstract

Pesticides management options for control of invertebrate pests in many parts of the world. Despite an increase in the use of pesticides, crop losses due to pests have remained largely unchanged for 30–40 years. Beyond the target pests, broad-spectrum pesticides may affect non-target invertebrate species, including causing reductions in natural enemy population abundance and activity, and competition between pest species. Assays of invertebrates against weathered residues have shown the persistence of pesticides might play an important part in their negative impacts on natural enemies in the field. A potential outcome of frequent broad-spectrum pesticide use is the emergence of pests not controlled by the pesticides but benefiting from reduced mortality from natural enemies and competitive release, commonly known as secondary pests.

Keywords: effect, insecticides, natural enemies

1. Introduction

Pesticides management options for control of invertebrate pests in many parts of the world [1, 2]. Despite an increase in the use of pesticides, crop losses due to pests have remained largely unchanged for 30–40 years [3]. Beyond the target pests, broad-spectrum pesticides may affect non-target invertebrate species [4], including causing reductions in natural enemy population abundance and activity [5, 6], and competition between pest species [7]. Assays of invertebrates against weathered residues have shown the persistence of pesticides might play an important part in their negative impacts on natural enemies in the field [8].

A potential outcome of frequent broad-spectrum pesticide use is the emergence of pests not controlled by the pesticides but benefiting from reduced mortality from natural enemies and competitive release, commonly known as secondary pests [9–11]. Secondary pest outbreaks are challenging as they may also be caused by other mechanisms, which inherently make it difficult to determine how frequently pesticide use results in this outcome [10]. In cotton fields, it was estimated that 20% of late-season pesticide costs were attributable to secondary pest outbreaks caused by early-season pesticide applications for *Lygus* pests [10]. Higher numbers of cotton aphids, *Aphis gossypii* Glover and spider mites, *Tetranychus urticae* Koch were found in cotton fields that received early-season applications of insecticides against *Helicoverpa* spp. [5, 6].

One standardized approach for assessing non-target impacts of pesticides is the International Organization for Biological and Integrated Control—Pesticides and Beneficial Organisms (IOBC) rating system [12–14]. Subsequently, more bioassays under field conditions are needed to incorporate the dynamic interaction between pest populations and their natural enemy communities [15] and the environmental context at the time of application [16–18].

In Australian broad-acre grains the pest management practitioners are primarily concerned with pesticide efficacy, crop phytotoxicity, and cost; seldom are broader impacts of pesticides included in decision-making [19–21]. Chlorpyrifos is applied for the control of pests such as earwigs, isopods, and millipedes (Portuguese millipede, *Ommatoiulus moreleti* Lucas, 1860) [15], despite not being registered specifically to control those pests. A reduced application rate of broad-spectrum pesticides may lessen the impact on natural enemies but remain efficacious against pests [5, 22]. Repeated applications of broad-spectrum pesticides to control typical pest species are common in broad-acre crops, in particular canola [21] and pulses [23], therefore growers cannot often relate the pest numbers observed in a field to likely yield losses and adjust pesticide application [24]. The outcome is that pesticides are often applied prophylactically or in response to some observed crop damage that may or may not result in yield loss.

2. Indirect effects of pesticides on natural enemies

The indirect effects of pesticides on natural enemies have not been studied as extensively compared to direct effects, this chapter presents the indirect effects of pesticides that have primarily involved evaluating fecundity and longevity [25–35].

Prey consumption is the most important to successfully integrating natural enemies with pesticides and prevents indirect consequences on population dynamics [36, 37].

Some factors affiliated with natural enemies that may influence the indirect effects of pesticides include natural enemy age, type of natural enemy, life stages exposed to pesticides, and sex [38, 39]. Indirect affect may be related to residues remaining after a foliar application [40, 41]. Residues remaining after application may indirectly affect parasitoids by inhibiting adult emergence [42].

Natural enemies, indirectly affected by feeding on contaminated honeydew excreted by phloem-feeding insect prey [43, 44]. Certain pesticides may also exhibit repellent activity [45, 46] or alter host plant physiology [47, 48] indirectly affecting the ability of natural enemies to regulate existing arthropod pest populations [49].

3. Systemic insecticides

Applied as granules have been promoted to be relatively non-toxic to natural enemies [49–51]. However, insecticides as systemic effect exhibit indirect effects against natural enemies via several mechanisms of prey floral parts contaminated with the active ingredients [52–54]. Systemic insecticides may indirectly influence natural enemies if the mortality of prey populations is high [55, 56].

Natural enemies decrease the populations during starvation or dispersal [55, 57–59]. This effect depends on the foraging efficiency of the specific natural enemy. Decrease quantity or density of available prey or decrease their quality such that they are not acceptable as a food source, indirectly affected on larvae and adults or female parasitoids not lay eggs. Reproduction, foraging, fecundity, and longevity [33].

The active ingredient of systemic insecticide is distributed into flower parts indirectly impact natural enemies that feed on plant pollen or nectar such as minute pirate bug, *Orius* spp., which feed on plants during their life cycle [60–62], After feeding on the nectar of buckwheat (*Fagopyrum esculentum*) plants adults of, *Anagyrus pseudococci* are indirect affect [53]. *Microplitis croceipes* after feeding on the extrafloral nectaries of cotton plants was decreased foraging ability and longevity [63]. The application method and possibly timing of application may influence any indirect effects on parasitoids that feed on flower pollen and nectar as a food source [63]. Translocation of systemic insecticides into flowers indirectly affect natural enemies by altering foraging behavior as has been shown with the pink lady beetle, the green lacewing, and the parasitoid, *A. pseudococci* [53, 62]. The ability of systemic insecticides, when applied to the soil or growing medium as a drench or granule, to move into floral parts contingent on water solubility, application rate, and plant type [38, 63].

4. Insect growth regulators

Insect growth regulators are active directly on immature stages of some insect pests, there are three types of insect growth regulators: juvenile hormone mimics, chitin synthesis inhibitors, and ecdysone antagonists [64–69].

4.1 Pyriproxyfen

Pyriproxyfen, a juvenile hormone mimic is not indirect harmful effects against adult female oviposition and egg viability of green lacewing, *C. carnea* [70–72]. Also, not indirect effects on development time, female longevity, and fertility of *Orius* sp. [72]; exposure to pyriproxyfen delayed development and decreased the rate of parasitism of, *Hyposoter didymator* [73], and demonstrated to substantially alter of development time on *Chrysoperla rufilabris* of immatures [74], also, did not indirect impact against *Delphastus catalinae* female fecundity [70].

Fifth instars of *Podisus maculiventris* exposure to pyriproxyfen did not an indirect effect against reproduction. *Encarsia pergandiella* and *Encarsia transvena* are not indirect affect after exposure while *Encarsia formosa* exhibited decreased rates of emergence [75].

4.2 Kinoprene

Kinoprene is indirectly harmful against natural enemies by inhibiting adult emergence of, *Opius dimidiatus* and *Aphidius nigripes* [76, 77]. Kinoprene did not indirectly affect parasitoid emergence from *Planococcus citri* mummies [78]. Also, it inhibits adult emergence against some parasitoids [79].

4.3 Fenoxycarb

It is a juvenile hormone analog [80, 81] that has shown to be indirectly harmful to some natural enemies. It is delay development time from of pupae and adult of *C. rufilabris* [81], also, delay development of third instar larvae but not first instar larvae. Also, reproduction of females is inhibiting when second and third instars were initially exposed to it [82, 83]. Also, the same result against third instar larvae of *C. carnea* [84]. Also, happened indirect affect against female longevity and fecundity of, *Micromus tasmaniae* [74].

4.4 Cyromazine

It is a growth regulator that disrupts molting, it is affecting cuticle sclerotization during increasing cuticle stiffness [65], and exhibits indirect effects on the reproduction of *Phytoseiulus persimilis* [74], no indirect effect, against rates of adult emergence, of *Chrysocharis parksi* [85]. Exposure to it did not indirectly affect on longevity and reproduction of, *Hemiptarsenus varicornis* and *Diglyphus isaea* [86].

4.5 Diflubenzuron

It is a chitin synthesis inhibitor [65], less indirect impact against natural enemies, both parasitoids and predators [87].

Exposure to it decreased female longevity and reduced the parasitization rate of, *Hyposoter didymator* [73] and reproduction of, *Eulophus pennicornis* [88].

M. tasmaniae, exposed to diflubenzuron, resulted in indirect affects on reproduction, sex ratio, and longevity [74]. Diflubenzuron exhibited no indirect effects on the reproduction of, *Podisus maculiventris* adults. Diflubenzuron displayed minimal indirect effects on the parasitoid, *Macrocentrus ancylivorus* [89].

4.6 Buprofezin

It is a chitin synthesis inhibitor [66, 90], sterilizes certain natural enemies [91], reduces the number of progeny per female and sex ratios [73]. Feeding on it decreases female fertility and fecundity, and sterilized the males of the predatory coccinellid, *Delphastus catalinae* [69]. It did not affect the development of *Orius tristicolor* [92] or inhibits the reproduction of females of, *P. persimilis* [74]. Also, no indirect affect on oviposition and foraging of some parasitoids as *Eretmocerus* sp., and *Encarsia luteola* [90, 93]. Insect growth regulators are susceptible to early instars [90, 94, 95].

Indirect effects on natural enemies due to the volatility of the compound as it is known to be volatile and display vapor activity on some insect pests [96].

4.7 Azadirachtin

It is an ecdysone antagonist [72, 97–101], indirect effects against natural enemies [102]. It inhibits oviposition of the green lacewing, *C. carnea* and indirect affect against fertility and fecundity [99, 100]. Reproduction of, *Aphidoletes aphidimyza* is not indirect affect after exposure to it [103], and did not indirectly affect on the fecundity of, *Aphidius colemani* [91]; longevity and foraging ability of the parasitoids, *Cotesia plutellae* and *Diadromus collaris*, and sex ratio of progeny [6]; nor a reproduction of, *Neoseiulus californicus* [104]. Also, do not inhibit prey consumption of, *Atheta coriaria* adults [105].

First larvae of *Harmonia axyridis*, exhibit increase of development time, also, no indirect effect on adult fecundity [106–108].

5. Selective feeding blockers

It is include flonicamid and pymetrozine, inhibits feeding activity of piercing-sucking insects after initial insertion of their stylets into plant tissues and interfere with neural regulation of fluid intake through the mouthparts resulting in starvation [102, 109–112]. Flonicamid and pymetrozine, did not affect the development time, fertility, and parasitism of natural enemies, *Episyrphus balteatus*, *Bembidion lampros*; *Aphidius rhopalosiphi*, *Adalia bipunctata*; and *Aleochara bilineata* [112]. Pymetrozine exhibited minimal indirect effects on the reproduction of *N. californicus* [104]. Flonicamid did not indirectly affect parasitism, the sex ratio, and adult emergence of the parasitoid, *L. dactylopii*. Overall, minimal research has been conducted to determine the indirect effects of these types of pesticides on natural enemies [113].

6. Microbials

Entomopathogenic fungi and bacteria are, in general, not indirectly harmful to natural enemies, this may vary depending on concentration, natural enemy type, life stage exposed, the timing of application, and environmental conditions [114, 115].

Indirect effect not be associated with entomopathogenic fungi or bacteria [116]. *B. thuringiensis* has been indirect effects on some parasitoids this is depended on the formulation [117].

Natural enemies ingest fungal spores during grooming or feeding on contaminated hosts [89]; also, indirect effects depend on the concentration of spores [118]. Entomopathogenic fungi indirectly affect some natural enemies during feeding on prey that have been sprayed. Larvae of, *Cryptolaemus montrouzieri* were killed (50% mortality) after consuming mealybugs that had been sprayed with *Beauveria bassiana* [115]. *B. bassiana* decreased the fecundity of *N. californicus* females [104]. Fungus *Cephalosporium lecanii* exhibited no indirect effects on the longevity of the leafminer parasitoid, *Diglyphus begini* [119]. Exposure to *Metarhizium anisopliae* had no indirect effect on prey consumption (fungus gnat larvae) of rove beetle, *A. coriaria* adults [101]. Exposure to *Isaria* (=*Paecilomyces*) *fumosoroseus* at low relative humidity (55%) resulted in no indirect effects on foraging behavior and longevity of the aphid parasitoid, *Aphelinus asychis* whereas both parameters were significantly reduced when exposed to a high (≥95%) relative humidity, which could impact the ability of the parasitoid to regulate aphid populations. Ovipositing females may avoid prey that is infected by entomopathogenic fungi [114].

Spinosad has been demonstrated to be indirectly harmful to a variety of predatory insects such as, *C. carnea* [120]; *Hippodamia convergens*; *Orius laevigatus*, *Geocoris punctipes*; and *Nabis* sp. [121, 122]. Exposure to spinosad extended development time from the first instar to adult and decreased fertility of *Harmonia axyridis* females. Nevertheless, exposure to spinosad did not inhibit foraging behavior and reproduction of *P. persimilis* females [123, 124]. Parasitoids may be indirectly affected by spinosad based on decreased reproduction and reduced longevity [125, 126].

7. Miticides

It is like other pesticides, demonstrate variability in regards to any indirect effects against natural enemies depending on the type of miticide and predatory mite species [127]. It did not affect *Neoseiulus* (=*Amblyseius*) *womersleyi* on *Tetranychus urticae*, eggs [127, 128]. Exposure to concentrations of fenpyroximate indirectly affect on longevity and fecundity of *P. plumifer* [129]. Pyridaben inhibited reproduction of *Galendromus occidentalis* [130]. No indirect effects associated with sex ratio and prey consumption of *P. persimilis* [131, 132].

Exposure to bifenazate did not reduce fecundity, longevity, or prey consumption of *P. persimilis* or *N. californicus* [133].

8. Fungicides

It is considered low harmful to natural enemies comparing with insecticides and miticides [134]. Mancozeb was negatively affected against fecundity and reproduction of, *Amblyseius andersoni*, *G. occidentalis* [135], and *Euseius victoriensis* and inhibited the reproduction of, *Amblyseius fallacis* [130, 136]. Also, it did not indirectly affect on longevity or reproduction of, *Hemiptarsenus varicornis* and *Diglyphus isaea* [55]. Fungicides did not indirectly affect the fecundity of both *E. victoriensis* and *G. occidentalis* [130].

9. Additional factors associated with indirect effects of pesticides on natural enemies

The methodology evaluates the indirect effects of pesticides on natural enemies that may influence the results obtained [136–144]. The indirect effects of pesticides against natural enemies not necessarily are affiliated with the active ingredient [136, [141–144]. It is can be formulations as emulsifiable concentrates (EC) and soluble powders (SP) contain additives as adjuvants, surfactants, solvents, or carriers that are indirectly harmful to natural enemies [145].

10. Summary

This chapter has demonstrated the feasibility of combining or integrating natural enemies with certain pesticides including systemic insecticides, insect growth regulators, selective feeding blockers, microbials, miticides, and fungicides. There are three primary means by which natural enemies integrated with pesticides including pesticide selection, spatial separation of natural enemies and pesticides, and temporal discontinuity between natural enemies and pesticides [114]. Indirect effects are evaluated to determine if pesticides are compatible with natural enemies [29]. Indirect effects depending on concentration, natural enemy species, pesticide exposure time, developmental life stage(s) evaluated, and the influence of residues and repellency [50].

Author details

Mohamed Abdel-Raheem
Pests and Plant Protection Department, Agricultural and Biological Research Division, National Research Centre, Cairo, Egypt

*Address all correspondence to: ma.abdel-raheem@nrc.sci.eg; abdelraheem_nrc@yahoo.com; abdelraheem_nrc@homail.com

References

[1] Thomson LJ, Hoffmann AA. Field validation of laboratory-derived IOBC toxicity ratings for natural enemies in commercial vineyards. Biological Control. 2006;**39**(3):507-515. DOI: 10.1016/j.biocontrol.2006.06.009

[2] Guedes RNC, Smagghe G, Stark JD, Desneux N. Pesticide-induced stress in arthropod pests for optimized integrated pest management programs. Annual Review of Entomology. 2016;**61**:43-62. DOI: 10.1146/annurev-ento-010715-023646

[3] Altieri MA, Nicholls CI. Biodiversity and Pest Management in Agrosystems. 2nd ed. Boca Raton, FL. CRC Press; 2004

[4] Readshaw JL. Biological control of orchard mites in Australia with an insecticide-resistant predator. Journal of the Australian Institute of Agricultural Science. 1975;**41**(3):213-214

[5] Wilson LJ, Bauer LR, Lally DA. Effect of early season insecticide use on predators and outbreaks of spider mites (Acari: Tetranychidae) in cotton. Bulletin of Entomological Research. 1998;**88**(4):477-488. DOI: 10.1017/S000748530004222X

[6] Wilson LJ, Bauer LR, Lally DA. Insecticide-induced increases in aphid abundance in cotton. Austral Entomology. 1999;**38**(3):242-243. DOI: 10.1046/j.1440-6055.1999.00100.x

[7] Zeilinger AR, Olson DM, Andow DA. Competitive release and outbreaks of non-target pests associated with transgenic Bt cotton. Ecological Applications. 2016;**26**(4):1047-1054. DOI: 10.1890/15-1314

[8] Grundy PR, Maelzer D, Collins PJ, Hassan E. Potential for integrating eleven agricultural insecticides with the predatory bug *Pristhesancus plagipennis* (Hemiptera: Reduviidae). Journal of Economic Entomology. 2000;**93**:584-589. DOI: 10.1603/0022-0493-93.3.584

[9] Dutcher JD. A review of resurgence and replacement causing pest outbreaks in IPM. In: General Concepts in Integrated Pest and Disease Management. Dordrecht, Netherlands: Springer; 2007. pp. 27-43

[10] Gross K, Rosenheim JA. Quantifying secondary pest outbreaks in cotton and their monetary cost with causal-inference statistics. Ecological Applications. 2011;**21**(7):2770-2780. DOI: 10.1890/11-0118.1

[11] Steinmann KP, Zhang M, Grant JA. Does use of pesticides known to harm natural enemies of spider mites (Acari: Tetranychidae) result in increased number of miticide applications? An examination of California walnut orchards. Journal of Economic Entomology. 2011;**104**(5):1496-1501. DOI: 10.1603/EC11168

[12] Hassan SA, Bigler F, Blaisinger P, Bogenschütz H, Brun J, Chiverton P, et al. Standard methods to test the side-effects of pesticides on natural enemies of insects and mites developed by the IOBC/WPRS Working Group 'pesticides and beneficial organisms'. Bulletin OEPP/EPPO. 1985;**15**:214-255. DOI: 10.1111/j.1365-2338.1985.tb00224.x

[13] Nash MA, Thomson LJ, Hoffmann AA. Effect of remnant vegetation, pesticides and farm management on abundance of the beneficial predator *Notonomus gravis* (Chaudoir) (Coleoptera: Carabidae). Biological Control. 2008;**46**:83-93. DOI: 10.1016/j.biocontrol.2008.03.018

[14] Stark JD, Banks JE, Acheampong S. Estimating susceptibility of biological control agents to pesticides: Influence of life history strategies and population

structure. Biological Control. 2004;**29**(3):392-398. DOI: 10.1016/j.biocontrol.2003.07.003

[15] Thomson LJ, Hoffmann AA. Ecologically sustainable chemical recommendations for agricultural pest control? Journal of Economic Entomology. 2007;**100**(6):1741-1750. DOI: 10.1093/jee/100.6.1741

[16] Naranjo SE. Long-term assessment of the effects of transgenic Bt cotton on the abundance of nontarget arthropod natural enemies. Environmental Entomology. 2005;**34**(5):1193-1210. DOI: 10.1093/ee/34.5.1193

[17] Whitehouse MA, Wilson LJ, Fitt GP. A comparison of arthropod communities in transgenic Bt and conventional cotton in Australia. Environmental Entomology. 2005;**34**(5):1224-1241. DOI: 10.1093/ee/34.5.1224

[18] Rose R, Dively GP. Effects of insecticide-treated and lepidopteran-active Bt transgenic sweet corn on the abundance and diversity of arthropods. Environmental Entomology. 2007;**36**(5):1254-1268. DOI: 10.1603/0046-225X(2007)36[1254:EOIALB]2.0.CO;2

[19] Van der Werf HM. Assessing the impact of pesticides on the environment. Agriculture, Ecosystems & Environment. 1996;**60**(2-3):81-96. DOI: 10.1016/S0167-8809(96)01096-1

[20] Umina PA, Jenkins S, McColl S, Arthur A, Hoffmann AA. A framework for identifying selective chemical applications for IPM in dryland agriculture. Insects. 2015;**6**(4):988-1012. DOI: 10.3390/insects6040988

[21] Gu H, Fitt GP, Baker GH. Invertebrate pests of canola and their management in Australia: A review. Australian Journal of Entomology. 2007;**46**:231-243. DOI: 10.1111/j.1440-6055.2007.00594.x

[22] Edwards OR, Franzmann B, Thackray D, Micic S. Insecticide resistance and implications for future aphid management in Australian grains and pastures: A review. Australian Journal of Experimental Agriculture. 2008;**48**(12):1523-1530. DOI: 10.1071/EA07426

[23] Murray DAH, Clarke MB, Ronning DA. Estimating invertebrate pest losses in six major Australian grain crops. Australian Journal of Entomology. 2013;**52**:227-241. DOI: 10.1111/aen.12017

[24] Valenzuela I, Hoffmann AA. Effects of aphid feeding and associated virus injury on grain crops in Australia. Austral Entomology. 2015;**54**:292-305. DOI: 10.1111/aen.12122

[25] Grosch DS. Reproductive performance of a braconid after heptachlor poisoning. Journal of Economic Entomology. 1970;**63**:1348-1349

[26] Grosch DS. Reproductive performance of *Bracon hebetor* after sublethal doses of carbaryl. Journal of Economic Entomology. 1975;**68**:659-662

[27] O'Brien PJ, Elzen GW, Vinson SB. Toxicity of azinophos methyl and chlordimeform to parasitoid *Bracon mellitor* (Hymenoptera: Braconidae): Lethal and sublethal effects. Environmental Entomology. 1985;**14**:891-894

[28] Hsieh CY, Allen WW. Effects of insecticides on emergence, survival, longevity, and fecundity of the parasitoid *Diaeretiella rapae* (Hymenoptera: Aphididae) from mummified *Myzus persicae* (Homoptera: Aphididae). Journal of Economic Entomology. 1986;**79**:1599-1602

[29] Rosenheim JA, Hoy MA. Sublethal effects of pesticides on the parasitoid *Aphytis melinus* (Hymenoptera:

Aphelinidae). Journal of Economic Entomology. 1988;**81**:476-483

[30] Borgemeister C, Poehling HM, Dinter A, Holler C. Effects of insecticides on life-history parameters of the aphid parasitoid *Aphidius rhopalosiphi* (Hym, Aphidiidae). Entomophaga. 1993;**38**:245-255

[31] Haseeb M, Liu TX, Jones WA. Effects of selected insecticides on *Cotesia plutellae*, endoparasitoid of *Plutella xylostella*. BioControl. 2004;**49**:33-46

[32] Elzen GW. Sublethal effects of pesticides on beneficial parasitoids. In: Jepson PC, editor. Pesticides and Non-Target Invertebrates. Wimborne, UK: Intercept; 1990. pp. 129-150

[33] Croft BA. Arthropod Biological Control Agents and Pesticides. New York, NY: John Wiley & Sons; 1990

[34] van de Veire M, Tirry L. Side effects of pesticides on four species of beneficials used in IPM in glasshouse vegetable crops: 'Worst case' laboratory tests. Bulletin of the Organization of International Biological Control. 2003;**26**:41-50

[35] Grafton-Cardwell EE, Lee JE, Stewart JR, Olsen KD. Role of two insect growth regulators in integrated pest management of citrus scales. Journal of Economic Entomology. 2006;**99**:733-744

[36] Sáenz-de-Cabzón Irigaray FJ, Zalom FG, Thompson PB. Residual toxicity of acaricides to *Galendromus occidentalis* and *Phytoseiulus persimilis* reproductive potential. Biological Control. 2007;**40**:153-159

[37] Desneux N, Decourtye A, Delpuech J-M. The sublethal effects of pesticides on beneficial arthropods. Annual Review of Entomology. 2007;**52**:81-106

[38] Cloyd RA, Bethke JA. Impact of Neonicotinoid insecticides on natural enemies in greenhouse and interiorscape environments. Pest Management Science. 2011;**67**:3-9

[39] Bartlett BR. Integration of chemical and biological control. In: DeBach P, editor. Biological Control of Insect Pests and Weeds. New York, NY: Chapman and Hall; 1964. pp. 489-511

[40] Dicke M, Vet LEM. Plant-carnivore interactions: Evolutionary and ecological consequences for plant, herbivore and carnivore. In: Oliff H, Brown VK, Drent RH, editors. Herbivores: Between Plant and Predators. Oxford, UK: Blackwell Science; 1999. pp. 483-520

[41] Teodoro AV, Pallini A, Oliveira C. Sub-lethal effects of fenbutatin oxide on prey location by the predatory mite *Iphiseiodes zuluagai* (Acari: Phytoseiidae). Experimental and Applied Acarology. 2009;**47**:293-299

[42] Rill SM, Grafton-Cardwell EE, Morse JG. Effects of two insect growth regulators and a neonicotinoid on various life stages of *Aphytis melinus* (Hymenoptera: Aphelinidae). BioControl. 2008;**53**:579

[43] Longley M, Jepson PC. Effects of honeydew and insecticide residues on the distribution of foraging aphid parasitoids under glasshouse and field conditions. Entomologia Experimentalis et Applicata. 1996;**81**:189-198

[44] Wang HY, Yang Y, Su JY, Shen JL, Gao CF, Zhu YC. Assessment of the impact of insecticides on *Anagrus nilaparvatae* (Pang et Wang) (Hymenoptera: Mymanidae), an egg parasitoid of the rice planthopper, *Nilaparvata lugens* (Hemiptera: Delphacidae). Crop Protection. 2008;**27**:514-522

[45] Bartlett BR. The repellent effects of some pesticides to hymenopterous

parasites and coccinellid predators. Journal of Economic Entomology. 1965;**58**:294-296

[46] Garcia P. Sublethal effects of pyrethroids on insect parasitoids: What we need further know. In: Stoytcheva M, editor. Pesticides: Formulations, Effects, Fate. Rijeka: Intech; 2011. pp. 477-494

[47] Elzen GW, O'Brien PJ, Powell JE. Toxic and behavioral effects of selected insecticides on the *Heliothis* Parasitoid *Microplitis croceipes*. Entomophaga. 1989;**34**:87-94

[48] Tran DH, Takagi M, Takasu K. Effects of selective insecticides on host searching and oviposition behavior of *Neochrysocharis formosa* (Westwood) (Hymenoptera: Eulophidae), a larval parasitoid of the American serpentine leafminer. Applied Entomology and Zoology. 2004;**39**:435-441

[49] Ripper WE, Greenslade RM, Hartley GS. Selective insecticides and biological control. Journal of Economic Entomology. 1951;**44**:448-459

[50] McClanahan RJ. Food-chain toxicity of systemic acaricides to predaceous mites. Nature. 1967;**215**:1001

[51] Mizell RF, Sconyers MC. Toxicity of imidacloprid to selected arthropod predators in the laboratory. Florida Entomologist. 1992;**75**:277-280

[52] Tillman PG, Mullinix BG. Comparison of susceptibility of pest *Euschistus servus* and predator *Podisus maculiventris* (Heteroptera: Pentatomidae) to selected insecticides. Journal of Economic Entomology. 2004;**97**:800-806

[53] Krischik VA, Landmark AL, Heimpel GE. Soil-applied imidacloprid is translocated to nectar and kills nectar-feeding *Anagyrus pseudococci* (Girault) (Hymenoptera: Encyrtidae). Environmental Entomology. 2007;**36**:1238-1245

[54] Szczepaniec A, Creary SF, Laskowski KL, Nyrop JP, Raupp MJ. Neonicotinoid insecticide imidacloprid causes outbreaks of spider mites on elm trees in urban landscapes. PLoS One. 2011;**6**(5):e20018. DOI: 10.1371/journal.pone.0020018

[55] Radcliffe EB. Population responses of green peach aphid in Minnesota on potatoes treated with various insecticides. Proceedings of the North Central Branch of the Entomological Society of America. 1972;**27**:103-105

[56] Kiritani K. Pest management of rice. Annual Review of Entomology. 1979;**24**:79-312

[57] Hardin MR, Benrey B, Coll M, Lamp WO, Roderick GK, Barbosa P. Arthropod pest resurgence: An overview of potential mechanisms. Crop Protection. 1995;**14**:3-18

[58] Flanders SE. Environmental resistance to the establishment of parasitic hymenoptera. Annals of the Entomological Society of America. 1940;**33**:245-253

[59] Ripper WE. Biological control as a supplement to chemical control of insects. Nature. 1944;**153**:448-551

[60] Kiman ZB, Yeargan KV. Development and reproduction of the predator *Orius insidiosus* (Hemiptera: Anthocoridae) reared on the diets of selected plant material and arthropod prey. Annals of the Entomological Society of America. 1985;**78**:464-467

[61] Hagen KS. Ecosystem analysis: Plant cultivars (HPR), entomophagous species and food supplements. In: Boethel DJ, Eikenbary RD, editors. Interactions of Plant Resistance and Parasitoids and Predators of Insects.

New York, NY: John Wiley & Sons, Inc.; 1986. pp. 151-197

[62] Stapel JO, Cortesero AM, De Moraes CM, Tumlinson JH, Lewis WJ. Effects of extrafloral nectar, honeydew, and sucrose on searching behavior and efficiency of *Microplitis croceipes* (Hymenoptera: Braconidae) in cotton. Environmental Entomology. 1997;**26**:617-623

[63] Stapel JO, Cortesero AM, Lewis WJ. Disruptive sublethal effects of insecticides on biological control: Altered foraging ability and life span of a parasitoid after feeding on extrafloral nectar of cotton treated with systemic insecticides. Biological Control. 2000;**17**:243-249

[64] Stenersen J. Chemical Pesticides: Mode of Action and Toxicology. Boca Raton, FL: CRC Press; 2004

[65] Yu SJ. The Toxicology and Biochemistry of Insecticides. Boca Raton, FL: CRC Press, Taylor & Francis Group; 2008

[66] Staal GB. Insect growth regulators with juvenile hormone activity. Annual Review of Entomology. 1975;**20**:417-460

[67] Ables JR, Jones SL, Bee MJ. Effect of diflubenzuron on beneficial arthropods associated with cotton. Southwestern Entomologist. 1977;**2**:66-72

[68] Keever DW, Bradley JR Jr, Ganyard MC. Effects of diflubenzuron (dimilin) on selected beneficial arthropods on cotton fields. Environmental Entomology. 1977;**6**:732-736

[69] Liu T-X, Stansly PA. Lethal and sublethal effects of two insect growth regulators on adult *Delphastus catalinae* (Coleoptera: Coccinellidae), a predator of whiteflies (Homoptera: Aleyrodidae). Biological Control. 2004;**30**:298-305

[70] Koehler PG, Patterson RJ. Incorporation of pyriproxyfen in German cockroach (Dictyoptera: Blattelidae) management program. Journal of Economic Entomology. 1991;**84**:917-921

[71] Ishaaya I, Barazani A, Kontsedalov S, Horowitz AR. Insecticides with novel modes of action: Mechanism, selectivity and cross-resistance. Entomological Research. 2007;**37**:148-152

[72] Nagai K. Effects of juvenile hormone mimic material, 4-phenoxyphenyl (*RS*)-2-(2-pyridyloxy) propyl ether, on *Thrips palmi* Karny (Thysanoptera: Thripidae) and its predator *Orius* sp. (Hemiptera: Anthocoridae). Applied Entomology and Zoology. 1990;**25**:199-204

[73] Schneider MI, Smagghe G, Pineda S, Viñuela E. Action of insect growth regulator insecticides and spinosad on life history parameters and absorption in third-instar larvae of the endoparasitoid *Hyposoter didymator*. Biological Control. 2004;**31**:189-198

[74] Blümel S, Stolz M. Investigations on the effect of insect growth regulators and inhibitors on the predatory mite *Phytoseiulus persimilis* A. H. with particular emphasis on cyromazine. Journal of Plant Diseases and Protection. 1993;**100**:150-154

[75] Liu T-X, Stansly PA. Effects of pyriproxyfen on three species of *Encarsia* (Hymenoptera: Aphelinidae), endoparasitoids of *Bemisia argentifolii* (Homoptera: Aleyrodidae). Journal of Economic Entomology. 1997;**90**(2):404-411

[76] Ware GW, Whitacre DM. The Pesticide Book. Willoughby, OH: MeisterPro Information Resources; 2005

[77] Lemma KM, Poe SL. Juvenile hormone analogues: Effects of ZR-777

on *Liromyza sativae* and its endoparasite. Florida Entomologist. 1978;**61**:67-68

[78] McNeil J. Juvenile hormone analogs: Detrimental effects on the development of an endoparasitoid. Science. 1975;**189**:640-642

[79] Rothwangl KB, Cloyd RA, Wiedenmann RN. Effects of insect growth regulators on citrus mealybug parasitoid *Leptomastix dactylopii* (Hymenoptera: Encyrtidae). Journal of Economic Entomology. 2004;**97**:1239-1244

[80] Grenier S, Grenier AM. Fenoxycarb, a fairly new insect growth regulator: A review of its effects on insects. Annals of Applied Biology. 1993;**122**:369-403

[81] Liu T-X, Chen T-Y. Effects of the insect growth regulator fenoxycarb on immature *Chrysoperla rufilabris* (Neuroptera: Chrysopidae). Florida Entomologist. 2001;**84**:628-633

[82] Celli G, Bortolotti L, Nanni C, Porrini C, Brenna GS. Effects of the IGR fenoxycarb on eggs and larvae of *Chrysoperla carnea* (Neuroptera: Chrysopidae) laboratory test. In: Haskell PT, McEwen PK, editors. New Studies in Ecotoxicology. Cardiff, UK: The Welsh Pest Management in Forum; 1997. pp. 15-18

[83] Grenier S, Plantevin G. Development modifications of the parasitoid *Pseudoperichaeta nigrolineata* (Dipt., Tachinidae) by fenoxycarb, an insect growth regulator, applied onto its host *Ostrinia nubilalis* (Lep., Pyralidae). Journal of Applied Entomology. 1990;**110**:462-470

[84] Rumpf S, Frampton C, Dietrich DR. Effects of conventional insecticides and insect growth regulators on fecundity and other life-table parameters of *Micromus tasmaniae* (Neuroptera: Hemerobiidae). Journal of Economic Entomology. 1998;**91**:34-40

[85] Parrella MP, Christie GD, Robb KL. Compatibility of insect growth regulators and *Chrysocharis parksi* (Hymenoptera: Eulophidae) for the control of *Liriomyza trifolii* (Diptera: Agromyzidae). Journal of Economic Entomology. 1983;**76**:949-951

[86] Bjorksten TA, Robinson M. Juvenile and sublethal effects of selected pesticides on the leafminer parasitoids *Hemiptarsenus varicornis* and *Diglyphus isaea* (Hymenoptera: Eulophidae) from Australia. Journal of Economic Entomology. 2005;**98**:1831-1838

[87] Gordon R, Cornect M. Toxicity of the insect growth regulator diflubenzuron to the rove beetle *Aleochara bilineata*, a parasitoid and predator of the cabbage maggot *Delia radicum*. Entomologia Experimentalis et Applicata. 1986;**42**:179-185

[88] Butaye L, Degheele D. Benzoylphenyl ureas effect on growth and development of *Eulophus pennicornis* (Hymenoptera: Eulophidae), a larval ectoparasite of the cabbage moth (Lepidoptera: Noctuidae). Journal of Economic Entomology. 1995;**88**(3):600-605

[89] Broadbent AB, Pree DJ. Effects of diflubenzuron and BAY SIR 8514 on beneficial insects associated with peach. Environmental Entomology. 1984;**13**:133-136

[90] Gerling D, Sinai P. Buprofezin effects on two parasitoid species of whitefly (Homoptera: Aleyrodidae). Journal of Economic Entomology. 1994;**87**:842-846

[91] Stara J, Ourednickova J, Kocourek F. Laboratory evaluation of the side effects of insecticides on *Aphidius colemani* (Hymenoptera: Aphidiidae), *Aphidoletes aphidimyza* (Diptera: Cecidomyiidae), and *Neoseiulus cucumeris* (Acari: Phytoseidae). Journal of Pesticide Science. 2011;**84**:25-31

[92] James DG. Effect of buprofezin on survival of immature stages of *Harmonia axyridis*, *Stethorus punctum picipes* (Coleoptera: Coccinellidae), *Orius tristicolor* (Hemiptera: Anthocoridae), and *Geocoris* spp. (Hemiptera: Geocoridae). Journal of Economic Entomology. 2004;**97**:900-904

[93] Liu T-X, Chen T-Y. Effects of the chitin synthesis inhibitor buprofezin on survival and development of immatures of *Chrysoperla rufilabris* (Neuroptera: Chrysopidae). Journal of Economic Entomology. 2000;**93**:234-239

[94] Retnakaran A, Wright JE. Control of insect pests with benzoylphenyl ureas. In: Wright JE, Retnakaran A, editors. Chitin and Benzoylphenyl Ureas. Netherlands: Dr. W. Junk Publishers; 1987. pp. 205-282

[95] Darvas B, Polgar LA. Novel type insecticides: Specificity and effects on non-target organisms. In: Ishaaya I, Degheele D, editors. Insecticides with Novel Modes of Action. Berlin, Germany: Springer; 1998. pp. 188-259

[96] De Cock A, Ishaaya I, Degheele D, Veierov D. Vapor toxicity and concentration dependent persistence of buprofezin applied to cotton foliage for controlling the sweet potato whitefly (Homoptera: Aleyrodidae). Journal of Economic Entomology. 1990;**83**:1254-1260

[97] Ascher KRS. Non-conventional insecticidal effects of pesticides available from the neem tree, *Azadirachta indica*. Archives of Insect Biochemistry and Physiology. 1993;**22**:433-449

[98] Mordue AJ, Simmonds MSJ, Ley SV, Blaney WM, Mordue W, Nasiruddin M, et al. Actions of azadirachtin, a plant allelochemical against insects. Pesticide Science. 1998;**54**:277-284

[99] Medina P, Smagghe G, Budia F, Tirry L, Vinuela E. Toxicity and absorption of azadirachtin, diflubenzuron, pyriproxyfen, and tebufenozide after topical application in predatory larvae of *Chrysoperla carnea* (Neuroptera: Chrysopidae). Environmental Entomology. 2003;**32**:196-203

[100] Medina P, Budia F, Del Estal P, Viñela E. Influence of azadirachtin, a botanical insecticide, on *Chrysoperla carnea* (Stephens) reproduction: Toxicity and ultrastructural approach. Journal of Economic Entomology. 2004;**97**:43-50

[101] Tedeschi R, Alma A, Tavella L. Side-effects of three neem (*Azadirachta indica* A. Juss) products on the predator *Macrolophus caliginosus* Wagner (Het., Miridae). Journal of Applied Entomology. 2001;**125**:397-402

[102] Hollingworth RM, Treacy MF. Classification and properties of commercial insecticides and acaricides. In: All JN, Treacy MF, editors. Use and Management of Insecticides, Acaricides, and Transgenic Crops. St. Paul, MN: The American Phytopathological Society; 2006. pp. 36-67

[103] Spollen KM, Isman MB. Acute and sublethal effects of a neem insecticide on the commercial biological control agents *Phytoseiulus persimilis* and *Amblyseius cucumeris* (Acari: Phytoseiidae) and *Aphidoletes aphidimyza* (Diptera: Cecidomyiidae). Journal of Economic Entomology. 1996;**89**:1379-1386

[104] Castagnoli M, Liguori M, Simoni S, Duso C. Toxicity of some insecticides to *Tetranychus urticae*, *Neoseiulus californicus* and *Tydeus californicus*. BioControl. 2005;**50**:611-622

[105] Cloyd RA, Timmons NR, Goebel JM, Kemp KE. Effect of pesticides on adult rove beetle *Atheta*

coriaria (Coleoptera: Staphylinidae) survival in growing medium. Journal of Economic Entomology. 2009;**102**:1750-1758

[106] Kraiss H, Cullen EM. Insect growth regulator effects of azadirachtin and neem oil on survivorship, development and fecundity of *Aphis glycines* (Homoptera: Aphididae) and its predator, *Harmonia axyrides* (Coleoptera: Coccinellidae). Pest Management Science. 2008;**64**:660-668

[107] Banken JA, Stark JD. Stage and age influence on the susceptibility of *Coccinella septempunctata* (Coleoptera: Coccinellidae) after direct exposure to neemix, a neem insecticide. Journal of Economic Entomology. 1997;**90**:1102-1105

[108] Schmutterer H. Side-effects of neem (*Azadirachta indica*) products on insect pathogens and natural enemies of spider mites and insects. Journal of Applied Entomology. 1997;**121**:121-128

[109] Harrewijn P, Kayser H. Pymetrozine, a fast-acting and selective inhibitor of aphid feeding. *In-situ* studies with electronic monitoring of feeding behavior. Pesticide Science. 1997;**49**:130-140

[110] Fuog D, Fergusson SJ, Fluckiger C. Pymetrozine: A novel insecticide affecting aphids and whiteflies. In: Ishaaya I, Degheele D, editors. Insecticides with Novel Modes of Action. New York, NY: Springer-Verlag; 1998. pp. 40-49

[111] Morita M, Ueda T, Yoneda T, Koyanagi T, Haga T. Flonicamid, a novel insecticide with a rapid inhibitory effect on aphid feeding. Pest Management Science. 2007;**63**:969-973

[112] Jansen JP, Defrance T, Warnier AM. Side effects of flonicamid and pymetrozine on five aphid natural enemy species. BioControl. 2011;**56**:759-770

[113] Cloyd RA, Dickinson A. Effect of insecticides on mealybug destroyer (Coleoptera: Coccinellidae) and parasitoid *Leptomastix dactylopii* (Hymenoptera: Encyrtidae), natural enemies of citrus mealybug (Homoptera: Pseudococcidae). Journal of Economic Entomology. 2006;**99**:1596-1604

[114] Lacey LA, Mesquita ALM, Mercadier G, Debire R, Kazmer DJ, Lecant F. Acute and sublethal activity of the entomopathogenic fungus *Paecilomyces fumosoroseus* (Deuteromycotina: Hyphomycetes) on adult *Aphelinus asychis* (Hymenoptera: Aphelinidae). Environmental Entomology. 1977;**26**:1452-1460

[115] Kiselek EV. The effect of biopreparations on insect enemies. Zashchita Rastenii Moskva. 1975;**12**:23

[116] Marchal-Segault D. Susceptibility of the hymenopterous braconids *Apanteles glomeratus* and *Phanerotoma glavitestacea* to the spore-crystal complex of *Bacillus thuringiensis* Berliner. Annales de Zoologie Ecologie Animale. 1975;**6**:521-528

[117] Thoms EM, Watson TF. Effect of dipel (*Bacillus thuringiensis*) on the survival of immature and adult *Hyposoter exiguae* (Hymenoptera: Ichneumonidae). Journal of Invertebrate Pathology. 1986;**47**:178-183

[118] James RR, Lighthart B. Susceptibility of the convergent lady beetle (Coleoptera: Coccinellidae) to four entomogenous fungi. Environmental Entomology. 1994;**23**:190-192

[119] Bethke JA, Parrella MP. Compatibility of the aphid fungus *Cephalosporium lecanii* with the leafminer parasite, *Diglyphus beginii* (Hymenoptera: Eulophidae). Pan-Pacific Entomologist. 1989;**65**:385-390

[120] Medina P, Budia F, Tirry L, Smagghe G, Vinuela E. Compatibility of spinosad, tebufenozide and azadirachtin with eggs and pupae of the predator *Chrysoperla carnea* (Stephens) under laboratory conditions. Biocontrol Science and Technology. 2001;**11**:597-610

[121] Thompson GD, Dutton R, Sparks TC. Spinosad—A case study: An example from natural products discovery programme. Pest Management Science. 2000;**56**:696-702

[122] Copping LG. The Biopesticide Manual. Bracknell, UK: BCPC Publishing; 2001

[123] Galvan TL, Koch RL, Hutchison WD. Effects of spinosad and indoxacarb on survival, development, and reproduction of the multicolored Asian lady beetle (Coleoptera: Coccinellidae). Biological Control. 2005;**34**:108-114

[124] Holt KM, Opit GP, Nechols JR, Margolies DC. Testing for non-target effects of spinosad on twospotted spider mites and their predator *Phytoseiulus persimilis* under greenhouse conditions. Experimental and Applied Acarology. 2006;**38**:141-149

[125] Miles M, Dutton R. Spinosad—A naturally derived insect control agent with potential for use in integrated pest management systems in greenhouses. In: Proceedings of the BCPC Conference—Pests and Diseases; 13-16 November 2000; Brighton, UK. Farnham, Surrey, UK: British Crop Protection Council; 2000. pp. 339-344

[126] Williams T, Valle J, Viñuela E. Is the naturally derived insecticide spinosad compatible with insect natural enemies? Biocontrol Science and Technology. 2003;**13**:459-475

[127] Park J-J, Kim M, Lee J-H, Shin K-I, Lee S-E, Kim J-G, et al. Sublethal effects of fenpyroximate and pyridaben on two predatory mite species, *Neoseiulus womersleyi* and *Phytoseiulus persimilis* (Acari, Phytoseiidae). Experimental and Applied Acarology. 2011;**54**:243-259

[128] Kim SS, Paik CH. Comparative toxicity of fenpyroximate to the predatory mite, *Amblyseius womersleyi* Schicha and the Kanzawa spider mite, *Tetranychus kanzawai* Kishida (Acarina: Phytoseiidae, Tetranychidae). Applied Entomology and Zoology. 1996;**31**:369-377

[129] Hamedi N, Fathipour Y, Saber M. Sublethal effects of fenpyroximate on life table parameters of the predatory mite *Phytoseius plumifer*. BioControl. 2010;**55**:271-278

[130] Alston DG, Thomson SV. Effects of fungicide residues on the survival, fecundity, and predation of the mites *Tetranychus urticae* (Acari: Tetranychidae) and *Galendromus occidentalis* (Acari: Phytoseiidae). Journal of Economic Entomology. 2004;**97**:950-956

[131] Ahn K, Lee SY, Lee KY, Lee YS, Kim GH. Selective toxicity of pesticides to the predatory mite, *Phytoseiulus persimilis* and control effects of the two-spotted spider mite, *Tetranychus urticae* by predatory mite and pesticide mixture on rose. Korean Journal of Applied Entomology. 2004;**43**:71-79

[132] Kim SS, Yoo SS. Comparative toxicity of some acaricides to the predatory mite, *Phytoseiulus persimilis* and the twospotted spider mite, *Tetranychus urticae*. BioControl. 2002;**47**:563-573

[133] Kim SS, Seo SG. Relative toxicity of some acaricides to the predatory mite, *Amblyseius womersleyi* and the twospotted spider mite, *Tetranychus urticae* (Acari: Phytoseiidae, Tetranychidae). Applied Entomology and Zoology. 2001;**36**:509-514

[134] Wright DJ, Verkerk RHJ. Integration of chemical and biological control systems for arthropods: Evaluation in a multitrophic context. Pesticide Science. 1995;**44**:207-218

[135] Angeli G, Ioriatti C. Susceptibility of two strains of *Amblyseius andersoni* Chant. (Acari: Phytoseiidae) to dithiocarbamate fungicides. Experimental and Applied Acarology. 1994;**18**:669-679

[136] Cowles RS, Cowles EA, McDermott AM, Ramoutar D. "Inert" formulation ingredients with activity: Toxicity of trisilioxane surfactant solutions to twospotted spider mites (Acari: Tetranychidae). Journal of Economic Entomology. 2000;**93**:180-188

[137] Hoy MA, Cave FE. Laboratory evaluation of avermectin as a selective acaricide for use with *Metaseiulus occidentalis* (Nesbitt) (Acarina: Phytoseiidae). Applied Entomology and Zoology. 1985;**39**:401-409

[138] Kogan M. Ecological Theory and Integrated Pest Management Practice. New York, NY: John Wiley & Sons; 1986

[139] Stark JD, Jeppson PC, Mayer DF. Limitations to use of topical toxicity data for predictions of pesticide side effects in the field. Journal of Economic Entomology. 1995;**88**:1081-1088

[140] Villanuéva-Jimenez JA, Hoy MA. Toxicity of pesticides to the citrus leaf miner and its parasitoid *Ageniaspis citricola* evaluated to assess their suitability for an IPM program in citrus nurseries. BioControl. 1998;**43**:357-388

[141] Dahl GH, Lowell JR. Microencapsulated pesticides and their effects on non-target insects. In: Scher HB, editor. Advances in Pesticide Formulation Technology. Washington, DC, USA: America Chemistry Society; 1984. pp. 141-150

[142] Stevens PJG. Organosilicone surfactants as adjuvants for agrochemicals. Pesticide Science. 1993;**38**:103-122

[143] Imai T, Tsuchiya S, Fujimori T. Aphicidal effects of silwet L-77, organosilicone nonionic surfactant. Applied Entomology and Zoology. 1995;**30**:380-382

[144] Wood BW, Tedders WL, Taylor J. Control of pecan aphids with an organosilicone surfactant. HortScience. 1997;**32**:1074-1076

[145] Alix A, Cortesero AM, Nénon JP, Anger JP. Selectivity assessment of chlorfenvinphos reevaluated by including physiological and behavioral effects on an important beneficial insect. Environmental Toxicology and Chemistry. 2001;**20**:2530-2536

Section 4

Integrated Pest Management

Chapter 11

Revolutionizing Integrated Pest Management Using Nanobiotechnology: A Novel Approach to Curb Overuse of Synthetic Insecticides

Muzafar Riyaz, Saqib Hassan and Gulzar Ahmed Rather

Abstract

Nanotechnology: A promising field of advanced interdisciplinary research has unlocked an extensive range of scenarios in the sectors like agriculture, electronics, pharmacy, healthcare, pest management and much more. In agriculture, the potential uses and benefits of nanotechnology are enormous. With the use of Nanotechnology, the management of insect pests through the formulations of nanomaterial-based insecticides have changed the course of Integrated Pest Management (IPM). Traditional strategies in Integrated Pest Management used in agriculture are insufficient and the application of chemical pesticides have caused adverse effects on animals, human health and the environment. With the utilization of nanotechnological approaches, the green and efficient alternatives would provide the management of insect pests without causing an impact on animals and the environment. The present study aims to focus on the management of insect pests utilizing modern nanotechnological approaches.

Keywords: insecticides, pests, IPM, nanotechnology, sustainability

1. Introduction

There has been a major spike in terms or global human population across the planet. Since global croplands are limited and to feed almost 8 billion people is a boiling issue both developed and developing nations are facing in the present times [1, 2]. Agriculture sector is a backbone for the economies of many farming dependent nations [3, 4]. However, cultivating large-scale crops and higher quantity of crop production requires time, money and energy and most importantly the care. As long as there have been farms, farmers have been battling pests. Pests, weeds and other fungal, bacterial or viral diseases are a natural result of ecological disturbance. Modern pesticides have now been available for over 60 years [5]. In the 1940's when pesticides became available, farmers gained powerful and easy to use weapons for defeating harmful organisms [6]. The new chemicals were so effective that research on ecological methods of pest control was largely abandoned. The next generation of farmers learned very little about nonchemical approaches

IntechOpen

to controlling pests. However, over the years mankind has discovered serious drawbacks to chemical pest control. A large number of insects developed resistance to pesticides and cropland weeds learned to tolerate herbicides [7]. Newer more expensive products were required to cope with the resistance and many pest control products contaminated the environment and caused unintended damage to beneficial insects including pollinators, predators and parasitoids, aquatic life, avifauna and wildlife [8–14]. The agricultural communities from different parts of the world were compelled to develop broad-based ecologically sound pest-fighting strategies. From their efforts, a series of practices emerged we now call Integrated Pest Management (IPM).

Integrated pest management is an approach to controlling pests that takes advantage of the broad variety of management practices that are available to farmers [15]. The strategies used in IPM can save both economy and crops of farmers as it offers alternatives to expensive pesticides and herbicides which most of the farmers are not able to buy. Integrated Pest Management is built on four main principles, often known as prevention, avoidance, monitoring and suppression (PAMS) approach [16]. Since it is easier to prevent pests and diseases from developing than to control them after they appear in our farm fields. In IPM, the most important aspect for controlling pests is that the pathways need to be interrupted that enables pests to reach the farm fields as it requires less energy and time. IPM is literally based on the initial pest control strategies which include:

1. Knowledge of pests (Life cycles of pests and their natural enemies).

2. Prevention strategies (Site selection, time of plantation, nutrition and hygiene).

3. Monitoring (Observation, monitoring and traps).

4. Intervention (Chemical, Mechanical and Biological control).

5. Evaluation and review

Integrated Pest Management has been Individuals initiating and utilizing the IPM strategies must be educated about each pest and the options that he is choosing for eliminating pests. The strategies initiated in the IPM must be closely monitored as methods differ for each crop and area. As pests have a higher reproducing rate in increasing temperature especially in spring and summer which is also the developing period of most of the crops. Avoiding the reproduction and development of such pests, conventional pesticide application usually synthetic pesticides take less time to eliminate pests from the farm fields. However, with the passage of time, the pests and weeds develop resistant behaviors to most of the chemical pesticides, herbicides and biopesticides as well. The large-scale utilization of chemical pesticides and low stability of biopesticides have cost us both the health of ecosystems and burden of crop losses.

Nanobiotechnology is the modern approach and can be used to overcome the large-scale utilization of synthetic pesticides. The novel nanopesticides and nanofertilizers aims at reducing the pesticide pollution which has innumerable impacts on our ecosystems, life and human health. The advancement in the field nanotechnology and the utilization of nanomaterials in the agriculture will change the course of human history and sustainable development. Nanobiotechnology in agriculture, if planned and utilized in a proper manner will lead to sustainable development (**Figure 1**).

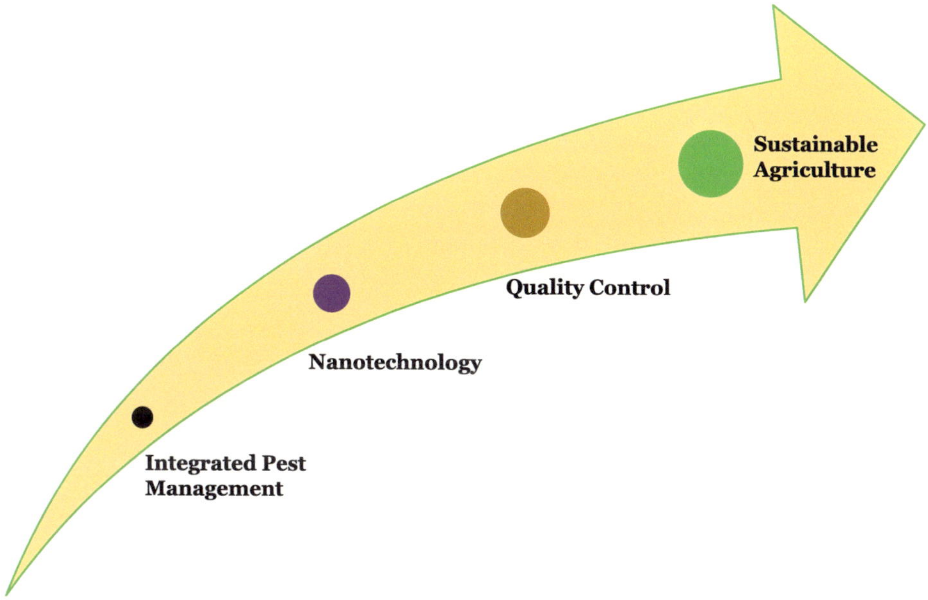

Figure 1.
Sustainable agriculture can be achieved by utilizing nanotechnology with quality control.

In this chapter we give an overview of agriculture and global food security, we also discuss about crop loss and rising global food demand, impact of synthetic pesticides on environment and finally we have comprehensively described the applications of nanobiotechnology/nanotechnology in agriculture especially in pest control.

2. Agriculture and global food security

Food security has always been a major concern for human civilizations throughout the history [17]. In 1974 World Food Summit, food security was defined as for the first time: "availability at all times of adequate world food supplies of basic foodstuffs to sustain a steady expansion of food consumption and to offset fluctuations in production and prices" [18]. From time to time the definitions have been modified. "Food security, as defined by the United Nations' Committee on World Food Security, means that all people, at all times, have physical, social, and economic access to sufficient, safe, and nutritious food that meets their food preferences and dietary needs for an active and healthy life". Concerns regarding the global food security and production have grown in recent decades. Climate change, population growth, rising food prices, and environmental stresses will all have substantial effects on food security in the future decades. Adaptation techniques and policy responses to global change are urgently needed, including alternatives for managing water allocation, land use patterns, food trade, postharvest food processing, food prices, and food safety. Furthermore, in developing countries, changing lifestyles and diets have impacted the demand for meat and dairy products [19]. Increasing demand for the food will inevitably increase land competition, food security and availability remains one of the most pressing issues on the public agenda [20]. The agriculture sector's challenge is no longer merely to maximize output and feed 9 billion people without large increases in food prices until 2050, but also to ensure food availability, distribution, and social justice for all people.

3. Factors affecting global food security

Overpopulation, climate change, and urbanization are three worldwide concerns that have an impact on food security. Hunger and disease are most prevalent in areas of the world with the largest birth rates and population, where demand typically exceeds supply. Local ecosystems provide resources for food production, health, environmental management, and water to the people. The local ecosystem has a limited carrying capacity, and when this capacity is surpassed, the ecosystem comes under stress and starts to break down. This condition has been termed as ecosystem vulnerability [21]. Such vulnerable condition of the ecosystem can give to over-farmed soils, denuded grazing lands and dried up or contaminated wells.

Climate change is affecting various regions of the world. Climate change can affect food security and hence human health in a variety of ways [22, 23]. One important route is through climate change affecting the amount of food produced, both directly and indirectly through implications on water availability and quality, pests and diseases, and pollination services. Another way involves altering CO_2 levels in the atmosphere, which has an impact on biomass and nutritional quality. Climate change can worsen food safety issues during transportation and storage. Limited food sources lead to nutritional deficiencies and these deficiencies can cause immune suppression and make the population more prone to illness [24]. Diseases lead to a greater demand on the body's energy reserves and necessitates the consumption of nutrients that aren't readily available. This is a cycle of hunger and disease that shortens one's life, decreases productivity, and limits one's potential.

Furthermore, due to the agricultural workforce's exposure to high temperatures, the direct effects of altering weather might have an influence on human health. There is the possibility for interactions with food availability due to changes in metabolic needs and physiological stress for persons exposed to severe temperatures; people may require more food to cope while simultaneously being unable to produce it [22]. All of these elements have the potential to affect both physical and cultural health by affecting the amount and quality of food available to people.

Food insecurity is aggravated by urbanization. Megacities, defined as cities with populations of 10 million or more, are particularly frequent in developing nations. In fact, low- and middle-income nations account for three-quarters of urban residents. The global economy has benefited from these nations' low-cost labour, which has fueled rural-to-urban migration. The majority of manufacturing jobs are in cities, and when overcrowded and depleted lands fail to provide basic economic requirements despite their ability to generate sufficient food, people migrate to these urban jobs and opportunities. Until now, rural farmers have been able to keep up with demand, but urban people' economic access to food remains a concern. The demand to feed these megacities, as well as the difficulties linked with economic food access, is enormous. This was seen in the years 2007 and 2008, when rising oil costs made food expensive for city residents all across the world [25].

Food security is not only about humans and their environment; it also involves animals, who are a key source of food for humans. As a result, under this situation, we cannot neglect infectious animal illness. Contagious animal illnesses can wipe out an animal population or drastically impair its productivity, which has an impact on food supply and access. Even if humans are not afflicted, the illness frequently has a significant impact on them. Any loss of animals means a loss of protein in the form of milk, meat, or blood, as well as a supply of labour to plow fields and deliver food to markets. Even illnesses with a low case-fatality rate but a high morbidity rate, such as foot and mouth disease, produce such severe sickness in the animal that its productivity is severely diminished. This can influence the food supply. Avian influenza, especially H5N1, decimated flocks across the Western Pacific, posing a food security issue [26]. Furthermore, zoonotic illnesses put people at danger when they ingest these animals, which happens often even when the animal is unwell. In terms of access and availability, food instability may push communities to seek alternative food sources, such as bush meat, thus exposing them to additional illnesses. This has been the case with illnesses such as sudden acute respiratory syndrome, acquired immunodeficiency syndrome (AIDS), and others [27].

Agriculture is directly or indirectly linked to the livelihood of a large percentage of the population in the developing countries. In recent years, agricultural productivity growth has slowed. Land is a limited resource, and many emerging nations are unable to expand their farmed areas [28, 29]. As a result, increasing agricultural production may be the only way to fulfill future food demand for a growing population. Because cultivable fertile land and related inputs are limited in most areas of the world, a new strategy to increasing future agricultural productivity development in most regions of the world may be intense agricultural expansion rather than widespread growth. As a result, diversification coupled with intensification of production and improving the resource utilization are important essential measures [30]. The disparity between feasible and actual yields for most crops implies a huge opportunity to boost food and agricultural output by increasing productivity [31]. According to the Food and Agriculture Organization, UN (FAO), in the developing world, around 80% of the growth in food production will have to come from increased yields and cropping intensity, with just 20% coming from arable land expansion [32]. As a result, intensification is critical not just to satisfy the growing demand for food grains, but also to reduce deforestation, environmental devastation, and global warming.

4. Global burden of crop loss and rising food demand

Crop loss is impeding efforts to achieve the Sustainable Development Goals in food security, nutrition, and livelihoods. Pests and disease may take away up to 40% of agricultural output, but the data available to confirm and illustrate trends is inadequate. The Global Burden of Crop Loss project will gather, validate, analyze, and disseminate data on the extent and causes of crop loss, with the goal of gathering enough and reliable data to enable prioritization of plant health research and policy, improving our ability to predict the impact of emerging diseases. One of the defining problems of our time is meeting rising food demand in the face of climate change and increasingly variable agricultural conditions. To ensure that adequate food is accessible for all of us for centuries to come, we will need to produce considerably more food while minimizing environmental damage. Pests and diseases are responsible for an estimated 20–40% of agricultural production loss worldwide. Losses of basic cereals (rice, wheat, maize) and tuber crops (potatoes and sweet potatoes) have a direct impact on food security and nutrition, but losses of essential commodity crops like banana and coffee have a significant impact on household incomes and national economies. Climate change also increases the threat of plant pests and diseases, impeding progress on several of the UN's Sustainable Development Goals. A big component of this will be reducing crop loss, and considerable efforts will be required to enhance pest control, particularly diseases and weeds.

Crop loss assessment research has traditionally been divided into three stages: exploratory, development, and implementation. These phases occurred at various times in agricultural research, with the goal of improving our understanding of the influence of diseases on crop output quantity and quality. The loss of biodiversity is speeding up all across the planet as this tendency is mostly driven by the global food system. The conversion of natural ecosystems for agricultural production or grazing has been the primary driver of habitat loss in the last 50 years, resulting in a reduction in biodiversity. The 'cheaper food' concept has influenced our food system over the last few decades. The goal of policies and economic systems has been to produce more food at cheaper costs. Intense agricultural production damages soils and ecosystems, reducing land productivity and forcing even more intensive food production to keep up with demand. These constraints are being exacerbated by the worldwide consumption of lower-calorie, resource-intensive foods. Food production today is largely reliant on unsustainable techniques such as monocropping and excessive tilling, as well as inputs such as fertilizer, pesticides, energy, land, and water. This has diminished the diversity of landscapes and habitats, posing a threat to or eliminating the breeding, feeding, and/or nesting of birds, animals, insects, and microbiological species, as well as pushing out many others. Our food system is a major source of global greenhouse gas emissions. Climate change is also causing habitat degradation and species extinction to spread to different places. Biodiversity loss will continue to increase until our food system is reformed. Our ability to survive will be jeopardized if ecosystems and habitats are further degraded.

By 2050, food consumption is predicted to grow by 59 to 98%. This will have a profound impact on agricultural markets. Farmers throughout the world will need to boost crop output, either by expanding the quantity of agricultural area available for crop production or by improving productivity on existing agricultural lands through fertilizer and irrigation, as well as adopting innovative technologies such as precision farming. Global demand for agricultural commodities is rising and may continue to do so for decades, fueled by a projected 2.3 billion rise in global population and higher per capita incomes through 2050 [19]. Both land clearing and more intensive use of existing croplands might help satisfy this need, but the environmental consequences and costs of these various agricultural expansion pathways remain unknown [33]. Agriculture already has significant global environmental impacts: land clearing and habitat fragmentation threaten biodiversity, land clearing, crop production, and fertilization account for about a quarter of global greenhouse gas (GHG) emissions [34], and fertilizers can harm marine, freshwater, and terrestrial ecosystems [35]. Quantitative estimates of future crop demand and how alternative production techniques affect yields and environmental factors are required to understand the future environmental consequences of global crop production and how to attain higher yields with fewer impacts. To fulfill increasing food demand while also halting and reversing environmental deterioration, significant improvements in resource-use efficiency and advances in resource conservation will be required globally. Despite certain technical advancements, crop growth has slowed substantially compared to earlier decades. The detrimental side effects of heavy usage of chemical inputs in agricultural production have become increasingly evident, raising major questions about long-term sustainability.

Agriculture, fisheries, and forestry investments, as well as research and development funding, must be increased, particularly in and for low-income nations. This is necessary to encourage the adoption of sustainable production systems and practices, such as integrated crop-livestock and aquaculture-crop systems, conservation agriculture, agroforestry, nutrition-sensitive agriculture, sustainable forest management, and sustainable fisheries management, among others. These and other types of climate smart agriculture will assist farms, ecosystems, and communities in adapting to, mitigating, and building resilience in the face of climate change.

5. Impact of synthetic insecticides on environment

Insecticides are generally chemical compounds developed for eliminating the insect pests from agricultural fields, storage warehouses, homes etc. Man has been utilizing pesticides from the time immemorial. Though, utilization of chemical pesticides brought a relief to farmers by expelling the pests from the farm lands. However, the large-scale usage of synthetic insecticides was proven to be incompatible for the environment. Impact of synthetic insecticides on various life forms and different ecosystems has been reported across different places of the world. The nations engaged in the different agricultural and allied sectors are mostly affected from it. With the rise in global population and the food requirements, there has been a parallel growth in large-scale cultivation of high yielding monocrops. Since, crop loss by the pests was controlled by the pesticides, however a long-lasting adverse impact on various life forms and natural environment end up being a major issue to be taken care of. On contrary, health of farmers has also been declined to the worst in some studies as the farmers are most exposed to the toxicity levels of synthetic insecticides. The synthetic insecticides have been proven to be most devastative on the beneficial insect diversity including pollinators such as honey bees, dipteran pollinators, predators, parasitoids and other useful insects which deliver several ecosystem services. In an agricultural field, a number of insects can be seen collecting nectar and pollinating the flowers such as bees, wasps, hoverflies, moths & butterflies and some species checking the populations of insect pests such as parasitic wasps, hornets, beetles, lacewings etc. While splashing synthetic insecticides on crops infested with pests, 15 to 40% of an estimated fraction of insecticides are scattered into the atmosphere by either volatilization or spray drift processes [36]. After spraying, the insecticides in atmospheric particulate phase will remain in the air for about 7–12 days and can thoroughly orbit many geographical locations across the globe. The orbiting of pesticides in the atmosphere can alter air quality and may add more events to the climate change [37]. The pesticide runoff from the agricultural lands into streams and lakes have a great impact in the aquatic life and water contamination. Though runoff can be the transportation of pesticides into aquatic ecosystem, the atmospheric dispersal of pesticides can travel to other places like grazing fields, human settlements potentially affecting other living organisms and human wellbeing. The impact of synthetic agro-chemicals on insect diversity has been well documented across the globe. There has been a massive decrease in insect pollinators and other beneficial class of insects from past few decades due to the large-scale utilization of insecticides and other agrochemicals [38, 39]. Alternative measures are needed to substitute the overuse of synthetic insecticides and other agrochemicals.

6. Nanotechnology and agriculture

Agriculture has traditionally been the most significant and stable sector in the economy since it generates and supplies raw materials for the food and feed industries. Due to the finite nature of natural resources and the world's growing population, agricultural expansion must be economically feasible, ecologically friendly, and efficient. This change will be critical for attaining several goals in the coming years [40–42]. Agricultural nutrient balances change considerably with economic expansion, and as a result of this assumption, the improvement of soil fertility in emerging nations is extremely important [43]. Recently, a wide range of possible uses of nanotechnology in agriculture have been proposed, prompting extensive research at both the academic and industry levels [44–46]. Indeed, the unique characteristics of nanoscale materials make them ideal candidates for the design and development of new agricultural equipment. As a result, research into nanotechnology's uses in agriculture has gained a lot of attention in recent years.

6.1 Nanopesticides

Pesticide use is common in commercial agriculture, and research and development of novel, effective, and target-specific pesticides is ongoing. As a result, each year a huge number of pesticides are spayed in the agroecosystems. Only 0.1% of pesticides sprayed reach the target pests, while the rest (99.9%) contaminates the environment [47]. This has significant repercussions for the food chain and human health. Pesticides' pervasive presence in the environment has led in the development of pesticide resistance in weeds, insects, and diseases, in addition to the impact on non-target species [48]. Biopesticides tend to decrease the harmful effects of synthetic pesticides, but their usage is limited due to their sluggish and environment-dependent pest-control efficacy. As a result, nanopesticides are critical for the successful and long-term control of many pests, and they have the potential to reduce the usage of synthetic chemicals and their related environmental hazards. To improve their efficacy, nanopesticides act differently than the conventional pesticides [49]. Nanoparticles may be transported in dissolved and colloidal phases, and this process explains why their behavior differs from that of ordinary solutes of the same particles [50]. The mobility and breakdown of active substances by soil-dwelling microbes may be aided by their solubility. Because nanoparticle-based pesticides improve the solubility of aluminum, they are also regarded to have a lower environmental effect than conventional pesticides [51].

Nanoparticles have been shown to have antimicrobial action against bacterial, fungal, and viral infections. Silver [52], copper [53], and aluminum are significant inorganic nanoparticles with pesticidal capabilities [54]. In a study, utilizing silica-silver nanoparticles to suppress pathogenic fungus (*Rhizoctonia solani, Magnaporthe grisea, Colletotrichum gloeosporioides*), the disease-causing pathogens vanished from diseased leaves after 3 days of spraying the product [55]. Silver nanoparticles were also found to have antifungal action against *Raffaelea sp*., a fungus that causes harm to oak trees [56]. The evaluation of a nanoformulated commercial fungicide (Trifloxystrobin 25% + Tebuconazole 50%) against the soil borne fungal pathogen *Macrophomina phaseolina* at various concentrations (5, 10, 15 and 25 ppm) demonstrated higher efficacy than a commercial product [57].

6.2 Nanofertilizers

A tremendous rise in agricultural yields, particularly grain yields, has played a key role in satisfying the world's nutritional needs during the previous five decades. Increased use of chemical fertilizers is one of the primary contributors to increased crop yield. Fertilizer-responsive crop cultivars have expanded the use of chemical fertilizers. Chemical fertilizers, on the other hand, have a low usage efficiency due to fertilizer loss (through volatilization and leaching), which pollutes the environment and raises production costs [33]. For example, there is a loss of about 50–70% of the nitrogen, when conventional fertilizers are applied. As a result, the scientific community is paying close attention to the development of alternative techniques to assure the long-term usage of nutrients. Nanotechnology is being utilized in this setting to decrease mobile nutrient losses, produce slow-release fertilizers, and increase the accessibility of nutrients that are currently unavailable. Nanofertilizers are nanomaterials that function as carriers/additives for nutrients (e.g., by compositing with minerals) or as nutrients themselves (micro- or macronutrients). Encapsulating nutrients inside a nanofertilizer can also be used to create nanofertilizers. Nanofertilizers boost crop production and quality by supplying more nutrients usage efficiency while lowering manufacturing costs and therefore help in the perspective of agricultural sustainability. It was found that nanofertilizers had a median effectiveness and lead to an increase in the crop yield by 18–29% when compared to traditional fertilizers [58, 59].

Phosphatic nanofertilizers have also been reported to increase the productivity of seeds by 32%. When compared to plants fed with ordinary fertilizer, soybean (*Glycine max* L.) exhibited an increase in yield by 20%. Nanofertilizers are also beneficial to the environment. Plant metabolism and nutrient absorption has been found to improve through nanometric holes molecular transporters or nanostructure cuticle pores. Nanotechnology in plant nutrition enables the creation of slow/controlled release fertilizers, which increase fertilizer efficiency and minimize nutrient losses to the environment, making them more environmentally friendly [60]. Conventional nitrogenous fertilizers have a fertilizer usage efficiency of 30–60%, but chemical bonding in soil causes 80–90% of conventional phosphatic fertilizers to be lost and inaccessible to plants [61]. Clay minerals, hydroxyapatite, chitosan, polyacrylic acid, zeolite, and other nanostructured materials are utilized to produce fertilizers for soil and/or foliar application. Because of hydroxyapatite's large surface area and strong interactions with urea, the release of nitrogen from urea is delayed [62]. Enhanced organic waste decomposition and compost generation might be an important element of nanotechnology in agriculture, although the research is still in its early stages, with no concrete results to yet. Based on the current data, nanofertilizers can minimize the quantity of fertilizer needed owing to their high usage efficiency, therefore reducing the environmental effect of nutrient losses. This discovery comes at an ideal moment to secure global food security.

6.3 Nanobiosensors in agriculture

Biosensors are a type of hybrid receptor-transducer system, a device that detects the physical and chemical characteristics of a medium in the environment to identify the presence of a biological or organic recognition element to detect the presence of a particular biological analyte [63]. Detection of a specific analyte at ultra-low concentrations using a sensitive element by means of a physicochemical transducer Nano-biosensor technology has the potential to revolutionize healthcare, aid in early identification and quick decisions to improve crop production through proper water, land, fertilizer, and pesticide management. A large surface area, fast electron-transfer kinetics, high sensitivity, and a high surface-to-volume ratio. Organophosphates, neonicotinoids, carbamates and atrazines are most common pesticides, and their residues can be found even at low concentrations. Because of the poor homogeneity of the soil, low concentrations last longer. These pesticides are identified using nano-biosensors [64]. Metals (gold, silver, cobalt, etc.) NPs, carbon nanotubes (CNTs), magnetic NPs, and quantum dots (QDs) have all been studied for their uses in biosensors, which have become a new multidisciplinary frontier between biological sensing and material research. As a result, a biosensor is a device that incorporates a biological recognition element as well as physical or chemical principles. It combines a biological and an electronic component to produce a quantifiable signal component, with biological recognition accomplished via the transducer process and signal processing accomplished by electronic accomplishment. The existence of a bioreceptor (biological element) paired with an appropriate transducer that creates a signal after contact with the target molecule of interest gives biosensor systems a greater specificity and sensitivity than conventional techniques. Enzymes, dendrimers, thin films, and other natural and artificial bioreceptors have recently been created and utilized. As a result, a biological reaction is converted into an electrical signal by a biosensor, an analytical instrument. It's about the components of biological elements including antibodies, enzymes, proteins, and nucleic acids. The transducer, as well as any accompanying electronics or signal processors, are in charge of detecting the functions. The AuNP-based micro cantilever-based DNA biosensor has been created and is commonly used to detect low levels of DNA during a hybridization procedure [65]. Although, usage of nanotechnology has created new revolution in smart farming and lowered related concerns, broad usage of nanomaterials -based agriculture and food items and less-likely immobilized nano-sensors have risen impacts on human and environmental health. Complexity of nanobio-eco-interactions restricts tracking their activity in soils. Therefore, a comprehensive approach is required to comprehend these connections in soil-plant-air and eventually in food chain.

6.4 Soil remediation using Nanomaterials

Soil is an important element of the ecosystem that has been under threat for decades owing to numerous forms of pollution. Soil recovery and regeneration has become a global issue. Nanotechnology has recently emerged as an effective, cost-effective, environmentally friendly, and promising soil remediation technique. This technique offers a lot of promise for removing pollutants from the environment through adsorption, redox reactions, conversion, stabilization, and other methods. To remove pollutants from soil, a variety of nanomaterials and devices are employed. As a result, soil might be efficiently remediated using nanotechnology-based concepts, techniques, and products, which are not possible to do using traditional approaches. Heavy metal is one of the most dangerous soil pollutants. Heavy metals, like other pollutants, might be removed from soil utilizing nanotechnology-based methods.

Bioremediation is an in situ, natural, environmentally friendly, cost-effective, and flexible approach to detoxify hazardous contaminants (organic and inorganic). However, due to extended treatment times, low pollutant availability, low remediation efficiency in highly polluted soils caused by pollutant toxicity to biological agents (bacteria, fungus, plants, and so on), and the generation of hazardous byproducts, its efficacy may be restricted [66]. The use of nanomaterials in conjunction with bioremediation provides a method for overcoming the constraints of this green technology. Although nanomaterials have been utilized for chemical decontamination of sites for the past two decades, their application in bioremediation is a relatively new area that is still in its early stages.

7. Future of nanopesticides and pest control

Excessive use of pesticides and fertilizers in agriculture to boost yields has shown to be ineffective since a substantial portion of them is wasted, causing harm to the environment and human health. As a result, farmers have a significant difficulty in replacing pesticides and fertilizers with nanopesticides [67]. Over the last few decades, considerable research into the application of nanotechnology to increase agricultural production has been undertaken. The use nanoparticles as nanopesticides have been proven to be successful in the agri-food production. Herbicides, fungicides, pesticides and fertilizers are encapsulated with different types of nanoparticles that aid in the delayed release of fertilizers and pesticides, which results in precise dose availability to crops [68]. The chemically designed fertilizers are inorganic compounds or materials with specified chemical composition and mostly synthetic in origin. They are primarily involved in agriculture sector to provide nutrients like potassium, nitrogen and phosphorous that is not present in soil. However, to the loss of these fertilizers from the soil due to leaching, volatilization and water runoff is a major problem associated with their use.

The release of these fertilizers causes environmental degradation and a significant quantity of nutrients is lost and therefore inaccessible to the plants [69]. In agriculture pesticides are used to control pests, rats, mice, ticks, mosquitoes and other disease carrier. In addition to those pesticides are being used to control disease, insect infection and weeds. Currently using of pesticides is the fastest way to control effects of various pests and diseases. The usage of pesticides in agriculture is associated with number of risks which includes harmful effects on pollinating and domestic animals, effects on human health, penetrate into the water and cause toxic effects on aquatic animals and finally effects on our ecosystem. The programmed and regulated usage of chemicals on a nanoscale basis is an acceptable and suitable solution of the difficulties described above. In a controlled manner way, these ingredients are injected directly into the plant portion that has been attacked by the disease or insect [70].

Nanofertilizers are the modified forms of traditional chemical fertilizers designed through nanotechnological interventions and a variety of biological and physiochemical techniques. Nanofertilizers have unique properties which differs it from bulk materials [71]. Nanofertilizers have various advantages over the traditional fertilizers such as higher solubility and bioavailability, release of nutrients in controlled and slow manner and less loss rate [72]. In present condition, meeting the dietary needs of the world's fastest growing population is pressing necessity. It has been estimated nearly one third of crops get damaged due to pest infestation by applying conventional methods. There is an immediate requirement of innovative methods to overcome these issues. In this aspect nanotechnology is playing a leading role to the agro technological revolution [73]. With the help of these cutting–edge materials, modern agriculture is evolving into smart farming allowing farmers to get the most out of their resources. Nanotechnology is not limited to the plant against pest control in addition to that it monitors plant growth, enhanced crop yield and minimizing waste. Nanomaterials in the form of nano insecticides, nano fungicides, nano herbicides and nanonematocides are the most often tested nano pesticides. The nanomaterials utilized as nanopesticides can be metal nanoparticles like silver nanoparticles, gold nanoparticles and copper nanoparticles or metal oxide nanoparticles such as zinc oxide nanoparticles, copper oxide nanoparticles, silicon dioxide, magnese dioxide or titanium dioxide etc. Studies conducted on silver nanoparticles at 30–150 mg/dl against *Meloidogyne spp*. have showed the 99% reduction of nematodes [74]. In another study silver

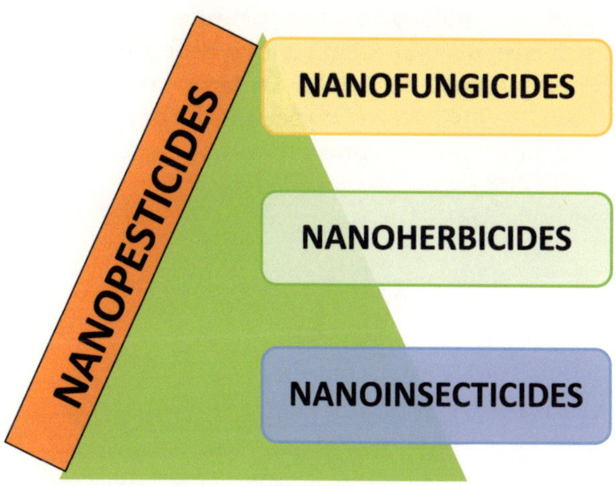

Figure 2.
Different types of nanomaterials used against plant protection.

nanoparticles used as nanopesticides against *Xanthomonas campestris pv. campestris* have shown the significant reduction of bacteria causing bacterial blight disease [75]. Copper nanoparticles when used against *Fusarium spp*. has shown excellent antifungal activity as nanopesticide. In another study conducted on copper oxide when used at a concentration of 10 mg/L against *S. littoralis* showed good insecticide on Bt-transgenic cotton [76, 77]. Iron oxide nanoparticles when used against fungal pathogen like *Rhophitulus solani, B. cinerea* and *F. oxyporium* have showed the reduction of fungal infection about 60–80% [78]. Magnesium oxide nanoparticles when applied against fungal pathogen *F. oxyporium* reduces the disease in tomato plants [79]. Gold nanoparticles-ferbam with a size of about 30 nm when applied in tea leaves promoted surface adhesion in tea plant [80]. In **Figure 2**, we show the different types of nanoparticles that are used as nanopesticides for crop protection.

8. Conclusion

Global food security is one of the major concerns that both developed and developing nations are undertaking from the time of global human population upsurge. Global crops lands are limited and global population is rising at an alarming rate. Since, the extension of agricultural activities has brought the increase in the production of synthetic agricultural pesticides and fertilizers that have been used by the farmers for decades all over the world. The large-scale utilization of these synthetic agrochemicals from the time of their evolution to the present era has cost the humans; their health, air pollution, water pollution, soil pollution, decline of the beneficial entomofauna, environmental contamination and so on. Implementation of alternative measures to overcome the use of synthetic pesticides and fertilizers is need of the hour. With the advancement of the technology, the novel approach of the nanotechnology can bring agriculture towards the sustainability. The novel nanopesticides and nanofertilizers are the best alternatives to save the crops, human health and the environment. The regulations must be employed and quality control ought to be instigated for the utilization of nanomaterials in the agriculture.

Acknowledgements

The authors thank Entomology Research Institute, Loyola College Chennai, Tamil Nadu, India; Department of Microbiology, School of Life Sciences, Pondicherry University, Puducherry, India; Department of Biomedical Engineering, Sathyabama Institute of Science and Technology, deemed to be University, India for the extended support.

Author details

Muzafar Riyaz[1], Saqib Hassan[2] and Gulzar Ahmed Rather[3]*

1 Entomology Research Institute, Loyola College, Chennai, Tamil Nadu, India

2 Department of Microbiology, School of Life Sciences, Pondicherry University, Puducherry, India

3 Department of Biomedical Engineering, Sathyabama Institute of Science and Technology, Deemed to be University, Chennai, Tamilnadu, India

*Address all correspondence to: gulzarrather05635@gmail.com

References

[1] Crist E, Mora C, Engelman R. The interaction of human population, food production, and biodiversity protection. Science. 2017;**356**(6335):260-264. DOI: 10.1126/science.aal2011

[2] Susman P, O'Keefe P, Wisner B. Global disasters, a radical interpretation. In: Hewitt K, editor. Interpretations of Calamity. Oxfordshire, UK: Routledge; 2019. pp. 263-283. DOI: 10.4324/9780429329579

[3] Harwood RR. A history of sustainable agriculture. In: Edwards CA, editor. Sustainable Agricultural Systems. London: CRC Press; 2020. pp. 3-19. DOI: 10.1201/9781003070474

[4] Jones AD, Ejeta G. A new global agenda for nutrition and health: The importance of agriculture and food systems. Bulletin of the World Health Organization. 2016;**94**(3):228-229. DOI: 10.2471/BLT.15.164509

[5] Saravi SSS, Shokrzadeh M. Role of pesticides in human life in the modern age: A review. Pesticides in the Modern World-Risks and Benefits. 2011;**5**:3-12

[6] Abubakar Y, Tijjani H, Egbuna C, Adetunji CO, Kala S, Kryeziu TL, et al. Pesticides, history, and classification. In: Egbuna C, Sawicka B, editors. Natural Remedies for Pest, Disease and Weed Control. Cambridge, MA: Academic Press; 2020. pp. 29-42. DOI: 10.1016/B978-0-12-819304-4.00003-8

[7] Bass C, Denholm I, Williamson MS, Nauen R. The global status of insect resistance to neonicotinoid insecticides. Pesticide Biochemistry and Physiology. 2015;**121**:78-87. DOI: 10.1016/j.pestbp.2015.04.004

[8] Aktar MW, Sengupta D, Chowdhury A. Impact of pesticides use in agriculture: Their benefits and hazards. Interdisciplinary Toxicology. 2009;**2**(1):1-12. DOI: 10.2478/v10102-009-0001-7

[9] Bernardes MFF, Pazin M, Pereira LC, Dorta DJ. Impact of pesticides on environmental and human health. In: Andreazza AC, Scola G, editors. Toxicology Studies-Cells, Drugs and Environment. Rijeka: IntechOpen; 2015. pp. 195-233. DOI: 10.5772/59710

[10] Madhun YA, Freed VH. Impact of pesticides on the environment. In: Cheng HH, editor. Pesticides in the Soil Environment: Processes, Impacts and Modeling. Vol. 2. Hoboken, NJ: John Wiley and Sons; 1990. pp. 429-466. DOI: 10.2136/sssabookser2.c12

[11] Jawale CA, Rajput KH, Ugale BJ. Assessing the impact of pesticides: An overview. International Journal of Life Sciences. 2017;**5**(3):474-479

[12] Mahmood I, Imadi SR, Shazadi K, Gul A, Hakeem KR. Effects of pesticides on environment. In: Hakeem KR, Akhtar MS, Abdullah SN, editors. Plant, Soil and Microbes. Switzerland: Springer; 2016. pp. 253-269. DOI: 10.1007/978-3-319-27455-3_13

[13] Riyaz M, Iqbal WA, Sivasankaran K, Ignacimuthu S. Impact on farmers' health due to the pesticide exposure in the Agrarian zones of Kashmir Valley: A review. Acta Scientific Agriculture. 2020;**4**(2):16-22

[14] Riyaz M, Shah RA, Sivasankaran K. Pesticide residues: Impacts on fauna and the environment. Rijeka: IntechOpen; 2021. DOI: 10.5772/intechopen.98379

[15] Ehler LE. Integrated pest management (IPM): Definition, historical development and implementation, and the other IPM. Pest Management Science. 2006;**62**(9):787-789. DOI: 10.1002/ps.1247

[16] Vreysen MJB, Robinson AS, Hendrichs J, Kenmore P. Area-wide integrated pest management (AW-IPM): Principles, practice and prospects. In: Vreysen MJ, Robinson AS, Hendrichs J, editors. Area-Wide Control Insect Pests. Dordrecht: Springer; 2007. pp. 3-33. DOI: 10.1007/978-1-4020-6059-5_1

[17] Wahlqvist ML. Why food in health security (FIHS)? Asia Pacific Journal of Clinical Nutrition. 2009;**18**(4):480-485

[18] United Nations. Report of the World Food Conference, Rome. New York: United Nations; 1975

[19] Godfray HC, Beddington JR, Crute IR, Haddad L, Lawrence D, Muir JF, et al. Food security: The challenge of feeding 9 billion people. Science. 2010;**327**(5967):812-818. DOI: 10.1126/science.1185383

[20] Erkan R. Food for thought: "four Ss with one F": Security, safety, sovereignty and shareability of food. British Food Journal. 2012;**114**(3): 353-371

[21] Ericksen PJ. What is the vulnerability of a food system to global environmental change? Ecology and Society. 2008;**13**(2):1-18

[22] Watts N, Amann M, Ayeb-Karlsson S, Belesova K, Bouley T, Boykoff M, et al. The lancet countdown on health and climate change: From 25 years of inaction to a global transformation for public health. Lancet. 2018;**391**(10120):581-630. DOI: 10.1016/S0140-6736(17)32464-9

[23] Choufani J, Davis C, McLaren R, Fanzo J. Climate Change and Variability: What are the Risks for Nutrition, Diets, and Food Systems? Vol. 1645. Washington, DC: International Food Policy Research Institute; 2017

[24] Wakabi W. 'Worst drought in a decade' leaves Kenya crippled. Lancet. 2006;**367**(9514):891-892. DOI: 10.1016/S0140-6736(06)68364-5

[25] Satterthwaite D, McGranahan G, Tacoli C. Urbanization and its implications for food and farming. Philosophical Transactions of the Royal Society of London. Series B, Biological Sciences. 2010;**365**(1554):2809-2820. DOI: 10.1098/rstb.2010.0136

[26] Basuno E, Yusdja Y, Ilham N. Socio-economic impacts of avian influenza outbreaks on small-scale producers in Indonesia. Transboundary and Emerging Diseases. 2010;**57**(1-2):7-10. DOI: 10.1111/j.1865-1682.2010.01121.x

[27] ProMED-mail. Avian influenza, human (16). Indonesia, Vietnam; 2008. Archive Number: 20080124.0302

[28] Wolfe ND, Daszak P, Kilpatrick AM, Burke DS. Bushmeat hunting, deforestation, and prediction of zoonotic disease emergence. Emerging Infectious Diseases. 2005;**11**(12):1822-1827. DOI: 10.3201/eid1112.040789

[29] ECG (Evaluation Cooperation Group). Evaluative Lessons for Agriculture and Agribusiness. Washington, DC: World Bank; 2011

[30] Dixon J, Gulliver A, Gibbon D. Farming Systems and Poverty: Improving Farmers' Livelihoods in a Changing World. Rome and Washington, DC: Food and Agriculture Organization and World Bank; 2001

[31] Zepeda L. Agricultural investment, production capacity and productivity. In: Zepeda L, editor. Agricultural Investment and Productivity in Developing Countries. Rome, Italy: FAO Economic and Social Development; 2001

[32] FAO. The Technology Challenge: How to Feed the World 2050 High-Level Export Forum. Rome, Italy: Food and

Agriculture Organization, United Nations; 2009

[33] FAO. The Future of Food and Agriculture Trends and Challenges Foresight, U.K. (2011). The Future of Food and Farming [Final Project Report]. London: The Government Office for Science; 2017

[34] Burney JA, Davis SJ, Lobell DB. Greenhouse gas mitigation by agricultural intensification. Proceedings of National Academy of Sciences. 2010;**107**(26):12052-12057. DOI: 10.1073/pnas.0914216107

[35] Vitousek PM, Aber JD, Howarth RW, Likens GE, Matson PA, Schindler DW, et al. Human alteration of the global nitrogen cycle: Sources and consequences. Ecological Applications. 1997;7(3):737-750

[36] Socorro J, Durand A, Temime-Roussel B, Gligorovski S, Wortham H, Quivet E. The persistence of pesticides in atmospheric particulate phase: An emerging air quality issue. Scientific Reports. 2016;**6**:33456. DOI: 10.1038/srep33456

[37] Miller GT, Spoolman S. Living in the Environment: Principles. Connections, and Solutions. Wodsworth: Melbourne; 1996

[38] Sánchez-Bayo F, Wyckhuys KA. Worldwide decline of the entomofauna: A review of its drivers. Biological Conservation. 2019;**232**:8-27. DOI: 10.1016/j.biocon.2019.01.020

[39] Tesfahunegny W. Impact of pesticides on birds from DDT to current fatality: A literature review. Journal of Zoology Studies. 2016;**3**(2):44-55

[40] Johnston BF, Mellor JW. The role of agriculture in economic development. The American Economic Review. 1961;**51**(4):566-593

[41] Yunlong C, Smit B. Sustainability in agriculture: A general review. Agriculture, Ecosystems and Environment. 1994;**49**(3):299-307. DOI: 10.1016/0167-8809(94)90059-0

[42] Mukhopadhyay SS. Nanotechnology in agriculture: Prospects and constraints. Nanotechnology, Science and Applications. 2014;**7**:63-71. DOI: 10.2147/NSA.S39409

[43] Pretty J, Sutherland WJ, Ashby J, Auburn J, Baulcombe D, Bell M, et al. The top 100 questions of importance to the future of global agriculture. International Journal of Agricultural Sustainability. 2010;**8**(4):219-236. DOI: 10.3763/ijas.2010.0534

[44] Chen H, Yada R. Nanotechnologies in agriculture: New tools for sustainable development. Trends in Food Science and Technology. 2011;**22**(11):585-594. DOI: 10.1016/j.tifs.2011.09.004

[45] Dasgupta N, Ranjan S, Mundekkad D, Ramalingam C, Shanker R, Kumar A. Nanotechnology in agro-food: From field to plate. Food Research International. 2015;**69**:381-400. DOI: 10.1016/j.foodres.2015.01.005

[46] Parisi C, Vigani M, Rodríguez-Cerezo E. Agricultural nanotechnologies: What are the current possibilities? Nano Today. 2015;**10**(2):124-127. DOI: 10.1016/j.nantod.2014.09.009

[47] Carriger JF, Rand GM, Gardinali PR, Perry WB, Tompkins MS, Fernandez AM. Pesticides of potential ecological concern in sediment from South Florida canals: An ecological risk prioritization for aquatic arthropods. Soil and Sediment Contamination. 2006;**15**(1):21-45. DOI: 10.1080/15320380500363095

[48] Rai M, Ingle A. Role of nanotechnology in agriculture with special reference to management of

insect pests. Applied Microbiology and Biotechnology. 2012;**94**(2):287-293. DOI: 10.1007/s00253-012-3969-4

[49] Kah M, Tufenkji N, White JC. Nano-enabled strategies to enhance crop nutrition and protection. Nature Nanotechnology. 2019;**14**(6):532-540. DOI: 10.1038/s41565-019-0439-5

[50] Kumpiene J, Lagerkvist A, Maurice C. Stabilization of As, Cr, Cu, Pb and Zn in soil using amendments—A review. Waste Management. 2008;**28**(1): 215-225. DOI: 10.1016/j.wasman.2006.12.012

[51] Kah M, Hofmann T. Nanopesticide research: Current trends and future priorities. Environment International. 2014;**63**:224-235. DOI: 10.1016/j.envint.2013.11.015

[52] Kim HJ, Phenrat T, Tilton RD, Lowry GV. Effect of kaolinite, silica fines and pH on transport of polymer-modified zero valent iron nano-particles in heterogeneous porous media. Journal of Colloid and Interface Science. 2012;**370**(1):1-10. DOI: 10.1016/j.jcis.2011.12.059

[53] Gogos A, Knauer K, Bucheli TD. Nanomaterials in plant protection and fertilization: Current state, foreseen applications, and research priorities. Journal of Agricultural and Food Chemistry. 2012;**60**(39):9781-9792. DOI: 10.1021/jf302154y

[54] Stadler T, Buteler M, Weaver DK, Sofie S. Comparative toxicity of nanostructured alumina and a commercial inert dust for Sitophilus oryzae (L.) and Rhyzopertha dominica (f.) at varying ambient humidity levels. Journal of Stored Products Research. 2012;**48**:81-90. DOI: 10.1016/j.jspr.2011.09.004

[55] Park HJ, Kim SH, Kim HJ, Choi SH. A new composition of nanosized silica-silver for control of various plant diseases. The Plant Pathology Journal. 2006;**22**(3):295-302. DOI: 10.5423/PPJ.2006.22.3.295

[56] Kim SW, Kim KS, Lamsal K, Kim YJ, Kim SB, Jung MY, et al. An in vitro study of the antifungal effect of silver nanoparticles on oak wilt pathogen Raffaelea sp. Journal of Microbiology and Biotechnology. 2009;**19**(8):760-764. DOI: 10.4014/jmb.0812.649

[57] Kumar GD, Natarajan N, Nakkeeran S. Antifungal activity of nanofungicide Trifloxystrobin 25%+ Tebuconazole 50% against Macrophomina phaseolina. African Journal of Microbiology Research. 2016;**10**(4):100-105. DOI: 10.5897/AJMR2015.7692

[58] DeRosa MC, Monreal C, Schnitzer M, Walsh R, Sultan Y. Nanotechnology in fertilizers. Nature Nanotechnology. 2010;**5**(2):91. DOI: 10.1038/nnano.2010.2

[59] Kah M, Kookana RS, Gogos A, Bucheli TD. A critical evaluation of nanopesticides and nanofertilizers against their conventional analogues. Nature Nanotechnology. 2018;**13**(8): 677-684. DOI: 10.1038/s41565-018-0131-1

[60] Liu R, Lal R. Synthetic apatite nanoparticles as a phosphorus fertilizer for soybean (Glycine max). Scientific Reports. 2014;**4**(1):5686. DOI: 10.1038/srep05686

[61] Giroto AS, Guimarães GG, Foschini M, Ribeiro C. Role of slow-release nanocomposite fertilizers on nitrogen and phosphate availability in soil. Scientific Reports. 2017;**7**(1):46032. DOI: 10.1038/srep46032

[62] Kottegoda N, Munaweera I, Madusanka N, Karunaratne V. A green slow-release fertilizer composition based on urea-modified hydroxyapatite

nanoparticles encapsulated wood. Current Science. 2011:73-78

[63] Sun YP, Li XQ, Cao J, Zhang WX, Wang HP. Characterization of zero-valent iron nanoparticles. Advances in Colloid and Interface Science. 2006;**120**(1-3):47-56. DOI: 10.1016/j.cis.2006.03.001

[64] Singh A, Prasad SM. Nanotechnology and its role in agro-ecosystem: A strategic perspective. International Journal of Environmental Science and Technology. 2017;**14**(10): 2277-2300. DOI: 10.1007/s13762-016-1062-8

[65] Brolo AG. Plasmonics for future biosensors. Nature Photonics. 2012;**6**(11):709-713. DOI: 10.1038/nphoton.2012.266

[66] Cecchin I, Reddy KR, Thomé A, Tessaro EF, Schnaid F. Nanobioremediation: Integration of nanoparticles and bioremediation for sustainable remediation of chlorinated organic contaminants in soils. International Biodeterioration and Biodegradation. 2017;**119**:419-428. DOI: 10.1016/j.ibiod.2016.09.027

[67] Bratovcic A, Hikal WM, Said-Al Ahl HAH, Tkachenko KG, Baeshen RS, Sabra AS, et al. Nanopesticides and nanofertilizers and agricultural development: Scopes, advances and applications. Open Journal of Ecology. 2021;**11**(4):301-316. DOI: 10.4236/oje.2021.114022

[68] Konappa N, Krishnamurthy S, Arakere UC, Chowdappa S, Akbarbasha R, Ramachandrappa NS. Nanofertilizers and nanopesticides: Recent trends, future prospects in agriculture. In: Jogaiah S, Singh HB, Fraceto LF, De Lima R, editors. Advances in Nano-Fertilizers and Nano-Pesticides in Agriculture. 2021. pp. 281-330. DOI: 10.1016/B978-0-12-820092-6.00012-4

[69] Ombódi A, Saigusa M. Broadcast application versus band application of polyolefin-coated fertilizer on green peppers grown on andisol. Journal of Plant Nutrition. 2000;**23**(10): 1485-1493

[70] Sharon M, Choudhary AK, Kumar R. Nanotechnology in agricultural diseases and food safety. Journal of Phytology. 2010;**2**(4):83-92

[71] Brunner TJ, Wick P, Manser P, Spohn P, Grass RN, Limbach LK, et al. In vitro cytotoxicity of oxide nanoparticles: Comparison to asbestos, silica, and the effect of particle solubility. Environmental Science and Technology. 2006;**40**(14):4374-4381. DOI: 10.1021/es052069i

[72] Miransari M. Soil microbes and plant fertilization. Applied Microbiology and Biotechnology. 2011;**92**(5):875-885. DOI: 10.1007/s00253-011-3521-y

[73] Mittal D, Kaur G, Singh P, Yadav K, Ali SA. Nanoparticle-based sustainable agriculture and food science: Recent advances and future outlook. Frontiers in Nanotechnology. 2020;**4**(2):10. DOI: 10.3389/fnano.2020.579954

[74] Cromwell WA, Yang J, Starr JL, Jo YK. Nematicidal effects of silver nanoparticles on root-knot nematode in Bermudagrass. Journal of Nematology. 2014;**46**(3):261-266

[75] Rajesh S, Raja DP, Rathi JM, Sahayaraj K. Biosynthesis of silver nanoparticles using Ulva fasciata (Delile) ethyl acetate extract and its activity against Xanthomonas campestris pv. malvacearum. Journal of Biopesticides. 2012;**5**:119

[76] Bramhanwade K, Shende S, Bonde S, Gade A, Rai M. Fungicidal activity of Cu nanoparticles against Fusarium causing crop diseases. Environmental Chemistry Letters.

2016;**14**(2):229-235. DOI: 10.1007/s10311-015-0543-1

[77] Ayoub HA, Khairy M, Elsaid S, Rashwan FA, Abdel-Hafez HF. Pesticidal activity of nanostructured metal oxides for generation of alternative pesticide formulations. Journal of Agricultural Food and Chemistry. 2018;**66**(22):5491-5498. DOI: 10.1021/acs.jafc.8b01600

[78] Chhipa H, Kaushik N. Development of nano-bio-pesticide using Iron and Eucalyptus plant extract and their application in pest management. In: Conference Proceeding of Symposium on Recent Advances in Biotechnology for Food and Fuel. New Delhi: TERI; 2015. pp. 19-20

[79] Elmer WH, White JC. The use of metallic oxide nanoparticles to enhance growth of tomatoes and eggplants in disease infested soil or soilless medium. Environmental Science: Nano. 2016;**3**(5):1072-1079. DOI: 10.1039/C6EN00146G

[80] Hou R, Zhang Z, Pang S, Yang T, Clark JM, He L. Alteration of the nonsystemic behavior of the pesticide ferbam on tea leaves by engineered gold nanoparticles. Environmental Science and Technology. 2016;**50**(12):6216-6223. DOI: 10.1021/acs.est.6b01336

Chapter 12

Integrated Management of the Cattle Tick *Rhipicephalus (Boophilus) microplus* (Acari: Ixodidae) and the Acaricide Resistance Mitigation

Rodrigo Rosario Cruz, Delia Inés Domínguez García, Saúl López Silva and Fernando Rosario Domínguez

Abstract

Resistance to insecticides is one of the major obstacles to the control of agricultural pests, as well as species important to human and veterinary health. The World Health Organization has called insecticide resistance "the great little obstacle against vector-borne diseases". *Rhipicephalus (Boophilus) microplus* is one of the most important vector, transmitting diseases to cattle such as anaplasmosis and babesiosis. These diseases cause great economic losses that significantly increased because of the appearance of tick populations resistant to acaricides, as a result of the intensive use of pesticides. Resistance to ixodicides in Latin America is a growing problem, since control of disease-transmitting ticks, depends heavily on the use of pesticides. In Mexico, the resistance of *R. microplus* to organophosphate compounds, pyrethroids, and recently amidines, has been detected in some areas, affected by multiple acaricide resistance to the three families of ixodicides. The cattle tick *R. microplus* in addition to the great ecological impact represents the most expensive pest for livestock in Mexico, since the producers are directly affected by this tick, due to the decrease in the production of meat, milk and damage to the skin, as well as the indirect damage, such as the transmission of diseases, including Anaplasmosis and Babesiosis, which, in turn, represents a serious limitation for the introduction of specialized cattle in endemic areas. Therefore, the use of integrated management programs is a mandatory issue that should be implemented in all those areas affected by this parasite.

Keywords: ticks, *R. microplus*, tick control, tick vaccines, insecticides, Acaricide resistance

1. Introduction

Parasitic diseases are a global problem for health and animal production performance due to endoparasites or ectoparasites, Among ectoparasites (ticks, mites, flies, fleas mosquitoes etc.), ticks have adapted to most of the terrestrial niches on

IntechOpen

the planet and have specialized in feeding on the blood of mammals, birds and reptiles around the world [1–3]. The evolutionary adaptation of ticks to hematophagy, is the major reason of the great economic losses caused by this group of parasites, however, the greatest impact of tick infestations to human and animal health is also related with the tick borne diseases.

Ticks are considered responsible for more than 100,000 cases of human diseases, and are the most important vectors of disease-causing pathogens in wild and domestic animals. Globally, they are the second most important disease vectors in humans only after the mosquitoes [4, 5], however they are considered to be the most important vector of pathogens in North America [6].

The families Argasidae and Ixodidae are two groups of thelmophagous ticks of great importance for human and animal health, since they act as reservoirs of a lot of pathogens including parasitic protozoos (*Babesia spp* and *Theileria spp*.), bacteria (*Rickettsia spp*., *Ehrlichia spp* and *Anaplasma spp*), viruses (Nairovirus, Flavivirus and Asfavirus) and nematodes (*Acanthocheilonema*) [5].

Ticks belong to the group of ectoparasites that cause important economic losses in the cattle industry in tropical and subtropical ecosystems all over the world. Specifically, *R. microplus* causes direct damage due to the action of bites [7] and indirect damage caused by the transmission of three etiological agents: *Babesia bovis*, *Babesia bigemina* and *Anaplasma marginale* [8]. In the US prior to the eradication of *R. microplus* and *R. annulatus*, indirect economic losses from babesiosis were estimated at $ 130.5 US million dollars (which today would be three US billion dollars). If ticks had not been eradicated from the US, the livestock industry's annual losses due to ticks would be approximately one billion US dollars [9, 10]. Currently, the Texas Animal Health Commission (TAHC) has expanded the preventive quarantine zone in South Texas, because of the presence of resistant ticks on livestock and wildlife in 139 grassland areas [11]. The aim of this review is to contribute to the discussion of the cattle tick issues, as well as to provide a reference, for all those interested in the current problem of acaricide resistance, the importance of vaccine development and the perspectives of tick genomic research in Mexico.

2. *Rhipicephalus (Boophilus) microplus* life cycle

Rhipicephalus (Boophilus) microplus, is an important endemic tick specie causing great loses and damages to livestock production in tropical and subtropical regions [12]. It is a one-host telmophagous ectoparasite, showing a parasitic and a free living stage and four different evolutionary ontological stages: Egg, larvae, nymph and the adult engorged female (**Figure 1**).

Bovine cattle is parasitized by *R. microplus* as a preferred host, however, it can sporadically infest horses, sheep and goats. Its life cycle is divided in two phases: the parasitic and the non-parasitic free living stages, as well as four ontological stages: egg, larva, nymph and the adult engorged female [13].

The parasitic phase (**Figure 1**), begins when the larvae overcome the climatic and host barriers, since its life cycle is influenced by climatic factors acting on the free living tick stage and the host response against the tick as a parasite. Larvae, then reaches the bovine skin, where they will start the physiological processes of feeding, molting and copulation of the larva, nymph and adult stages respectively [14]. The duration of the parasitic phase is relatively constant, it has been estimated that the duration of this stage from larvae to the adult engorged female, occurs approximately from 18 to 22 days, including feeding, molting and change to the next stage; the whole process takes place all the time on the bovine. The mortality

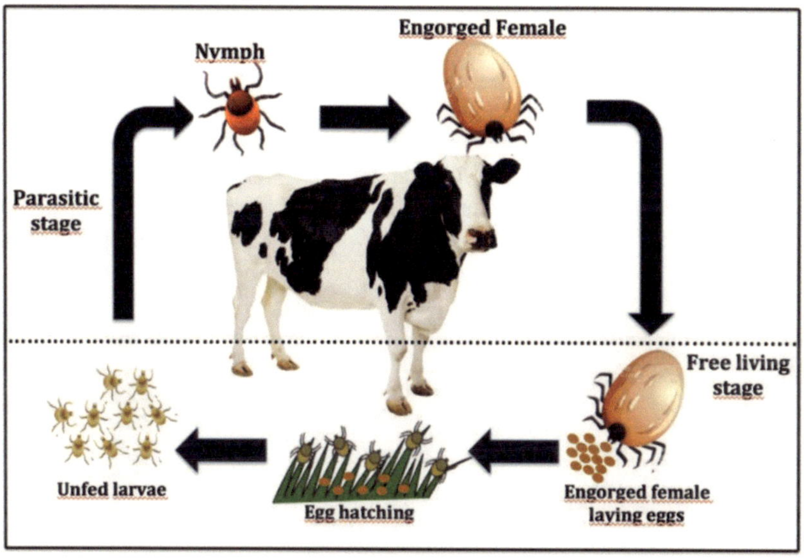

Figure 1.
Life cycle of the one host thelmophagous cattle tick Rhipicephalus (Boophilus) microplus, *showing a parasitic and a free living stage and four different ontological forms: Egg, larvae, nymph and adult or engorged female. (Artwork composed by Fernando Rosario & Delia Inés Dominguez 2021).*

rate of ticks in this phase is determined by the resistance of the host, since, larvae as we already mentioned, is influenced by changing climatic factors as well as the host response against the different tick parasitic stages, the larvae, nymph and the adult engorged female [14].

The non-parasitic phase (**Figure 1**), begins when the engorged female detaches from the host in search of suitable places for oviposition, the eggs laid remain under the grass, until the larvae hatches and appear on the grassland ready to infest the bovine host. Under the grass, the non-parasitic cycle goes through several stages: pre-oviposition, oviposition, incubation and larval hatching. The intervals duration for the completion of each stage, are variable and greatly conditioned by factors such as season, host abundance, selection of host species, and the climatic conditions mainly humidity and temperature [15]. Biologically, these processes involve the physiological and behavioral response of ticks to temperature, moisture stress and day length that result in specific patterns of seasonal population dynamics and hosts availability [14].

R. microplus changes to the juvenile adult stage approximately 13 days after the larva attaches to the bovine host, in this stage male and female become sexually dimorphic. Once the exubia is lost, the male is ready to copulate the next day. The male is very mobile and walks around the host looking for females to mate, regularly male ticks are found below of semi-engorged females. The female is not as mobile as the male and remain attached to the host throughout her life cycle. The female ends her cycle as soon as it finishes laying eggs on the grass (**Figure 1**) [16].

3. Global importance of ticks

Undoubtedly Ticks are the most important group of pathogen vectors causing diseases in wild and domestic animals [5]. Its great economic and sanitary importance is due to its wide distribution, vectorial capacity, hematophagous habits and the number of cattle it affects [17].

The control of tick populations and the diseases they transmit in countries with emerging economies in Latin America, is a prevailing need due to the millionaire economic losses they cause. On the other hand acaricides with a tick-killing effect is the main tool available to control ticks [18].

The cattle fever tick *R. (B.) microplus* and *R. (B.) annulatus* are two of the known vectors of *Babesia bovis* and *Babesia bigemina*, the causative agents of bovine babesiosis [19]. These ticks are invasive livestock parasites (**Figure 2**) in the trans-boundary region between United States (U.S.) and northern Mexico [19], affecting

Figure 2.
Infested cow from the Northern transboundary region between Mexico and the United States, shows the tick infestations resulting from the intensive use of acaricides based on a regular and systematic chemical application approach. (The photograph has been kindly provided by Dr. Martin Ortiz Estrada).

Figure 3.
Map showing the distribution of the cattle tick Rhipicephalus (Boophilus) microplus *and the current state of tick control officially recognized by the National Tick Campaign Office from the National Center for Verification Services on Animal Health SENASICA from the Mexican Government. Consulted and taken from the official SENASICA web site on August 19, 2021. (https://www.gob.mx/senasica/documentos/situacion-actual-del-control-de-la- garrapata-boophilus-spp).*

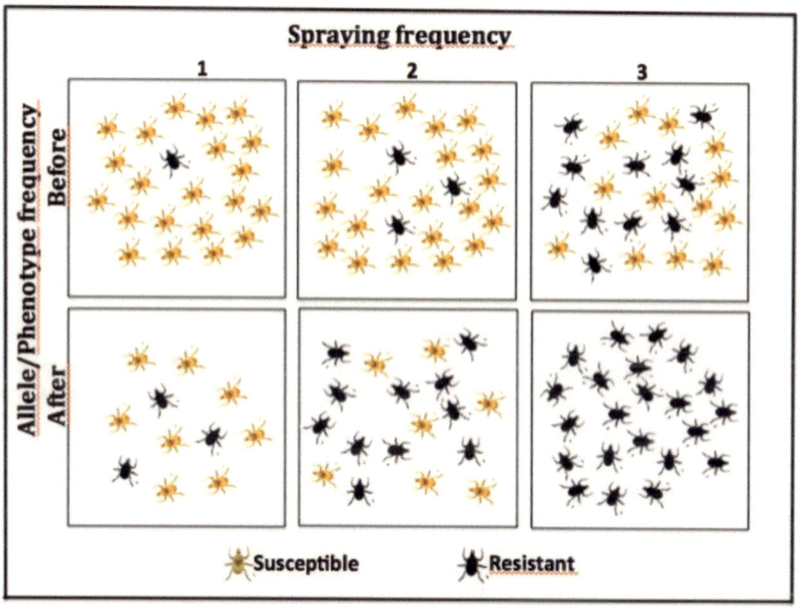

Figure 4.
Theoretical illustration showing the increase of acaricide resistance phenotypes and/or allele frequency levels in a tick population. Some individuals (black) with genetic traits allowing them to survive the acaricide applications can reproduce; if the selection pressure is frequent, they progressively become the preponderant part of the tick population. (Artwork composed by Fernando Rosario & Delia Inés Dominguez 2021).

32 million heads of cattle (**Figure 3**), and are considered to be the most economically important ectoparasites of cattle worldwide [20, 21]. The use of acaricides has an environmental impact as well, originated by the contamination of the soil, and water, besides the killing effect on other beneficial arthropod species and the contamination of food products for human consumption such as milk and meat, derived from this type of livestock operations.

Acaricides have been intensively used for tick control; as a result the use of chemicals combined with the plasticity of tick genomes, has increased progressively the appearance of resistance to different families of acaricides unavoidably (**Figures 2** and **4**) [22], and in many cases to the appearance of multiple resistance. The appearance of resistance in the cattle industry has highlighted the greater inconvenience of the use of acaricides, the selection of resistant tick populations due to the use of acaricides or mixtures of acaricides elaborated based on the ignorance of the resistance mechanisms [22]. There are some critical and basic concepts that allow us to make a road map, on how acaricide resistance occurs, after the continuous and frequent application of chemical treatments. After the continuous exposure, the acaricide kills a fraction of the susceptible ticks, and an increase of tick resistant phenotypes gradually occurs, as illustrated in **Figure 4**. As a consequence, the half-life of pesticides in some regions of northern Mexico has been reduced to such a degree that they no longer represent an alternative to control ticks (**Figure 2**), and the interest of looking for new approaches is currently focused to search for new potentially useful immunogenic vaccine candidates to control resistant tick populations [23].

4. Acaricide resistance in Mexico

Acaricide resistance is a genetic condition driven by randomly arise genetic traits that can be inherited to the progeny and spread throughout the population along time, promoted by natural or artificial selective pressure on a toxic environment, contaminated with synthetic or natural acaricides.

Parasitic diseases have become a global problem due to free trade agreements or commercial exchange of goods and services, because the geographical borders between countries have disappeared from the political geography. One of the biggest issues associated with exportation and importation of animals and products, is the free movement of vector and vector borne diseases associated with animal health and food safety [24], as well as the tick genomes, encoding the acaricide resistance traits that will be transferred to the progeny.

Two general mechanisms of acaricide resistance have been described in *R. microplus*: The enhanced metabolic detoxification, mediated by multigenic families of enzymes (**Figure 5**) such as: esterases, Glutathion-S-Transferases, Mix function Oxidases (Cytochrome P-450) [25, 26], the recently proposed mechanism mediated by the ATP Binding Cassette (ABCt) (**Figure 5**), which is a transporter group of proteins [27], and the target site modification [28–31].

However, the most common mechanism in pyrethroid resistant tick populations in the field, is the target site modification, mediated by a substitution occurring on gene sequences as it was demonstrated for the occurrence of a point mutation located at the segment six domine III (S6III) of the sodium channel gene [30], which encodes the substitution of a Phenylalanine by an Isoleucine in Mexican field samples (**Figure 6**).

Figure 6, show the association between genotypes and phenotypes of nine tick strains that were grouped based on three different phenotypes: very resistant, moderately resistant and susceptible to pyrethroids as measured by the larval packet test (LPT) and later analyzed by the allele specific PCR amplification test in order to identify the three different genotypes (RR, RS and SS) in samples collected from Yucatán, Mexico.

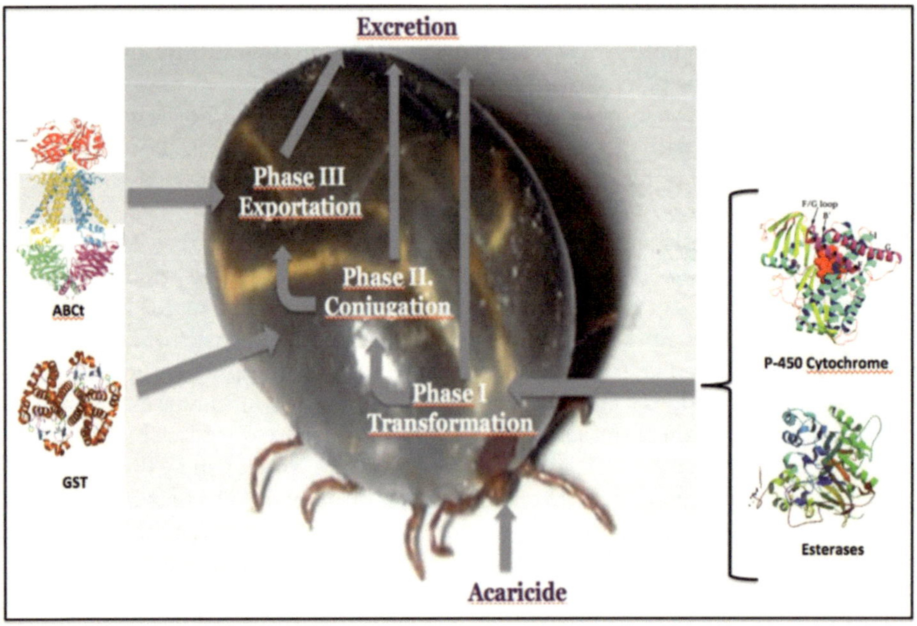

Figure 5.
Illustration showing the different phases of detoxification mechanisms and the participation of multigenic families of hydrolyzing (Esterases and cytochrome P450), modifying enzymes (GST) or the group of transporter proteins ATP binding Cassette (ABCt) at different levels of the metabolic detoxification process (transformation, conjugation and exportation). (Artwork composed by Fernando Rosario & Delia Inés Dominguez 2021).

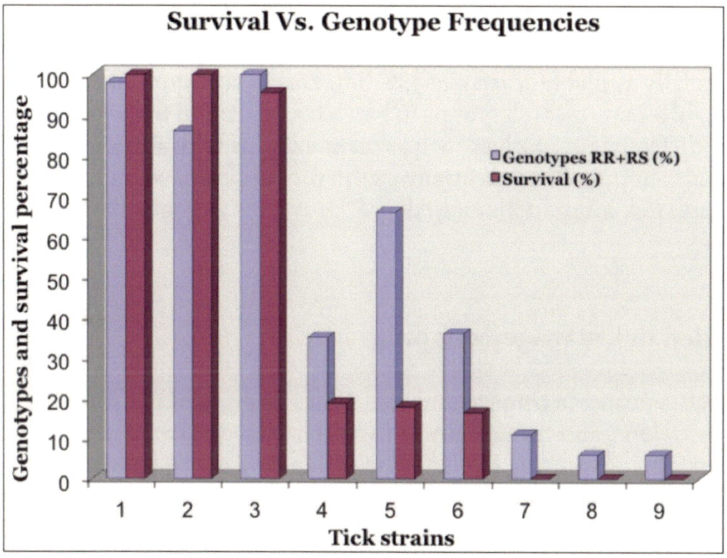

Figure 6.
Genotype and larval survival percentages obtained by the allele specific PCR and LPT respectively, were plotted in order to show the statistical correlation between the presence of the mutation on the sodium channel gene and the resistance to pyrethroids in samples collected from Yucatán, Mexico ($p < 0.05$). Tick strains were classified by their phenotype as: Very resistant (strains 1, 2 and 3), intermediate resistance (strains 4, 5 and 6) and very susceptible (7, 8 and 9). Genotypes are represented by blue bars and phenotypes by red bars. (Graph from Dr. R. Rosario-Cruz archives).

Results demonstrated that PCR test can be used as a molecular tool to detect and predict the appearance of pyrethroid resistance phenotypes, since a resistant allele frequency lower than ten percent, started showing up in the susceptible strains while they were still susceptible as measured by the LPT [30]. According to this results the genotype and phenotype frequency (**Figure 3**) increases in parallel with the continuous application of acaricides ($p < 0.05$) (**Figure 4**).

Up to date, numerous studies have been reported in order to predict the mode of inheritance of pyrethroid resistance in various insects such as mosquitoes (31), the horn fly *Haematobia irritans* (L), [32], *Plutella xylostella* [33] and *Cydia pomonella* [34]. However, the conclusions obtained from these studies have been made based on measurements of the phenotypic response to toxicological bioassays, and did not take into account the genotypes present in the strains analyzed; therefore, no general conclusions can be made, based on these data, obtained from the phenotypes of these arthropod species.

There are evidences demonstrating that in *R. microplus* the sodium channel substitution, encoding the pyrethroid resistance trait, is inherited to the progeny by a completely recessive mode when the male is resistant and female susceptible (RS), since the heterozygous RS genotype behave as a susceptible phenotype [35], suggesting that resistance to flumethrine (Flu-R) and deltamethrine (Del-R) is due to a single gene recessively inherited, while the Cypermethrin resistant RS genotype (Cyp-R), show a residual 36% of larval survival, suggesting that pyrethroid resistance in Cyp-R SR heterozygous strain, is probably due to different mechanisms. On the other hand pyrethroids as a class show a residual maternal effect for all Flu-R, Del-R and Cyp-R for the SR heterozygous genotype strain obtained from the cross of a susceptible male and a resistant female, since, approximately 30% of the heterozygous progeny behave as a resistant phenotype probably due to a mechanism of resistance different than the sodium channel gene substitution [35] inherited by the resistant female, so called maternal effect.

The use of acaricides has been the most important tool against ticks; however, the abuse of chemical control has led to multiple resistance to different classes of commercially available acaricides [36, 37]. Due to recent problems of multiple resistance, different research groups in Mexico and around the world, have focused on finding different alternatives such as plant extracts with acaricide activity [38] and recombinant proteins, potentially useful for vaccine development as an alternative to control tick infestations caused by *R. microplus* [39] as well as transmission of pathogens.

5. Integrated tick management program

Livestock industry in tropical and subtropical regions of the world, is affected by tick infestations, and the economic losses it causes due to the direct effects associated to their blood-sucking habits, such as skin damage and pathogens transmission in tropical and subtropical areas [5, 12, 40]. The producers also have important losses associated with decreased weight gain and low production of milk and meat due to the economical impact on cattle by pathogens causing Babesiosis (*B. bovis and B. bigmina*) and Anaplasmosis (*A. marginale*) [41, 42]. However, there are no precise information on the contribution of each of these components to the complex network of interactions between the host, the tick and the tick borne pathogens.

The cattle tick control has been traditionally based on application of acaricide strategies, by dip bath, sprying, pour on or injection, ignoring the consequences of frequent acaricide applications and the biology of the vector. In some states of Mexico, multi-resistance is a constant and current threat, which affects the National livestock production and therefore the economy of producers, since they depend completely on the use of acaricides, but do not have access to technical advice from any public or private office, in order to design a control program to prevent the sudden loss of efficacy of the acaricides used for tick control [43].

The most reliable information on global economic losses, date from the 1980s, these figures estimate that one billion head of cattle are exposed to tick infestations in tropical and subtropical regions of the world and in 1984, economic losses were estimated in eight US billion dollars [44].

Reported data in the literature does not include the loss of human life due to ticks and tick borne diseases, such as the thousands of cases of Lyme disease that occur annually in Europe and North America [45, 46], tick-borne encephalitis cases in Europe [47], and tick-borne rocky spotted fever cases in the United States [48].

In the context of animal health, the most important tick is *R. microplus* to which losses in productivity were attributed and quantified in 1987, in more than one US billion dollars annually in South America [49], and in 1974 in Australia, annual losses were estimated at 62 million dollars [50]. Recent studies reported that Brazil losses were quantified in two US billion dollars [51].

Recent experimental trials of an integrated tick management (ITM) program in Mexico, suggest that ITM programs should included the combined use of acaricides with an anti-tick vaccine against *R. microplus* [52]. The application of the ITM program in field facilities, decreased the frequency of acaricide applications by period of time from 27 to 155 days. This extension on the average application time, decreased the number of total annual applications from 14 to 2.8, which mean a reduction in the use and purchase of acaricides of 80%. It is predictable that a proportional reduction of the environmental contamination can be expected by including an anti-tick vaccine within the ITM program. The reduction of costs and use of acaricides was attributed to the effects of the vaccine on the tick reproductive parameters, for instance, the tick weight was reduced in vaccinated cattle, from

Figure 7.
Comparison of 21 days old adult engorged females, collected from vaccinated and unvaccinated cattle under an integrated tick management program in the coastal state of Guerrero, Mexico. (Photo from Dr. R. Rosario-Cruz archives).

166 mg/tick to 25 mg (**Figure 7**), for example, this tick reduction in size, meant an average egg mass reduction of 84%.

These data was used to calculate the reduction of production costs due to the purchase of pesticides but did not include the purchase of antibiotics to control babesiosis and anaplasmosis, for calculation purposes [52]. The cost of chemical tick control in this study was $ 408.3 Mexican pesos per animal, while the combined program was only $ 128 pesos, which mean a reduction of 68.63% for the purchase of ixodicides.

The extrapolation of these data to the national livestock herd estimated in 30 million head of cattle, is equivalent to 12,248.7 million Mexican pesos. The use of a combined control program would reduce these losses from 12,248.7 million of Mexican pesos to 3,843.7, that is a reduction of 68.63% of the losses applied to the Mexican livestock herd [52].

The data published in this paper shows an estimated annual loss of 942 million USA dollars, (considering a current exchange rate of 20 Mexican pesos/US dollar, the equivalent annual losses would be 612 US million Dollar) it is worth mentioning that this and other previously published papers, does not include the loss of animals dead by ticks and tick borne diseases such as Babesiosis and Anaplasmosis, nor the expenses produced by the costs of the medication used to control these tick borne pathogens which can double the annual losses due to tick infestations.

6. The perspective of immunological approaches

Edward Jenner was the first to scientifically prove in studies carried out in 1796, a method to protect against smallpox, thereby laying the foundations of vaccinology, and although the invention is not directly attributed to him, he is often considered the father of vaccines due to his scientific approach that proved that the "vaccination" method worked, and from then until today, more than 200 years after its discovery, new biotechnological tools have substantially improved not only the application of vaccines, but the way to produce them.

A vaccine is a biological preparation that provides an active acquired immunity to a particular pathogen. The vaccine preparation stimulates the immune system to recognize a foreign threat and thus destroys and remembers it, so that the immune system can easily destroy any of these pathogens when they later invade into the body. The vaccine characteristics can be enhanced by Biotechnology.

Vaccines have been the most significant advance in public health, and its preventive prophylactic treatment has been demonstrated as we mentioned, for over 200 years for bacterial and viral diseases preventing morbidity and mortality in millions of people annually [53]. Vaccine development during the pre-genomic era was based on the use of dead, live or attenuated organism or on the use of subunitarian proteins purified from total extracts from organisms of interest [54].

These subunitarian proteins may contain one or more antigens combined. To develop such vaccines is a critical necessary step, identification of proteins of interest and eliminating others that are not useful. In this particular case, in order to be recognized as protector, an antigen must be able to limit the development or reproduction of the organism, parasite or pest in question in subsequent exposure challenges [55].

The empirical approach to the development of subunit vaccines includes several steps: a) culturing the parasite, microorganism or pest to be controlled, b) the analysis, and identification of its components, c) purification of antigens having immunogenic properties required for product development and d) the subsequent challenge with the infectious agent or parasite against which we want to develop the vaccine, in an appropriate animal model to evaluate the immunogenic characteristics of this technology [56, 57].

This methodology has difficulties inherent in the process of identification and purification of the fractions possessing the optimal antigenic characteristics for vaccine development and the availability of macro or microorganism to be controlled by this biotechnological tool because the vaccine production It is severely limited when the target organism cannot easily grow [58]. There are other drawbacks that have to do also with the biology of the target organism, since in some cases the most abundant proteins are not necessarily immunoprotective, or may be the case that the antigens expressed during in vivo or infestation infection as the case are not the same as those expressed during cultivation in vitro latter may not be the case in ticks [59].

Difficult as it may seems, the hard work has already made great progresses, the number of cloned and analyzed genes are already a big list, and experiments have shown that genes as Bm86, subolesin, ferritin, aquaporin and a growing number of orthologous genes can be used to control ticks. The future in the field of vaccine development is becoming shorter, and the scope of modern technology in the field is increasingly longer. Landing knowledge regarding the tick vaccines development for tick control, is very close to pays off as seen by the growing list of new antigens discovered, although the tick control still represents a challenge for the scientific community.

7. Conclusive remarks

Ticks and tick-borne pathogens constitute a growing problem for human and animal health worldwide, since, they are considered, the most important vectors of disease-causing pathogens in wild and domestic animals and the second most important vector of pathogens causing diseases in humans, only after the mosquitoes.

Resistance to acaricides impacts directly the economy and the competitiveness of producers, and its presence within the ranches implies the expenses associated to control of ticks and tick borne diseases. Efficient integrated control programs are required to mitigate the direct effects on cattle infested with resistant ticks, and to keep a low prevalence of tick borne diseases.

The use of an integrated tick management program in Mexico, including a combined control strategy (acaricides and tick vaccine), reduced the use of acaricides for tick control by 80% approximately, with a cost–benefit ratio of 3:1, lowering the environmental and food products contamination derived from this activity, reducing the mortality attributed to Babesiosis and Anaplasmosis and contributing to the development of a sustainable, and environmentally friendly livestock production system.

New candidate protective antigens and research on tick vaccine development need to be addressed to establish and design better strategic control programs, since vaccines have demonstrated to be the most effective and an environmentally friendly intervention for the control of tick infestations and tick-borne diseases.

The hard work, difficult as it may seems, has already made great progresses, the future in the field of tick vaccine development is becoming shorter, and very close to pays off as seen by the growing list of new antigens discovered, although the tick control still represents an innovation challenge for the scientific community in Mexico and all over the world.

Conflict of interest

The authors declare having no relevant affiliations or financial involvement with any organization or entity with a financial interest in or financial conflict with the subject matter or materials discussed in the manuscript.

Author details

Rodrigo Rosario Cruz[1*], Delia Inés Domínguez García[1], Saúl López Silva[2] and Fernando Rosario Domínguez[3]

1 Biotechnology in Health and Environmental Research Lab (BioSA), Natural Sciences College, Autonomous Guerrero State University, Chilpancingo, Guerrero, Mexico

2 Genomol Laboratory S.A. de C.V. Costera Miguel Aleman Ave. Acapulco, Guerrero, Mexico

3 Veterinary College, Universidad Mesoamericana, Cuernavaca, Morelos, Mexico

*Address all correspondence to: rockdrig@yahoo.com.mx

References

[1] Furman, D.P., Loomis, E.C., 1984. The Ticks of California (Acari: Ixodida). University of California Publications, Bulletin of the California Insect Survey, Vol. 25. University of California Press, California, p. 1-35. Library of Congress Catalog Card Number 83-9265. ISBN 0-520-09685-1.

[2] Tahir, D., Meyer, L., Fourie, J., Jongejan, F., Mather, T., Choumet, V., Blagburn, B., Straubinger, R.K., & Varloud, M. Interrupted blood feeding in Ticks: causes and consequences. Microorganisms. 2020;8(6)910. https//doi.org/10.3390/microorganisms8060910.

[3] Sharifah, N., Heo, C. C., Ehlers, J., Houssaini, J., & Tappe, D. Ticks and tick-borne pathogens in animals and humans in the island nations of Southeast Asia: A review. Acta tropica, 2020;209:105527. https://doi.org/10.1016/j.actatropica.2020.105527.

[4] FAO. The state of food and agriculture. Climate Change, Agriculture and Food Security. FAO. Roma. 2016. http://www.fao.org/3/i6030e/i6030e.pdf.

[5] de la Fuente, J., Estrada-Pena, A., Venzal, J.M., Kocan, K.M., Sonenshine, D.E. Overview: Ticks as vectors of pathogens that cause disease in humans and animals. Frontiers in Bioscience. 2008;13:6938-6946. doi: 10.2741/3200.

[6] Parola P., and Raoult D. Ticks and tick borne bacterial diseases in humans: an emerging infectious threat. Clinical Infectious Diseases. 2001:32: 897-928. doi: 10.1086/319347.

[7] Buczek A, Bartosik K. [Tick-host interactions]. Przeglad Epidemiologiczny. 2006; 60 Suppl 1:28-33.

[8] Vieira, L.L., Canever, M.F., Cardozo, L:L: Cardozo, C.P., Herkenhoff, M.E., Neto, A.T., Vogel, C., & Milleti, L.C. Prevalence of *Anaplasma marginale*, Babesia bovis and Babesia bigemina in the Cattle in the campos de Lages región Santa Catarina state, Brazil estimated by multiplex-PCR. Parasite epidemiology and control. 2001;6:e00114. https://doi.org/10.1016/j.parepi.2019.e00114.

[9] Shyma, K.P., Gupta, J.P., & Singh, V. Breeding strategies for tick resistance in tropical cattle: a sustainable approach for tick control. Journal of parasitic diseases: Official organ of the Indian Society for Parasitology. 2015;39(1): 1-6. http://doi.org/10.1007/s12639-013-0294-5.

[10] Raijput, Z.I., Hu, S.H., Chen, W.J., Arijo, A.G., Xiao, C.W. Importance of ticks and their chemical and immunological control in livestock. Journal of Zhejiang University-Science B (Biomedicine & Biotechnology). 2006;7(11):912-921. doi:10.1631/jzus.2006.B0912.

[11] George J.E. The effects of global change on the threat of exotic arthropods and arthropod-borne pathogens to livestock in the United States. Ann. New York Academy of Sciences. 2008;1149: 249-254. doi: 10.1196/annals.1428.084.

[12] Estrada-Peña, A, García, Z, Sánchez, H.F. The distribution and ecological preferences of *Boophilus microplus* (Acari: Ixodidae) in Mexico. Experimental and Applied Acarology. 2006;38 (4):307-316. https://doi.org/10.1007/s10493-006-7251-2.

[13] Troughton, D.R. and Levin, M.L. Life cycle of seven Ixodid Tick Species (Acari: Ixodidae) Under Standardized Laboratory Conditions, Lournal of Medical Entomology. 2007;44:732-740. https://doi.org/10-1093/jmedent/44.5.732.

[14] Randolph, S.E., Tick Ecology: Processes and patterns behind the epidemiological risk posed by ixodid ticks as vectors. Parasitology. 2004;129Suppl:37-65. doi: 10.1017/ soo31182004004925.

[15] Foldvari, G. Life cycle and ecology of *Ixodes ricinus*: The roots of public Health importance. In: Braks MAH, Van Wieren, S.E., Takken, W., Sprong, H. Editors. Ecology and prevention of lyme borreliosis. Ecology and control of vector-borne diseases, vol 4. Wageningen: Wageningen Academic Pubishers; 2016.

[16] Senbill, H., Hazarika, L., Baruah, A., Borah, D.K., Bhattacharyya, B., & Rahman, S. Life cycle of the southern cattle tick, *Rhipicephalus (Boophilus) microplus* Canestrini 1888 (Acari: Ixodidae) under Laboratory conditions. Systematic and Applied Acarology. 2018. 23;6:1169-1179. https://doi.org/10.11158/saa.23.6.12.

[17] Guerrero, F.D., Bendele, K.G., Chen, A.C., Li, A.Y., Miller, R.J., Pleasance, E., Varhol, R., Rousseau, M.E., Nene, V.M. Serial analysis of gene expression in the southern cattle tick following acaricide treatment of larvae from organophosphate resistant and susceptible strains. Insect Molecular Biology. 2007;16:49-60.

[18] George, J.E., Pound, J.M., Davey, R.B. Chemical control of ticks on cattle and the resistance of these parasites to acaricides. Parasitology. 2004;129 Suppl: 353-366. doi: 10.1017/ s0031182003004682

[19] Giles, J.R., Peterson, A.T., Busch, J.D. et al. Invasive potential of cattle fever ticks in the southern United States. Parasites Vectors. 2014;7:189. https://doi.org/10.1186/1756-3305-7-189

[20] Perez de Leon, A.A. and Mitchell, R.D. Ectoparasites of cattle. Veterinary Clinics of North America: Food Animal Practice. 2020;36:173-185. doi: 10.1016/j.cvfa.2019.12.004.

[21] Grisi, L. Romário Cerqueira Leite, João Ricardo de Souza Martins, Antonio Thadeu Medeiros de Barros, Renato Andreotti, Paulo Henrique Duarte Cançado, Adalberto Angel Pérez de León, Jairo Barros Pereira, Humberto Silva Villela. Reassessment of the potential economic impact of cattle parasites in Brazil. Brazilian Journal of Veterinary Parasitology 2014;23(2): 150-156. https://doi.org/10.1590/ S1984-29612014042.

[22] Rosario-Cruz, R., Almazan, C., Miller, R.J., Domínguez-Garcia, D.I., Hernandez-Ortiz, R., de la Fuente, J. Genetic basis and impact of tick acaricide resistance. Frontiers in Bioscience. 2009;14:2657-2665.

[23] Almazán, C., Lagunes R, Villar, M., Canales, M., Rosario-Cruz, R., Jongejan, F., de la Fuente, J. Identification and characterization of *Rhipicephalus (Boophilus) microplus* candidate protective antigens for te control of cattle tick infestations. Parasitology Research. 2010;106(2):471-479. doi: 10.1007/s00436-009-1689-1.

[24] FAO, Food and Agricultural Organization of the United States Nations. La ganaderísa a Examen. Estado Mundial de la Agricultura y la Alimentación. Roma. 2009. http//www.fao.org/docrep/012/io68os/io68os.pdf.

[25] Gaudencio, F.N., Klafke, G.M., Tunholi-Alves, V.M., Ferreira, T.P., Coelho, C.N., da Fonseca, A.H., da Costa Angelo, I., Pinheiro, J. Activity of carboxylesterases, glutathione-S-Transferase and monooxygenase on *Rhipicephalus microplus* exposed to fluazuron. Parasitology International. 2017;66(5):584-587. doi: 10.1016/j.parint.2017.04.006.

[26] Miller, J.R., Davey, B.R., George, E.J. Characterization of Pyrethroid

Resistance and susceptibility to Coumaphos in Mexican *Boophilus microplus* (Acari: Ixodidae). Journal of Medical Entomology. 1999;36(5):533-538. https://doi.org/10.1093/jmedent/36.5.533.

[27] Lara, F.A., Pohl, P.C., Gandara, A.C., Ferreira, J., Nascimento-Silva, M.C., Bechara, G.H., Sorgine, M.H., Almeida, I.C., Vaz, I. Jr. & Oliveira, P.L. ATP binding Cassette Transporter mediates both Heme and Pesticide detoxification in Tick Midgut Cells. Plos one. 2015;10(8):e0134779. https://doi.org/10.1371/journal.pone.0134779.

[28] Jamroz, R.C., Guerrero, F.D., Pruett, J.H., Oehler, D.D., Miller, R.J. Molecular and biochemical survey of acaricide resistance mechanisms in larvae from Mexican strains of the southern cattle tick, *Boophilus microplus*. Journal of insect physiology. 2000;46:685-695. doi: 10.1016/s0022-1910(99)00157-2

[29] Hernandez, O.R., Guerrero, F.D., George, J.E., Wagner G.G. Allele frecuency and gene expression of a putative carboxylesterase encoding gene in a pyrethroid resistant strain of the tick *Boophilus microplus*. Insect Biochemistry and Molecular Biology 2002;32:1009-1016. doi: 10.1016/s0965-1748(02)00037-1.

[30] Rosario-Cruz, R., Guerrero, D.F., Miller, J.R., Rodriguez-Vivas, R.I., Domínguez-García, D.I., Cornel, J.A., Hernandez-Ortiz, R., George, E.J. Roles Played by esterase activity and by a sodium channel mutation envolved in pyrethroid Resistance in populations of *Boophilus microplus* (Acari: Ixodidae) collected from Yucatán, Mexico. Journal of Medical Entomology. 2005;42(6):1020-1025. doi: 10.1603/0022-2585(2005)042[1020,rpbeaa]2.0.co;2.

[31] Halliday, W.R., Georghiou, G.P. Cross-resistance and dominance relationships of pyrethorids in a permethrin-selected strain of *Culex quinquefasciatus* (diptera: Culiscidae). Journal of Economic Entomology. 1985;78(6):1227-1232. doi: 10.1093/jee/78.6.1227.

[32] Roush, R.T., Miller, G.L. Considerations for design of insecticide Resistance Monitoring Programs, Journal of Economic Entomology. 1986;79,2(1):293-298. https://doi.org/10-1093/jee/85.4.1046.

[33] Tabashnik, B.E., Schwartz, J.M., Finson, N., Johnson, M.W. Inheritance of Resistance to *Bacillus thuringiensis* in Diamondback Moth (Lepidoptera: Plutellidae). Journal of Economic Entomology. 1992;85(4):1046-1055. https://doi.org/10.1093/jee/85.4.1046.

[34] Bouvier, J.C., Buès, R., Boivin, T., et al. Deltamethrin Resistance in the codling moth (Lepidoptera: Tortricidae): inheritance and number of genes involved. Heredity. 2001;87: 456-462. https://doi.org/10.1046/j.1365-2540.2001.00928.x.

[35] Aguilar-Tipacamu, G., Rosario-Cruz, R., Miller, J.R., Guerrero D.F., Rodriguez-Vivas, R. I., Garcıa-Vazquez, Z. Phenotype changes inherited by crossing pyrethroid susceptible and resistant genotypes from the cattle tick *Riphicephalus (Boophilus) microplus*. Experimental and Applied Acarology 2011;54:301-311. doi: 10.1007/s10493-011-9441-9.

[36] Abbas, R.Z., Zaman, M.A., Colwell, D.D., Gilleard J., Iqbal, Z. Acaricide resistance in the cattle ticks and approaches to its management: The state of Play. Veterinary Parasitology. 2014;203(1-2):6-20. doi: 10.1016/j.vetpar.2014.03.006.

[37] Vudriko, P., Okwee-Acai, J., Tayebwa, D.S., Byaruhanga, J., Kakooza, S., Wampande, E., Omara, R., Muhindo, J.B., Tweyongyere, R., Owiny, D.O., Hatta, T., Tsuji, N., Unemiya-Shirafuji,

R., Xuan, X., Kanameda, M., Fujisaki, K., &Suzuki, H. Emergence of multi-acaricide resistant *Riphicephalus* ticks and its implication on chemical tick control in Uganda. Parasites & Vectors. 2016;9:4. https://doi.org/10.1186/s13071-015-1278-3.

[38] Martinez, V.M., Rosario-Cruz, R., Castillo Herrera, G., Flores Fernandez, J.M., Alvarez, A.H., Lugo Cervantez, E. (2011). Acaricidal effect ofr essential oils from *Lipia graveolens* (Lamiales: Vervenaceae), *Rosmarinus officinalis* (Lamiales: Lamiaceae), and *allium sativum* (Liliales: Liliaceae) against *Rhipicephalus (Boophilus) microplus* (Acari:Ixodidae). Journal of Medical Entomology. 48(4):822-827. doi: 10.1603/me10140

[39] Prudencio, C.R., Marra, A.O., Cardoso, R., Goulart, L.R. Recombinant peptides as new immunogens for the control of the bovine tick, *Rhipicephalus (Boophilus) microplus*. Veterinary Parasitology. 2010;172(1-2):122-31. doi: 101016/j.vetpar.2010.04.012.

[40] Snelson, J.T. 1975. Animal ectoparasites and disease vectors causing major reduction in world supplies. FAO Plant Protection Bulletin. 1975;13: 103-114.

[41] L´Hostis, M., Seegers, H. Tick Borne parasitic diseases in Cattle: current Knowledge and prospective risk analysis related to the ongoing evolution in French Catlle Farming System. Veterinary Research. 2002;33(5):599-611. doi: 10.1051/vetres:2002041.

[42] Peter, R.J., Van den Bossche, P., Penzhorn, B.L., Sharp, B. Tick, Fly and mosquito control: lessons from the past, solutions for the future. Veterinary Parasitology. 2005;30;132(3-4):205-15. doi: 10.1016/j-vetpar.2005.07.004.

[43] George, J.E., Present and future technologies for tick control. Annals of the New York Academy of Sciences. 2000;916:583-588. https://doi.org/10.1111/j.1749-6632.2000.tb05340.x

[44] Brown, S.J. and askenase, P.W., (1984). Analysis of host components mediating immune resistance to ticks. In Acarology VI, Vol 2, Griffiths, D.A. and Bowman, C.E. Eds., 1040.

[45] Jaenson, T.G. The epidemiology of Lyme borreliosis. Parasitology Today. 1991;7(2):39-45. Doi:10.1016/0169-4758(91)90187-s.

[46] Pieceman, J., & Gern, L. Lyme Borreliosis in Europe and North America. Parasitology. 2004;129(S1):191-220. DOI: 10.1017/S0031182003004694.

[47] Amicizia, D., Domnich, A., Panatto, D., Lai, P.L., Cristina, M.L., Avio, U., & Gasparini, R., Epidemiology of tick-borne encephalitis (TBE) in Europe and its prevention by available vaccines. Human vaccines & immunotherapeutics. 2013;9(5):1163-1171. https://doi.org/10.4161/hv.23802.

[48] Openshaw, J.J., Swerdlow, D.L., Krebs, J.W., Holman, R.C., Mandel, E., Harvey, A., Haberling, D., Massung, R.F., & McQuiston, J.H. Rocky Mountain spotted fever in the United States, 2000-2007: interpreting contemporary increases in incidence. The American journal of Tropical Medicine and Hygiene. 23010;83(1):174-182. https//doi.org/10.4269/ajtmh.2010.09-0752.

[49] Horn, S. (1987), Ectoparasites on animals and their impact on the economy of South America, in Proc. 23rd World Veterinary Congress, Montreal, August 1987.

[50] Springel, P.H. The cattle tick in relation to Animal Production in Australia. Wild Animal Review. FAO. 1983;36:1-5.

[51] Grisi, L., Massard, C.L., Moya, B.G.E., Pereira, J.B. Impacto economico

das principais ectoparasitoses em bovinos no Brasil. A Hora Veterinaria. 2002;21:8-10.

[52] Dominguez-García, D.I., Torres-Agatón, F., and Rosario-Cruz, R. Economic evaluation of tick (*Rhipicephalus microplus*) control in Mexico. Revista Iberoamericana de las ciencias Biológicas y Agropecuarias. 2016;5(9):1-10. ISSN 2007-9990.

[53] Stutzer, C., Richards, S.A., Ferreira, M., Baron, S., & Maritz-Olivier, C. Metazoan Parasite Vaccines: Present Status and Future Prospects. Frontiers in cellular and infection microbiology. 2018;8:67. https://doi.org/10.3389/fcimb.2018.00067.

[54] Karch, P. C., and Burkhard, P. Vaccine technologies: From whole organisms to rationally designed protein assemblies. Biochemical Pharmacology. 2016;120:1-14. https://doi.org/10.1016/j.bcp.2016.05.001

[55] Willadsen, P. (2008). Anti-tick vaccines. In a Bowman & P. Nuttall (Eds.), Ticks: Biology, Disease and Control. (pp.424-446). Cambridge: Cambridge University Press. doi:10.1017/CBO9780511551802.020

[56] Mulenga, A., Sugimoto, C., Onuma, M. Issues in tick vaccine development: identification and characterization of potencial vandidate vaccine antigens. Microbes and Infection. 2000:23(11):1353-1361. doi: 10.1016/s1286-4579(00)01289-2.

[57] Willadsen, P. Anti-tick vaccines. Parasitology. 2004;129S:367-387. doi: 10.1017/s0031182003004657.

[58] Willadsen P. Tick control: thoughts on a research agenda. Vet Parasitol. 2006 May 31;138(1 2):161-8. doi: 10.1016/j.vetpar.2006.01.050.

[59] Nuttall, P.A., Trinmell, A.R., Kazimirova, M., Labuda, M. Exposed and concealed antigens as vaccine targets for controlling ticks and tick-borne diseases. Parasite immunology. 2006;28(4):155-163. doi: 10.1111/j.1365-3024.2006.00806.x.

Section 5

Insecticides Effects

Chapter 13

Effect of Biodegradable Multiple Pesticides on Aquatic Biospecies

Kenneth Ojotogba Achema and Charity Jumai Alhassan

Abstract

The subject of pesticides usage has become a serious threat to sound ecological sustainability. In this regard, the effects of biodegradable multiple pesticides on aquatic biospecies have been discussed in detail. They are always different forms of pesticides in the aquatic environment. These pesticides are bioavailable in both water body and sediments, and the aquatic species do feed on water and sediment materials. The pesticides are also capable to bioaccumulate and biomagnify along the food chain. These attributes pose serious risks to human health and the sound ecological system that is needed for life sustainability. Cancer, infertility, lesions, headache, dizziness, eye irritation, vomiting, dermal diseases, and gastrointestinal problems have been observed as the direct pesticides effects on biological populations in several countries. The needs for different safety guidelines required for pesticides manufacturing and usage have been recommended.

Keywords: pesticide, biodegradability, bioaccumulate, biomagnify, water, sediments, aquatic biospecies

1. Introduction

Aquatic ecosystems are complex systems that are the compass of nutrients, biotic pelagic and benthic communities, pools of detritus and that have the bulk of both water and sediment [1, 2]. Anthropogenic activities lead to multiple types of stresses, including emissions of pesticides and nutrients into the environment [3, 4]. These pesticides are capable of affecting species in the aquatic ecosystems and the nutrients can cause eutrophication. Emissions of pesticides can also lead to accumulation in the environmental compartments of water and the complex matrix that forms the sediment. Simultaneous pesticides bioaccumulation into organic substances such as biota and detritus prove to have an adverse effect on aquatic bio-species. Various toxicokinetic models have described these types of accumulations [5].

1.1 Environmental pollution

Recently, the cry for environmental pesticides pollution is heard from every nook and crannies of the world. Pollution of pesticides has now become a distinct threat to the very existence of mankind and animals on this earth. It is a problem challenge for our days. In the past, man has been disturbing the balance of nature for comfort, wealth, and ego, but now nature has started disturbing the balance of nature. In the late century, there has been growing concern in developing countries and developed countries over the pollution effects from sources, such as sewage, pesticides, and trade effluents discharged from domestic habitations and by the industrial units [6].

The immediate catastrophic effects of pesticides pollution by some industrial units and agricultural application of pesticides have pointed out the essence of its environmental effects: prevention and control. There is now a global awareness that pesticides production and utilization activities in the future time need to be assessed for their environmental hazard or effects without any form of compromise to the said assessment. Too much rise in pesticides application and their industrial production activities in the past have led to the emission of harmful pesticides into human and aquatic habitats and have led to various ecological issues.

Disturbance of pesticides emission has become a serious threat to both human and animals life and it puts the ecosystem out of balance. Maintenance of ecological balance and environmental purity due to the sudden increase in the production of pesticides and their applications in both the home and agricultural sector should be the inclusive concern of each member of society. This situation could be improved through awareness program creation, which must gain the support of people from all works of life with the aim of enlighten them on its pros and cons and their responsibilities that will meet up with the global standard. Pesticides pollution under discussion here has a different meaning and environmental disorder to different biological organisms. Human beings feed on aquatic animals, fishes and drink untreated water. Thus, they are more susceptible to multiple pesticides effects than aquatic animals that feed on prey and uptake water only. This discussion cannot exclude how water is important to all living organisms as it sustains life as the human body depends on water for about seventy percent (70%) to function normally.

Yichen et al. [5] also pointed out the indisputable dependability of water by living organisms as it functions in every living organism cell and cell is said to be the smallest unit of life. In order to reduce and prevent the issues of pesticides pollution in an aquatic habitat, law enforcement agencies have primary responsibilities of ensuring that laws and implantation of pesticides used must be seriously put in place. Companies or industries operating along rivers, seas, and lakes banks, to mention a few, need to redirect their discharge wastes formally channeled into water bodies to a sustainable environment (**Figure 1**) [7].

1.2 Pesticides and environmental pollution

Pesticides pollution have different routes into aquatic organism domesticated habitat such as leaching from the agricultural farm, erosion from farmlands during

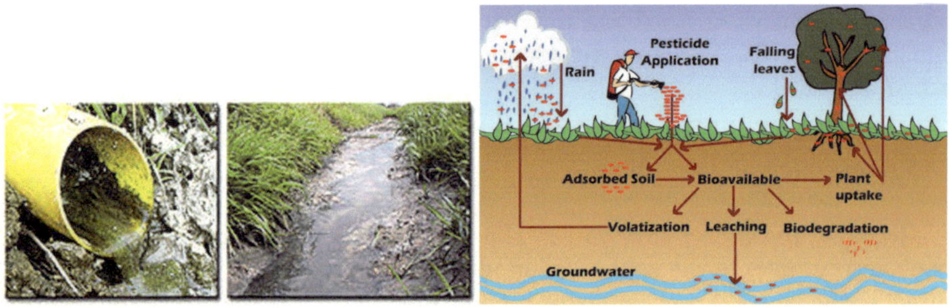

Figure 1.
Environmental pollution.

rainfall, and become available in the water body through spills from industrial effluents and discharges of environmental wastes into water bodies [8]. Pesticides have different meanings to different people. Generally, pesticides are large classification or group of chemical substances that are developed to model and thereafter substitute for a unique molecule in a particular biological process. This implies that the mode of action of pesticides is peculiar to specific organism, plant, or grass such as pests, weeds to mention a few [9].

The issue of poor quality of water is of the utmost environmental problem in the human health-related issues [10]. Pollution from pesticides has contributed to the said threat and it will continue to gain more ground in as much as they are present in water bodies. The primary aim of pesticides production is mostly for specific targeted organisms or plants but their effects are non-targeted as it affects humans, animals, and plants and produces range of toxicity effects which include carcinogenicity [11] and have the ability to disrupt endocrine [12].

The wide usage of pesticides are detected across different nations such as Europe's freshwater. In Ireland, pesticides are present beyond the European Union (EU) permission limit bound on different numbers of inspections [8, 13]. In Europe, freshwaters in the UK were found to be susceptible to pesticide pollution and in Germany, groundwater and sediment materials were specifically polluted due to pesticides applications and discharge [14].

EC legislators have provided different legislation or laws in place to prevent or minimize the discharge of pesticides and their applications in their environment. The legislation includes Water Framework Directive [15], the strategy for the prevention of endocrine-disrupting compounds [16], and the Stockholm convention [17]."

The normal water purification tools or equipment have proved inefficient to remove toxic pesticides substances from water bodies [16]. The need to have an efficient method is of utmost importance to be researched. Up to now, the best efficient pesticides removal methods include photocatalysis and adsorption [18, 19].

Human activities like land cover change, urbanization and industrialization have impaired ecosystems for several decades in order to increase the access to natural resources for an exponentially growing population [20, 21]. The activities of humans have led to the impairment of the planet earth's boundaries and are causing biodiversity loss and climate changes [22, 23]. Keeping the account of human activities on the global freshwater, land use, acidification of oceans, rivers, lakes, and streams are already tending to a threshold value [22]. Rockström et al. [22] in their study have noted that the human population is facing un-quantifiable threats due to freshwater contamination by different forms of contaminants that are unknown in an aquatic environment.

More than 14 million different chemical compounds exist, out of which above 100,000 synthetic chemical compounds are frequently used in consumer products in different countries of the world for different purposes [24, 25]. Thus, an uncertain number of chemicals may potentially be released into the aquatic environment by diverse routes like point sources, remobilization from contaminated sediments, and groundwater input (**Figure 2**) [26].

1.2.1 Routes of pesticides into aquatic environment

A lot of literature has identified different routes of pesticides to the aquatic environment [27, 28]. It was observed mainly that the rate at which pesticides enter the aquatic environment is not the same with all pesticides. The routes mainly depend on the physiochemical properties, that is, the ability of pesticides to persist for several years or a decade in an environment without totally losing their

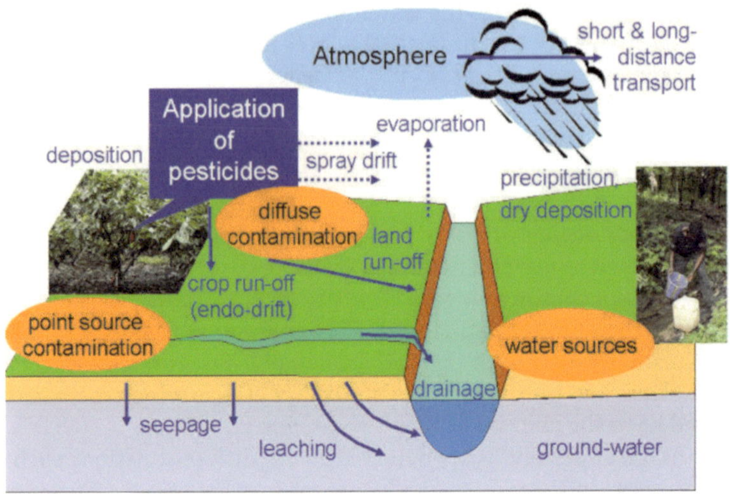

Figure 2.
Pesticides pollution.

concentrations. The land use and climate changes facilitate pesticides entering into an aquatic environment where it was not originally applied. Majorly, pesticides enter into aquatic environment through wind drift during pesticides applications, erosion due to rainfall immediately after pesticides applications to agricultural farmland, migration of living organisms that were affected by the pesticides concentration into the aquatic environment, and through drainage to mention a few.

1.3 Mixture of pesticides effect

In an aquatic environment, different contaminants especially pesticides pollution have been found with different names and concentrations. These pesticides have been found to integrate with each other and form different pesticides with different concentration levels in aquatic habitat [29].

In this century, many agricultural investors (e.g., farmers) have priority for growing high-yielding species of different crops to meet the increasing population demand for food. However, one of the essential phenomena of this subject of discussion is that varieties are that most of those crops are highly prone to different diseases and pests [30], which was evidence to cause about 40–50% of crop loss [31]. As a result of this information, the use of pesticides to protect crops from those pests and diseases, and reduce crop lost, herewith improving the yield quality as well as quantity became necessary [32–34]. In Bangladesh, pesticides were introduced in 1951 but their uses were negligible until the end of the 1960s [35]. An exponential increase in their uses have occurred from 7350 metric tons of active ingredient of pesticides in 1992 to about 45,172 metric tons in 2010 [34].

One of the major reasons for the high rate of pesticides application in some countries such as Bangladesh is due to the adoption of the government policy to increase the control of pests and diseases of crops through a chemical measure process in order to increase overall crop yield and to prevent and control crops losses [34].

More so, about eighty-four (84%) of pesticides significant materials are in the family of 242 trade names of a different group of chemicals namely: organochlorine compound, carbamates, organophosphate, neonicotinoids, pyrethroids, nitro compound, heterocyclic pesticides that are registered in Bangladesh and other parts of the world and are used in the agricultural sector and household applications [35]. However, organochlorine pesticides have been banned in Bangladesh in 1993 [36] and in many countries of the world because of the nature of their toxicities in both human and aquatic environments and they are capable to bioaccumulate and biomagnify in the biological process of feeding such as food chain [37, 38]. Considering other group of pesticides available, the organophosphorus pesticide has gained popularity in the application by farmers in Bangladesh. In addition, more than 35% of their farmers use organophosphorus pesticide to treat varieties of crops for protection reasons [39]."

Pesticides applied on agricultural land have the capability to reach the aquatic environment through several ways which may include but are not limited to leaching of groundwater, spray drift, runoff of surface water, disposal of pesticides containers nearby or inside rivers, cleaning of pesticides equipment in rivers or lakes [40–42]. The indiscriminate use of pesticides and their disposal methods constitute a major threat to the aquatic organism and have led to eco-toxicological risk. More than sixty percent (60%) of animal protein emanated from fish [43]. Since fish serves as the major source of protein in man's food, the indiscriminate use of pesticides in an aquatic environment needs to be reviewed as their toxic effects on fishes are harmful to their normal behavior, physiology and sometimes lead to their deaths [30, 44–54].

Different studies [55, 56] have shown the adverse effects of pesticides on fish species which include but not limited to histopathological alterations such as kidney, gonad, liver, and gill tissue. A study by Dutta and Maxwell (2003) reported that bluegill fish shows histopathological alteration in the ovary namely; cytoplasmic reaction, cytoplasm, and karyoplasmic clumping, necrosis, and thinning of follicular lining exposed to diazinon, atretic oocytes, and adhesion (**Figure 3**).

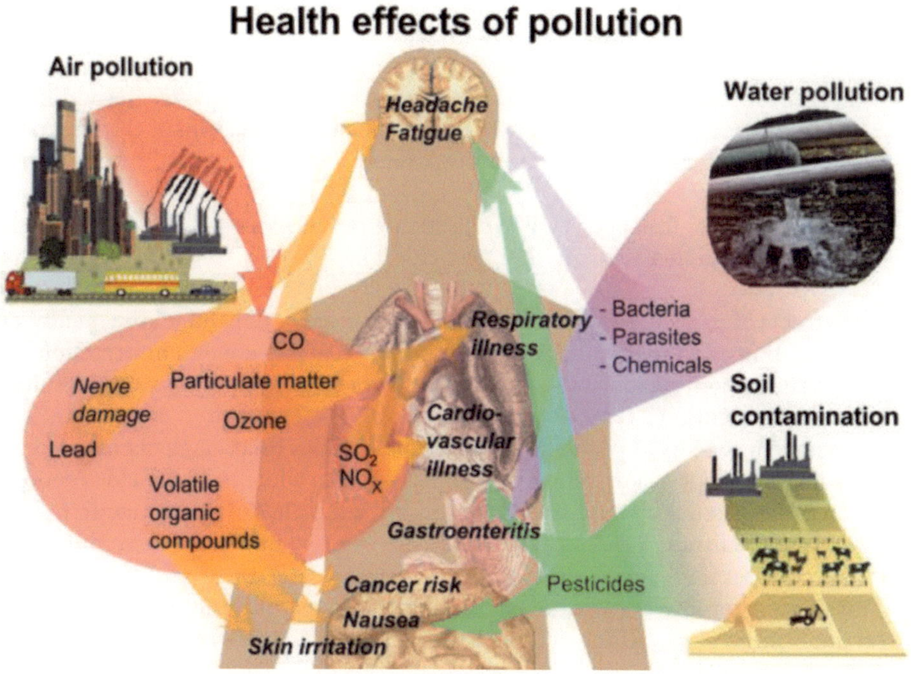

Figure 3.
Mixtures of pesticides.

1.4 Effect of pesticides on aquatic organisms

Manufacturers of pesticides especially those that are licensed by the legislators of their respective countries always ensure their products to be selective and specific to targeted organisms in terms of their toxicity effects. Yet, some of them are totally specific, a few of them are relatively specific while the majority of them are not due to their biodegradable process. Pesticides toxicity most times always depends on the mode of applications. Pesticides applied during wind action and those applied closely to rainfall time always drift from the specific area of application to the non-targeted area. Biodegradation and degradation of pesticides vary from one pesticides compound to another. This is always due to their respective elements that are made up of a particular chemical organic compound. Some are more toxic than their original parent compounds while others are less toxic during the splitting of the individual elements that are made up of the compounds [57].

The susceptibility of humans and animals to pesticides mixtures always produces toxicological interactions effects [58]. Exposure effects of multiple pesticides may be toxic or less toxic or have the same effects with exposure to their individual component. Most times, it is more poisonous to be susceptible to pesticides mixture than exposure to their respective individual element at different time period due to their effect of synergy [58].

Some of the related terminologies used for explaining toxicological interactions of exposure to multiple pesticides include but are not limited to:

i. Effect of synergy: This happens due to the greater effects of two pesticides affecting biological organism at the same time than the sum of their individual effects when they are applied individually or separately.

ii. Effect of antagonist: This exists whenever two pesticides are applied at the same time and each of the pesticides action interferes with the other pesticide that was mixed or combined with.

iii. Effect of additive: It usually takes place or happens whenever the addition of two pesticides has the same toxicity effects with the sum of the effects of the individual pesticides applied separately.

iv. Potentiating: It happens when a pesticide produces a toxic effect anytime is applied together with another pesticide(s).

The effects of exposure to the interactions of pesticides depend on the quality of their application and the prediction of their effects requires enough information regarding the factors responsible for pesticides exposures such as magnitude, time, and toxicity to mention a few.

Antagonism process of pesticides interaction includes functional, chemical, dispositional, and receptor. Functional antagonism occurs when two pesticides counterbalance one another by opposite effects on the same physiological function. Chemical antagonism is a chemical reaction between two pesticides to produce a less toxic product [59].

Pesticides enter into the aquatic environment through different routes such as direct application of pesticides to rivers, seas, lakes, or any other water source to prevent or control weeds, pests, or diseases of crops [59]. Atmospheric nature may truly take place due to the movement of the spray of pesticides from crops surfaces or soil surfaces to the aquatic environment. Yeo et al. [60] noted from their work

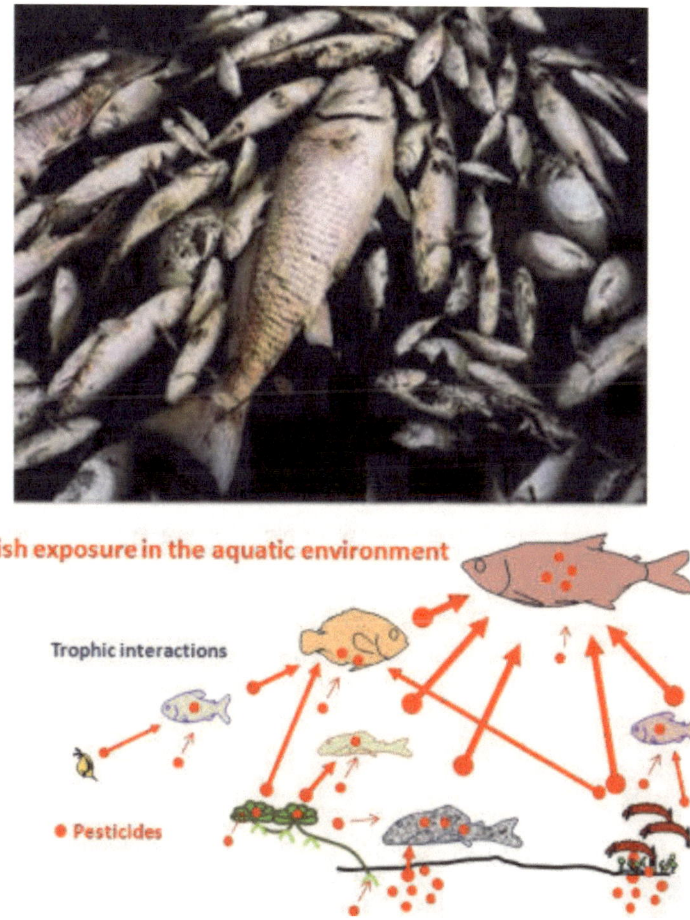

Figure 4.
Effect of pesticides on aquatic organisms.

the effects of atmospheric concentration of pesticides compounds such as organochlorine pesticides which include: Dichlorodiphenyltrichloroethane (DDTs), heptachlor, endosulfan, chlordane, hexachlorocyclohexanes have reported to have minimum and maximum seasonal variation in a rural setting [61]. Furthermore, plantations have been shown to help in reducing surface soil erosion and influence pesticides biodegradation over time, and could lead to the concentration of pesticides in water [62] (**Figure 4**).

1.4.1 Volatilization

Pesticides have the potential to change their state from one form of state of matter into another [63]. The process of the term pesticide volatilization is motivated and catalyzed by pesticides transportation such as soil, water, plant, and surface matrix sorption [64], transportation by air, and diffusion through the boundary layer. The following factors are said to influence the volatilization of pesticides which include physicochemical properties of pesticides [65] including vapor pressure, water solubility, Henry's law constant, adsorption properties, and some environmental factors including soil moisture and soil/air temperature.

1.4.2 Photolysis

Photolysis of pesticides occurs whenever pesticide compounds have adequate energy from light and causes decomposition of the compound molecules through either direct or indirect process [66].

Photolysis is also known to involve in some form of organic pesticides compounds reaction such as carbon bond, isomerization, decarboxylation, and ester cleavage [67]. Different type of photolysis rate of reaction always depends on the absorption spectrum of the pesticides involved (Dureja, 2012; [68]). Direct photolysis is said to take place when pesticide absorbs directly from light energy and result in some form of chemical reaction [69].

Pesticides mixture toxicity is always difficult or complex to predict their toxic effects. Various models that are used to predict a toxic mixture of pesticides effects are always based on their structures activity composition and are always formulated for the complex organic compound of heavy pesticides.

However, different pesticides mixtures are expected to change the behavior of biological species from their combined effects than those effects from the concentration of a single compound.

A lot of studies have been carried out on the pesticides effects in relation to different ecological settings which mainly focused on the restricted compound known as organochlorines [70–72]. This said organic compound is capable of assimilating into crops, animals, and entire ecosystem at high rates of pesticides concentration emission [73–74].

1.4.3 Degradation of pesticides

Several researchers were able to point out that "the degradation of pesticides rates are faster and higher under hot weather. In addition, pesticides solubility is temperature-dependent, that is, the higher the temperature the more soluble it becomes and light intensity was also found to be responsible for the high rate of pesticides degradation [75]." More so, hydrolysis has been found as one of the factors that are responsible for the speedy degradation of pesticides especially when combined with changes in pH and aerobic/anaerobic conditions. The transformation process is mediated by living organisms such as plants, algae, bacteria, or fungi as a result of biodegradation. Complex pesticides like carbon compounds such as synthetic pesticides are used for crops growth as the nutrient substrate is capable of degrading into other compounds or elements [76–78].

Degradation process of the majority of pesticides is mostly affected by bacteria and fungi such as DDT pesticide, chlophyrifos, and cypermethrin. Some factors such as plants, animals (e.g., earthworms), soil moisture, temperature, pH, soil organic matter, carbon source concentration of pesticides greatly influence microbial degradation of pesticides [79–85]. On the whole, most of the microbial activities are found during warm temperatures and in moist soil [9].

During the application of pesticides to a specific area of concern, for instance, crops farm, such pesticides concentration may be watered down by irrigation, runoff water, leaching, rainfall, drainage to the non-targeted environment like groundwater usually pollute aquatic habitant. In addition, pesticides present in the atmosphere, water, soil, or sediment can be degraded via photolysis, hydrolysis, microbial degradation, and biotic uptake [9].

Few workers of the Environmental Protection Agency (EPA) have researched on the possible effects of combined pesticides on aquatic biological species, most especially fish. They reported from their studies that the combined effects of pesticides are often determined as a simulation of their separate effects. Their findings may

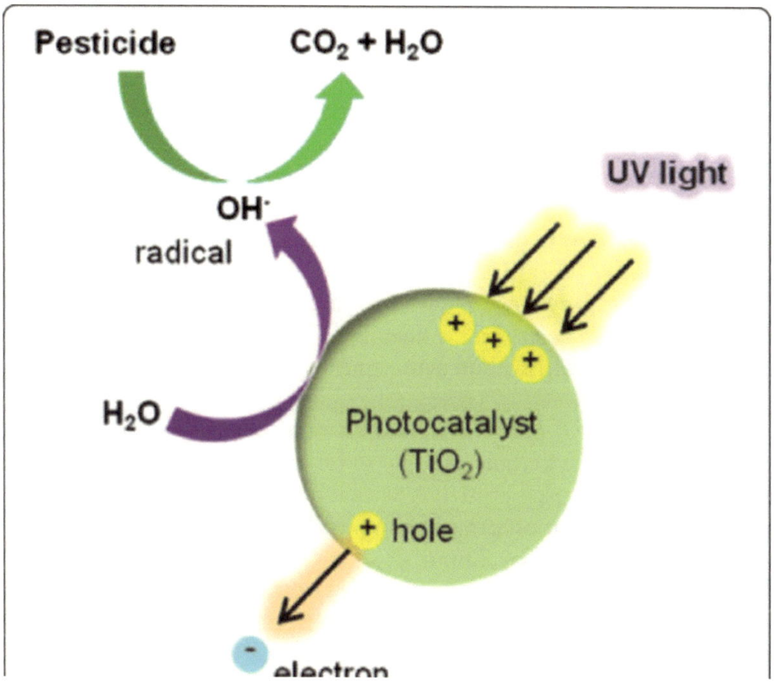

Figure 5.
Degradation of pesticides.

not mirror the combined effects of pesticides as one may expect because pesticides exist in synergistic form. After their finding, much work has been published in respect of the toxicity effects of combined pesticides effects on fishes and other aquatic organisms. They later found out that accurate and comprehensive data are required to model the effects of pesticides mixture on aquatic and any other living organisms' population.

The toxicity of pesticides on aquatic organisms can be measured in a number of ways. The World Health Organisation [86] measures the toxicity of pesticides under the following headings.

i. Toxicity effect of pesticide(s) on microorganisms;

ii. Toxicity effect of pesticide(s) on aquatic organisms;

iii. Toxicity effect of pesticide(s) on terrestrial organisms.

The World Health Organisation only reports the toxicity of individual pesticides in their study on the effects of pesticides on the targeted organism [86]. Most of the harmful pesticides such as LC50, LD50, and the physiochemical pesticides properties were considered (**Figure 5**).

1.5 Effects of pesticides on human health

The use and improper handling of pesticides during their application cause a lot of problems to human health in developing countries. Many studies have pointed out the occupational health hazards of farmers posed by the unsafe use of pesticides. The adverse effects usually observed by farmers after the usage of pesticides on their farms include but are not limited to eye irritation, vomiting, and headache.

About eighty percent (80%) of farmers are aware of the adverse health symptoms poses by pesticides as a result of their intoxication at the time of their applications [42].

The outcomes of their study on the adverse effects of pesticides have similar results with other research works carried out in other developing countries. For instance, Dasgupta et al. [87] reported "negative health effects such as headache, dizziness, eye irritation, vomiting, dermal diseases and gastrointestinal problems after pesticide application in different parts of developing countries." More so, a study by Miah et al. [88] found some similar negative health symptoms. Also, nausea in farmers that grow vegetables in south-east Bangladesh was attributed to pesticides effects. The majority of the negative health effects or signs were reported after the application of pesticides in some countries in South Asia like Pakistan, India, and Nepal [89–91].

However, most of the negative health issues reported by the farmers after pesticides application were due to their inability to follow safety measures on the labels of the pesticides such as spraying pesticides without the use of a nose and mouth mask, covering shoes, and without covering other part of their body effectively [87, 88].

Although, a report issued by Sumon et al. [42] stated that about 82% of farmers normally cover their faces and body with clothes during pesticides application. However, mere covering of the face and body are not enough preventive measures to observe during pesticides application. The more advanced ways of pesticides effects preventive measures during application require farmers or pesticides users to follow the guidelines stipulated by Kabir and Rainis [92]. In their study, they gave some preventive measures required by every pesticides applicator to observe during pesticides application which include: wearing gum-boots, hand gloves, masks during pesticides application and washing of spraying equipment, and taking bath immediately after application.

Furthermore, to reduce or eliminate the dilemma of pesticides risks, it must be the primary responsibility of both governmental and non-governmental organizations to shoulder the responsibility of creating awareness programs in the communities where the pesticides applicators lived. More so, the government must be responsible for the training of farmers to ensure that agricultural workers have good knowledge of the protective guidelines or measures. To achieve this, pesticides industries can bring about product stewardship programs making the industries themselves co-responsible for their products during usage in the field, and the storage. Furthermore, the public sector that is, the government needs to ensure basic training among the agricultural workers that use pesticides for farming to gather knowledge and to build awareness on the safe use and handling of pesticides and subsequently can introduce laws on the use of pesticides and the license for pesticide spraying only for the trained farmers.

In order to create a sound ecological environment, the total removal of harmful pesticides (Chlorfenvinphos, Diuron, Atrazine, Endosulfan, Alachlor, Pentachlorophenol to mention a few) that were banned as reported in their studies as a result of their persistence in the environment should be implemented without any form of compromise to the said assessments [15, 93–95].

Pesticide-related pollution has been causing a persistent and continuous environmental problem [96]. Pesticides pose potential risks to air and water quality, crops, animal health, and human health, to mention but just a few. Significant issues related to pesticide use and application, include over-application, contamination of surface and underground water [97], and drift to unintended targets environment thereby affecting non-target organisms.

Pesticide drift which signifies the amount of pesticide active ingredient that is deflected out of the treated area by the action of air currents has the potential to affect

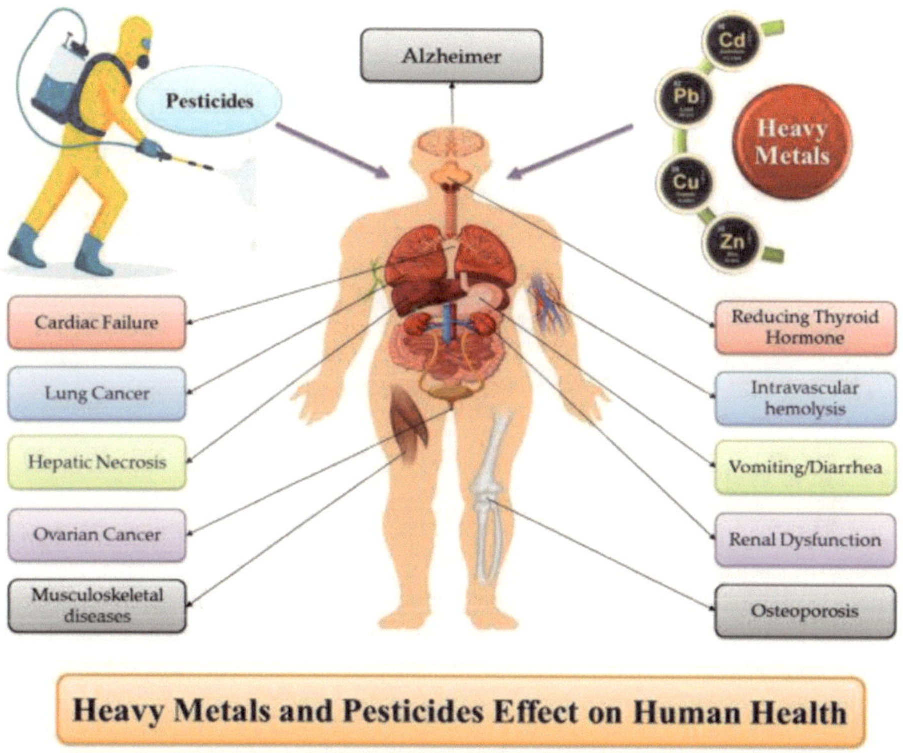

Figure 6.
Effect of pesticides on human health 1.

non-target organisms and the environment [98]. Greater proportions of pesticides concentration were unable to reach the targeted area [99]. However, the presence of pesticides concentration in an unwanted area always leads to the loss of some crops, wildlife population and sometimes cause chaos in natural environment [100]. In the past, the EU has experienced serious concerns about the dispersal of pesticides categorized as persistent organic pollutants (POPs). These POPs are capable of transporting across international boundaries far from their original sources, even to regions where they have never been used or produced (**Figures 6** and 7) [101].

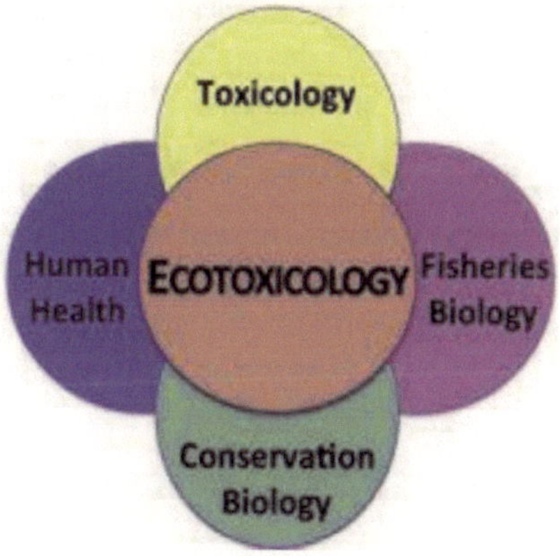

Figure 7.
Effect of pesticides on human health 2.

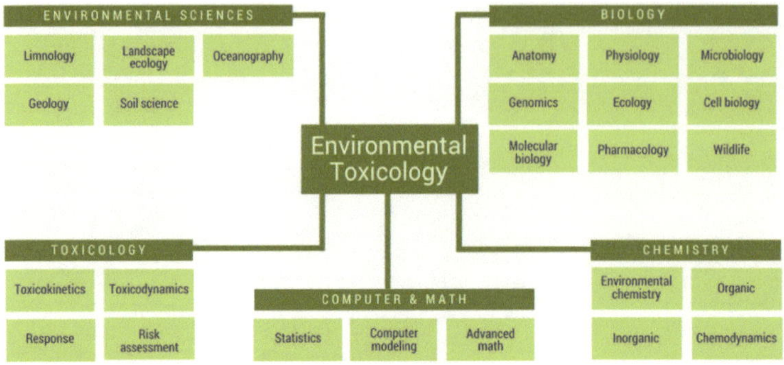

Figure 8.
Field of ecotoxicology.

1.6 Field of ecotoxicological study

The increase in usage of pesticides on agricultural farms has led to ecosystems disturbance. As a result of this incident, the need for research on how pesticides should be applied to increase agricultural productivity without compromising the ecological standard became necessary over the years. The study of the effect of chemicals/pesticides on biological species is known as ecotoxicology. The science of ecotoxicology and environmental toxicology [102] includes:

i. Mathematics

ii. Environmental biology and

iii. Chemistry

Due to the involvement of different disciplines, they are different aims of ecotoxicology research [102] which include the following:

i. **As a scientific discipline:** It involves understanding the fundamentals of the interactions between chemicals and biological systems on different levels of complexity and curiosity-driven.

ii. **As a technological field:** It connotes the development of bio-assays on various levels of complexity, for different compounds and different environmental compartments; development of models of distribution, fate and effects of chemicals in the biosphere; and chemical and analytic techniques.

iii. **As an input provider for environmental regulation:** It provides the scientific basis for environmental quality standards that ensure ecosystem services; sustainable development and ecosystem health; provides clean-up goals and strategies; and provide options (**Figure 8**).

DOI: http://dx.doi.org/10.5772/intechopen.104626

Author details

Kenneth Ojotogba Achema[1*] and Charity Jumai Alhassan[2]

1 Department of Mathematics and Statistics, Confluence University of Science and Technology, Osara, Kogi State, Nigeria

2 Department of Mathematics and Computer Science, Edo State University, Uzairue, Edo State, Nigeria

*Address all correspondence to: adonaipoly34@gmail.com

References

[1] Tongo S, Lawrence E, Kingsley A. Levels, distribution and characterization of polyclclic aromatic hydrocarbon. Journal of Environmental Chemical Engineering. 2016;**5**:504-512

[2] Achema KO, Okuonghae D, Tongo I. Dual-level toxicity assessment of biodegradable pesticides to aquatic species. Ecological Complexity. 2021;**45**: 1-15

[3] Noyees PD, McElwee MK, Miller HD. The toxicology of climate change: Environmental contaminants in a worming world. Environmental International. 2009;**35**:971-986

[4] United State Geological Survey Fact Sheet. Pesticides in stream sediment and aquatic biota. 2019. Available from: http://water.usgs.gov/nawqa

[5] Yichen H, Lijuan X, Feiyu L, Mengshi X, Derong L, Xraomei L, et al. Microbial degradation of pesticide residue and an emphasis on the degradation of cypermethrin and 3-phenoxy benzoic acid: A review. Molecules. 2018;**23**:1-23

[6] Reda F, Bakr A, Ahmad M, Kamel S, Sheba A, Doaa R. A mathematical model for estimating the LC50 or LD50 among an in seet life cycle. Egyptian Academic Journal of Biological Sciences. 2010;**32**: 75-81

[7] Dugan R. Biochemical Ecology of Water Pollution. New York: Plenum Publishing Co. Ltd.; 1972

[8] McGarrigle M, Lucy J, O'Cinneide M. Water Quality in Ireland 2007–2009. Wexford: Environmental Protection Agency; 2010

[9] Gavrilescu M. Review: Fate of pesticides in the environment and its bioremediation. Engineering in Life Sciences. 2005;**5**(6):497-526

[10] European Environment Agency. Europe's Environment: An Assessment of Assessments. Luxembourg: Publications Office of the European Union: EEA; 2011a

[11] Mathur V, Bhatnagar P, Sharma RG, Acharya V, Sexana R. Breast cancer incidence and exposure to pesticides among women originating from Jaipur. Environment International. 2002;**28**(5): 331-336

[12] McKinlay R, Plant JA, Bell JNB, Voulvoulis N. Endocrine disrupting pesticides: Implications for risk assessment. Environment International. 2008;**34**(2):168-183

[13] Gibs J, Stackelberg PE, Furlong ET, Meyer M, Zaugg SD, Lippincott RL. Persistence of pharmaceuticals and other organic compounds in chlorinated drinking water as a function of time. Science of the Total Environment. 2007; **373**(1):240-249

[14] European Environment Agency. Hazardous Substances in Europe's Fresh and Marine Waters—An Overview. Luxembourg: European Environment Agency; 2011b

[15] European Commission. Water Framework Directive 2000/60/EC. 2000

[16] EC. Community Strategy for Endocrine Disruptors: A Range of Substances Suspected of Interfering with the Hormone Systems of Humans and Wildlife. Brussels: Commission of the European Communities; 1999

[17] EC. Council Decision 2006/507/EC concerning the conclusion, on behalf of the European Community, of the Stockholm Convention on Persistent Organic Pollutants. Decision ed. Brussels: EU. Official Journal of the

European Communities L. 2004b; 209:1-2

[18] Ahmed N, Garnett ST. Sustainability of freshwater prawn farming in rice fields in Southwest Bangladesh. Journal of Sustainable Agriculture. 2010;**34**(6):659-679

[19] Devipriya S, Yesodharan S. Photocatalytic degradation of pesticide contaminants in water. Solar Energy Materials and Solar Cells. 2005;**86**(3): 309-348

[20] Carpenter KD, Kuivila KM, Hladik ML, Haluska T, Cole MB. Storm-event transport of urban-use pesticides to streams likely impairs invertebrate assemblages. Environmental Monitoring and Assessment. 2016;**188**:345. DOI: 10.1007/s10661-016-5215-5

[21] Vörösmarty CJ, McIntyre P, Gessner M, Dudgeon D, Prusevich A, Green P, et al. Global threats to human water security and river biodiversity. Nature. 2010;**467**(7315):555-561

[22] Rockström J, Falkenmark M, Karlberg L, Hoff H, Rost S. The potential of green water for increasing resilience to global change. Water Resources Research. 2009;**45**:W00A12. DOI: 10.1029/2007WR006767

[23] Steffen W, Richardson K, Rockström J. Planetary boundaries: Guiding human development on a changing planet. Science. 2015;**347**:7-36

[24] Hartung T, Rovida C. Chemical regulators have overreached. Nature. 2009;**460**:1080-1081

[25] Schwarzenbach RP, Egli T, Hofstetter TB, von Gunten U, Wehrli B. Global water pollution and human health. Annual Review of Environment and Resources. 2010;**35**:109-136

[26] Ritter L, Solomon KR, Forget J, Sterneroff M. A Review of Selected Persistent Organic Pollutants, Aldrin, Chlordane, DDT, Dieldrine, Dioxin and Furans, Endrin, Heptachlor, Hexachlorobenzene, Mirex, Polychlorinated Biphenyls, and Toxaphene. Gevena: WHO; 1995

[27] Ashauer R, Boxall A, Brown C. Predicting effects on aquatic organisms from fluctuating or pulsed exposure to pesticides. Environmental Toxicology and Chemistry. 2006;**25**:1899-1912

[28] Brock TCM, Lahre J, Van Den Brick PJ. Ecological Risks of Pesticides in Freshwater Ecosystem Part 1: Herbicides. Green World Research: Alterra; 2006

[29] Altenburger R, Ait-Aissa S, Antcak P, Backhaus T, Barcel D, Seiler TB, et al. Future water quality monitoring: Adapting tools to deal with mixtures of pollutants in water resource management. Science of the Total Environment. 2015;**512–513**:540-551

[30] Ali MH, Sumon KA, Sultana M, Rashid H. Toxicity of cypermethrin on the embryo and larvae of Gangetic mystus, Mystus cavasius. Environmental Science and Pollution Research. 2018;**25**:3193-3199

[31] Uddin MA, Saha M, Chowdhury M, Rahman M. Pesticide residues in some selected pond water samples of Meherpur Region of Bangladesh. Journal of The Asiatic Society of Bangladesh (Science). 2013;**39**(1):77-82

[32] Ansara-Ross TM, Wepener V, Van den Brink PJ, Ross MJ. Pesticides in South African fresh waters. African Journal of Aquatic Science. 2012;**37**(1): 1-16

[33] Peluso F, Dubny S, Othax N, Castelain JG. Environmental risk of pesticides. Applying the Del Azul Pest risk model to freshwater of an agricultural area of Argentina. Human and Ecological Risk Assessment: An

International Journal. 2014;**20**(5): 1177-1199

[34] Rahman S. Pesticide consumption and productivity and the potential of IPM in Bangladesh. Science of The Total Environment. 2013;**445**:48-56

[35] Ara AG, Haque W, Hasanuzzaman M. Detection of organochlorine and organophosphorus pesticides residues in water samples of Taragong thana in Rangpur district in Bangladesh. Research Journal of Environmental and Earth Sciences. 2014;**6**(2):85-89

[36] Matin MA, Malek MA, Amin MR, Rahman S, Khatoon J, Rahman M, et al. Organochlorine insecticide residues in surface and underground water from different regions of Bangladesh. Agriculture, Ecosystems and Environment. 1998;**69**(1):11-15

[37] Sun Q, Zhu L, Dong M. Risk assessment of organic pesticides pollution in surface water of Hangzhou. Environmental Monitoring and Assessment. 2006;**117**(1):377-385

[38] Teklu BM, Adriaanse PI, Van den Brink PJ. Monitoring and risk assessment of pesticides in irrigation systems in Debra Zeit, Ethiopia. Chemosphere. 2016;**161**:280-291

[39] Chowdhury MAZ, Fakhruddin ANM, Islam MN, Moniruzzaman M, Gan SH, Alam MK. Detection of the residues of nineteen pesticides in fresh vegetable samples using gas chromatography: Mass spectrometry. Food Control. 2012;**34**:457-465

[40] Hossain MS, Chowdhury MAZ, Pramanik MK, Rahman MA, Fakhruddin ANM, Alam MK. Determination of selected pesticides in water samples adjacent to agricultural fields and removal of organophosphorus insecticide chlorpyrifos using soil bacterial isolates. Applied Water Science. 2015;**5**(2):171-179

[41] Sankararamakrishnan N, Sharma AK, Sanghi R. Organochlorine and organophosphorous pesticide residues in ground water and surface waters of Kanpur, Uttar Pradesh, India. Environmental International. 2005; **31**(1):113-120

[42] Sumon KA, Rico A, Ter Horst MM, Van den Brink PJ, Haque MM, Rashid H. Risk assessment of pesticides used in rice-prawn concurrent systems in Bangladesh. Science of the Total Environment. 2016;**568**:498-506

[43] FRSS. Yearbook of Fisheries Statistics of Bangladesh. Vol. 33. Bangladesh: Fisheries Resources Survey System (FRSS), Department of Fisheries; 2017. p. 124

[44] Daam MA, Crum SJ, Van den Brink PJ, Nogueira AJ. Fate and effects of the insecticide chlorpyrifos in outdoor plankton-dominated microcosms in Thailand. Environmental Toxicology and Chemistry. 2008;**27**(12):2530-2538

[45] Kumar KS, Dahms HU, Lee JS, Kim HC, Lee WC, Shin KH. Algal photosynthetic responses to toxic metals and herbicides assessed by chlorophyll a fluorescence. Ecotoxicology and Environmental Safety. 2014;**104**:51-71

[46] Liu SS, Wang CL, Zhang J, Zhu XW, Li WY. Combined toxicity of pesticide mixtures on green algae and photobacteria. Ecotoxicology and Environmental Safety. 2013;**95**:98-103

[47] Malev O, Klobučar RS, Fabbretti E, Trebše P. Comparative toxicity of imidacloprid and its transformation product 6-chloronicotinic acid to non-target aquatic organisms: Microalgae *Desmodesmus subspicatus* and amphipod *Gammarus fossarum*. Pesticide Biochemistry and Physiology. 2012; **104**(3):178-186

[48] Maltby L, Blake N, Brock T, Van den Brink PJ. Insecticide species

sensitivity distributions: Importance of test species selection and relevance to aquatic ecosystems. Environmental Toxicology and Chemistry. 2005;**24**(2): 379-388

[49] Palma P, Palma VL, Fernandes RM, Bohn A, Soares AMVM, Barbosa IR. Embryo-toxic effects of environmental concentrations of chlorpyrifos on the crustacean *Daphnia magna*. Ecotoxicology and Environmental Safety. 2009;**72**(6):1714-1718

[50] Roessink I, Merga LB, Zweers HJ, Van den Brink PJ. The neonicotinoid imidacloprid shows high chronic toxicity to mayfly nymphs. Environmental Toxicology and Chemistry. 2013;**32**(5):1096-1100

[51] Rubach MN, Crum SJ, Van den Brink PJ. Variability in the dynamics of mortality and immobility responses of freshwater arthropods exposed to chlorpyrifos. Archives of Environmental Contamination and Toxicology. 2011; **60**(4):708-721

[52] Sumon KA, Saha S, van den Brink PJ, Peeters ET, Bosma RH, Rashid H. Acute toxicity of chlorpyrifos to embryo and larvae of banded gourami *Trichogaster fasciata*. Journal of Environmental Science and Health, Part B. 2017;**52**(2):92-98

[53] Tillitt DE, Papoulias DM, Whyte JJ, Richter CA. Atrazine reduces reproduction in fathead minnow (*Pimephales promelas*). Aquatic Toxicology. 2010;**99**(2):149-159

[54] Van den Brink PJ, Blake N, Brock TCM, Maltby L. Predictive value of species sensitivity distributions for effects of herbicides in freshwater ecosystems. Human and Ecological Risk Assessment. 2016;**12**:645-674

[55] Deneer JW. Toxicity of mixtures of pesticides in aquatic systems. Pest Management Science. 2000;**56**:516-520

[56] Rand GM, Wells PG, McCarty LS. Introduction to aquatic toxicology. In: Rand GM, editor. Fundamentals of Aquatic Toxicology: Effects, Environmental Fate, and Risk Assessment. North Palm Beach: Taylor & Francis; 1995. pp. 3-67

[57] Hayes WJ, Laws ER. Handbook of Pesticide Toxicology. New York: Academic Press Inc.; 1990

[58] Alabaster JS, Lioyd F. Water Quality Criteria for Fresh Water Fish. London: Butterworths; 1980

[59] Carter A. How pesticides get into water—And proposed reduction measures. Pesticide Outlook. European Crop Protection Association. 2000;**11**: 149-156. DOI: 10.1039/b006243j

[60] Yeo HG, Choi M, Sunwoo Y. Seasonal variations in atmospheric concentrations of organochlorine pesticides in urban and rural areas of Korea. Atmospheric Environment. 2004;**38**:4779-4788

[61] Matocha MA, Krutz LJ, Reddy KN, Senseman SA, Locke MA, Steinriede RW, et al. Foliar washoff potential and simulated surface runoff losses of trifloxysulfuron in cotton. Journal of Agricultural and Food Chemistry. 2006; **54**:5498-5502

[62] Riise G, Lundekvam H, Wu QL, Haugen LE, Mulder J. Loss of pesticides from agricultural fields in SE Norway: Runoff through surface and drainage water. Environmental Geochemistry and Health. 2004;**26**:269-276

[63] Racke KD, Skidmore M, Hamilton DJ, Unsworth JB, Miyamoto J, Cohen SZ. Pesticide fate in tropical soils. Pesticide Science. 1999;**55**:219-220

[64] Van Wesenbeeck I, Driver J, Ross J. Relationship between the evaporation rate and vapor pressure of moderately and highly volatile chemicals. Bulletin of

Environmental Contamination and Toxicology. 2008;**80**:315-318

[65] Guth JA, Reischmann FJ, Allen R, Arnold D, Hassink J, Leake CR, et al. Volatilisation of crop protection chemicals from crop and soil surfaces under controlled conditions-prediction of volatile losses from physico-chemical properties. Chemosphere. 2004;**57**:871-887

[66] Hemond HF, Fechner-Levy EJ. Chemical Fate and Transport in the Environment. 2nd ed. Cambridge, MA: Academic Press; 1999

[67] Katagi T. Photo-degradation of pesticides on plant and soil surfaces. In: Ware GW, editor. Reviews of Environmental Contamination and Toxicology. New York, NY: Springer; 2004

[68] Bhattacharjee S, Fakhruddin ANM, Chowdhury MAZ, Rahman MA, Alam MK. Monitoring of selected pesticides residue levels in water samples of paddy fields and removal of cypermethrin and chlorpyrifos residues from water using rice bran. Bulletin of Environmental Contamination and Toxicology. 2012; **89**(2):348-353

[69] Burrows HD, Canle LM, Santaballa JA, Steenken S. Reaction pathways and mechanisms of photodegradation of pesticides. Journal of Photochemistry and Photobiology B: Biology. 2002;**67**: 71-108

[70] Plianbangchang P, Jetiyanon K, Wittaya-areekul S. Pesticide use patterns among small-scale farmers: A case study from Phitsanulok, Thailand. Southeast Asian Journal of Tropical Medicine and Public Health. 2009;**40**: 401-410

[71] Poolpak T, Pokethitiyook P, Kruatrachue M, Arjarasirikoon U, Thanwaniwat N. Residue analysis of organochlorine pesticides in the Mae Klong River of Central Thailand. Journal of Hazardous Materials. 2008;**156**: 230-239

[72] Samoh ANH, Ibrahim MS. Organochlorine pesticide residues in the major rivers of Southern Thailand. Malaysian Journal of Analytical Sciences. 2008;**1**:1-9

[73] Jaipieam S, Visuthismajarn P, Sutheravut P, Siriwong W, Thoumsang S, Borjan M, et al. Organophosphate pesticide residues in drinking water from Artesian Wells and health risk assessment of agricultural communities, Thailand. Human and Ecological Risk Assessment. 2009;**15**:1304-1316

[74] Panuwet P, Siriwong W, Prapamontol T, Ryan PB, Fiedler N, Robson MG, et al. Agricultural pesticide management in Thailand: Status and population health risk. Environmental Science & Policy. 2012;**17**:72-81

[75] Klein W. Mobility of environmental chemicals, including abiotic degradation. In: Bordeau P, Haines JA, Klein W, Krishna Murti CR, editors. Ecotoxicology and Climate SCOPE 38. Chichester, UK: John Wiley & Sons Ltd.; 1989. pp. 65-78

[76] Hussain S, Siddique T, Arshad M, Saleem M. Bioremediation and phytoremediation of pesticides: Recent advances. Critical Reviews in Environmental Science and Technology. 2009;**39**:843-907

[77] Murthy HMR, Manonmani HK. Aerobic degradation of technical hexachlorocyclohexane by a defined microbial consortium. Journal of Hazardous Materials. 2007;**149**:18-25

[78] Navarro-Ortega A, Tauler R, Lacorte S, Barcelö D. Occurrence and transport of PAHs, pesticides and alkylphenols in sediment samples along the Ebro River Basin. Journal of Hydrology. 2010;**383**(1–2):5-17

[79] Benoit P, Perceval J, Stenrod M, Moni C, Eklo OM, Barriuso E, et al. Availability and biodegradation of metribuzin in alluvial soils as affected by temperature and soil properties. European Weed Research Society. 2007; **47**:517-526

[80] Briceno G, Palma G, Duran N. Influence of organic amendment on the biodegradation and movement of pesticides. Critical Reviews in Environmental Science and Technology. 2012;**37**:233-271

[81] Caceres TC, Megharaj M, Naidu R. Degradation of fenamiphos in soils collected from different geographical regions: The influence of soil properties and climatic conditions. Journal of Environmental Science and Health, Part B, Pesticides, Food Contaminants, and Agricultural Wastes. 2008;**43**:314-322

[82] Chiu TC, Yen JH, Hsieh YN, Wang YS. Reductive transformation of dieldrin under anaerobic sediment culture. Chemosphere. 2005;**60**:1182-1189

[83] Druzina B, Stegu M. Degradation study of selected organophosphorus insecticides in natural waters. International Journal of Environmental Analytical Chemistry. 2007;**87**:1079-1093

[84] Rani R, Juwarkar A. Biodegradation of phorate in soil and rhizosphere of *Brassica juncea* (L.) (Indian Mustard) by a microbial consortium. International Biodeterioration and Biodegradation. 2012;**71**:36-42

[85] Xie L, Flippin JL, Deighton N, Funk DH, Dickey DA, Buch walter, D.B. Mercury(II) bioaccumulation and antioxidant physiology in four aquatic insects. Environmental Science & Technology. 2011;**43**:934-940

[86] Sheffer M. Environmental Helath Criteria No. 240 Principles and Methods for the Risk Assessment of Chemicals in Food. Geneva: World Health Organisation; 2009

[87] Dasgupta S, Meisner C, Huq M. A pinch or a pint? Evidence of pesticide overuse in Bangladesh. Journal of Agricultural Economics. 2007;**58**(1): 91-114

[88] Miah SJ, Hoque A, Paul A, Rahman A. Unsafe use of pesticide and its impact on health of farmers: A case study in Burichong upazila, Bangladesh. IOSR Journal of Environmental Science, Toxicology and Food Technology. 2014; **8**:57-67

[89] Chitra GA, Muraleedharan VR, Swaminathan T, Veeraraghavan D. Use of pesticides and its impact on health of farmers in South India. International Journal of Occupational and Environmental Health. 2006;**12**(3): 228-233

[90] Khan B, Lee LS, Sassman SA. Degradation of synthetic androgens 17α- and 17β-trenbolone and trendione in agricultural soils. Environmental Science and Technology. 2010;**42**(10): 3570-3574

[91] Mohanty MK, Behera BK, Jena SK, Srikanth S, Mogane C, Samal S, et al. Knowledge attitude and practice of pesticide use among agricultural workers in Puducherry, South India. Journal of Forensic and Legal Medicine. 2013;**20**(8):1028-1031

[92] Kabir MH, Rainis R. Farmers' perception on the adverse effects of pesticides on environment: The case of Bangladesh. International Journal of Sustainable Agricultural Research. 2012; **4**(2):25-32

[93] EC. Commission Decision 2004/247/EC concerning the non-inclusion of simazine in Annex I to Council Directive 91/414/EEC and the withdrawal of authorisations for plant protection

products containing this active substance. Decision ed. Brussels: EU. Official Journal of the European Communities L. 2004a;78:50-52

[94] EC. Commission Decision 2005/864/EC concerning the non-inclusion of endosulfan in Annex I to Council Directive 91/414/EEC and the withdrawal of authorisations for plant protection products containing this active substance. Decision ed. Brussels: EU. Official Journal of the European Communities L. 2005;317:25-27

[95] European Commission. Directive on Environmental Quality Standards 2008/105/EC. 2008

[96] De Schampheleire M, Spanoghe P, Brusselman E, Sonck S. Risk assessment of pesticide spray drift damage in Belgium. Crop Protection. 2007;**26**(4): 602-611

[97] Reichenberger S, Bach M, Skitschak A, Frede HG. Mitigation strategies to reduce pesticide inputs into groundand surface water and their effectiveness; A review. Science of the Total Environment. 2007;**384**:1-35

[98] Oerke EC, Dehne HW. Safeguarding production losses in major crops and the role of crop protection. Crop Protection. 2004;**23**:275-285

[99] Reimer AP, Prokopy LS. Environmental attitudes and drift reduction behaviour among commercial pesticide applicators in a US agricultural landscape. Journal of Environmental Management. 2012;**113**:361-369

[100] Pimentel D, Acquay H, Biltonen M, Rice P, Silva M, Nelson J, et al. Environmental and economic costs of pesticide use. Bioscience. 1992;**42**:750-760

[101] UN Economic and Social Council. The 1998 Protocol on Porganic Pollutants, Including the Amendments. Geneva: UN; 2009

[102] Thomas B. Ecotoxicology and Environmental Toxicology: An Introduction. Sweden: University of Gothenburg; 2012

Chapter 14

The Effect of Neonicotinoid Insecticides on the Structure and Stability of Bio-Macromolecules

Valéria Verebová and Jana Staničová

Abstract

Insecticides are among the most widely used pesticides in the world. They are preparations of chemical and biological origin used to control insects, which means its killing or preventing its destructive activity. Majority are used in forestry, agriculture, and households. Neonicotinoids represent the class of insecticides that is most frequently used in the world and replaced by more dangerous pyrethroids, organophosphates, and carbamates. In recent years, the focus has been mainly on the ecological and environmental risks caused by the use of neonicotinoids. These insecticides pose a very high risk to bees and also to soil and aquatic organisms. It is therefore highly topical to address the impact of neonicotinoids on biological systems on individual bio-macromolecules (DNA and serum albumins). Monitoring the impact of neonicotinoids on the structure and stability of biological macromolecules may contribute to reducing the use of these insecticides, as well as to considering and adjusting the tolerances of insecticides and their residues in food.

Keywords: insecticide, neonicotinoids, DNA, serum albumin, structure, stability

1. Introduction

The most widely applied pesticides in common practice are insecticides. They are preparations of chemical and biological origin used to control insects, which means their killing or preventing its destructive activity (**Table 1**).

Insecticides are mostly used in forestry, agriculture, and households. The use of insecticides has increased agricultural productivity, quantity, quality, and prolonged the lifetime of food and fodder plants. Insecticides have also revolutionized the fight against endemic diseases in developing countries. Unfortunately, these compounds pose the risk to humans and animals due to the presence of their residues in the food [4]. Insecticides are divided according to their mode of action into the ovicides (destroy eggs), larvicides (destroy larvae), and imagocides (destroy adult insects) [5]. The effect of insecticides is either immediate or slow acting and the insects die after a longer period of time [6].

1.1 Neonicotinoids

Neonicotinoids are a class of insecticides that belong to the most widely used in the world [7] because they allow for a rational approach to agricultural pest control [8, 9].

Insecticides class	Mechanism of action on pests	
Organochlorines	Dichlorodiphenyl-trichloroethane (DDT)	Destroying the delicate balance of Na^+ and K^+ ions in the axons of the neuron, which prevents the normal transmission of nerve impulses
	Hexachloro-cyclohexane (HCH)	Similar to DDT, only the effect is faster
	Cyclodienes	Act on the gamma-aminobutyric (GABA) acid receptor, which causes increased permeability of neurons to Cl^- ions
	Polychloroterpenes	Similar to cyclodienes
Organophosphates	Inhibition of nervous system enzymes, resulting in the accumulation of acetylcholine at neuron–neuron or neuron-muscle junctions	
Carbamates	Inhibition of aliesterase, which promotes hydrolysis of aliphatic ester bonds	
Formamidines	Inhibition of monoamine oxidase, which is responsible for the degradation of the neurotransmitters norepinephrine and serotonin	
Dinitrophenols	Inhibition of oxidative phosphorylation	
Pyrethroids	Influencing the peripheral and central nervous system	
Nicotinoids	Postsynaptic acetylcholine receptor blockade	
Neonicotinoids	Strong bond to nicotinic acetylcholine receptors in the central nervous system	

Table 1.
The major classes of insecticides and mechanism of action on pests [1–3].

Such insecticides tend to be referred to as "bio-rational" [10]. They replaced by more dangerous pyrethroids, organophosphates, and carbamates [11]. Discovery of imidacloprid and its subsequent introduction to the market in 1991 ushered in the era of neonicotinoids [12], which in 2014 accounted for more than 25% of the world's insecticide market [13]. We know the neonicotinoids of four generations. The first generation consists of chloropyridyls, which include imidacloprid, acetamiprid, thiacloprid, nitenpyram, cycloxaprid and paichongding. These are divided into three classes according to pharmacophore groups: N-nitroimine, N-cyanoimine, and nitromethylene [14]. Chlorothiazoles, as thiamethoxam, imidaclothiz and clothianidin, form the second-generation of neonicotinoids. Furanyls (e.g., dinotefuran) belong to the third generation. Fourth generation is made up of sulfoximines, such as sulfoxaflor [7, 15–18]. Neonicotinoids are used on all types of crops in temperate and tropical regions as well as in forestry, gardens, urban parks, and as veterinary control products ectoparasites in domestic animals [19]. The flexibility of their use is due to their system properties, which allow their application in the form of direct sprays on crops, soil granules or seed coating [20]. They have unique biological and chemical properties such as broad-spectrum insecticidal activity, low application rates and mode of action [21]. Due to their octanol/water partition coefficient and dissociation constant (pKa) values, they readily enter plant tissues and translocate to all parts of the plant regardless of the method of application. This will automatically become toxic to any insect and potentially other organisms feeding on plants [22, 23]. However, some of their characteristics increase their negative impact on the environment and non-target organisms [23, 24]. Recent studies show that neonicotinoids are already ubiquitous in the environment because of their versatile use, high mobility and relatively long half-life in water and soil [25, 26]. In the USA, for example, their presence was confirmed in twelve out of nineteen different fruits monitored and vegetables, eleven of which contained more than two neonicotinoids. Value of thiamethoxam even exceeded the maximum residue limit. The most frequent occurrence was paradoxically found in foods primarily intended for infants and toddlers (prevalence ranging 6–31%) [27]. Neonicotinoid contamination has also been demonstrated in drinking water [28, 29], vegetables and fruit [30], milk [31], and in honey [32].

1.1.1 Mechanism of neonicotinoid action

Neonicotinoid's mode of action is by blocking the nicotinacetylcholine receptor (nAChR) leading to paralysis and death of the pests [33, 34]. NAChR is an ion channel responsible for immediate neurotransmission and belongs to the group of neurotransmitter ion channels along with gamma-aminobutyric acid, glycine, 5-HT3 and serotonin receptors. It consists of ten α, four β, γ, and δ subunits, which combine to form three basic types of receptors (muscle, neuronal, and ganglionic) with different structures. Different combinations of subunits result in differences in sensitivity to acetylcholine and other pharmacological systems. The most potent nAChR agonist is the nicotinic derivative epibatidine [35].

Neonicotinoids contain a negatively charged, electronegative cyano or nitro group that reacts with the positively charged nAChR site of the insect. In vertebrates, this interaction is blocked by the protonation of nitrogen in their organism [35].

1.1.2 Neonicotinoid toxicity

In recent years, the focus has been mainly on ecological and environmental risks caused by the use of neonicotinoids.

N-nitromine neonicotinoids show a very high risk to wild bees and honey bees [36]. Exposure to already low doses of insecticides occurs sublethal effects such as reduced immunity, disorientation and behavioral changes and reproduction of bees [37–40]. Based on these facts, the use of imidacloprid, thiamethoxam, clothianidin have been completely banned in the European Union since May 2018 [41].

N-cyanoimine insecticides in turn pose a high risk to soil and aquatic organisms [36, 42, 43]. Because of their long persistence in soil and high water solubility, they tend to pass into groundwater, surrounding rivers [44], lakes, and seas [45, 46]. Studies focusing on neonicotinoid content in watercourses have shown concentrations ranging from 0 to 380 ng/L depending on the region [47–49]. Because of adverse effect of neonicotinoids on aquatic organisms such as in food intake [50, 51], changes in swimming and nesting [52, 53], growth inhibition [54], changes in reproduction [53] and acute and chronic mortality [53, 55, 56], this contamination poses a major risk to aquatic ecosystem and for the supply water for its consumption [34, 57].

Several other studies of the adverse impact of neonicotinoids on non target organisms have demonstrated the development of immunosuppression in birds, bats, fish and amphibians [58]. Neonicotinoids can also adversely affect mammalian nAChRs, leading to neurobehavioral deficits and increased glial fibrillary acidic protein expression in the hippocampus [59–61], further affect the reproductive cycle, liver function and have genotoxic and neurological effects [11, 61–65].

Neonicotinoids are capable of disrupting the endocrine system. The study focused on exposure to thiacloprid in rats showed an increase in triiodothyronine and thyroxine hormones [66]. They are further associated with the development of oxidative stress [67], which in various cases leads to changes in the levels of ovarian and antioxidant hormones [64], but also to increased germ cell apoptosis and DNA fragmentation in rat testes [68].

Next, we will discuss the most widely used representatives of first-generation neonicotinoid insecticides belonging to the group of N-cyanoimines (thiacloprid, acetamiprid) and N-nitroimines (imidacloprid).

1.1.3 Thiacloprid

Thiacloprid ((Z)-3-(6-chloro-3-pyridylmethyl)-1,3-thiazolidine-2-ylidenecyanamide) is an insecticide belonging to the class of neonicotinoids (**Figure 1**), which is used for protection of vegetables, orchards, tea, maize and oilseed rape seeds [69].

Due to quite good solubility of thiacloprid in water and its low potential in groundwater, contamination of mainly surface water bodies of water, but its presence has also been detected in drinking water [28]. Thiacloprid occurs as a white to yellow powder [70] and it is polar, slightly soluble in water (0.185 g/L) and in organic solvents: dimethyl sulfoxide (150 g/L), acetone (64 g/L), ethyl acetate (9.4 g/L), acetonitrile (52 g/L). Partition coefficient octanol/water has a logP = 1.26 (pH = 7, 20°C), indicating poor solubility in fats, low absorption and distribution in the body and its pH is in the range of 4–9 stable [71].

In general, thiacloprid shows higher toxicity to aquatic organisms compared to other neonicotinoids studied [35]. This toxicity is probably related to its resistance to degradation in water at neutral and acidic pH values. WHO classifies it as moderately hazardous (Class II) [72] while the Environmental Protection Agency in the United States (US EPA) characterizes it as a potential carcinogen, based on the occurrence of thyroid tumors in male rats and uterine and ovarian tumors in rats, and mice [73]. Thiacloprid is metabolized and excreted in the urine within 24 hours after oral administration. The target organ is primarily the liver; a toxic effect on the liver has also been observed in dog prostate [41, 74]. Other studies have shown that it causes fetal 25 resorption, skeletal retardation, changes in motor activity in rats, thyroid adenomas, and uterine adenocarcinogens in mice [74]. In fish, after exposure to TCLs, there is growth retardation, delayed fetal development, and changes in antioxidant enzyme levels [75]. Moreover observed were reduced cell proliferation associated with higher levels of chromosomal aberrations in bovine lymphocytes [76] and genotoxic and cytotoxic effects on human [77] and bovine lymphocytes [78].

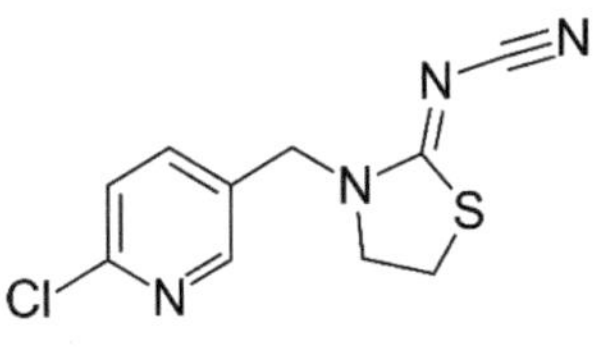

Figure 1.
Structural formula of thiacloprid.

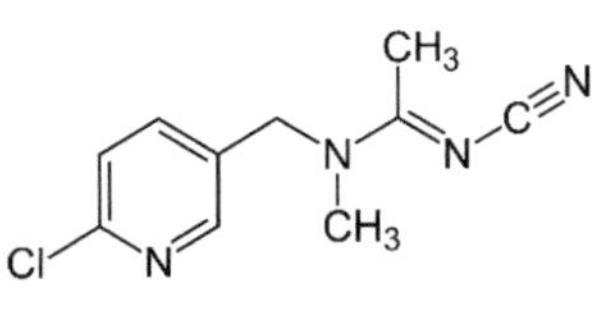

Figure 2.
Structural formula of acetamiprid.

1.1.4 Acetamiprid

Acetamiprid ((E)-N1-[(6-chloro-3-pyridyl)methyl]-N2-cyano-N1-methylacetamidine) is a new broad-spectrum neonicotinoid (**Figure 2**) which is used to control sucking insects by interfering with insects'nervous system [79, 80]. It protects of crops such leafy vegetables, citrus fruits, pome fruits, grapes, cotton, cole crops, and ornamental plants. Acetamiprid plays a key role in commercial cherry farming due to its effectiveness against the larvae of the cherry fruit fly [20].

Acetamiprid occurs as a white crystalline substance, it is polar and water soluble (4.2 g/L), therefore it can be transported to surface waters and may be toxic to aquatic organisms and life. It is also soluble in organic solvents: acetone, chloroform and dichloromethane (200g/L) and no very frothy in hexane (0.005g/L). Partition coefficient octanol/water has a logP = 0.80 (pH = 7, 25°C) and its dissociation constant is pKa = 0.7 [81].

Acetamiprid contains a 6-chloro-pyridine motif in its molecule, as does the animal alkaloid epibatidine from the poisonous South American frog *Epipedobates tricolor* [82]. An important property of acetamiprid as a representative of N-cyanoimine neonicotinoids is that, unlike N-nitroimines such as clothianidin, dinotefuran, imidacloprid, thiamethoxam or nitenpyram, it is of little toxicity to bees [83]. It does not accumulate in soil, is mobile and rapidly degrades by aerobic mechanisms [9] and microorganisms are involved in its degradation [84]. Its half-life in soil ranges from <1 to 8.2 days [85] and its content in vegetables and fruits is low [86, 87]. The application of acetamiprid in greenhouses and agricultural farms is safe and not associated with major health risks [88]. Human poisonings are known only in cases where acetamiprid is used as a suicide agent. Two such cases of acute suicidal poisoning are described in the medical literature. In both cases, nausea and vomiting, muscle weakness, hypothermia, convulsions and other clinical manifestations, including tachycardia, hypotension, ECG changes, hypoxia and thirst, occurred. These symptoms are partly similar to acute organophosphate intoxication. Supportive treatment of the clinical symptoms was sufficient and both patients were discharged without complications 2 days after ingestion of acetamiprid [89]. It is not yet sufficiently clear whether acetamiprid is genotoxic to mammals [90].

1.1.5 Imidacloprid

Imidacloprid (N-{1-[6-chloro-3-pyridyl)methyl]-4,5-dihydroimidazol-2-yl} nitramide) is a systemic insecticide (**Figure 3**) that acts as an insect neurotoxin, used for pest control in agriculture. In the year 1999, it was the most widely applied insecticide in the world [91]. Imidacloprid can be used by soil injection, broadcast foliar, application to the skin of the plant, tree injection, ground application as a liquid or granular formulation, or as a pesticide-coated seed treatment [92]. It is extra effective against sucking insects and mining pests such as mealybugs, aphids, thrips, and rice leafhoppers and against whitefly [93].

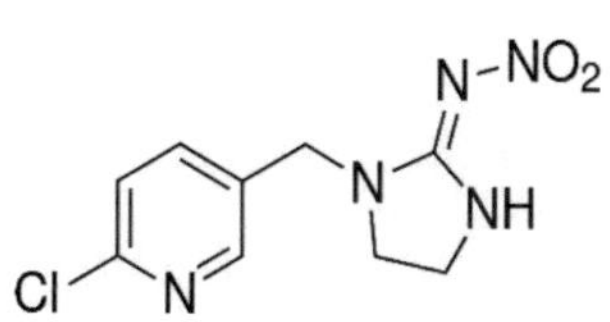

Figure 3.
Structural formula of imidacloprid.

Imidacloprid is a colorless crystalline substance with slight characteristic odor. It is weakly soluble in water (0.61 g/L) but better soluble in organic solvents: dichloromethane (67 g/L), isopropanol (23 g/L) and toluene (0.69 g/l). Partition coefficient octanol/water has a log P = 0.57 (pH = 7, 21°C) and its dissociation constants are $pKa_1 = 1.56$ and $pKa_2 = 11.12$ [94].

Imidacloprid has been classified by WHO as moderately hazardous (Class-II) based on animal studies [95]. The use of imidacloprid has a devastating impact on biodiversity, in particular on rivers and watercourses, not only on crustaceans [96], mollusks [97] and non-target species (insects), but also on soil organisms [36], as well as on bird populations [98]. Acute intoxication by imidacloprid and its metabolites resulted in the fast appearance of neurotoxicity symptoms, such as hyperactivity, hyperresponsiveness and trembling and led to hypoactivity and hyporesponsiveness [99]. It has been mentioned having harmful effects (oral toxicity) on honeybees in fields [100]. There is known case of imidacloprid poisoning with suicidal intent that developed various manifestations including hypokalaemia, central nervous system depression, respiratory arrest and paroxysmal atrial fibrillation [101]. Imidacloprid is metabolized by photodegradation from soil surface and water [102]. It is proven that plant metabolites of imidacloprid, the imidazolidine derivative, the olefin metabolite and nitroso-derivative were more toxic to aphids than imidacloprid itself [103].

2. Interaction of neonicotinoids with bio-macromolecules

Monitoring the interaction of neonicotinoid insecticides with biological macromolecules (DNA and serum albumins), determining the binding constant and the interaction mode provides a more comprehensive view of their distribution, toxicity and metabolism in the organism. It is also of great importance in adjusting the tolerance of these insecticides and their residues in food. The presented summarized results could contribute positively to the reduction of the use of the group of insecticides discussed by us.

2.1 Methods and analysis

The interaction of neonicotinoids with bio-macromolecules can be studied most often using UV–Vis, fluorescence and IR spectroscopy, circular dichroism, monitoring of bio-macromolecules melting, viscosity assays and last but not least molecular docking.

2.1.1 UV–Vis measurements

UV–Vis spectroscopy is a very effective method for investigating structural changes in the formation of DNA or protein and ligand/DNA or ligand/protein complexes. These measurements have been made on absorption spectrophotometer using quartz cuvettes of 1 cm path length at laboratory temperature. Usually the concentration of DNA or protein is constant while the concentration of neonicotinoid varies. A common practice is to add the same concentration of insecticide to the reference sample (with respect to the absorption maximum of the neonicotinoid). By monitoring the changes in absorbance intensity and the shift of the DNA/protein absorption maximum, it can be predicted how the neonicotinoid binds to the DNA/protein structure.

2.1.2 Fluorescence measurements

Fluorescence measurements were performed on spectrofluorimeter in a 1 cm quartz cuvette. Upon interaction, an extinguishing mechanism can be noted.

In general, extinguishing can be qualified as dynamic and static. The quenching mechanism is accurately described by the Stern-Volmer Eq. (1)

$$\frac{F_0}{F} = 1 + K_q \tau_0 = 1 + K_{SV}[Q] \tag{1}$$

where F_0 and F are the fluorescence intensities in the absence and in the presence of the quencher (Q) respectively. K_q and K_{SV} are the bio-macromolecule quenching rate constant, and Stern-Volmer constant respectively. τ_0 is the average lifetime of the bio-macromolecule without quencher and $[Q]$ is the concentration of quencher [104, 105].

Subsequently the binding parameters such as binding constant (K_a) and the number of binding sites (n) can be calculated according to Hill Eq. (2).

$$\log\left[\frac{F_0 - F}{F}\right] = \log K_a + n\log[Q] \tag{2}$$

The number of binding sites and association constant have been obtained by the plot of $\log[(F_0\text{-}F)/F]$ versus $\log[Q]$ [106–108].

2.1.3 Melting studies

The values of thermodynamic parameters can be determined from the Van't Hoff equation. This equation has been used in the form (3) for the interaction of neonicotinoid insecticides with proteins

$$\ln K_a = \frac{-\Delta H}{RT} + \frac{\Delta S}{R} \tag{3}$$

while ΔH is the enthalpy change, ΔS the entropy change, T is the temperature, and R is the universal gas constant. The binding studies from fluorescence measurements were usually realized at three different temperatures. By plotting $\ln K_a$ versus $1/T$ it can be possible to determine the values of ΔH and ΔS. Gibbs free energy change ΔG could be calculated using Eq. (4) [109, 110].

$$\Delta G = \Delta H - T\Delta S = -RT\,ln\,K_a \tag{4}$$

The thermal stability of DNA was studied by an absorption spectrophotometer with a Peltier module. A special type of Van't Hoff Eq. (5) has fitted melting curves

$$A = A_{\min} + \frac{A_{\max} - A_{\min}}{1 + e^{\left[\frac{\Delta H}{R} x \left(\frac{1}{T} - \frac{1}{T_m}\right)\right]}} \tag{5}$$

where A is the absorbance, $A_{\min}$, $A_{\max}$ are the minimal and maximal measured absorbance respectively, H is the enthalpy of transition, T, T_m are actual and melting temperature [111].

2.1.4 CD studies

CD spectroscopy was used to record changes in the secondary structure of optically active substances, including proteins and DNA. The CD results of complexes neonicotinoid/protein have been expressed in terms of mean residue ellipticity (MRE) (6)

$$MRE = \frac{Observed\ CD\left(mdeg\right)}{10\,c\,n\,l} \tag{6}$$

where c is the concentration of protein, n is the number of amino acid residues of protein, and l is the path length. The α-helix contents of free protein and neonicotinoid/protein complexes can be calculated from MRE values at 208 nm using the following Eq. (7) [112].

$$\alpha - helix\left(\%\right) = \frac{-MRE_{208nm} - 4{,}000}{33{,}000 - 4{,}000} x100 \tag{7}$$

CD spectra of DNA are very sensitive to the interaction mode of DNA with small molecules like neonicotinoid insecticides [113]. The characteristic spectrum of DNA in the B-form helix is characterized by a negative band at 245 nm resulting from right-handed helicity and a positive band at 278 nm resulting from base stacking [114]. By forming a neonicotinoid/DNA complex, a decrease in the intensity of the negative band at 245 nm was usually observed with accompanying red shift, while the intensity of the positive band at 276 nm increased. This increase in intensity was associated with a slight red shift. These changes of CD spectra represent the transformation from a B-form DNA structure to an A-conformation [115].

2.1.5 FT-IR spectra analysis

Infrared spectra have been collected approximately after 2h incubation of solutions containing neonicotinoid insecticides and bio-macromolecule at various rations. The amount of DNA or protein was constant. The intensity and shifting changes in infrared spectra could be applied to accurately specify the site of incorporation of neonicotinoids into the structure of bio-macromolecules (DNA and proteins) [116].

2.1.6 Viscosity tests

Viscosity tests can provide further information about the nature of the interaction neonicotinoid insecticides and DNA. This type of experiment consists in measuring the relative viscosity (t/t_0) of ethidium bromide/DNA and neonicotinoid/DNA complexes. A significant increase in viscosity was detected during the interaction of a typical intercalator ethidium bromide and DNA. The task was to compare the changes in DNA viscosity after adding neonicotinoids to the solution. If the same increase in viscosity was observed as for ethidium bromide, then the intercalation mode of binding between insecticides and DNA could be predicted [117].

2.1.7 Molecular docking studies

The trend of today's research is to confirm the results obtained from experiments by molecular docking studies. Serum albumin and DNA crystallographic data can be used from the Brookhaven Protein Data Bank. The species selected were 1AO6 (HSA), 4F5S (BSA), and 453D (DNA) [116, 118]. The molecular structures of neonicotinoid insecticides could be taken from The PubChem database. Most often AutoDockVina has been used for molecular docking and molecular simulation of the interaction between neonicotinoids and bio-macromolecules [119].

2.2 Interaction of neonicotinoids with DNA

Neonicotinoid insecticides are highly reactive compounds that form complexes with a variety of cellular biomolecules, including DNA. DNA plays an important role in cell proliferation, protein synthesis and genetic transcription. It is necessary to protect DNA from the harmful effects of insecticides because they will damage the genetic structure of cells and disrupt metabolic processes [120]. The study of the interaction between insecticides and nucleic acids is considered to be important to allow screening for carcinogenic properties of pesticides, especially insecticides [121, 122]. Studies addressing the specific mode of interaction and binding sites of neonicotinoid insecticides with DNA are few. Several techniques have been used to study the binding properties between neonicotinoid insecticides and DNA, including UV–visible absorption, fluorescence, circular dichroism, Fourier transform infrared spectroscopy, coupled with DNA melting investigations and viscosity measurements in physiological buffer. To predict the possible binding site and binding mode was used the molecular docking study [120–122].

Neonicotinoid insecticides belong to small molecules known to bind to the double helix of DNA by two dominant modes, especially such as groove binding and intercalation. Groove binding presents docking the thin ribbon-like molecules in the DNA minor groove, in close proximity to the sugar-phosphate backbone. Conversely, their intercalation into the helix includes the insertion of a drug – usually a planar aromatic cation – into the base reservoir of the helix [123].

It has been demonstrated that thiacloprid interacts with DNA, which influences on the length and denaturation of DNA. The binding strength of thiacloprid and DNA is expressed by a binding constant K_a whose value is $9.3 \cdot 10^3$ L/mol at laboratory temperature [117]. The presence of DNA significantly affects the emission spectrum of thiacloprid. The quenching of its fluorescence intensity was analyzed using the Stern-Volmer method [114]. The quenching constant K_{SV} was determined to be $2.8 \cdot 10^4$ L/mol and correlation coefficient for K_{SV} was $R = 0.998$. Not considerable changes in viscosity behavior were observed as a result of its addition to the DNA solution. The presence of thiacloprid causes a decrease in melting temperature T_m and Van't Hoff enthalpy ΔH of DNA. These changes in thermodynamic parameters lead to destabilization of the DNA molecule. In addition, it was recorded a two-phase character melting curve of the complex DNA-thiacloprid with an expression destabilization of AT regions in DNA [117]. It is generally assumed small molecules that consist of at least two aromatic rings coupled by a no rigid bond enabling their torsional flexibility, like thiacloprid bind preferentially to the minor groove of DNA, specially to the regions rich in AT-base pairs [124, 125]. Considering all the above results, it can be concluded that thiacloprid does not interact with DNA via an intercalating binding mode. Increasing of the length and influencing of DNA denaturation points to the groove-binding mode of interaction. Probably incorporation of this insecticide occurs into DNA minor groove by hydrophobic or hydrogen bonds [117].

Similarly, the effect of acetamiprid on DNA structure and stability was investigated. As shown by the UV–visible spectra, after the addition of acetamiprid to DNA solution, the absorption peak of DNA at 260 nm (associated with strong purine and pyrimidine base absorption in DNA [126]) increased markedly and there was significant blue shift. The fluorescence emission spectra of acetamiprid upon addition of DNA revealed that increasing concentration of DNA gradually decreased fluorescence emission peak of acetamiprid and a new band developed at approximately 370 nm. These facts can be attributed to the formation of the acetamiprid-DNA complex. The quenching mechanism between acetamiprid and DNA was analyzed by use of the Stern–Volmer method. The obtained K_{SV} value was $1.86 \cdot 10^4$ L/mol and the relevant correlation coefficient for K_{SV} was $R = 0.997$. The binding constant value K_a was determined $5.27 \cdot 10^3$ L/mol. Both constants were found at laboratory temperature [116]. Small molecules intercalating into the DNA causes stabilization of base stacking and leads to significant increase of *Tm*, whereas non-intercalation binding leads to no obvious increase in melting temperature [127]. In acetamiprid binding, T_m DNA increased by approximately 3°C. It can therefore be deduced that acetamiprid interacted with DNA via no classic rather than classical intercalation. This result was supported by evidence obtained from molecular docking, namely, the intercalation of acetamiprid into the double helix DNA from one side. Measurement of the relative viscosity of DNA shows a significant decrease in its viscosity, which excludes the classical intercalation mode of binding. The model in which acetamiprid binds to DNA in a partial intercalation manner explained this phenomenon. This means that partial intercalation can occur where acetamiprid can act as a 'wedge' that pushes apart one side of the base pair stack (but in contrast to the classical intercalation model, does not completely separate the stack), and thus induce static bending or kinking in the double helix [128]. Thermodynamic data for the interaction of acetamiprid and DNA were calculated. The Gibbs free energy value ΔG has a negative sign, indicating that the process is happening spontaneously [116]. Positive values of enthalpy ΔH and entropy ΔS changes suggest that hydrophobic interactions play an important function in the formation of the bond between acetamiprid and DNA and stabilize the complex [129]. Significant changes, in position and intensity, were observed in FTIR spectra for the GC base pair than for the AT base pair. It can be predicted that the specific binding site of acetamiprid to DNA is a site rich in GC base pairs. In addition, it is confirmed that acetamiprid binding leads to a change in the secondary structure of DNA from the B conformation to the A [116].

The interaction of acetamiprid with DNA is the basis for the development of the DNA probe, which is used in practice to detect the presence of acetamiprid in environmental samples and agricultural products. The results of the detection of acetamiprid using the DNA probe are in almost complete agreement with the results obtained using the HPLC technique [130].

At present, the interaction of imidacloprid with DNA is poorly studied, as its mode of incorporation into DNA, exact binding site, and binding constant are not known. Nevertheless, it is proven that imidacloprid can induce oxidative stress and DNA damage in zebrafish [131] and bees [132]. The study of the interaction between imidacloprid (also other neonicotinoid insecticides) and DNA is important for us to understand the insecticidal mechanism of neonicotinoids and their side effects, such as carcinogenesis, teratogenesis and mutagenesis [133]. Therefore, the mechanism of the reaction between imidacloprid and DNA is necessary to examine in detail.

2.3 Interaction of neonicotinoids with serum albumins

The binding study of neonicotinoids with proteins has toxicological importance [134]. The results of these interactions may cast some light on the future study of the interaction between neonicotinoid insecticides and other proteins such as enzymes and have toxicological to ecotoxicology importance. Therefore, it is essential to investigate the effect of neonicotinoids on the structural and optical properties of serum albumins, especially human serum albumin (HSA), the thermodynamic aspects in the binding process, and characters of the binding sites. The binding of insecticides, but also pesticides in general, to proteins has been exploited in the construction of pesticide biosensors. Biosensor assays can bring measures of the toxic effects of chemicals on the target organism and of the molecular mechanisms that are the basis of toxicity [135]. Another important direction in studies of the biological properties of neonicotinoids is directly related to protection of the health of humans and agricultural animals, and includes study of the interactions between these compounds and proteins, enzymes, and receptors in blood plasma and tissues. The structure of neonicotinoid insecticides contains sections and groups capable of forming electrostatic, hydrophobic, and hydrogen bonds as well as other types of bonds typical of endogenous ligands in their complexes with proteins. It is the reason many insecticides can play the important role of exogenous ligands and change their own properties within the composition of protein complexes, as is true for natural bio-regulators such as hormones. These alternations can involve metabolic parameters and biological effects of the pesticides in the human and animal bodies. The mechanism for the interaction of neonicotinoid with serum albumin probably includes "recognition" and initial binding of the ligand because of its polar group. Followed by adaptation of the ligand-binding site of the HSA molecule for binding by means of conformational transitions. This mechanism is terminated by subsequent interaction of the hydrophobic core of the ligand with the nonpolar side chains of serum albumin in the cavity of the binding site. The new insights into the interaction of serum albumins and neonicotinoid insecticides, about the forms in which neonicotinoids exist in the body and the approaches developed for studying interactions of these compounds with the major transport protein HSA are a necessary basis for estimating the biological effects of pesticides in this class when they enter human blood [136].

It has been shown that thiacloprid interacts with HSA, its binding properties have been characterized at the molecular level under physiological conditions. The intensity of HSA absorption maximum decreased after the addition of thiacloprid, and its little red shift occurred at the same time. This indicated the probable formation of a complex between thiacloprid and HSA. With a gradual increase in thiacloprid concentration, a period decrease in fluorescence intensity was also observed. The Stern-Volmer plots were used for analyses of the quenching mechanism. The obtained K_{SV} value was $3.304{\cdot}10^4$ L/mol and the binding constant K_a was found $3.07{\cdot}10^4$ L/mol. Both constants were determined at laboratory temperature. The linear Van't Hoff equation was applied to track changes in the thermodynamic parameters. The entropy, enthalpy and Gibbs free energy values indicate that the coupling of thiacloprid to HSA is an exothermic process due to the positive value of ΔS and negative values of ΔH and ΔG [134]. It is well known that thermodynamic parameters play an important role in determining the type of binding by which a ligand (insecticide) binds to an HSA. Positive value of ΔH and ΔS is the result of hydrogen bonding. Negative values of these quantities (ΔH, ΔS) are corresponded with hydrogen bonds and van der Waals interaction in a low dielectric solution. The electrostatic interaction in aqueous solution between ionic species is associated with a positive change of ΔS and very small negative change of ΔH, almost zero [137]. Hydrophobic force and electrostatic force interactions are characterized by negative value of ΔH and positive ΔS values. In view of the above facts, it is not easy to interpret the results obtained from the thermodynamic analysis of the thiacloprid-HSA interaction. It is difficult to explain this interaction mode by a single intermolecular force. It has been published that the interaction force acting between small molecules and proteins is generally not just a single force. Probably may be there are variety of forces existing in the interaction forces which contain the electrostatic force. The hydrophobic molecule embedded in the internal hydrophobic region of proteins can be responsible for the fluorescence quenching [138, 139]. One can hypothesize that the incorporation of thiacloprid into HSA can be realized by hydrophobic interaction, as evidenced by the positive ΔS value, but it cannot be excluded the influence of electrostatic interaction. The given conclusions are consistent with molecular modeling. This suggests that thiacloprid could be located on the surface of the binding pocket of subdomain IIA in the HSA molecule. It also confirms that the hydrogen bonding also plays a significant role [134]. Study of HSA secondary structure revealed the decrease of the percentage α-helix structure the effect of thiacloprid binding on the amino acid residue of the main polypeptide chain of HSA. Thiacloprid upset their hydrogen bonds [140].

Hemoglobin is essential protein in the blood plasma. It is working as a transporter of oxygen. Bovine hemoglobin (BHb) is used as a model protein to study the binding properties of drugs and insecticides [141]. The results obtained from the study of the influence of thiacloprid on this model protein show that their interaction is a static process. Two types of bonds play an important role in this binding, namely, hydrogen bonding and hydrophobic interactions. Increasing concentration of thiacloprid causes a marked decrease in BHb fluorescence intensity. The strength of cross-linking is characterized by the binding constant K_a, the magnitude of which was determined to be $8.04 \cdot 10^4$ L/mol at laboratory temperature. Tracking changes in thermodynamic parameters suggests that hydrogen bonding forces are most important for a given interaction. This is evident from the fact that negative values of the enthalpy change ΔH and entropy change ΔS were calculated. Changes in the secondary structure of BHb in the presence of thiacloprid show that there is a 3.7% decrease in α-helix content. Molecular modeling was used to determine the amino acid residues involved in thiacloprid-BHb binding. The thiacloprid pyridine ring interacts with Leu105, Pro95, Trp37 by hydrophobic interactions. The incorporation of thiaclopride into BHb is not exclusively hydrophobic, as several ionic and polar residues (Thr137, Tyr35) are present near the bound ligand. These polar residues stabilize the neonicotinoid insecticide via H-bonding and electrostatic interactions. Trp37 is involved in the formation of H-bonds with side chain imino groups [142].

Similarly, the effect of imidacloprid on HSA structure was investigated. The absorbance of the imidacloprid-HSA complex decreased with increasing imidacloprid concentration. A shift of the absorption maximum to the red region was also observed as in the thiacloprid interaction [135]. The HSA molecule contains 585 amino acid residues forming a single polypeptide of known sequence [118]. The protein response to conformational transitions, subunit association, ligand binding, or denaturation are changes in tryptophan emission spectra [143]. The study of the intrinsic fluorescence of Trp HSA is important for a better understanding of the specific changes that occur in the macromolecule [135]. Tyrosine fluorescence in HSA is mostly quenched due to the presence of nearby amino acids or efficient energy transfer from tyrosine fluorescence to Trp214 [144]. Imidacloprid caused a decrease in the fluorescence intensity of tryptophan residues. Fluorescence resonance energy transfer determined the distance between the donor (HSA) and acceptor (imidacloprid). Its size is 2.10 nm [135]. It is generally accepted that the average distances between the donor fluorophore and the acceptor fluorophore are 2–8 nm and indicate that energy transfer from HSA to imidacloprid occurs with a high probability [145]. It represents static quenching. During the measurement, a change in the structure of the surroundings of the Trp and Tyr residues was observed. Imidacloprid affected the physiological function of HSA. The fluorescence maximum of HSA shifted to higher wavelengths. The above results suggest that imidacloprid formed a specific bond in subdomain IIA, close to the tryptophan residue at position 214 of the HSA polypeptide chain like thiacloprid [135, 136]. The binding constant K_a was determined to the value $1.51 \cdot 10^4$ L/mol [135]. Thermodynamic parameters allowed predicting that imidacloprid incorporates into HSA via hydrophobic interactions. These occur in the presence of an aromatic pyridine imidacloprid ring. Its nitroimine group could provide further enhancement of the affinity of the above neonicotinoid insecticide for HSA [135, 136].

A study looking at the interaction of imidacloprid with bovine serum albumin (BSA) gave the same results as in the interaction with HSA. From the fluorescence quenching spectra, the binding constant K_a was calculated, the magnitude of which reaches a value of $3.42{\cdot}10^4$ L/mol. Static quenching of emission without energy transfer by radiation within a single BSA molecule was also detected. Variations in thermodynamic parameters ($\Delta H > 0, \Delta G < 0, \Delta S > 0$) made it possible to predict the type of interaction. It is probably a hydrophobic interaction. Among the effects, that imidacloprid has on BSA can include increasing the polarity of the microenvironment in which Trp and Tyr are found, increasing the compaction of peptide bonds, and modifying the conformation of BSA [146].

The effect of acetamiprid on the structure of serum albumin is currently not sufficiently studied (not enough relevant results have been published). Considering the above results obtained by studying the interaction of similar neonicotinoid insecticides (thiacloprid and imidacloprid), it can be assumed that acetamiprid also affects the structure and conformation of serum albumins as well as their thermodynamic parameters. Therefore, it is essential to study and characterize in detail the incorporation of this insecticide into albumin.

3. Conclusion

In conclusion, we have to state that the topic discussed by us is still insufficiently studied and it is necessary to further continuously address the issue of the interaction of neonicotinoid insecticides with bio-macromolecules. Using of neonicotinoid insecticides leads to serious environmental problems, including contamination of soil and ground water, which can cause them to accumulate in the human and animal bodies and subsequently damage DNA. Several studies published in this chapter show that molecules of thiacloprid, acetamiprid, and imidacloprid are able to incorporate into important biological macromolecules and disturb their structure and function. All of them are honey bee killers and harmful to pollinators. Only thiacloprid is allowed to use in EU, imidacloprid and acetamiprid are prohibited which are considered to be one of the causes of bee colony decline in the world [41, 147]. On the other hand, they can still stay in the US market by Environmental Protection Agency decision [148].

However, the published results clearly indicate that neonicotinoid insecticides such as thiacloprid, acetamiprid, and imidacloprid, should be used very sparingly and cautiously in practice, especially in densely populated countries of the world where insecticides are overused.

Acknowledgements

This work was also supported with grant from the Slovak Research Grant Agency VEGA No. 1/0242/19.

Conflict of interest

The authors declare no conflict of interest among themselves.

Author details

Valéria Verebová[1*] and Jana Staničová[1,2]

1 Department of Chemistry, Biochemistry and Biophysics, University of Veterinary Medicine and Pharmacy, Košice, Slovakia

2 First Faculty of Medicine, Charles University in Prague, Prague, Czech Republic

*Address all correspondence to: valeria.verebova@uvlf.sk

References

[1] Metcalf RL, Luckmann WH. Introduction to insect pest management. John Willey & Sons; 1994. 650 p.

[2] Grundlingh J, Dargan PI, El-Zanfaly M, Wood DM. 2,4-dinitrophenol (DNP): a weight loss agent with significant acute toxicity and risk of death. Journal of Medical Toxicology. 2011;**7**(3):205-212. DOI: 10.1007/s13181-011-0162-6

[3] Radcliffe ED, Hutchison WD, Cancelado RE. Integrated Pest Management. Cambridge University Press, Cambridge UK; 2009. 530 p.

[4] Elbert A, Buchholz A, Ebbinghaus-Kintscher U, Erdelen C, Nauen R, Schnorbach HJ. The biological profile of thiacloprid-A new chloronicotinyl insecticide. Pflanzenschutz- Nachrichten Bayer. 2001;**54**:185-208.

[5] Van Emden HF, Peakall DB. Beyond Silent Spring. Integrated pest management and chemical safety. Springer; 1996; **XVIII**: 320 p.

[6] Sánchez-Bayo F, Tennekes HA, Goka K. Impact of Systemic Insecticides on Organisms and Ecosystems, Insecticides. 2013; IntechOpen. DOI: 10.5772/52831. Available from: https://www.intechopen.com/books/insecticides-development-of-safer-and-more-effective-technologies/impact-of-systemic-insecticides-on-organisms-and-ecosystems [Accessed: 2021-06-15] DOI: 10.5772/52831

[7] Simon-Delso N, Amaral-Rogers V, Belzunces LP, Bonmatin JM, Chagnon M, Downs C, FurlanL, Gibbons DW, Giorio C, Girolami V, Goulson D, Kreutzweiser DP, Krupke CH, Liess M, Long E, McField M, Mineau P, Mitchell EAD, Morrissey CA, Noome DA, Pisa L, Settele J, Stark JD, Tapparo A, Van Dyck H, Van Praagh J, Van der Sluijs JP, Whitehorn PR, Wiemers M. Systemic insecticides (neonicotinoids and fipronil): trends, uses, mode of action and merabolites. Environmental Science and Pollution Research. 2015;**22**(1):5-34. DOI: 10.1007/s11356-014-3470-y

[8] Ishaaya I, Kontsedalov S, Mazirov D, Horowitz AR. Biorational agents--mechanism and importance in IPM and IRM programs for controlling agricultural pests. Mededelingen Rijksuniversiteit te Gent Fakulteit van de Landbouwkdundige en Toegepaste Biologishe Wetenschappen. 2001;**66**(2a):363-374

[9] Foster SP, Denholm I, Thompson R. Variation in response to neonicotinoid insecticides in peach-potato aphids, *Myzus persicae* (Hemiptera: Aphididae). Pest Management Science. 2003;**59**(2): 166-173. DOI: 10.1002/ps.570

[10] Ishaaya I, Kontsedalov S, Horowitz AR. Biorational insecticides: mechanism and cross-resistance. Archives of Insect Biochemistry and Physiology. 2005;**58**(4):192-129. Available from: Biorational insecticides: Mechanism and cross-resistance (wiley.com) [Accessed: 2020-03-12]

[11] Han W, Tian Y, Shen X. Human exposure to neonicotinoid insecticides and the evaluation of their potential toxicity: An overview. Chemosphere. 2018;**192**:59-65. DOI: 10.1016/j.chemosphere.2017.10.149

[12] Tomizawa M, Casida JE. Neonicotinoid insecticides: highlights of symposium on strategic molecular designs. Journal of Agricultural and Food Chemistry. 2010;**59**(7):2883-2886. DOI: 10.1021/jf103856c

[13] Bass Ch, Denholm I, Williamson MS, Nauen R. The global status of insect resistance to

neonicotinoid insecticides. Pesticide Biochemistry and Physiology. 2015;**121**:78-87. DOI: 10.1016/j.pestbp.2015.04.004

[14] Casida JE. Neonicotinoid metabolism: compounds, substituents, pathways, enzymes, organisms, and relevance. Journal of Agricultural and Food Chemistry. 2010;**59**(7):2923-2931. DOI: 10.1021/jf102438c

[15] Elbert A, Overbeck H, Iwaya K, Tsuboi S. Imidacloprid, a novel systemic nitromythylene analog insecticide for crop protection. The Brighton Crop Protection Conference: Pets and Diseases. 1990; Brighton. p. 21-28

[16] Maienfisch P, Angst M, Brandl F, Fischer W, Hofer D, Kayser H, Kobel W, Rindlisbacher A, Senn R, Steinemann A, Widmer H. Chemistry and biology of thiamethoxam: a second generation neonicotinoid. Pest Management Science. 2001;**57**:906-913. DOI: 10.1002/ps.365

[17] Meredith RH, Heatherington PJ, Morris E. Clothianidin – a new chloronicotinyl seed treatment for use on sugar beet and cereals: field trial experiences from Northern Europe. The Brighton Crop Protection Conference: Pets and Diseases. Brighton. 2002; p. 691-696

[18] Wang X, Anadon A, Wu Q, Qiao F, Ares I, Martiney-Larranaga MR, Yuan Z, Martinez MA. Mechanism of neonicotinoid toxicity: Impact on oxidative stress and metabolism. Annual Review of Pharmacology and Toxicology. 2018;**58**:471-507. DOI: 10.1146/annurev-pharmtox-010617-052429

[19] Sanchez-Bayo F. Systemic insecticides and their environmental repercussions. Dellasala DA, Goldstein MI (eds.) Encyclopedia of the Anthropocene. Elsevier, Oxford. 2018; p. 111-117. Available from: https://doi.org/ 10.1016/B978-0-12-409548-9.09895-X [Accessed: 2020-09-21]

[20] Jeschke P, Nauen R, Schindler M, Elbert A. Overview of the status and global strategy for neonicotinoids. Journal of Agricultural and Food Chemistry. 2011;**59**(7):2897-2908. DOI: 10.1021/jf101303g

[21] Maienfisch P, Huerlimann H, Rindlisbacher A, Gsell L, Dettwiler H, Haettenschwiler J, Sieger E, Walti M. The discovery of thiamethoxam: a second-generation neonicotinoid. Pest Management Science. 2001;**57**(2):165-176: DOI: 10.1002/1526-4998(200102)57:2<165::AID-PS289>3.0.CO;2-G

[22] Bromilow RH, Chamberlain K. Principles overning uptake and transport of chemicals. Trapp S, McFarlane JC (eds.) Plant contamination: modelling and simulation of organic chemical process. CRC Press, Boca Raton, Florida, 1995; 37-68

[23] Bonmatin JM, Giorio C, Girolami V, Goulson D, Kreutzweiser DP, Krupke C, Liess M, Long E, Marzaro M, Mitchell EAD, Noome DA, Simon-Delso N, Tapparo A. Environmental fate and exposure; neonicotinoids and fipronil. Environmental Science and Pollution Research. 2015;**22**(1):35-67. DOI: 0.1007/s11356-014-3332-7

[24] Magalhaes LC, Hunt TE, Siegfries BD. Efficacy of neonicotinoid seed treatments to reduce soybean aphid populations under field and controlled conditions in Nebraska. Journal of Economic Entomology. 2009;**102**(1):187-195. DOI: 10.1603/029.102.0127

[25] Goulson D. An overview of the environmental risks posed by neonicotinoid insecticides. Journal of Applied Ecology. 2013;**50**(4):977-987. DOI: 10.1111/1365-2664.12111

[26] Hladik M, Main AR, Goulson D. Environmental risks and challenges associated with neonicotinoid insecticides. Environmental Science and Technology. 2018;**52**:3329-3335. DOI: 10.1021/acs.est.7b06388

[27] United States Department of Agriculture (USDA). Pesticide Data Program: Annual Summary. Calendar Year 2014. 2016. Available from: https://www.ams.usda.gov/sites/default/files/media/2014%20PDP%20Annual%20Summary.pdf [Accessed: 2020-11-24]

[28] Seccia S, Fidente P, Barbini DA, Morrica P. Multiresidue determination of nicotinoid insecticide residues in drinking water by liquid chromato graphy with electrospray ionization mass spectrometry. Analytica Chimica acta. 2005;**553**(1-2):21-26. DOI: 10.1016/j.aca.2005.08.006

[29] Klarich KL, Pflug NC, DeWald EM, Hladik ML, Kolpin DW, Cwiertny DM, LeFevre GH. Occurrence of neonicotinoid insecticides in finished drinking water and fate during drinking water treatment. Environmental Science and Technology Letters. 2017;**4**(5):168-173. DOI: 10.1021/acs.estlett.7b00081

[30] Xie W, Han C, Qian Y, Ding H, Chen X, Xi J. Determination of neonicotinoid pesticides residues in agricultural samples by solid-phase extraction combined with liquid chromatography-tandem mass spectrometry. Journal of Chromatography A. 2011;**1218**(28):4426-4433. DOI: 10.1016/j.chroma.2011.05.026

[31] Seccia S, Fidente P, Montesano D, Morrica P. Determination of neonicotinoid insecticides residues in bovine milk samples by solid-phase extraction clean-up and liquid chromatography with diode-array detection. Journal of Chromatography A. 2008;**1214**(1-2):115-120. DOI: 10.1016/j.chroma.2008.10.088

[32] Mitchell EAD, Mulhauser B, Mulot M, Mutabazi A, Glauser G, Aebi A. A worldwide survey of neonicotinoids in honey. Science. 2017;**358**(6359):109-111. DOI: 10.1126/science.aan3684

[33] Matsuda K, Buckingham SD, Kleier D, Rauh JJ, Grauso M, Sattelle DB. Neonicotinoids: insecticides acting on insect nicotinic acetylcholine receptors. Trends in Pharmacological Sciences. 2001;**22**(11):573-580. DOI: 10.1016/s0165-6147(00)01820-4

[34] Morrissey CA, Mineau P, Devries JH, Sanchez-Bayo F, Liess M, Cavallaro MC, Liber K. Neonicotinoid contamination of global surface waters and associated risk to aquatic invertebrates: a review. Environment International. 2015;**74**:291-303. DOI: 10.1016/j.envint.2014.10.024

[35] Tomizawa M, Casida JE. Neonicotinoid insecticide toxicology: mechanism of selective action. Annual Review of Pharmacology and Toxicology. 2005;**45**:247-268. DOI: 10.1146/annurev.pharmtox.45.120403.095930

[36] E Silva CL, Brennan N, Brouwer JM, Commandeur D, Verweij RA, Van Gestel CAM. Comparative toxicity of imidacloprid and thiacloprid to different species of soil invertebrates. Ecotoxicology. 2017;**26**(4):555-564. DOI: 10.1007/s10646-017-1790-7

[37] Whitehorn PR, O'Connor S, Wackers FL, Goulson D. Neonicotinoid pesticide reduces bumble bee colony growth and queen production. Science. 2012;**336**(6079):351-352. DOI: 10.1126/science.1215025

[38] Sandrock C, Tanadini LG, Pettis JS, Biesmeijer JC, Potts SG, Neumann P. Sublethal neonicotinoid insecticide exposure reduces solitary bee reproductive success. Agricultural and Forest Entomology. 2014;**16**(2):119-128. DOI: 10.1111/afe.12041

[39] Williams GR, Troxler A, Retschnig G, Roth K, Zaney O, Shutler D, Neumann P, Gauthier L. Neonicotinoid pesticides severely affect honey bee queens. Scientific Report. 2015;**5**:14621. DOI: 10.1038/srep14621

[40] Brandt A, Gorenflo A, Siede R, Meixner M, Büchler R. The neonicotinoids thiacloprid, imidacloprid, and clothianidin affect the immunocompetence of honey bees (*Apis mellifera* L.). Journal of Insect Physiology. 2016;**86**:40-47. DOI: 10.1016/j.jinsphys.2016.01.001

[41] European Food Safety Authority (EFSA), Abdourahime H, Anastassiadou M, Arena M, Auteri D, Barmaz S, Brancato A, Brocca D, Bura L, Cabrera LC, Chiusolo A, Civitella C, Marques DC, Crivellente F, Ctverackova L, De Lentdecker C, Egsmose M, Fait G, Ferreira L, Gatto V, Greco L, Ippolito A, Istace F, Jarrah S, Kardassi D, Leuschner R, Lostia A, Lythgo C, Magrans JO, Medina P, Messinetti S, Mineo D, Miron I, Nave S, Molnar T, Padovani L, Morte JMP, Pedersen R, Raczyk M, Reich H, Ruocco S, Saari KE, Sacchi A, Santos M, Serafimova R, Sharp R, Stanek A, Streissl F, Sturma J, Szentes C, Tarazona J, Terron A, Theobald A, Vagenende B, Vainovska P, Van Dijk J, Verani A, Villamar-Bouza L. Peer review of the pesticide risk assessment of the active substance thiacloprid. EFSA Journal. 2019;**17**(3):e05595. DOI: 10.2903/j.efsa.2019.5595

[42] Ellis C, Park KJ, Whitehorn P, David A, Goulson D. The neonicotinoid insecticide thiacloprid impacts upon bumblebee colony development under field conditions. Environmental Science and Technology. 2017;**51**(3):1727-1732. DOI: 10.1021/acs.est.6b04791

[43] Zhang P, Sun H, Min L, Ren C. Biochars change the sorption and degradation of thiacloprid in soil: insights into chemical and biological mechanisms. Environmental pollution. 2018;**236**:158-167. DOI: 10.1016/j.envpol.2018.01.030

[44] Chrétien F, Giroux I, Thériault G, Garnon P, Corriveau J. Surface runoff and subsurface tile drain losses of neonicotinoids and companion herbicides at edge-of-field. Environmental pollution. 2017;**224**:255-264. DOI: 10.1016/j.envpol.2017.02.002

[45] Smith R, Middlebrook R, Turner R, Huggins R, Vardy S, Warne M. Large-scale pesticide monitoring across Great Barrier Reef catchments-paddock to reef integrates monitoring, modelling and reporting program. Marine Pollution Bulletin. 2012;**64**(4-9):117-127. DOI: 10.1016/j.marpolbul.2011.08.010

[46] Gonzalez-Rey M, Tapie N, Le Menach K, Dévier MH, Budzinski H, Bebianno MJ. Occurrence of pharmaceutical compounds and pesticides in aquatic systems. Marine Pollution Bulletin. 2015;**96**(1-2):384-400. DOI: 10.1016/j.marpolbul.2015.04.029

[47] Yamamoto A, Terao T, Hisatomi H, Kawasaki H, Arakawa R. Evaluation of river pollution of neonicotinoids in Osaka City (Japan) by LC/MS with dopant-assisted photoionization. Journal of Environmental Monitoring. 2012;**14**(8):2189-2194. DOI: 10.1039/c2em30296a

[48] Main AR, Headley JV, Peru KM, Michel NL, Cessna AJ, Morrissey CA. Widespread use and frequent detection of neonicotinoid insecticides in wetlands of Canada's Prairie Pothole Region. Plos One. 2014;**9**(3):e92821. DOI: 10.1371/journal.pone.0092821

[49] Sanchez-Bayo F, Hyne RV. Detection and analysis of neonicotinoids in river waters-development of a passive sampler for three commonly used insecticides. Chemosphere. 2014;**99**:143-151. DOI: 10.1016/j.chemosphere.2013.10.051

[50] Alexander AC, Culp JM, Liber K, Cessna AJ. Effects of insecticide exposure on feeding inhibition in mayflies and oligochaetes. Environmental Toxicology and Chemistry. 2007;**26**(8):1726-1732. DOI: 10.1897/07-015r.1

[51] Agatz A, Cole TA, Preuss TG, Zimmer E, Brown CD. Feeding inhibition explains effects of imidacloprid on the growth, maturation, reproduction, and survival of *Daphnia magna*. Environmental Science and Technology. 2013;**47**(6): 2909-2917. DOI: 10.1021/es304784t

[52] Barletta M, Lima ARA, Costa MF. Distribution, sources and consequences of nutrients, persistent organic pollutants, metals, and microplastics in South America estuaries. Science of the Total environment. 2019;**651**(1):1199-1218. DOI: 10.1016/j.scitotenv.2018.09.276

[53] Raby M, Nowierski M, Perlov D, Zhao X, Hao C, Poirier DG, Sibley PK. Acute toxicity of 6 neonicotinoid insecticides to freshwater invertebrates. Environmental Toxicity and Chemistry. 2018;**37**(5):1430-1445. DOI: 10.1002/etc.4088

[54] Cavallaro MC, Morrissey CA, Headley JV, Peru KM, Liber K. Comparative chronic toxicity of imidacloprid, clothianidin, and thiamethoxam to Chironomus dilutus and estimation of toxic equivalency factors. Environmental Toxicity and Chemistry. 2017;**36**(2):372-382. DOI: 10.1002/etc.3536

[55] Stoughton SJ, Liber K, Culp J, Cessna A. Acute and chronic toxicity of imidacloprid to the aquatic invertebrates *Chironomus tentans* and *Hyalella azteca* under constant- and pulse-exposure conditions. Archives of Environmental Contamination and Toxicology. 2008;**54**(4):662-673. DOI: 10.1007/s00244-007-9073-6

[56] Roessink I, Merga LB, Zweers HJ, Van den Brink PJ. The neonicotinoid imidacloprid shows high chronic toxicity to mayfly nymphs. Environmental Toxicity and Chemistry. 2013;**32**(5):1096-1100. DOI: 10.1002/etc.2201

[57] Basley K, Goulson D. Neonicotinoids thiamethoxam and clothianidin adversely affect the colonisation of invertebrate populations in aquatic microcosms. Environmental Science and Pollution research. 2018;**25**(10):9593-9599. DOI: 10.1007/s11356-017-1125-5

[58] Mason R, Tennekes H, Sánchez-Bayo F, Jepsen PU. Immune suppression by neonicotinoid insecticides at the root of global wildlife declines. Journal of Environmental Immunology and Toxicology. 2013;**1**(1):3-12. DOI: 10.7178/jeit.1

[59] Abou-Donia MB, Goldstein LB, Bullman S, Tu T, Khan WA, Dechkovskaia AM, Abdel-Rahman AA. Imidacloprid induces neurobehavioral deficits and increases expression of glial fibrillary acidic protein in the motor cortex and hippocampus in offspring rats following in utero exposure. Journal of Toxicology and Environmental Health, Part A. 2008;**71**(2):119-130. DOI: 10.1080/15287390701613140

[60] Li P, Ann J, Akk Gustav. Activation and modulation of human $\alpha4\beta2$ nicotinic acetylcholine receptors by the neonicotinoids, clothianidin and imidacloprid. Journal of Neuroscience Research. 2011;**89**(8):1295-1301. DOI: 10.1002/jnr.22644

[61] Kimura-Kuroda J, Komuta Y, Kuroda Y, Hayashi M, Kawano H. Nicotine-like effects of the neonicotinoid insecticides acetamiprid and imidacloprid on cerebellar neurons from neonatal rats. Plos One. 2012;**7**(2):e32432. DOI: 10.1371/journal.pone.0032432

[62] Cimino AM, Boyles AL, Thayer KA, Perry MJ. Effects of neonicotinoid pesticide exposure on human health: a systematic review. Environmental Health Perspectives. 2016;**125**(2):155-162. DOI: 10.1289/EHP515

[63] Gu Y, Li Y, Huang X, Zheng J, Yang J, Diao H, Yuan Y, Xu Y, Liu M, Shi H, Xu W. Reproductive effects of two neonicotinoid insecticides on mouse sperm function and early embryonic development in vitro. Plos One. 2013;**8**(7):e70112. DOI: 10.1371/journal.pone.0070112

[64] Kapoor U, Srivastava MK, Srivastava LP. Toxicological impact of technical imidacloprid on ovarian morphology, hormones and antioxidant enzymes in female rats. Food and Chemical Toxicology. 2011;**49**(12):3086-3089. DOI: 10.1016/j.fct.2011.09.009

[65] Toor HK, Sangha GK, Khera KS. Imidacloprid induces histological and biochemical alterations in liver of female albino rats. Pesticide Biochemistry and Physiology. 2013;**105**(1):1-4. DOI: 10.1016/j.pestbp.2012.10.001

[66] Şekeroğlu V, Şekeroğlu ZA, Demirhan ES. Effects of commercial formulations of deltamethrin and/or thiacloprid on thyroid hormone levels in rat serum. Toxicology and Industrial Health. 2014;**30**(1):40-46. DOI: 10.1177/0748233712448114

[67] Aydin B. Effects of thiacloprid, deltamethrin and their combination on oxidative stress in lymphoid organs, polymorphonuclear leukocytes and plasma of rats. Pesticide Biochemistry and physiology. 2011;**100**(2):165-171. DOI: 10.1016/j.pestbp.2011.03.006

[68] Bal R, Naziroğlu M, Türk G, Yilmaz Ö, Kuloğlu T, Etem E, Baydas G. Insecticide imidacloprid induces morphological and DNA damage through oxidative toxicity on the reproductive organs of developing male rats. Cell Biochemistry and Function. 2012;**30**(6):492-499. DOI: 10.1002/cbf.2826

[69] Sharma N, Banerjee H, Pal S, Sharma KK. Persistence of thiacloprid and deltamethrin residues in tea grown at different locations of North-East India. Food Chemistry. 2018;**253**:88-92. DOI: 10.1016/j.foodchem.2018.01.132

[70] O'Neil MJ. The Merck Index: An encyclopedia of chemicals, drugs and biologicals. Whitehouse Station: Merck Research Laboratories. Division of Merck and Co. 2006. p. 2708

[71] National Center for Biotechnology Information. PubChem Compound Summary for CID 115224. Thiacloprid. 2021. Available from: https://pubchem.ncbi.nlm.nih.gov/compound/Thiacloprid [Accessed: 2020-10-11]

[72] World Health Organization & International Programme on Chcemical Safety. The WHO recommended classification of pesticides by hazard and guidelines to classification 2009. World Health Oraganization. 2010. p. 78. Available from: https://apps.who.int/iris/handle/10665/44271 [Accessed: 2021-06-15]

[73] U.S. Environmental Protection Agency. Office of Pesticide Programs. Chemicals evaluated for carcinogenic potential annual cancer report 2018. p. 40. Available from: Chemicals Evaluated for Carcinogenic Potential Annual Cancer Report 2018 (apublica.org) [Accessed: 2020-09-10]

[74] EPA U.S. Thiacloprid. Pesticide Tolerances. Federal Register. 2013;**70**:8410-8416. Available from: Federal Register:: Thiacloprid; Pesticide Tolerances [Accessed: 2020-08-13]

[75] Velisek J, Stara A. effect of thiacloprid on early life stages of common carp (*Cyprinus carpio*).

Chemosphere. 2018;**194**:481-487. DOI: 10.1016/j.chemosphere.2017.11.176

[76] Galdíková M, Šiviková K, Holečková B, Dianovský J, Drážovská M, Schwarzbacherová V. The effect of thiacloprid formulation on DNA/chromosome damage and changes in GST activity in bovine peripheral lymphocytes. Journal of Environmental Science and Health, Part B. 2015;**50**(10):698-707. DOI: 10.1080/03601234.2015.1048102

[77] Kocaman AY, Rencüzoğullari E, Topaktaş M. In vitro investigation of the genotoxic and cytotoxic effects of thiacloprid in cultured human peripheral blood lymphocytes. Environmental Toxicology. 2014;**29**(6):631-641. DOI: 10.1002/tox.21790

[78] Schwarzbacherová V, Wnuk M, Deregowska A, Holečková B, Lewinska A. In vitro exposure to thiacloprid-based insecticide formulation promotes oxidative stress, apoptosis and genetic instability in bovine lymphocytes. Toxicology in Vitro. 2019;**61**:104654. DOI: 10.1016/j.tiv.2019.104654

[79] Fan LF, Zhao GH, Shi HJ, Liu MC, Li ZX. A highly selective electrochemical impedance spectroscopy-based aptasensor for sensitive detection of acetamiprid. Biosensors and Bioelectronics. 2013;**43**(1):12-18. DOI: 10.1016/j.bios.2012.11.033

[80] Fogel MN, Schneider MI, Desneux N, Gonzalez B, Ronco AE. Impact of the neonicotinoid acetamiprid on immature stages of the predator Eriopis connexa (Coleoptera: Coccinellidae). Ecotoxicology. 2013;**22**(6):1063-1071. DOI: 10.1007/s10646-013-1094-5

[81] National Center for Biotechnology Information. PubChem Compound Summary for CID 213021. Acetamiprid. 2021. Available from: https://pubchem.ncbi.nlm.nih.gov/compound/Acetamiprid [Accessed: 2021-07-15]

[82] Patočka J, Ardila MC, Vázquez MV. Poisons of fogs of the family Dendrobatidae-inspiration for bioorganic chemistry. Chemické Listy. 2000;**94**:230-233. Available from: http://chemicke-listy.cz/docs/full/archiv/2000-PDF/04-PDF/230-233.pdf [Accessed: 2020-04-07]

[83] Decourtye A, Devillers J. Ecotoxicity of Neonicotinoid Insecticides to Bees. In: Thany S.H. (eds) Insect Nicotinic Acetylcholine Receptors. Advances in Experimental Medicine and Biology, Springer, New York. 2010.**683**: 85-95. DOI: 10.1007/978-1-4419-6445-8_8

[84] Liu Z, Dai Y, Huang G, Gu Y, Ni J, Wei H, Yuan S. Soil microbial degradation of neonicotinoid insecticides imidacloprid, acetamiprid, thiacloprid and imidaclothiz and its effect on the persistence of bioefficacy against horsebean aphid *Aphis craccivora* Koch after soil application. Pest Management Science. 2011;**67**(10):1245-1252. DOI: 10.1002/ps.2174

[85] Gupta S, Gajbhiye VT. Persistence of acetamiprid in soil. Bulletin of Environmental Contamination and Toxicology. 2007;**78**(5):349-352. DOI: 10.1007/s00128-007-9097-7

[86] Obana H, Okihashi M, Akutsu K, Kitagawa Y, Hori S. Determination of neonicotinoid pesticide residues in vegetables and fruits with solid phase extraction and liquid chromatography mass spectrometry. Journal of Agricultural and Food Chemistry. 2003;**51**(9):2501-2505. DOI: 10.1021/jf0261102

[87] Ferrer I, Thurman EM, Fernández-Alba AR. Quantitation and accurate mass analysis of pesticides in vegetables by LC/TOF-MS. Analytical

Chemistry. 2005;**77**(9):2818-2825. DOI: 10.1021/ac048458x

[88] Marín A, Martínez Vidal JL, Egea Gonzalez FJ, Garrido Frenich A, Glass CR, Sykes M. Assessment of potential (inhalation and dermal) and actual exposure toacetamiprid by greenhouse applicators using liquid chromatography-tandem massspectrometry. Journal of Chromatography B: Analytical Technologies in the Biomedical and Life Sciences. 2004;**804**(2):269-275. DOI: 10.1016/j.jchromb.2004.01.022

[89] Imamura T, Yanagawa Y, Nishikawa K, Matsumoto N, Sakamoto T. Two cases of acute poisoning with acetamiprid in humans. Clinical Toxicology (Philadelphia, Pa.). 2010;**48**(8):851-853. DOI: 10.3109/15563650.2010.517207

[90] Kocaman AY, Topaktaş M. In vitro evaluation of the genotoxicity of acetamiprid in human peripheral blood lymphocytes. Environmental and Molecular Mutagenesis. 2007;**48**(6):483-490. DOI: 10.1002/em.20309

[91] Yamamoto, I. Nicotine to Nicotinoids: 1962 to 1997. In: Nicotinoid insecticides and the nicotinic acetylcholine receptor. Springer-Verlag, Tokyo. 1999. p. 3-27. DOI: 10.1007/978-4-431-67933-2_1

[92] USDA Forest Service. Imidacloprid: Human Health and Ecological Risk Assessment-Final Report. 2005. Available from: ForestServiceImidacloprid.pdf (umn.edu) [Accessed: 2021-04-22]

[93] Elbert A, Becker B, Hartwig J, Erdelen C. Imidacloprid – a new systemic insecticide. Pflanzenschutz-Nachrichten Bayer. 1991;**44**(2):113-136

[94] National Center for Biotechnology Information. PubChem Compound Summary for CID 86287518. Imidacloprid. 2021. Available from: https://pubchem.ncbi.nlm.nih.gov/compound/Imidacloprid [Accessed: 2021-07-17]

[95] Mohamed F, Gawarammana I, Robertson TA, Roberts MS, Palangasinghe C, Zawahir S, Jayamanne S, Kandasamy J, Eddleston M, Buckleyn A, Dawson AH, Roberts DM. Acute human self-poisoning with imidacloprid compound: a neonicotinoid insecticide. Plos One. 2009;**4**(4):e5127. DOI: 10.1371/journal.pone.0005127

[96] Smit CE, Posthuma-Doodeman JAM, van Vlaardingen PLA, de Jong FMW. Ecotoxicity of imidacloprid to aquatic organisms: derivation of water quality standards for peak and long-term exposure. Human and Ecological Risk Assessment: An International Journal. 2015;**21**(6):1608-1630. DOI: 10.1080/10807039.2014.964071

[97] Ewere EE, Reichelt-Brushett A, Benkendorff K. Impacts of neonicotinoids on molluscs: what we know and what we need to know. Toxics. 2021;**9**(2):21. DOI: 10.3390/toxics9020021

[98] Hallmann CA, Foppen RPB, van Turnhout CAM, de Kroon H, Jongejans E. Declines in insectivorous birds are associated with high neonicotinoid concentrations. Nature. 2014;**511**:341-343. DOI: 10.1038/nature13531

[99] Suchail S, Guez D, Belzunces LP. Discrepancy between acute and chronic toxicity induced by imidacloprid and its metabolites in *Apis mellifera*. Environmental Toxicology and Chemistry. 2009;**20**(11):2482-2486. DOI: 10.1002/etc.5620201113

[100] Bonmatin JM, Moineau I, Charvet R, Colin ME, Fleche C, Bengsch ER. Behaviour of Imidacloprid

in Fields. Toxicity for Honey Bees. In: Lichtfouse E, Schwarzbauer J, Robert D. (eds). Environmental Chemistry. Springer, Berlin, Heidelberg; 2005. p. 483-494. DOI: 10.1007/3-540-26531-7_44

[101] Mundhe SA, Birajdar SV, Chavan SS, Pawar NR. Imidacloprid poisoning: an emerging cause of potentially fatal poisoning. Indian Journal of Critical Care Medicine. 2017;**21**(11):786-788. DOI: 10.4103/ijccm.IJCCM_152_17

[102] Roberts TR, Hutson DH. Imidacloprid. Metabolic pathways of agrochemicals – Part 2: Insecticides and Fungicides. The Royal Society of Chemistry. Cambridge, UK. 1999. p. 111-120

[103] Nauen R, Tietjen K, Wagner K, Elbert A. Efficacy of plant metabolites of imidacloprid against *Myzus persicae* and *Aphis gossypii* (Homoptera: Aphididae). Pesticide Science. 1998;**52**(1):53-57. DOI: 10.1002/(SICI)1096-9063(199801)52:1<53::AID-PS621>3.0.CO;2-6

[104] Turro NJ. Modern molecular photochemistry. Benjamin-Cummings. Menlo Park, CA. 1978. pp. 305

[105] Eftink MR, Ghiron CA. Fluorescence quenching studies with proteins. Analytical Biochemistry. 1981;**114**(2):199-227. DOI: 10.1016/0003-2697(81)90474-7

[106] Tian J, Liu J, He W, Hu Z, Yao X, Chen X. Probing the binding of scutellarin to human serum albumin by circular dichroism, fluorescence spectroscopy, FTIR, and molecular modeling method. Biomacromolecules. 2004;**5**(5):1956-1961. DOI: 10.1021/bm049668m

[107] Huang Y, Zhang Z, Zhang D, Lv J. Flow-injection analysis chemiluminescence detection combined with microdiaysis sampling for studying protein binding of drug. Talanta. 2001;**53**(4):835-841. DOI: 10.1016/S0039-9140(00)00569-5

[108] Zhang J, Zhuang S, Tong C, Liu W. Probing the molecular interaction of triazole fungicides with human serum albumin by multispectroscopic techniques and molecular modeling. Journal of Agricultural and Food Chemistry. 2013;**61**(30):7203-7211. DOI: 10.1021/jf401095n

[109] Ross PD, Subramanian S. Thermodynamics of protein association reactions: forces contributing to stability. Biochemistry. 1981;**20**(11):3096-3102. DOI: 10.1021/bi00514a017

[110] Némethy G, Scheraga HA. The structure of water and hydrophobic bonding in proteins. III. The thermodynamic properties of hydrophobic bonds in proteins 1,2. Journal of Physical Chemistry. 1962;**66**(10):1773-1789. DOI: 10.1021/j100816a004

[111] Pace CN. Measuring and increasing protein stability. Trends in Biotechnology. 1990;**8**:93-98. DOI: 10.1016/0167-7799(90)90146-O

[112] Greenfield NJ. Using circular dichroism spectra to estimate protein secondary structure. Nature Protocols. 2006;**1**(6):2876-2890. DOI: 10.1038/nprot.2006.202

[113] Mitsopoulou CA, Dagas CE, Makedonas C. Synthesis, characterization, DFT studies and DNA binding of mixed platinum(II) complexes containing quenuxaline and 1,2-dithiolate ligands. Journal of Inorganic Biochemistry. 2008;**102**(1):77-86. DOI: 10.1016/j.jinorgbio.2007.07.002

[114] Zhang YL, Zhang X, Fei XC, Wang SL, Gao HW. Binding of bisphenol A and acrylamide to BSA and

DNA: insights into the comparative interactions of harmful chemicals with functional biomacromolecules. Journal of Hazardous Materials. 2010;**182** (1-3):877-885. DOI: 10.1016/j.jhazmat.2010.06.131

[115] Shahabadi N, Hadidi S. Spectroscopic studies on the interaction of calf thymus DNA with the drug levetiracetam. Spectrochimica Acta Part A: Molecular and Biomolecular Spectroscopy. 2012;**96**:278-283. DOI: 10.1016/j.saa.2012.05.045

[116] Zhang Y, Zhang G, Zhou X, Li Y. Determination of acetamiprid partial-intercalative binding to DNA by use of spectroscopic, chemometrics, and molecular docking techniques. Analytical Bioanalytical Chemistry. 2013;**405**:8871-8883. DOI: 10.1007/s00216-013-7294-2

[117] Verebová V, Želonková K, Holečková B, Staničová J. The effect of neonicotinoid insecticide thiacloprid on the structure and stability of DNA. Physiological Research. 2019;**68**(Suppl.4):459-466. DOI: 10.33549/physiolres.934385

[118] Sugio S, Kashima A, Mochizuki S, Noda M, Kobayashi K. Crystal structure of human serum albumin at 2.5 A resolution. Protein Engineering, *Design and Selection.* 1999;**12**(6):439-446. DOI: 10.1093/protein/12.6.439

[119] Lee P, Wu X. Modifications of human serum albumin and their binding effect. Current Pharmaceutical Design. 2015;**21**(14):1862-1865. DOI: 10.2174/1381612821666150302115025

[120] Chen ZG, Zhang GM, Chen ZX, Gao WH. A resonance light-scattering off–on system for studies of the selective interaction between adriamycin and DNA. Analytical and Bioanalytical Chemistry. 2012;**402**(6):2163-2171. DOI: 10.1007/s00216-011-5672-1

[121] Ahamdi F, Jamali N, Moradian R, Astinchap B. Binding studies of pyriproxyfen to DNA by multispectroscopic atomic force microscopy and molecular modeling methods. DNA and Cell Biology. 2012;**31**:259-268. DOI: 10.1089/dna.2011.1303

[122] Ahmadi F, Jafari B, Rahimi-Nasrabadi M, Ghasemi S, Ghanbari K. Proposed model for in vitro interaction between fenitrothion and DNA, by using competitive fluorescence, 31P NMR, 1H NMR, FT-IR, CD and molecular modeling. Toxicology in Vitro. 2013;**27**(2):641-650. DOI: 10.1016/j.tiv.2012.11.004

[123] Kumar CV, Turner RS, Asuncion EH. Groove binding of styrylcyanine dye to DNA double helix: the salt effect. Journal of Photochemistry and Photobiology A: Chemistry. 1993;**74**(2-3):231-238. DOI: 10.1016/1010-6030(93)80121-O

[124] Lavery R, Pullman B. Molecular electrostatic potential on the surface envelopes of macromolecules: B-DNA. International Journal of Quantum Chemistry. 1981;**20**(1):259-272. DOI: 10.1002/qua.560200123

[125] Ihmels H, Faulhaber K, Vedaldi D, Dall'Acqua F, Viola G. Intercalation of organic dye molecules into double-stranded DNA. Part 2: The annelated quinolizinium ion as a structural motif in DNA intercalators. Photochemistry and Photobiology. 2005;**81**(5):1107-1115. DOI: 10.1562/2005-01-25-IR-427

[126] Wang F, Huang W, Su L, Dong ZJ, Zhang S. Spectrofluorimetric study of the binding of codeine to nucleic acids. Journal of Molecular Structure. 2009;**927**(1-3):1-6. DOI: 10.1016/j.molstruc.2009.01.022

[127] Sun YT, Zhang HQ, Bi SY, Zhou XF, Wang L, Yan YS. Studies on the arctiin and its interaction with DNA by spectral

methods. Journal of Luminescence. 2011; **131**(11):2299-2306. DOI: 10.1016/j.jlumin.2011.04.036

[128] Shahabadi N, Kashanian S, Ahmadipour Z. DNA binding and gel electrophoresis studies of a new silver(I) complex containing 1,9-dimethyl-1,10-phenanthroline ligands. DNA and Cell Biology. 2011;**30**(3):187-194. DOI: 10.1089/dna.2010.1104

[129] Temerk YM, Ibrahim MS, Kotb M, Schuhmann W. Interaction of antitumor flavonoids with dsDNA in the absence and presence of Cu(II). Analytical and Bioanalytical Chemistry.2013;**405**: 3839-3846. DOI: 10.1007/s00216-012-6675-2

[130] Sun N, Ding Y, Tao Z, You H, Hua X, Wang M. Development of an upconversion fluorescence DNA probe for the detection of acetamiprid by magnetic nanoparticles separation. Food Chemistry. 2018;**257**:289-294. DOI: 10.1016/j.foodchem.2018.02.148

[131] Ge W, Yan S, Wang J, Zhu L, Chen A, Wang J. Oxidative stress and DNA damage induced by imidacloprid in Zebrafish (*Danio rerio*). Journal of Agricultural and Food Chemistry. 2015;**63**(6):1856-1862. DOI: 10.1021/jf504895h

[132] Bebane PSA, Hunt BJ, Pegoraro M, Jones ARC, Marshall H, Rosato E, Mallon EB. The effects of the neonicotinoid imidacloprid in gene expression and DNA methylation in the buff-tailed bumblebee *Bombus terrestris*. Proceeding of the Royal Society B. 2019;**286**(1905):20190718. DOI: 10.1098/rspb.2019.0718

[133] Hou S, Jiang X, Du F, Cong H, Li Y, Jie N. Study on the interaction between nucleic acids and imidacloprid and its application. Nucleosides, Nucleotides and Nucleic Acids. 2009;**28**(11-12):989-997. DOI: 10.1080/15257770903362156

[134] Wang C, Chu Q, Chen C, Bo Z. Investigation of the mechanism of binding of thiacloprid to human serum albumin using spectroscopic techniques and molecular modeling methods. Spectroscopy. 2011;**25**:113-122. DOI: 10.3233/SPE-2011-0498

[135] Wang Y, Tang B, Zhang H, Zhou Q, Zhang G. Studies on the interaction between imidacloprid and human serum albumin: Spectroscopic approach. Journal of Photochemistry and Photobiology B: Biology. 2009;**94**:183-190. DOI: 0.1016/j.jphotobiol.2008.11.013

[136] Mikhailopulo KI, Serchenya TS, Kiseleva EP, Chernov YG, Tsvetkova TM, Kovganko NV, Sviridov OV. Interaction of molecules of the neonicotinoid imidacloprid and its structural analogs with human serum albumin. Journal of Applied Spectroscopy. 2008;**75**(6):857-863. DOI: 10.1007/s10812-009-9120-3

[137] Sjoholm I, Ekman B, Kober A, Ljungstedt-Pahlman I, Seiving B, Sjodin T. Binding of drugs to human serum albumin: XI. The specificity of three binding sites as studied with albumin immobilized in microparticles. Molecular Pharmacology. 1979;**16**(3):767-777

[138] Huang B, Zou GL, Yang YM. Studies on the interaction between adriamy and bovine serum albumin. Acta Chimica Sinica. 2002;**60**(10): 1867-1871

[139] Liu XF, Xia YM, Fang Y, Zou L, Liu LL. Interaction between natural pharmaceutical homologued of coumarin and bovine serum albumin. Acta Chimica Sinica. 2004;**64**(16): 1484-1490

[140] Kang J, Liu Y, Xie MX, Li S, Jiang M, Wang YD. Interactions of human serum albumin with chlorogenic acid and ferulic acid. Biochimica

et Biophysica Acta (BBA) – General Subjects. 2004;**1674**(2):205-214. DOI: 10.1016/j.bbagen.2004.06.021

[141] Bao XY, Zhu ZW, Li NQ, Chen JG. Electrochemical studies of rutin interacting with hemoglobin and determination of hemoglobin. Talanta. 2001;**54**(4):591-596. DOI: 10.1016/S0039-9140(00)00667-6

[142] Chen CY, Zhao B, Wang ZW. Interaction of thiacloprid with bovine hemoglobin using spectroscopic and molecular modeling methods. Spectroscopy. 2010;**24**:559-566. DOI: 10.3233/SPE-2010-0477

[143] Sulkowska A, Równicka J, Bojko B, Sulkowski W. Interaction of anticancer drugs with human and bovine serum albumin. Journal of Molecular Structure. 2003;**651**:133-140. DOI: 10.1016/S0022-2860(02)00642-7

[144] Abou-Zied OK, Al-Shihi OIK. Characterization of subdomain IIA binding site of human serum albumin in its native, unfolded, and refolded states using small molecular probes. Journal of the American Chemical Society. 2008;**130**(32);10793-10801. DOI: 10.1021/ja8031289

[145] Ashoka S, Seetharamappa J, Kandagal PB, Shaikh SMT. Investigation of the interaction between trazodone hydrochloride and bovine serum albumin. Journal of Luminiscence. 2006;**121**(1):179-186. DOI: 10.1016/j.jlumin.2005.12.001

[146] Yan CN, Mei P, Guan ZJ, Liu Y. Studies on thermodynamics features of the interaction between imidacloprid and bovine serum albumin. Chinese Journal of Chemistry. 2007;**25**:1085-1089. DOI: 10.1002/cjoc.200790202

[147] Ranz RR. Introductory Chapter: Actuality and trends of beekeeping worldwide, Beekeeping - New Challenges. IntechOpen. 2020. DOI: 10.5772/intechopen.86890. Available from: https://www.intechopen.com/chapters/67687 [Accessed: 2021-08-21]

[148] Erickson BM. Neonicotinoid pesticides can stay in the US market, EPA says. Chemical and Engineering News. 2021; American Chemical Society. ISSN 0009-2347

Section 6

Insecticides of Plant Origin

Chapter 15

Valorization of Olive Mill Wastewater in the Control of *Aphis pomi* De Geer 1773 (Hemiptera, Aphididae) Infesting Apple Plants in Nurseries

Nahid Haouache, Soukaina El Asri, Adil Asfers, Abdelhadi Ait Houssa, Bouchra Tazi and Ahmed Boughdad

Abstract

Olive mill wastewater (OMW), are the liquid residues generated during the extraction of oil by traditional and modern three-phase type crushing units. These effluents are characterized by an acidic pH and composition rich in water, organic matter, minerals and polyphenols. In general, they are directly discharged into natural ecosystems. Their danger is linked to the enormous quantities produced in a short period between October and March. To mitigate the effects of vegetable waters on the environment, their valorization in different areas is discussed. As biopesticides, crude OMW have been shown to be very toxic to *Aphis pomi*; the LC50 and LC95 varied respectively from 27.17 to 45.59 and from 77.19 to 134.57 mg of OMW/L of water; they vary according to the stage of the aphid considered. The young stages of *A. pomi* were more sensitive than the elderly are. Therefore, the OMW can be used as a means of controlling aphids. However, before operating on a large scale, it is necessary to repeat the trials in field and assess their impact on non-target organisms and treated crops.

Keywords: Olive Mill wastewater, Biopesticides, *Aphis pomi*, Apple plants, Nursery

1. Introduction

In Morocco, with an area of 49731 ha and an annual production of 809762 t [1], apple cultivation is exposed to the pressure of various harmful biological agents; approximately 182 synthetic pesticides are registered against these organisms [2]. Aphids are small, soft-bodied insects with long, slender mouthparts used to pierce stems, leaves and other tender parts of plants and, suck up sap from the host plant. They are among the most dangerous pests of crops; they directly weaken plants by sucking their sap; these results in curling and deformation of the leaves of young shoots, which affects the photosynthetic function of the attacked plant. Among the indirect damage, aphids are vectors of many phytopathogenic viruses and the secretion of honeydew favoring the development of sooty mold on leaves and fruits [3, 4]. The green apple aphid, *Aphis pomi* De Geer (Hemiptera, Aphididae) is 1.3–2.3 mm long and light green or yellowish green in color, with short antennae and black or dark brown siphunculi; asexual development goes through 4 Nymphs and an adult (**Figure 1**). It is a monoecious holocyclic species, i.e., the aphid has one sexual generation and several asexual (parthenogenetic) wingless and /or winged generations; they grow on the same plant species or on plants of related species. The aphid is widely distributed in the northern hemisphere [5]. This species is very harmful to pome fruit (Rosaceae), especially apple trees; its infestations are rife regularly. The species is particularly harmful in nurseries and young orchards. To control aphids, apple growers only use synthetic insecticides; thus, 82 pesticides are registered against aphids [2]; these pesticides are broad spectrum and effective against many pests other than aphids; they mainly belong to the groups of organophosphates, carbamates, pyrethroids and neonicotinoids. However, the intensive use of these products raises health, environmental and ecotoxicological problems (e.g., [6–8]). The use of these pesticides also generates resistance phenomena in pests [9–11]. In addition, they can cause the resurgence of secondary pests [12]. This latter phenomenon is characterized by a reversal of the biological response such as the shortening of the duration of the development, the increase in fecundity with fertility and longevity due to the application of the sublethal doses of the pesticides used [13]. Besides the unwise use of pesticides increases the mortality of natural enemies that contribute to pest control [14, 15]; which increases the cost of production and affects the efficiency of the techniques applied and the environmental sustainability of the agroecosystem [16].

To mitigate the ecotoxicological, environmental and social consequences of synthetic pesticides; the research for effective, economical, safe and ecological alternative methods compatible with sustainable development is therefore imperative. In other words, adopt the concept of integrated pest management (IPM) [17, 18]. Among the products likely to replace synthetic pesticides and, at the same time reduce pollution of natural ecosystems; valorization of OMW in plant protection responds well to this dilemma.

Around the world, there are more than 800 million productive olive trees, occupying an area of 10 million hectares; olives are used either as table olives or for the production of olive oil. Global table olive production was 2900000 tons,

Figure 1.
Colony of Aphis pomi on an apple plant (A) and magnified 35 times under a stereomicroscope (B).

while olive oil production exceeded 3300000 tons [19]. The Mediterranean region alone provides 97% of the total world olive oil production, being close to 3 million tons [20]. The oil extraction is carried out by three processes, namely the traditional process (press process), the two-phase process and the three-phase process [21]. However, olive oil production is also accompanied by the generation of huge amounts of by-products and waste that leave a cluttered environmental footprint [22]. Indeed, during the extraction of oil by press or three-phase systems, enormous quantities of vegetable water are produced annually, the reported volumes are estimated at around 440 L of OMW for 100 kg of processed olives [23], i.e., from 10 to 30 million m^3 per year [24]. These effluents are acidic and very rich in organic compounds including polyphenols [24–26]. The management of OMW is generally unregulated; they can be discharged directly into terrestrial and aquatic ecosystems without prior treatment. This results in disastrous environmental consequences due to their high polluting capacity (e.g., [27–30]). Thus, non-target organisms in these natural compartments are negatively affected (e.g., [31–35]). In addition, the presence of organic and inorganic pollutants makes this waste flow toxic for bacteria and other microorganisms used in biological treatments [32]. In Morocco, where olive tree occupies an area of 1073493 ha and produces 1912238 t of olives [1], oil extraction is carried out mainly by traditional press and, in few cases by the three-phase or two-phase system; thus, it generates enormous quantities of OMW. Like other countries, the dumping of vegetable waters is also disposed mainly on soils and in rivers. Indeed, as indicated in **Figure 2**, during the extraction of oils from the olives, the OMW are evacuated in a channel (**Figure 2A**) discharging them either in a watercourse (**Figure 2B** and **D**) either in a terrestrial environment (**Figure 2C**).

Figure 2.
Olive mill wastewater (OMW) disposed into ecosystems ((A) discharge channel, (C) soil receptacle, (B and D) Oued receptacle).

With regard to the harmful effects of OMW and to avoid or mitigate the issues associated with them, the scientific community has felt the need to promote them in several areas. This is because vegetable waters contain valuable components, such as water, organic compounds and a wide range of nutrients that could be recycled. Then, OMW could be recycled and used as a feedstock to generate cost effective compounds such as antioxidants, enzymes, biogas, soil conditioners, feeds, foods, fertilizer, etc. (e.g., [36–42]).

In crop pest management, OMW have been tested against weeds, fungi and nematodes [43], olive psyllid [44], green apple aphid [45], date palm white worm [46], tomato broomrape [47], plant pathogens [48], Mediterranean fruit fly fruits [49]. For full details, bibliographic reviews provide a synthetic overview on the use of OMW in plant protection [50, 51]. For our part, after having evaluated the toxicity of OMW with respect to *Gammarus gauthieri* (Amphipoda, Crustaceae) [31] and earthworms [33], we tried to valorize crude OMW in the control of pests by testing them against *A. pomi*. This work aimed at replacing synthetic pesticides used in aphid control and, at the same time reducing pollution of natural ecosystems owing to OMW. To this end, the valorization of OMW in plant protection responds well to this dilemma. In this chapter, the results relating to the efficacy of crude olive mill wastewater with respect to the green apple aphid are presented, as well as their physicochemical composition.

2. Materials and methods

2.1 Location of the study nursery

The experiment was carried out in summer in a 2-year-old apple tree nursery (*Malus communis var*. Golden Delicious, Rosaceae) located near the National School of Agriculture in Meknes (GPS coordinates: 33° 50′ 37″ N, 5° 28′ 39″ W) in plastic tunnel under ambient conditions. It is installed inside an orchard of the same culture.

2.2 Origin and physicochemical composition of olive mill wastewaters

The vegetable water samples were collected from a traditional olive oil extraction unit located in the Sefrou-Morocco region (GPS coordinates: 33° 49′ 50″ N, 4° 50′ 15″ W). Sampling was carried out immediately after olive oil extraction and OMW production; the effluents were taken, on three times with an interval of 15 min between two successive sampling. The samples were brought back to the Laboratory of the National School of Agriculture in Meknes in airtight boxes and stored at 4°C to avoid any alteration of their chemical composition.

OMW analysis (pH, electrical conductivity or EC, crude protein level, potassium, phosphorus, magnesium, calcium, copper, manganese, iron, zinc, sodium, cadmium and nickel) was carried out at the Official Laboratory of Chemical Analysis and Research, Casablanca Morocco (www.loarc.ma) according to standard methods. The content of suspended solids, the chemical oxygen demand (COD), the biological oxygen demand (BOD), concentrations of nitrites, ammonium, sulphate, nitrate, orthophosphat and chlorine were determined at the Laboratory of the Department of Basic Sciences of the National School of Agriculture in Meknes as methods described in Rodier et al. [52]. The total polyphenol content was determined according to the method of Folin Ciocalteu [53, 54]; to this end, 200 μL of OMWW were mixed with 1 ml of a freshly prepared Folin–Ciocalteu reagent (10 times diluted) and 0.8 ml of 7.5% sodium carbonate (Na2CO3). The whole mixture was incubated for 30 minutes at room temperature (20° C). The reading was carried out using a UV–visible spectrophotometer (Shimadzu 1601, Europe) at a wavelength of 765 nm. The concentration of polyphenols is expressed in milligrams of gallic acid equivalent per gram of OMW).

2.3 Strain of *Aphis pomi* and biotest

The different stages of the green apple aphid (L1, L2, L3, L4, Wingless adult) (**Figure 1**) were sampled with apple leaves infested in the nursery where the experiment occurred; the nursery has not undergone any insecticide treatment before. The leaves with the insects were then sorted by stage and placed in Petri dishes 9 cm in diameter and 1.5 cm in height ventilated and closed to prevent the aphids from escaping. Each box contained 10 individuals belonging to the same stage. Leaves bearing aphids were kept turgid by wrapping their peduncle with cotton wool soaked in water. The experiment was carried out in the nursery growing under a plastic tunnel in ambient conditions.

To determine the quantity of the crude OMW to be recommended to the farmer, i.e., the dose necessary to control aphids, we sought to determine the content of crude vegetable water to be diluted in water, i.e., the concentration of these effluents to be dissolved in the water required to kill a percentage of aphids. To this end, the concentrations were determined beforehand according to the formula of [55], i.e., concentrations capable of causing the mortality of 5% and 95% of the insects used. Then, the concentrations used are 5, 10, 20, 40 and 80 mg of OMW/liter of distilled water. The solutions of the vegetable waters were prepared using a magnetic stirrer for 10 min. At the same time, two control treatments were carried out, the negative one consisting of distilled water; while the positive one using Imidacloprid (Confidor (Active matter content 200 g/L) according to the recommended dose against aphids on apple trees (50 cc/hL) [2]. For each concentration or controls, 5 replicates were performed with 10 individuals each according to a randomized complete block design. The application of OMW to the leaves bearing the aphids was performed using 1-liter hand sprayer maintained at 0.3om from the targets. The count of dead insects was carried out 48 hours after spraying; each individual shaken by a paintbrush and not moving was considered dead. Aphid mortality was assessed under a stereomicroscope.

2.4 Statistical analysis of data

The lethal concentrations (LC50 and LC95 i.e., concentrations capable of killing 50 and 95% of the population treated) of the products tested against *A. pomi* were determined by Probit analysis [55] using SPSS version 21 (IBM Corp. Released 2015. Armonk, NY: IBM Corp.). The values were considered significantly different, when their 95% confidence intervals did not overlap. To identify the toxicity caused exclusively by OMW, the mortalities recorded with each concentration were corrected with Abbott formula [56]. The comparison of the lethal effects of OMW and the pesticide on aphid stages was performed by two-factor variance analysis (Stages* products) followed by the Tukey test (HSD) at 5% as post-hoc for the multiple comparison of means. The homogeneity and normality of the variance of the dependent variables were verified by Levene and Shapiro–Wilk tests, respectively; the dataset was transformed into arsin $\sqrt{}$ (percentage) before analysis of variance according to Sokal and Rohlf [57]. Linear models relating stage mortality to OMW concentration have been established; their choice was made based on the low values of the Akaike Information Criterion (AIC) and standard errors as well as on the high values of R^2 and their significance by the analysis of variance (F). To classify the different stages of the aphid according to their responses to OMW, a hierarchical classification was carried out on the LC50 and LC95 using Statistica version 7 software (Statsoft Inc. USA). The data are summarized as graphs or tables.

3. Results

3.1 Physicochemical properties of OMW

The physicochemical characteristics of OMW obtained from an extraction unit by the traditional press process are presented in **Table 1**. The effluents are reddish-black in color and darken during storage (**Figure 2**); they have a cloudy appearance and a strong odor of olive oil. It is an acidic liquid (pH = 4.90) with an electrical conductivity of 13.5mS/cm which is characteristic of wastewater. The contents of solid matter, chemical oxygen demand (COD) and biological oxygen demand (BOD5) weigh 40, 160 and 90 g/L of OMW, respectively, they are high. Crude proteins represent approximately 1% of the composition of vegetable waters; while

OMW properties	Units	Values
pH		4.90
Electrical conductivity	mS/cm	13.5
COD	g/L	160
BOD5		90
Solid matter		40
Crude protein	%	1.4
Potassium		1.0
Calcium		0.1
Sodium		0.06
Magnesium		0.045
Phosphorus	mg/kg	641.8
Iron		40.0
Manganese		4.5
Zinc		10.0
Copper		2.2
Nickel		˂ 0.75
Cadmium		˂ 0.015
Sulfate	mg/L	1591.84
Ammonium		1342.53
Nitrate		743.39
Nitrite		260.57
Chlorine		184.6
Orthophosphate		12.43
Polyphenols	mg of GAE/g	66,19

Table 1.
Physicochemical properties of OMW collected from a traditional processing unit in Sefrou region (GAE: gallic acid equivalent).

potassium, calcium, sodium and magnesium represent from 1 to 0.05% of the total (**Table 1**). The samples of vegetable water analyzed are relatively rich in inorganic ions (P, Mn, Cu, Fe, Zn, Ni, Cd), their contents vary from approximately 642 to 0.015 mg/Kg. Sulfates, ammonium, nitrates, nitrites, chlorine and orthophosphate

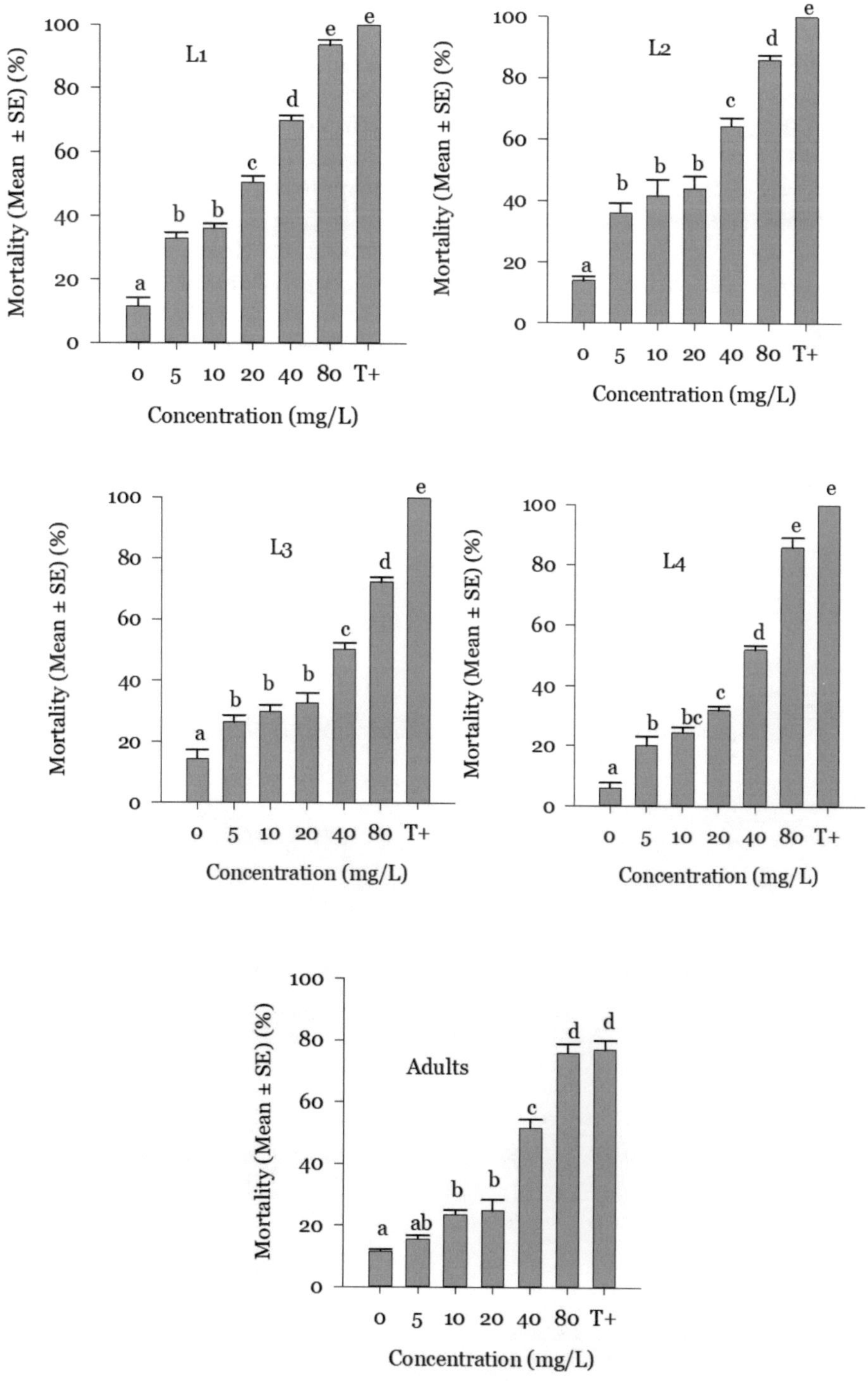

Figure 3.
Mortality of the different stages of Aphis pomi *treated with crude OMW (SE: standard error; T +: imidacloprid; the bars with the same letter do not differ statistically from each other, ANOVA2F followed by the Tukey HSD test at 5%).*

are also dissolved in vegetable waters, their concentrations ranging between approximately 1592 and 12 mg/L; while the total polyphenol content is around 66 mg /L of OMW (**Table 1**).

3.2 Aphid toxicity

Crude vegetable water from olives has been shown to be very toxic to *A. pomi*, their effects are amplified with concentration (**Figure 3**). The response of the treated individuals varies significantly between stages (F = 59.93, df = 4, 141, P < 0.001) and according to the concentration (F = 871.75, df = 6, 141, P < 0.001). Compared to the negative control, crude OMW significantly affect the survival of the different stages of the green apple aphid, the percentages of mortalities are approximately 4 to 38 times higher than the control. Whereas all stages have a comparable mortality rate in untreated lots (P > 0.05). On the other hand, compared to the positive control (Imidacloprid), only the high concentration (80 mg/L) which makes it possible to induce statistically comparable mortalities, the other concentrations cause mortalities markedly lower than this,

Stages	Models	R^2	$Df^{Residual}$	F	P
L1	y = 0.93x + 24.98	0.92	4	47.19	0.0024
L2	y = 0.78x + 27.42	0.91	4	38.11	0.0035
L3	y = 0.67x + 20.32	0.97	4	126.11	0.0004
L4	y = 0.94x + 12.45	0.98	4	252.45	<0.0001
Adults	y = 0.82x 12.70	0.98	4	171.21	0.0002

Table 2.
Relationship between concentration (x in mg of OMW/L of distilled water) and mortality (y in %) of Aphis pomi treated with crude olive mill wastewater during 48 h.

Stages	Number	Slope[b] ± SE	Constante ± SE	LC50 (CI[c]) mg/L	LC95 (CI) mg/L	χ^2	Df[d]	P[e]
L[a]1	300	0.03 ± 0.004	−0.75 ± 0.11	24.17 (19.08, 29.78)	77.19 (65.18, 96.25)	4.92	4	0.30
L2	300	0.02 ± 0.003	−0.62 ± 0.11	27.33 (20.61, 34.75)	100.00 (82.42, 130.04)	6.44	4	0.17
L3	300	0.02 ± 0.003	−0.82 ± 0.11	44.67 35.93, 57.10)	134.57 (108.30, 183.10)	2.26	4	0.69
L4	300	0.03 ± 0.003	−1.10 ± 0.12	39.08 (33.10, 46.50)	97.56 (83.28, 119.65)	3.40	4	0.49
Adults	300	0.02 ± 0.003	−1.06 ± 0.12	45.59 (38.27, 55.35)	116.37 (97.48, 147.32)	2.30	4	0.68

[a]*L = nymph.*
[b]*Probit model = slope – constante, SE: standard error.*
[c]*95% lower and upper confidence limits are shown in parenthesis expressed in mg of OMW/L of distilled water.*
[d]*Df: degree of freedom.*
[e]*Probaility.*

Table 3.
Toxicity parameters calculated for Aphis pomi exposed during 48 hours to crude OMW.

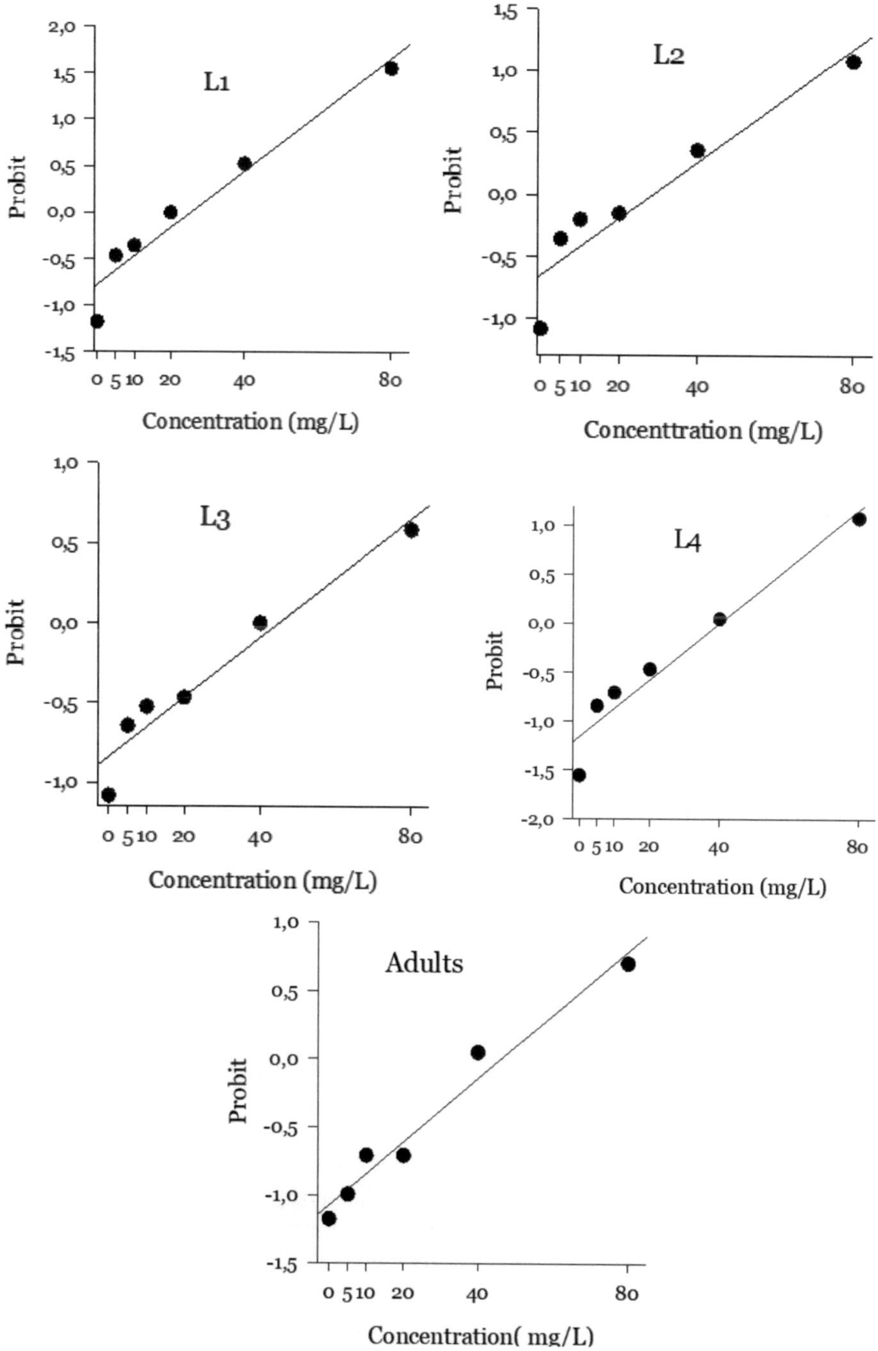

Figure 4.
Toxic activity of crude OMW against the different stages of Aphis pomi after 48 hours of treatment (on the x-axis are presented the concentrations instead of their logarithms).

$P < 0.05$ (**Figure 3**). With regard to the same concentration of OMW, the aphid's response varies depending on the stage. With 5 mg /L, the L1 and L2 are affected by the same mortality rate ($P > 0.05$); the same is true for L3 and L4 or L4 and adults compared in pairs. The same pattern is obtained with 10 mg/L. Treated with 20, 40 or 80 mg/L, the L1 and L2 were shown to be much more vulnerable than the other stages, $P < 0.05$.

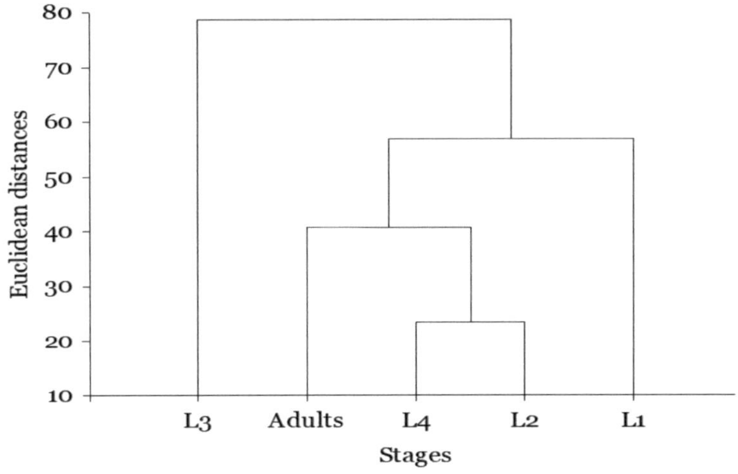

Figure 5.
Classification of the different stages according to their sensitivity to OMW (unweighted averages of lethal concentrations).

Moreover, the relationship linking the mortality of the stages of the aphid to the concentrations of OMW makes it possible to note that for each stage of the green apple aphid, the mortality therefore depends linearly and positively on the concentration of OMW (**Table 2**). In terms of corrected mortality, i.e., due exclusively to crude OMW, the percentages of dead insects vary from around 2 to 98% depending on the concentration and the stage considered with strong variations (Coefficients of variation = 4–70%).

Furthermore, the observed data fit well to the log-probit model and no statistically significant deviation from the regression equation was detected, $P > 0.05$ (**Table 3** and **Figure 4**). Lethal concentrations vary depending on the stage of the aphid considered; in fact, the extreme LC50 and LC95 vary from approximately 19 to 55 and 65 to 147 mg of OMW/L of distilled water, respectively. With regard to the values of the LC50, the tolerance of *A. pomi* increases with age, the sensitivity of the young stages is greater than that of the older ones. At high concentration (LC95), values increase with stage; while for L4 and adults, lethality is variable (**Table 3** and **Figure 4**).

With regard to lethal concentrations, the hierarchical classification of the stages of *A. pomi* according to their sensitivity to crude OMW allows them to be classified in decreasing order L1 ≤ L2 ≤ L4 < Adults <L3. The nymphs L1, L2 and L4 were therefore more vulnerable to crude OMW than adults and L3 (**Figure 5**). In terms of efficiency, the application of 135 mg of crude OMW/L of water allows to reduce the level of aphid populations below the threshold adopted by producers (15–20% of infested shoots).

4. Discussion

4.1 Physicochemical characteristics of OMW

Olive oil extraction by traditional and three-phase processes generates huge amounts of vegetable waters rich in various organic and inorganic compounds during a short period of the year (October–March). OMW have a high biological and chemical oxygen demand values, as well as high contents of organic matter, suspended matter, inhibitor substances (phenolic compounds) and minerals, especially potassium, phosphorus, magnesium and calcium [58]. The values of the physicochemical parameters presented in this work are similar to those reported in Morocco (eg, [26, 59–62]). However, the chemical composition of OMW depends on the olive variety with the stage of maturity of the olives, the harvest period and the extraction techniques [63, 64]. Their physicochemical characteristics confer them the status of polluting substances posing serious environmental problems. Indeed, certain elements present in OMW are responsible for their harmful effects. Thus, the high concentration of phenolic compounds in vegetable waters generates phytotoxic effects [63]. Moreover, according to [65], heavy metals (Cu, Ni, Pb, Cr, Zn) affect the performance of anaerobic digestion by inhibiting microorganisms; indeed, during the production of biogas by the anaerobic digestion process from of the olive mill waste, the methanogenic bacteria are inhibited. The recovery, recycling, and reuse of these by-products are considered the best options for a sustainable water management program. Thus, their use as bio-pesticides is one of the valorization modalities that we have tried to apply in this work.

4.2 Valorization of OMW in aphid control

The damage caused by *A. pomi*, especially in nurseries and young plantations, force producers to treat frequently the pest with synthetic insecticides. The misuse of synthetic pesticides, however, raises safety and environmental issues [6–8]. At the same time, olive oil extraction by traditional and three-phase systems generates enormous volumes of vegetable waters, which are generally discharged into natural ecosystems [24–26]. To mitigate the undesirable effects of synthetic pesticides and at the same time those of OMW, it has proved to be imperative to replace synthetic aphicides by valorizing vegetable water in the management of *A. pomi*; this will contribute to crop protection while solving the environmental and health problems raised by the two categories of products. The treatment of the aphid with crude OMW has made it possible to prove their effectiveness. Thus, the crude vegetable waters tested against *A. pomi* were toxic to aphids. Indeed, the applied concentrations cause a variable mortality according to the stage and dependent on the concentration. The LC50 and LC95 vary from 27.17 to 45.59 and from 77.19 to 134.57 mg of crude OMW /L of water, respectively; the young stages were more vulnerable to vegetable waters than the older ones and adults. By high concentration, crude OMW equal the efficacy of the reference product, imidacloprid, used at the recommended dose in the field. In terms of efficiency, to reduce the level of populations below the threshold adopted by producers, ie, 15–20% of infested shoots, an average dose of 135 mg of crude OMW/L of water can satisfy the farmer to control aphids.

Various studies have evaluated the effectiveness of the OMW or their polyphenols in the management of insect pests on different rops. Thus, for example, the treatment of *Euphyllura olivina* Costa (Hemiptera, Psyllidae), olive psyllid, with polyphenols from OMW at a rate of 2 g of hydroxytyrosol/L of water in an olive grove, allowed to control 41.1% of larvae and 72% of adults. In contrast, with regard to eggs, the products was ineffective [44]. Against *A. pomi*, crude OMW or their polyphenols cause significantly higher mortalities than the controls used; their toxicity depends both on the stages of the insect and on the concentration of the products tested; the average LC50s are around 25.10 ml of crude OMW and 42.8 mg of polyphenols/L of distilled water. The toxicity caused by crude OMW depends in 67% to approximately 100% of cases on that of the polyphenols contained in them [45]. Spraying the larvae of *Potosia opaca* (Coleoptera, Scarabeidae), a date palm pest in Morocco, with crude OMW, allowed to obtain the same efficacy as with the reference insecticides (chlorpyriphos-ethyl or chlorpyriphos-ethy + cypermethrin) 19 days after treatment; vegetable waters are both toxic and affect the weight of survivors; their toxicity depends directly on their polyphenol content [46]. Tested on the Mediterranean fruit fly (*Ceratitis capitata* (Wiedemann) (Diptera, Tephriditae), the polyphenolic fractions of OMW inhibit egg hatching and female fecundity without affecting larval development [49]. Overall, from all the studies cited in this paragraph, it emerges that the toxicity caused by OMW depends mainly on their polyphenol content (*op. Cit.*). In addition, although, the biochemical modes of action of OMW have not yet been elucidated in insects, the high levels of phenols present in vegetable water could block the transmission of nerve impulses [66, 67]. However, in this case, it is not excluded that the vegetable waters contained insecticides, in this case organophosphates and/or carbamates, used against the olive fly and which inhibit acetylcholinesterase (eg, [68]).

5. Conclusions and perspectives

Rejected agricultural by-products offer multiple opportunities for recovery and have significant potential not only in the agricultural and agrifood sectors but also in plant protection. In fact, in this work, crude OMW tested against *A. pomi* were effective in reducing the level of their populations to economically tolerable levels. However, the effect of products tested in nursery pest management must be compatible with integrated pest management (IPM) concept. Since, some plant producers also carry out augmentative releases of natural enemies (Unpublished data). Therefore, like conventional pesticides, risk assessment of side effects of OMW is still necessary [17, 18]; the evaluation of the effects of OMW on non-target organisms must include both lethal and sublethal effects (e.g., [14, 15]). In the event that the natural enemies bred massively and purchased by plant producers, their releases must be carried out outside the treatment periods. It is also possible to spray against pests with OMW outside the activity of natural enemies; preferably during vegetative rest against overwintering forms.

Moreover, knowing that OMW can also show phytotoxicity [69], an evaluation in this direction is planned. Our work can help to enhance the use of MOW to control the green apple aphid among other pests while integrating the ecological services provided by beneficial organisms in agroecosystems, and at the same time avoid the harmfulness of OMW. At the industrial level, the large-scale direct extraction of polyphenols for the production of biopesticides would result in high added-value. The identification and quantification of the constituents of polyphenols with their biochemical modes of action in treated pests should precede the economic estimation of pest control based on OMW and their polyphenols.

Acknowledgements

We express our sincere thanks to the National School of Agriculture of Meknes, to the Moroccan Association for the Protection of Plants and to Louata Field -Providence Verte Group for their valuable material assistance, and to the farmer for the domiciliation of the study. We also thank the entire Intech Open team for their invitation and for their judicious and relevant reviews of the manuscript.

Author details

Nahid Haouache[1], Soukaina El Asri[2], Adil Asfers[3], Abdelhadi Ait Houssa[3], Bouchra Tazi[4] and Ahmed Boughdad[2]*

1 Plant Protection Service, Menara Garden, Marrakech, Morocco

2 Department of Plant Protection and Environment, National School of Agriculture, Meknes, Morocco

3 Louata Field – Providence Verte Group, Sefrou, Morocco

4 Department of Basic Sciences, National School of Agriculture of Meknes, Meknes, Morocco

*Address all correspondence to: boughdad@enameknes.ac.ma

References

[1] FAOSTAT, 2019. Crop statistics, Food and Agriculture Organization of the United Nations. Available from: http://www.fao.org/faostat/fr/#data/QC (Acceced: 2019-07-19)

[2] ONSSA, 2019. http://eservice.onssa.gov.ma/IndPesticide.aspx (Acceced June 2021)

[3] Beers E. H., D. M. Suckling, R. J. Prokopy and J. Avilla. Ecology and Management of Apple Arthropod Pests. In Ferree D.C. and I.J. Warrington (eds), Apples: Botany, Production and Uses, CAB International 2003, pp: 489-519.

[4] Alford D.V. A Colour Handbook Pests of Fruit Crops, Second Edition. CRC Press Taylor & Francis Group; 2014. 434 p.

[5] Van Emden H. F. and Harrington R. (eds),. Aphids as Crop Pests. CABI, Second Edition. www.cabi.org; 2017. 686 p.

[6] Guedes R. N. C., Smagghe G., Stark J. D., & Desneux N. Pesticide-Induced Stress in Arthropod Pests for Optimized Integrated Pest Management Programs. Annual Review of Entomology. 2016; 61(1), 43-62. https://doi:10.1146/annurev-ento-010715-023646

[7] Kim KH, Kabir E, Jahan SA. Exposure to pesticides and the associated human health effects. Science of the Total Environment. 2017. 575:525-235. https://DOI:10.1016/j.scitotenv.2016.09.009

[8] Carvalho F. P. Pesticides, environment, and food safety. Pesticides, Food and Energy Security. 2017; 6(2): 48-60. https://doi:0.1002/fes3.108

[9] Tamaš N., Dojnov B., Margetić A., Vujčić M., Špirović B., Miletić N., Stević M. and Vujčić Z. Resistance to common organophosphate and carbamate insecticides in Aphis pomi (Hemiptera: Aphididae). Fruits. 2015; 70(3): 135-142. https://doi.org/10.1051/fruits/2015005

[10] Guedes R. N. C, Walse S. S. and Throne J. E. Sublethal exposure, insecticide resistance, and community stress. Current Opinion in Insect Science. 2017; 21:47-53. http://dx.doi.org/10.1016/j.cois.2017.04.010

[11] Paula, D. P., Menger, J., Andow, D. A., & Koch, R. L. Diverse patterns of constitutive and inducible overexpression of detoxifying enzyme genes among resistant Aphis glycines populations. Pesticide Biochemistry and Physiology. 2020; 164:100-114. https://doi.org/10.1016/j.pestbp.2019.12.012

[12] Cutler C., G., Ramanaidu, K., Astatkie, T., & Isman, M. B. Green peach aphid, *Myzus persicae* (Hemiptera: Aphididae), reproduction during exposure to sublethal concentrations of imidacloprid and azadirachtin. Pest Management Science. 2009; 65(2):205-209. https://doi.org/10.1002/ps.1669

[13] Guedes, R. N. C., & Cutler, G. C. Insecticide-induced hormesis and arthropod pest management. Pest Management Science. 2014; 70(5): 690-697. https://doi.org/10.1002/ps.3669

[14] Stark JD, Banks JE. Population-level effects of pesticides and other toxicants on arthropods. Annual Review of Entomology. 2003; 48: 505-519. https://doi.org/10.1146/annurev.ento.48.091801.112621

[15] Desneux, N., Decourtye, A., Delpuech, J.M. The sublethal effects of pesticides on beneficial arthropods. Annu. Rev. Entomol. 2007; 52:81-106. https://doi.org/10.1146/annurev.ento.52.110405.091440

[16] Ribeiro, L.P., Zanardi, O.Z., Vendramim, J.D., Yamamoto, P.T. Comparative toxicity of an acetogenin-based extract and commercial pesticides against citrus red mite. Exp. Appl. Acarol. 2014; 64, 87-98.

[17] Radcliffe, E. B., Hutchison, W. D., Cancelado, R. E. (Eds.), Integrated pest management: concepts, tactics, strategies and case studies. Cambridge University Press. 2009.

[18] Abrol D.P. Integrated pest management: current concepts and ecological perspective. Academic Press, San Diego, Calif; 2014. 576p.

[19] IOOC, 2018. International Olive Oil Council Webpage, World Olive Oil Figures. Available from: http://www.internationaloliveoil.org. (Accessed: 2021-08-05)

[20] Vossen, P. Growing olives for oil. In: R. Aparicio & J. Harwood (Eds.). Handbook of olive oil: Analysis and properties. Springer; 2013. (pp. 19-56). https://doi:10.1007/978-1-4614-7777-8_2

[21] Azbar, N., Bayram, A., Filibeli, A., Muezzinoglu, A., Sengul, F., Ozer, A. A review of waste management options in olive oil production. Critical Reviews in Environmental Science and Technology. 2004; 34 (3): 209-247. https://doi.org/10.1080/10643380490279932

[22] Galanakis C. M. (ed.). Olive Mill Waste Recent Advances for Sustainable Management. Academic Press; 2017. 284p.

[23] Alu'datt, M.H., Alli, I., Ereifej, K., Alhamad, M., Al-Tawaha, A.R., Rababah, T. Optimisation, characterisation and quantification of phenolic compounds in olive cake. Food Chemistry. 2010; 123(1):117-122. https://doi.10.1016/j.foodchem.2010.04.011

[24] El-Abbassi, A., Kiai, H., Hafidi, A. Phenolic profile and antioxidant activities of olive mill wastewater. Food Chem. 2012; 132(1), 406-412

[25] Jail A., Boukhoubza F., Nejmeddine A., Sayadic S., Hassani L. Co-treatment of olive-mill and urban wastewaters by experimental stabilization ponds. Journal of Hazardous Materials. 2010; 176: 893-900. https://doi:10.1016/j.jhazmat.2009.11.120

[26] El Yamani, M. E., Sakar, E. H., Boussakouran, A., Ghabbour, N., & Rharrabti, Y. Physicochemical and microbiological characterization of olive mill wastewater (OMW) from different regions of northern Morocco. Environmental Technology. 2020; 41(23): 3081-3093. https://doi.org/10.1080/09593330.2019.1597926

[27] Salomone, R., Ioppolo, G. Environmental impacts of olive oil production: a Life Cycle Assessment case study in the province of Messina (Sicily). Journal of Cleaner Production. 2012; 28:88-100. https://doi.org/10.1016/j.jclepro.2011.10.004

[28] Uyar, B., Kars, G., Eroğlu, İ., Yücel, M., Gündüz, U. Hydrogen production via photofermentation, in: Azbar, N., Levin, D.B. (Eds.), State of the art and progress in production of biohydrogen. Bentham Science Publishers, United Arab Emirates; 2012. pp. 54-77

[29] Dermeche S., M. Nadoura, C. Larroche, F. Moulti-Matia, P. Michaud. Olive mill wastes: Biochemical characterizations and valorization strategies Process Biochemistry. , 2013; 48(10): 1532-1552. https://doi.org/10.1016/j.procbio.2013.07.010

[30] Haddad, K., Jeguirim, M., Jellali, S., Thevenin, N., Ruidavets, L., & Limousy, L. Biochar production from Cypress sawdust and olive mill wastewater: Agronomic approach. Science of The

Total Environment. 2020; 141713. https://doi:10.1016/j.scitotenv.2020.141713

[31] Haouache N. A. Bouchaleta. Olive mill wastewater: Effect on the survival and reproduction of the ecological indicator Gammarus gauthieri. Journal of Materials and Environmental Science. 2016; 7 (7): 2288-2294

[32] Souilem, S., El-Abbassi, A., Kiai, H., Hafidi, A., Sayadi, S., Galanakis, C.M. Olive oil production sector: environmental effects and sustainability challenges. In: Galanakis C.M., .ed. Olive Mill Waste: Recent Advances for Sustainable Management. Academic Press. 2017. Pp: 1-28. https://doi.org/10.1016/B978-0-12-805314-0.00001-7

[33] Haouache N. Bouchelta A., Boughdad A., Bouabid R., Hannour K. Effect and Chemical composition of olive mill wastewater on the survival and the fecundity of *Eisenia foetida* (Haplotaxida, Lumbricidae, Savigny, 1826). International Journal of Academic Studies. 2018; 4(02): 36-49

[34] Mekersi N., Kadi K., Casini S., Addad ., Bazri K., Marref S., Lekmine S., Amari A. Effects of single and combined olive mill wastewater and olive mill pomace on the growth, reproduction, and survival of two earthworm species (*Aporrectodea trapezoides, Eisenia fetida*). Applied Soil Ecology. 2021; 168, 10412. https://doi.org/10.1016/j.apsoil.2021.104123

[35] Patsios S. I., Kontogiannopoulos K. N. and Banias G. F. Environmental impact assessment in agri-production: a comparative study of olive oil production in two European countries In: Bochtis D, Achillas C., Banias G. & M. Lampridi (eds). Bio-Economy and Agri-production Concepts and Evidence. Academic Press; 2021. Pp:83-116. https://doi.org/10.1016/B978-0-12-819774-5.00005-9

[36] Muktadirul Bari Chowdhury, A. K. M., Akratos, C. S., Vayenas, D. V., & Pavlou, S. Olive mill waste composting: A review. International Biodeterioration & Biodegradation. 2013; 85, 108-119. https://doi:10.1016/j.ibiod.2013.06.019

[37] Kourmentza, C., Koutra, E., Venetsaneas, N., Kornaros, M. Integrated biorefinery approach for the valorization of olive mill waste streams towards sustainable biofuels and bio-based products, in: Kalia, V., Kumar, P. (Eds.), Microbial Applications Vol. 1. Springer; 2017. pp. 211-238.

[38] Gullón, B., Gullón, P., Eibes, G., Cara, C., De Torres, A., López-Linares, J.C., Ruiz, E., Castro, E. Valorisation of olive agro-industrial by-products as a source of bioactive compounds. Sci. Total Environ. 2018; 645: 533-542

[39] Sarris, D., Rapti, A., Papafotis, N., Koutinas, A. A., & Papanikolaou, S. Production of added-value chemical compounds through bioconversions of olive-mill wastewaters blended with crude glycerol by a Yarrowia lipolytica strain. Molecules. 2019 ; 24(2), 222. https://doi.org/10.3390/molecules24020222

[40] Ahmed P.M., Fernández P.M., Figueroa L.I.C., Pajot H.F. Exploitation alternatives of olive mill wastewater: production of value added compounds useful for industry and agriculture. Biofuel Research Journal. 2019; 22:980-994. https://doi:10.18331/BRJ2019.6.2.4

[41] Zema, D.A.; Esteban Lucas-Borja, M.; Andiloro, S.; Tamburino, V.; Zimbone, S.M. Short-Term Effects of Olive MillWastewater Application on the Hydrological and Physico-Chemical Properties of a Loamy Soil. Agric. Water Manag. 2019; 221: 312-321.

[42] Ahmad, T., Belwal, T., Li, L., Ramola, S., Aadil, R. M., Abdullah, Zisheng, L. Utilization of wastewater from edible oil industry, turning waste

into valuable products: A review. Trends in Food Science & Technology. 2020; 99:21-33. https://doi.org/10.1016/j.tifs.2020.02.017

[43] Cayuela, M. L., Millner, P. D., Meyer, S. L. F., & Roig, A. Potential of olive mill waste and compost as biobased pesticides against weeds, fungi, and nematodes. Science of The Total Environment. , 2008 ; 399(1-3): 11-18. https://doi:10.1016/j.scitotenv.2008.03.031

[44] Debo, A., Yangui, T., Dhouib, A., Ksantini,M., Sayadi, S. Efficacy of a hydroxytyrosolrich preparation from olive mill wastewater for control of olive psyllid, Euphyllura olivina, infestations. Crop. Protection. 2011; 30(12): 1529-1534. http://dx.doi.org/10.1016/j.cropro

[45] Haouache N, Boughdad A. Utilisation des margines contre Aphis pomi (De Geer, 1773) (Homoptera, Aphididae). Neuvième Congrès de l'Association Marocaine de Protection des Plantes –Rabat Maroc. 2014. 15p.

[46] Boutaj, H., Boutasknit, A., Anli, M., Ait Ahmed, M., El Abbassi, A., & Meddich, A. Insecticidal Effect of Olive Mill Wastewaters on Potosia opaca (Coleoptera: Scarabeidae) Larva. Waste and Biomass Valorization. 2020 ; 11 (7): 3397-3405. https://doi:10.1007/s12649-019-00682-1

[47] Qasem, J. R. Control of branched broomrape (*Orobanche ramosa* L.) in tomato (*Lycopersicon esculentum* mill.) by olive cake and olive mill wastewater. Crop Protection. 2020; 129:105021. https://doi.org/10.1016/j.cropro.2019.105021

[48] Sciubba F, Chronopoulou L, Pizzichini D, Lionetti V, Fontana C, Aromolo R, Socciarelli S, Gambelli L, Bartolacci B, Finotti E, Benedetti A, Miccheli A, Neri U, Palocci C, Bellincampi D. Olive Mill Wastes: A Source of Bioactive Molecules for Plant Growth and Protection against Pathogens. Biology. 2020 ; 9(12):450. https://doi.org/10.3390/biology9120450

[49] Di Ilio1 V., & Cristofaro M. Polyphenolic extracts from the olive mill wastewater as a source of biopesticides and their effects on the life cycle of the Mediterranean fruit fly *Ceratitis capitata* (Diptera, Tephriditae). International Journal of Tropical Insect Science. 2021; 41:359-366. https://doi.org/10.1007/s42690-020-00224-6

[50] El-Abbassi A, Saadaoui N, Kiai H, Raiti J, Hafidi A. Potential applications of olive mill wastewater as biopesticide for crops protection. Science of the Total Environment. 2017 ; 576 (15): 10-21. https://doi.org/10.1016/j.scitotenv.2016.10.032

[51] Bailón-Salas, A. M., Ordaz-Diaz, L. A., López-Serrano, P. M., Flores-Villegas, M. Y., and Dominguez-Calleros, P. A. Wastewater as a resource for pest control: An overview. BioResources. 2021 ; 16(3): 6401-6425. https://doi:10.15376/biores.16.3.Bailon-Salas

[52] Rodier J., Legube B., Merlet N. Analyse de l'eau. Dunod, Paris. 2009. 1526 p

[53] Blainski, A., Lopes, G., & de Mello, J. Application and Analysis of the Folin Ciocalteu Method for the Determination of the Total Phenolic Content from *Limonium Brasiliense* L. Molecules. 2013 ; 18 (6): 6852-6865. https://doi:10.3390/molecules18066852

[54] Rover, M. R., & Brown, R. C. Quantification of total phenols in bio-oil using the Folin–Ciocalteu method. Journal of Analytical and Applied Pyrolysis. 2013; 104: 366-371. https://doi:10.1016/j.jaap.2013.06.011

[55] Finney, D.J. Probit Analysis. 3rd ed. Cambridge University Press. Cambridge. 1971. 333p.

[56] Abbott, W.S. A method of computing the effectiveness of an insecticide. J. Econ. Entomol. 1925; 18, 265-267.

[57] Sokal R. R., F. J. Rohlf, Biometry: The Principles and Practices of Statistics in Biological Research. Third edition. Publisher: W. H. Freeman. 1995. 899p

[58] Patsios, S.I., Kontogiannopoulos, K.N., Mitrouli, S.T, Plakas, K.V., Karabelas, A.J. Characterisation of agricultural waste co- and by-products, Report, AgroCycle project Deliverable 1.2. 2016. Available online at: http://www.agrocycle.eu/files/2017/10/D1.2_AgroCycle.pdf (Accessed: 2021-7-20)

[59] Ben Sassi, A., Boularbah, A., Jaouad, A., Walker, G., & Boussaid, A. A comparison of Olive oil Mill Wastewaters (OMW) from three different processes in Morocco. Process Biochemistry. 2006; 41(1): 74-78. https://doi:10.1016/j.procbio.2005.03.074

[60] Zbakh, H., El Abbassi, A. Potential use of olive mill wastewater in the preparation of functional beverages: a review. Journal of Functional Foods. , 2012; 4(1):53-65. https://doi:10.1016/j.jff.2012.01.002

[61] Belaqziz, M., El-Abbassi, A., Lakhal, E. K., Agrafioti, E., & Galanakis, C. M. Agronomic application of olive mill wastewater: Effects on maize production and soil properties. Journal of Environmental Management. 2016; 171: 158-165. https://doi:10.1016/j.jenvman.2016.02.006

[62] Elabdouni, A., Haboubi, K., Merimi, I., & El Youbi, M. S. Olive mill wastewater (OMW) production in the province of Al-Hoceima (Morocco) and their physico-chemical characterization by mill types. Materials Today: Proceedings. 2020; 27(4):3145-3150. https://doi.org/10.1016/j.matpr.2020.03.806

[63] Magdich, S., Jarboui, R., Rouina, B.B., Boukhris,M. and Ammar, E. A yearly spraying of olive mill wastewater on agricultural soil over six successive years: Impact of different application rates on olive production, phenolic compounds, phytotoxicity and microbial counts. Science of the Total Environment. 2012; 430(15): 209-216. https://doi.org/10.1016/j.scitotenv.2012.05.004

[64] Bene C. D., Pellegrino E., Debolini M., Silvestri N., Bonari E. Short- and long-term effects of olive mill wastewater land spreading on soil chemical and biological properties. Soil Biology & Biochemistry. 2013; 56: 21 – 30

[65] Alcrudeashdeh, K.A.b.; Gul, E.; Yang, Q.; Yang, H.; Bartocci, P.; Fantozzi, F. Effect of Heavy Metals in the Performance of Anaerobic Digestion of Olive Mill Waste. Processes. 2020; 8, 1146. https://doi.org/10.3390/pr8091146

[66] Danellakis, D., Ntaikou, I., Kornaros, M., Dailianis, S. Olive oil mill wastewater toxicity in the marine environment: alterations of stress indices in tissues of mussel *Mytilus galloprovincialis*. Aquatic Toxicology. 2011; 101(2): 358-366. https://doi:10.1016/j.aquatox.2010.11.015

[67] Campani, T., Caliani, I., Pozzuoli, C., Romi, M., Fossi, M.C., Casini, S. Assessment of toxicological effects of crude and bioremediated olive mill waste in the earthworm Eisenia fetida: a biomarker approach for sustainable agriculture. Applied Soil Ecology. 2017; 119:18-25. https://doi:10.1016/j.apsoil.2017.05.016

[68] Casida, J. E., & Durkin, K. A. Neuroactive Insecticides: Targets, Selectivity, Resistance, and Secondary Effects. Annual Review of Entomology.

2013; 58(1): 99-117. https://doi:10.1146/annurev-ento-120811-153645

[69] Enaime, G., Baçaoui, A., Yaacoubi, A. et al. Phytotoxicity assessment of olive mill wastewater treated by different technologies: effect on seed germination of maize and tomato. Enviro nmental Science and Pollution Research. 2020; 27: 8034-8045. https://doi.org/10.1007/s11356-019-06672-z

Section 7

Resistance and Residue to Insecticides

Chapter 16

Insecticide Resistance in Vectors of Medically Important Parasitic Infections

Taruna Kaura, Nikita Sharma and Abhishek Mewara

Abstract

Insecticide resistance is a major threat to vector control programmes as insecticides still remain the most effective method to control the vector-borne diseases. For effective management of insecticide resistance, a knowledge of the insecticides used along with their mode of action is a prerequisite to optimize their use. Nowadays, different detection methods, *viz.*, phenotypic, genotypic and proteomic assays are used for assessment of insecticide resistance in vectors. An understanding of the phenotypic and genotypic variations present in the vectors help in implementation of these techniques to evaluate the usefulness of insecticides in an area and to determine the efficacy of an ongoing vector control programmes. The understanding of different factors involved in emergence of insecticide resistance and the alternative solutions to control this problem by the use of rotational, mixture of insecticides and use of piperonyl butoxide to increase the efficacy of indoor residual spray and insecticide treated bed nets are some of the steps taken to tackle the problem of insecticide resistance in vectors.

Keywords: insecticides, resistance, vectors, parasitic infections, bioassays

1. Introduction

Many fatal disease-causing pathogens are transmitted to humans by insects which belong to the phylum "Arthropoda". These insects are known as vectors when they harbor the causative organisms in them and transmit it to other humans and animals. Most of the vectors have blood-sucking mouth parts and can transmit pathogens like parasites, viruses and bacteria. The vector-borne diseases are one of the significant causes of morbidity and mortality, particularly in the endemic regions of the tropical and subtropical nations [1], and affect more than 80% of world population. Numerous parasitic infections such as malaria, babesiosis, trypanosomiasis, leishmaniasis and filariasis which affect vast human populations are transmitted by these vectors. For most of these diseases, there is still no effective vaccine is available and a significant strategy to prevent and control these diseases is the control of their vectors by using different methods [2].

Among the various methods used for the control of vector-borne diseases, the most effective and common method is the use of insecticides. All the vector control programmes depend upon the use of insecticides in the form of larvicides, adulticides and insecticide treated nets [3]. It is not easy to say when insecticides were first used for vector control, but at least since 1000 BC people have been using natural chemicals, i.e., inorganic sulfur against the pest insects [4]. The first chemical insecticide synthesized for the control of medically important vector mosquito, which transmits malaria, was DDT in 1874. DDT was continuously use for the control of different pest insects until the first half of the 20th century, when due to development of resistance it was replaced by other insecticides such as organophosphates and carbamates [5]. The main hindrance to achieve success in vector control programmes is the development of insecticide resistance due to their overuse. Such resistance has a direct effect on the vector in terms of its longevity, infectiousness and on the management of disease [6].

2. Insects as vectors of parasitic infections

Most of the harmful parasitic infections are transmitted to humans by insects which have blood feeding behavior [7]. The parasitic infections such as malaria, lymphatic and non-lymphatic filariasis, leishmaniasis, sleeping sickness or the Human African Trypanosomiasis (HAT), and Chagas' disease or the American Trypanosomiasis, are a great burden to human health and life, especially in the poorest countries [8]. Malaria is a human parasitic disease with a very high burden. It is now especially important due to the widespread drug-resistant malaria and is at a risk of reemergence in many places worldwide. Malaria is transmitted by the bite of the mosquito species belonging to genus *Anopheles*, and filariasis is transmitted by the bite of *Culex* mosquito (**Table 1**). The parasitic zoonotic disease leishmaniasis is transmitted via the bite of phlebotomine sandflies in the Old and New World, tsetse fly is the vector of HAT, and triatomine kissing bugs are the vectors of Chagas' disease [8].

The vast expansion of these vector populations has become a growing concern and their control by different classes of insecticides is the most common method for their control, as insecticides suppress the insect populations by targeting insect metabolism in specific ways.

Insect	Disease	Insecticide use for their control
Mosquito	Malaria, filariasis	DDT, malathion, pyrethrum, deltamethrin, cyfluthrin,
Sandfly	Leishmaniasis	DDT, alpha cypermethrin, deltamethrin, deltamethrin + PBO
Tsetse fly	Human African Trypanosomiasis (HAT)	DDT, deltamethrin, HCB, dieldrin
Triatomine *bugs*	Chagas' disease/American trypanosomiasis	Pyrethroid, deltamethrin, fluralaner or afoxolaner

DDT: dichlorodiphenyltrichloroethane; HCB: hexachlorobenzene; PBO: piperonyl butoxide.

Table 1.
Major group of insects causing human diseases and insecticides used for their control [7, 8].

3. Types of insecticides and their modes of action

There are many classes of insecticides with varying modes of action. The insecticides used in different vector control programmes are generally classified into four classes on the basis of their chemistry, toxicological action, or their mode of penetration, *viz.*, organochlorines, organophosphates, carbamates, synthetic pyrethroids insect growth regulators and bacterial larvicides (**Figure 1**).

3.1 Organochlorines

These are chlorinated hydrocarbons which represent diverse group of compounds with carbon, hydrogen and chlorine in their structure. They comprise of three subgroups, namely, dichlorodiphenylethanes (dichlorodiphenyltrichloroethane [DDT], dicofol, methoxychlor, and perthane), chlorinated cyclodienes (aldrin, endrin, dieldrin, chlordane, endosulfan, and heptachlor), and hexachlorocyclohexanes (benzene hexachloride [BHC], chlordane, lindane, mirex, and toxaphene) [9]. The only organochlorine compound being used in residual spraying is DDT with 82% of organochlorines compounds being used in Southeast Asian region, mainly India, for vector control. DDT causes the alteration of the sodium and potassium ion transport across axonal membranes; this results in increased negative after-potential and prolonged action potentials, which consequently leads to repeated firing and occurrence of sequential action potentials, thus causing spasms and death of the insect [10]. DDT was first used during World War II for the control of mosquitoes and was extensively used in the period from 1940s to 1960s, and was banned in 1972 by the Environmental Protection Agency (U.S.A.).

3.2 Organophosphates

These are a group of synthetic compounds produced by the reaction of alcohols and phosphoric acid. These inhibit the enzyme acetylcholinesterase (AChE), which is responsible for the degradation of acetylcholine. The organophosphate binds to the enzyme, causing it to undergo a conformational change at its binding site to acetylcholine [11]. Their application is mainly by three methods: residual spraying, space spraying and to a lesser extent as larvicides. The organophosphates are extensively used in Southeast Asia, followed by the Americas and the Western Pacific.

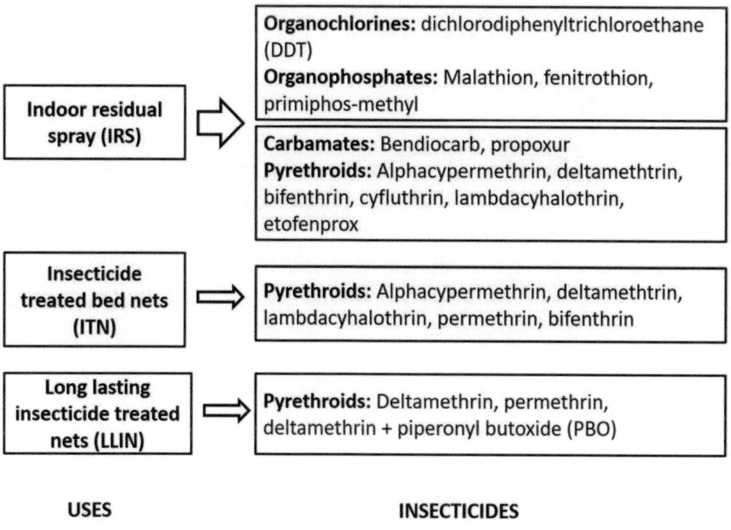

Figure 1.
Insecticides used for indoor residual spray (IRS), insecticide treated nets (ITN) and long-lasting insecticide treated nets (LLIN) [9, 11–14].

3.3 Carbamates

These are esters of carbamic acid and structurally and mechanistically similar to organophosphate (OP) insecticides. These insecticides work by inhibiting AChE and are commonly used to control agricultural pests. Compared with the other classes of insecticides, the use of carbamates is limited, and mainly used for residual spraying in the African Region [12].

3.4 Pyrethroids

These compounds are organic, and similar to the naturally occurring pyrethrins produced by the flowers of pyrethrums. On the basis of their biological response, they are divided into two groups – Type I and Type II pyrethroids. Pyrethroids are used in all the four major methods of application: about 70%for residual spraying, 25% for space spraying, and the remainder for treatment of nets and larvicidal purposes. In terms of the weight of active ingredient, pyrethroids are not the most used insecticides, but in terms of spray coverage they are by far the most used insecticides. The large-scale usage of pyrethroid insecticides for vector control is worrisome because it exerts a high selection pressure for the development of resistance in vector populations. The genes conferring resistance against pyrethroids have been spreading in vector populations, particularly in the populations of malaria and dengue vectors [13, 14]. This is particularly concerning because the use of long-lasting insecticidal nets (LLINs), a major tool in malaria control, depend solely on the action of pyrethroids. It is critical that the susceptibility of malaria vectors to pyrethroids is preserved. Therefore, it has been recommended to not use pyrethroids for indoor residual spraying where there is a high coverage of its use with treated nets [13].

3.5 Insect growth regulators (IGRs)

These are diverse group of chemical compounds, the use of which dates back to 1980s [15]. These compounds are effective against the larval stages of insects. They are divided into juvenile hormone analogs and chitin synthesis inhibitors. They mimic insect hormones, such as juvenile hormone and ecdysone, and interfere with the normal growth and development of the insect.

3.6 Bacterial larvicides

These larvicides are based on the bacteria of the species *Bacillus sphaericus* (Bs), and *Bacillus thuringiensis* serovar *israelensis* (Bti), which are entomopathogenic. In early 1960s the first strain of *Bs* with its larvicidal activity was discovered but in 1976 the Bti subspecies was discovered which was highly toxic to larvae of many species of mosquitoes. By mid-1980s, the use of bacterial larvicides started in different vector control programmes.

4. Insecticide resistance in disease vectors

The insecticide act in many different ways, to which insects have also developed different mechanisms by which they develop resistance against the toxic effects of these insecticides. Broadly, there are three different mechanisms of development of resistance, i.e., metabolic resistance, target site and resistance to penetration of the insecticide. Correspondingly, different biological, biochemical and molecular methods have been developed to detect these mechanisms in different vectors.

Status of insecticide resistance in vectors of parasitic diseases

4.1 Malaria

Malaria remains one of the deadliest vector-borne disease. Repeated exposure of the malaria vectors to insecticides over several decades has resulted in resistance to many of them. The insecticide resistance in vectors of malaria was first reported in 1950s for DDT [3]. Till date, the vectors of malaria are known to have developed resistance to the four major insecticide classes, *viz.*, pyrethroids, organochlorines, carbamates and organophosphate. Sub-Saharan Africa contributes to 90% and Southeast Asia contributes 7% of the malaria cases reported worldwide, whereas in the Latin Americas, malaria cases have significantly declined, the major contribution of cases now being from the Amazon region [16]. From Africa, all the four major species of *Anopheles,* which are the vector for malaria, have developed resistance to pyrethroids, except in the south-western Africa. In Southeast Asia, resistance to PY has been reported for *An. minimus*, *An. vagus* and *An. sinensis* in China, Thailand and Vietnam [17]. In India, *An. stephensi*, the prime urban vector and *An. culicifacies*, the rural vector of malaria, have developed resistance to PY, DDT and OP in the states of Goa, Tamil Nadu, Orrisa and Chhattisgarh, whereas only one population in western Columbia in the Amazon region has developed resistance to only PY and DDT and not OPs [18]. The resistance to DDT has also been reported in Thailand and Vietnam [19], and in India, the resistance has been reported from Gujarat and Rajasthan [20]. The major vector of malaria in Orrisa, *An. fluviatlis* (S form), and in the Amazon region, *An. darlingi,* are reported to be susceptible to the all classes of insecticides [21]. In addition, the resistance to bendiocarb, a CA which is commonly used in IRS, has also been reported across Africa. The resistance to OP has been limited to only west and east Africa. Multiple resistance to insecticides has been reported from different parts of all these regions [22].

4.2 Lymphatic filariasis

The prominent vector of lymphatic filariasis is *Culex quinquefasciatus*. The resistance against synthetic insecticides, i.e., DDT and malathion was reported from filariasis endemic states of India, Uttar Pradesh, Bihar and Kerala [23]. In a study carried in different localities of Brazil, the resistance to DDT, pyrethroids and carbamates have been recorded [24]. This mosquito was found resistant to malathion, permethrin in Kuala Lumpur and to pyrethroid, deltamethrin, permethrin in Zambia, only to pyrethroids in Zanzibar, and highly resistance to permethrin in Central Java [25, 26].

4.3 Leishmaniasis

It is the second most prominent parasitic infection after malaria in terms of fatalities caused by it globally. The phlebotomine sandflies are the prominent vectors of leishmaniasis, mainly found in tropics and subtropics [27]. For its vector control, the chemical interventions used are IRS and ITNs. Since 1944, DDT-based IRS was mainly used for the control of sandfly, but after 1970s, due to its toxic effect and reports on DDT resistance in sandfly from the disease endemic regions in India resulted in the use of alternative insecticide, i.e., pyrethroids [28, 29]. The deltamethrin, lambda-cyhalothrin and alpha-cypermethrin resulted in more than 70% of reduction in both *Lutzomyia* spp. and *Phlebotomus argentipes* sandflies [30]. Insecticide-treated durable wall lining (DWL) is being used as an alternative type of indoor residual intervention to increase the residual effect of insecticides used in IRS [31] and it drastically reduced the abundance of *P. argentipes* in south Asian countries [32]. Presently, IRS spraying and use of ITNs are the most common methods implemented in vector control programmes for the control of sandfly.

4.4 Human African Trypanosomiasis (HAT)

HAT is transmitted by the bite of the tsetse fly, i.e., *Glossina* spp., across most of the 38 countries of the sub-Saharan Africa [33]. In 1945, DDT and BHC were the only synthetic insecticides used for their control, but later on pyrethroids like deltamethrin were used. In Ethiopia, deltamethrin impregnated nets were used in 1990 for the control of *Glossina pallidipes* [34], but later on resistance was reported to deltamethrin [35].

4.5 Chagas' disease/American trypanosomiasis

The Chagas' disease, transmitted by triatomine bugs, is a major disease affecting millions of people in the Latin America countries. In 1950s, DDT was used to control the triatomine density [36], following which HCB was considered to be more effective, and later, dieldrin was also used for its control. In 1999, resistance in triatomine bugs was not a serious problem except for some reports where *Rhodnius prolixus* showed resistant to pyrethroids in Venezuela, and *Triatoma infestans* in Brazil [37, 38]. Some reports of deltamethrin resistance in *T. infestans* are also there from Argentina [37]. In a study carried out in Bolvia also, *T. infestans* populations were found resistant to deltamethrin [39]. The expression of resistance to pyrethroids during the early phase of embryonic development in *T. infestans* has also been reported [40].

5. Targets and techniques for detection of insecticide resistance:

5.1 Metabolic resistance

It is the most common mechanism of development of resistance to insecticides. In this type of resistance, either the enzymes which detoxify the insecticide are over expressed or there is an altered affinity of the enzyme for the compound used, mainly caused by substitution of amino acids, mainly in the three major enzyme families (cyrochrome P450 monooxygenase, glutathione S-transferase [GSTs] and esterase) which are involved in metabolism of the insecticide compound [41].

In mosquito species, the vectors of malaria and lymphatic filariasis, several CYP450 genes has been documented to be involved in resistance to pyrethroids, *viz.*, CYP6P_3 and CYP6M_2, CYP6P9a, CYP6Z_3 [42], while CYP6Z_1 is attributed for conferring resistance to both carbamates and pyrethroids [43]. CYP9M_{10} and CYP6AA_7 confer resistance to the insecticide permethrin in *Culex* species [44, 45]. The GSTs confer resistance to DDT in mosquitoes, i.e., *Anopheles*, and GSTe2 GSTe3, GSTe4 have been reported to be involved in pyrethroid resistance in *An. funestus* in Uganda and Kenya [46]. Several P450s and GST genes were reported to be overexpressed in a deltamethrin-resistant *An. sinensis* in China and Southeast Asian countries [47]. In a recent study carried out by Yan et al., several genes have been identified which confer permethrin resistance to *Culex pipiens quinquefasciatus,* particularly, 2 CCEs, 6 GSTs, and 7 P450s gene were highly expressed [49].

The detection of the enzymatic activity in biochemical assays is the most commonly used method to assess development of resistance to insecticides in insects. In several studies done to detect pyrethroid resistance, increased P450 monooxygenases and esterase activity have been recorded in the resistant strains of *Anopheles*, *Culex*, and triatomine bugs [40]. For sandfly, metabolic resistance is not adequately studied and only some studies of insecticide resistance related to bioassays on *P. argentipes* and *P. papatasi* populations from India and on *Lutzomyia* populations in South America have been reported [29].

5.2 Target site resistance

Here, the target site of the action of the insecticide may be modified genetically. As a result, the binding or interaction of the insecticide at its site of action is thereby prevented which decreases the efficacy of the insecticide.

Mechanism of resistance	Molecular determinants	Known point mutations	Type of vector
Metabolic resistance	Glutathione S-transferase gene (GSTe2)	L119F-GSTe2 mutation	*An. funestus* [46, 50]
	Cytochrome P450 monooxygenases	CYP6M2, CYP6P3 CYP4G16, CYP4G17 CPAP3-E and CPLCX1 *CYP9K1, CYP6M7, CYP4H18, CYP4H17, CYP4C36,* CYP6Z1, CYP6M2 and CYP6P3 in bendiocarb resistance Overexpression of Esterase A and Esterase B genes CYP9M10 (cyp450); CYP6AA7 overexpressed in pyrethroid resistance; $CYP_{5122}A1$	*An. gambiae*; *An. funestus, Cx. quinquefasciatus* [45, 48, 49]
Target site resistance	Voltage gated sodium channel gene (*VGSC*) gene (*kdr* mutations)	S6 segment of domain II- L1014F/S; N1575Y L925I mutation L1014F and L925I in the second domain of the sodium channel gene protein *VGSC*-1014F and *VGSC*-1014S, *ace-1* (G119S) and *rdl*-A296S or *rdl*-A296G	*An. funestus*; *An. sinensis*; *An. gambiae*; *An. culicifacies*); *An. vagus*; *Phlebotomus argentipes;* triatomne bugs [40, 43, 51–54]
	Acetylcholinesterase gene	G119S; N485I	*An. gambiae*; *An. funestus*; *An. coluzzii; Phlebotomus papatasi* [55, 56]
Penetration resistance		CPLCG3 gene	*An. gambiae*; *Cx. pipiens pallens* [18, 57]
	Aminobutyric acid	Alanine to serine at position 302 or 296	*An. gambiae*; *An. funestus* [58]

Table 2.
Genetic determinants and point mutations associated with insecticide resistance in various insect vectors [18, 40, 43, 45, 46, 48–58].

Various target-site mutations in the voltage-gated sodium channel (*VGSC*) gene were recorded in response to pyrethroids in different species of mosquitoes, *viz.*, *Anopheles gambiae* complex, *An. funestus* and *An. culicifacies*, *Phlebotomus argentipes* and triatomine bugs (**Table 2**). The most common resistance associated mutation reported in the AChE gene is G119S in species of *Anopheles* and *Phlebotomus*.

6. Methods of detection of insecticide resistance

The development of resistance to insecticides in insects is a complex phenomenon which depends on many direct and indirect factors. The direct factors include the natural variations in the genetic, biochemical, physiological, ecological and behavior of insects, while the indirect factors are the operational factors such as categories of insecticides used, the timing and method used for their application [59]. There are different bioassays which exploit some of these factors and help in detecting insecticide resistance, i.e., by phenotypic, genotypic and proteomic analysis.

The laboratory bioassays are useful in detecting susceptibility, tolerance and resistance in vectors against insecticides [60]. The phenotypes of vectors are utilized in detecting the insecticide resistance in methods such as by bottle bioassays. In this, a range of concentrations of the insecticide is used to study the target site resistance followed by knocking certain genes in the insect population [61]. The WHO cone tests, wireball assays, tube tests [62] and the Centers for Disease Control and Prevention (CDC) bottle bioassays take long exposure times, therefore, an alternative method – mosquito contamination device (MCD) bottle bioassay – has been developed for the control of malarial vectors in resource poor settings [63]. In vector control programmes, it is recommended that bottle bioassays should be routinely use in laboratory to measure the phenotypic resistance of a vector against a particular insecticide so as to determine whether it is still effective [64].

In addition to the traditional bioassays for determining resistance to insecticides, currently, various molecular markers and techniques to detect target site mutations have been developed. These genetic mutations can be identified by various PCR based methods such as random amplified polymorphic DNA (RAPD), restriction fragment length polymorphism (RFLP), amplified fragment length polymorphism (AFLP), microsatellites and single nucleotide polymorphisms (SNP). To observe changes at the RNA level, molecular techniques such as real-time polymerase chain reaction (RT-PCR), differential display reverse transcription PCR, northern blot and microarrays may be used. For protein estimation enzyme assays, various techniques such as enzyme linked enzyme-linked immunosorbent assay (ELISA), western blot, sodium dodecyl sulphate–polyacrylamide gel electrophoresis (SDS-PAGE), and matrix-assisted laser desorption ionization time-of-flight mass spectrometry (MALDI-TOF) may be used [65, 66]. In addition to this, to study the mechanism of resistance at the molecular level through proteomics, the identification of resistance-related proteins with their expression profiling can provide knowledge of the activity of proteins related to insecticide resistance in insects [67].

6.1 Detection of insecticide resistance in malaria vectors

6.1.1 Phenotypic assays

The WHO and CDC bioassays are the most common assays carried out to detect the insecticide resistance in malarial vectors till now. The WHO bioassays have been carried out in various studies, such as in detection of *Anopheles culicifacies* resistance to DDT, malathion and deltamethrin in India (60), *An. stephensi* in eastern

Ethiopia [68], and *An. arabiensis* susceptibility to bendiocarb, lambda-cyhalothrin and deltamethrin in Yemen [69]. The CDC bottle bioassay had been carried out to detect metabolic resistance as well as biochemical resistance against permethrin in *An. arabiensis* in Tanzania [70], and to quantify the resistance of insecticides of deltamethrin, lambda-cyhalothrin, alpha-cypermethrin, permethrin and DDT in *An. darlingi, An. nuneztovari* and *An. albimanus* in Colombia [71]. Another bioassay used for detecting the phenotypic resistance in the malarial vectors is MCD bottle bioassay. These bioassays are helpful and significant as they detected the effect on the behavior of *An. stephensi* and *An. gambiae* after the exposure to insecticides [63].

6.1.2 Genetic assays

To detect target site mutations, due to resistance to insecticides such as pyrethroids, carbamates and organophosphates in different *Anopheles* species, PCR assay is most commonly used [72]. It helps in detecting kdr mutations in different species of *Anopheles* due to overuse of pyrethroids/DDT [73, 74]. This assay also detected the resistance in vector population at the metabolic level which involves genes such as cytochrome P450 genes, carboxylesterases and glutathione S-transferases in *Anopheles coluzzii.* Profiling of gene expression was carried out by multiplexing followed by qRT-PCR, the findings from these studies proved it as another effective method for detection of insecticide resistance in malarial vectors. Moreover, study had been carried for the polymorphic genes: P450 genes CYP6Z1, CYP6Z3 and CYP6M7 against pyrethroid resistance in the malaria vector *An. funestus.* The analysis reported the changes in amino acids by QTL, which showed the contribution of these polymorphic genes in the insecticide resistance [75]. Thus, the characterizations and genetic profiling methods can improve the understanding regarding the target site mutations and metabolic resistance in malarial vectors.

6.1.3 Proteomics assays

The major driving factors which contribute to the malaria transmission are the age, method of blood feeding and way of infection spread. These factors could be promising target for detecting the insecticide resistance through proteomics. In *Anopheles* mosquitoes, with the help of artificial neural networks (ANNs) and MALDI-TOF/MS, the effect of insecticide resistance on these factors were described [76]. Proteomics detected the contribution of age as one of significant factor to insecticide resistance by using matrix-assisted laser desorption ionization tandem time-of-flight mass spectrometry or capillary high-pressure liquid chromatography with linear ion-trap (LTQ)-Orbitrap XL hybrid mass spectrometer which was further quantified by Western Blot leading to detection of protein biomarkers [77]. Proteomics study also detected metabolic resistance and target sited mutations by NCBInr/Protein BLAST and MS/MS-FTMS in *An. gambiae* in Burkina Faso [78]. Multiple resistance, which is emerging as the big issue in insecticide resistance management has also been detected with the help of proteomics like 2D electrophoresis and MALDI TOF in *An. stephensi* [79]. Thus, proteomics analysis can be a promising tool to tackle various obstacles in insecticide resistance management.

6.2 Detection of insecticide resistance in filarial vectors

6.2.1 Phenotypic assays

WHO bioassays are most commonly used diagnostic assay for detecting insecticide resistance in *Culex* phenotypically which is proved by extensive experimental studies such as to detect the knockdown resistance in *Cx. quinquefasciatus* in Sri Lanka [80]; as the diagnostic assay in *Cx. pipiens* against organochlorine, pyrethroid, organophosphate and carbamate insecticides. Larval bioassay by microplate method in the *Cx. pipens* larvae has also been carried out to study the knockdown and metabolic resistance in filariasis endemic areas of Egypt [81]. In India, following the WHO diagnostic methods for phenotypic detections, susceptibility studies have been carried out against DDT and Deltamethrin in *Cx. quinquefasciatus* from northeastern India to study the target sited resistance in this vector [82]. Some other studies have also been carried out using WHO bioassays for detecting the insecticide resistance such as metabolic resistance in *Cx. pipiens pallens* in China [83]; target sited mutations of genes G119S ace-1 and L1014F in *Cx. pipiens* complex and their hybrids in Morocco [84].

6.2.2 Genotypic assays

PCR based methods such as amplification of specific gene targets by PCR followed by sequencing of the amplified product can be used to detect specific mutations associated with resistance. For example, L1014F mutation on the VGSC domain IIS6 had been reported in *Cx. quinquefasciatus* in Sri Lanka associated with resistance to insecticides [85]. Various resistance genes have also been discovered through transcriptome profile by whole-transcriptome microarrays in *Culex* species. With the help of pyrosequencing, the mutations and metabolic insecticide resistance due to deltamethrin in *Cx. quinquefasciatus* from Zanzibar was determined [86]. These assays provided evidence for insecticide resistance caused due to target site mutations and metabolic alterations at the genetic level [87].

6.2.3 Proteomics assays

A global protein profile among insecticide resistance strains and susceptible strains of *Culex* may be obtained through proteomics. The alterations at the proteomics level can be determined by quantification of certain tags using liquid chromatography/tandem mass spectrometric analysis. The protein profiles have also been detected using isobaric tag for relative and absolute quantitation (iTRAQ) data analysis method which has confirmed the susceptibility status of strains of *Culex* in studies [88]. Such studies assure that proteomics can be considered as promising tools for studying insecticide resistance in mosquito populations.

6.3 Detection of insecticide resistance in the vector of *Leishmania*: Sandfly

6.3.1 Phenotypic assays

The WHO bioassays detected the knockdown resistance type mutations and certain metabolic enzyme resistances due to exposure of DDT, malathion, deltamethrin and propoxur in *Phlebotomus argentipes* [89]. Another study detected the resistance in *Paralongicollum sergenti* and *P. papatasi* on exposure to DDT and lambdacyhalothrin in Morocco [90]. The WHO and CDC bottle bioassays have also been conducted to study the diagnostic time and dose in *P. papatasi* to study insecticide resistance for pyrethroids, organophosphates, carbamates and DDT [91].

6.3.2 Genotypic assays

The detection of molecular alterations such as the mutations in *VGSC* gene have been detected in sandflies [92]. The pyrethroid resistance mutations (kdr) were also studied in the sandfly vector population worldwide with PCR as the major tool for detection [93]. Target site resistance of voltage gated sodium channel gene (kdr mutations) at position L1014F/S in *Phlebotomus papatasi* has also been determined through PCR [91].

6.4 Detection of insecticide resistance in the vector of Chagas' disease: Triatomine bug

6.4.1 Phenotypic assays

The WHO bioassays have been conducted in many species of Triatominae against DDT resistance [94], where behavioral responses were observed in *T. infestans* strains to deltamethrin, on the basis of which the insect populations were differentiated into susceptible and resistant [95]. The other behavioral and environmental factors such as dispersal or migration of the vector populations leading to the changes in feeding behavior, nutritional status and reproductive behavior also contribute to the alterations in the response to insecticides [96]. Therefore, observation of such changes can be utilized in detection of insecticide resistance in vector populations.

6.4.2 Genotypic assays

Target site mutations and metabolic resistance are the major pathways of insecticide resistance in insect vectors. In case of *Triatomine* bugs, studies have utilized the detection of these altered effects on certain enzymes and genes. The qRT-PCR was used to compare three triatomine species: *Triatoma dimidiata, T. infestans* and *T. pallidipennis* by studying their resistance associated metabolic and the target mutations in Latin America [97]. The transcriptome analysis of the genes encoding various proteins played a significant role in *T. infestans* to analyze the metabolism of essential compounds in the vector's body and its effect on insecticide resistance [98]. The detection of pyrethroid induced alterations in the developing eggs of *Triatomine* at the molecular level has proved to be another promising tool [99, 100].

6.4.3 Proteomics assays

Some insecticides interfere with the physiological functions in insects such as affecting the neuro-endocrine hormones. This property may be used to study the development of resistance to these insecticides and proposing candidate molecular targets responsible for resistance. LC/MS–MS has been utilized to validate certain post translational modifications in the neuro-endocrine factors in *Triatomine* [101]. The proteins essential for the structural and the metabolic constructs in Triatomine can prove to be good targets for detecting insecticide resistant populations. The transcriptome analysis, BLAST analysis and other sequencing platforms can pave the way for detecting the resistance effects on the basis of transcription factors, cell signaling pathways and cellular biology of the vector [98].

7. Alternative strategies to combat insecticide resistance

7.1 Rotation and mixture

This method involves use of insecticides of one or different classes with different target sites or distinguishable mechanism of actions in rotation or in sequence. The hypothetical basis of this strategy is that if the resistance to a particular insecticide is rare, then the use of multiple insecticides may decrease the chances of resistance to the least possible. This strategy should be allowed to run for prolonged periods of time so that there is no reversal of the effect [102]. Thus, annual rotation programmes are carried out extensively with the rotation of multiple classes of insecticides with different target of actions for vector populations. Usually, this practice is carried out during the growing season in agricultural practices. The rotation and mixture method of insecticide resistance may further be modified by using insecticide mosaics and combinations. This includes use of two different insecticides in two different areas thus lowering down the probability of development of resistance against one particular insecticide [103].

7.2 Bioinsecticides

Insecticides create a huge bioburden due to their chemical nature and pose a threat to human health and environment. Hence, there is a need to replace insecticides with biologically friendlier methods, such as by using bioinsecticides such as bacterial and fungal species against insecticides. Following this approach, the production of bioinsecticide has gained momentum, but have yet not been put to common use due to their high production cost. There is also a need to more research regarding their use and reliability. Many species of *Bacillus* are used in agriculture and have showed effective results. Plant extracts such as nicotine, pyrethrum, and neem oils are also are also been increasingly used with the green revolution. Another promising biocontrol agent is *Androctonus australis* anti-insect toxin (AaIT), which targets the neurological system of the insect vectors. Thus, exploitation of certain characteristics of the naturally occurring organisms in nature may potentially lead to the development of useful biocontrol agents against vector populations.

8. Future perspectives

Insecticide resistance management is the only way to reduce the selection pressure of insecticides. Newer tools are needed to be designed to detect the resistance apart from the existing phenotypic and genotypic methods. The lack of adequate research and development of methods for resistance detection is a major obstacle in insecticide management. Advancements in these areas hold the potential to eradicate vector borne diseases in the future and must be promoted. There are many approaches of genetic and proteomics studies for resistance detection which are yet unutilized for certain vectors such as *Leishmania*, *Triatomine*, and *Glossina*, indicating that much work needs to be done in this area. The development of these techniques can pave the way to study alterations in the physiology of vectors at genetic level due to insecticide resistance. The various other factors influencing the resistance development in vectors such as seasonality, distribution of vectors in given geographical area, alterations at allelic level which may lead to modified phenotypic traits are yet to be explored. An overall understanding of the vector species, their genotypes, phenotypes and proteomes involved in insecticide resistance for various vectors is required. Such models will help in deploying various vector control strategies optimally, and will also help in innovating newer methods to combat insecticide resistance.

9. Conclusions

The key component of vector control programmes is the use of insecticides, however, a timely emergence of resistance in insects is very crucial for the success of this component. Large number biological, environmental, and geographical factors are responsible for emergence of insecticide resistance in vectors. The two most common mechanisms by which vectors develop resistance to insecticides are metabolic and target site. In metabolic resistance, the three major enzymes involved are cytochrome P450 oxygenases, GSTs and esterases. The activity of these enzymes can be studied for different insecticides by standard biochemical tests. For target site resistance, the known or unknown mutations in *VGSC* and AChE genes can be studied by various techniques like PCR, RT-PCR and AFLP. To combat the emergence of insecticide resistance, various alternative strategies which involve the use of rotational insecticides, bioinsecticides and ITNs/LLINs are used. Yet many more strategies like extensive and regular surveillance of insecticide resistance, development of more sensitive techniques for the detection and area wise mapping of insecticide resistance of individual vectors may help to overcome the development of resistance in vectors.

Conflict of interest

None to declare.

Author details

Taruna Kaura, Nikita Sharma and Abhishek Mewara*
Department of Medical Parasitology, Postgraduate Institute of Medical Education and Research, Chandigarh, India

*Address all correspondence to: abhishekmewara@gmail.com

References

[1] Rivero A, Vézilier J, Weill M, Read AF, Gandon S. Insecticide Control of Vector- Borne Diseases: When Is Insecticide Resistance a Problem? PLoS Pathog. 2010;6(8):e1001000. DOI: 10.1371/journal.ppat.1001000.

[2] Cholewiński M, Derda M, Hadaś E. Parasitic diseases in humans transmitted by vectors. Ann Parasitol. 2015;61(3):137-57. DOI: 10.17420/ap6103.01.

[3] Corbel V, Guessan RN. Distribution, Mechanisms, Impact and Management of Insecticide Resistance in Malaria Vectors: A Pragmatic Review. Manguin S, editor. Rijeka: In Tech; 2013. p. 579-633. DOI: 10.5772/56117.

[4] Oberemok, V, Aleksei, Z, Levchenko, NN, Nyadar P. A Brief Review of Most Widely used modern insecticides and prospects for the creation of DNA insecticides. Entomol Rev. 2015;97(7): 507-518. DOI: 10.1134/S00138738 15070027.

[5] van den Berg H, Manuweera G, Konradsen F. Global trends in the production and use of DDT for control of malaria and other vector-borne diseases. Malar J. 2017; 5;16(1):401. DOI: 10.1186/s12936-017-2050-2.

[6] Karunamoorthi K, Sabesan S. Insecticide Resistance in Insect Vectors of Disease with Special Reference to Mosquitoes: A Potential Threat to Global Public Health, Health Scope. 2013;2(1):4-18. DOI: 10.17795/jhealthscope-9840.

[7] Hurd H. Manipulation of medically important insect vectors by their parasites. Annu Rev. Entomol. 2003;48:141-161. DOI: 10.1146/annurev.ento.48.091801.112722

[8] Cholewiński M, Derda M, Hadaś E. Parasitic diseases in humans transmitted by vectors. Ann Parasitol. 2015;61(3):137-57. DOI: 10.17420/ap6103.01.

[9] Jayaraj R, Megha P, Sreedev P. Organochlorine pesticides, their toxic effects on living organisms and their fate in the environment. Interdiscip Toxicol. 2016;9(3-4):90-100. DOI: 10.1515/intox-2016-0012.

[10] O'Reilly AO, Khambay BPS, Williamson MS, Field LM, Wallace BA, Davies TGE. Modeling insecticidebinding sites in the voltage-gated sodium channel. Biochem. J. 2006;396:255-263. DOI: 10.1042/BJ20051925.

[11] Mladenović van den Dries MA, Pronk A, Guxens M, Spaan S, Voortman T, Jaddoe VW, Jusko TA, Longnecker MP, Tiemeier H. Determinants of organophosphate pesticide exposure in pregnant women: A population-based cohort study in the Netherlands. Int J Hyg Environ Health. 2018;221(3):489-501. DOI: 10.1016/j.ijheh.2018.01.013.

[12] Tangena JA, Hendriks CMJ, Devine M, Tammaro M, Trett AE, Williams I, DePina AJ, Sisay A, Herizo R, Kafy HT, Chizema E, Were A, Rozier J, Coleman M, Moyes CL. Indoor residual spraying for malaria control in sub-Saharan Africa 1997 to 2017: an adjusted retrospective analysis. Malar J. 2020;10:19(1):150. DOI: 10.1186/s12936-020-03216-6.

[13] WHO (World Health Organization) DDT in Indoor Residual Spraying: Human Health Aspects. Environmental Health Criteria 241. 2011b. Available: http://www.who.int/ipcs/publications/ehc/ehc241.pdf

[14] Ranson H. Current and Future Prospects for Preventing Malaria Transmission via the Use of Insecticides.

Cold Spring Harb Perspect Med. 2017;7(11):a026823. DOI: 10.1101/cshperspect.a026823.

[15] Mulla MS. The future of insect growth regulators in vector control. J Am Mosq Control Assoc. 1995;11 (2 Pt 2):269-273.

[16] Al-Awadhi, M.; Ahmad, S.; Iqbal, J. Current Status and the Epidemiology of Malaria in the Middle East Region and Beyond. Microorganisms. 2021;9:338. DOI: 10.3390/microorganisms9020338.

[17] Qin Q, Li Y, Zhong D, et al. Insecticide resistance of Anopheles sinensis and An. vagus in Hainan Island, a malaria-endemic area of China. Parasit Vectors. 2014;7:92. DOI: 10.1186/1756-3305-7-92.

[18] Riveron MJ, Magellan T, Leon M, Benjamin M, Mu-Chun C, Charles W. Insecticide Resistance in Malaria Vectors: An Update at a Global Scale. 2018;10.5772/intechopen.78375. DOI: 10.5772/intechopen.78375.

[19] Van- Bortel W, Trung HD, Thuan le K, Sochantha T, Socheat D, Sumrandee C, Baimai V, Keokenchanh K, Samlane P, Roelants P, Denis L, Verhaeghen K, Obsomer V, Coosemans M. The insecticide resistance status of malaria vectors in the Mekong region. Malar J. 2008;7:102. DOI: 10.1186/1475-2875-7-102.

[20] Tikar SN, Mendki MJ, Sharma AK, et al. Resistance status of the malaria vector mosquitoes, Anopheles stephensi and Anopheles subpictus towards adulticides and larvicides in arid and semi-arid areas of India. J Insect Sci. 2011;11:85. DOI: 10.1673/031.011.8501.

[21] Sahu SS, Gunasekaran K, Raju HK, Vanamail P, Pradhan MM, Jambulingam P. Response of malaria vectors to conventional insecticides in the southern districts of Odisha State, India. Indian J Med Res. 2014;139(2): 294-300.

[22] Keïta M, Nafomon S, Fousseyni K, Boissé T, Francis Z, Boubacar C, Ambiélè BS, Sekou FT, Djouaka R, Doumbia S. Multiple Resistance Mechanisms to Pyrethroids Insecticides in A. gambiae sensu lato Population From Mali, West Africa. J Infect Dis. 2021 223:(2)1:S81–S90. DOI: 10.1093/infdis/jiaa190.

[23] Kumar K, Sharma AK, Kumar S, Patel S, Sarkar M, Chauhan LS. Multiple insecticide resistance/susceptibility status of Culex quinquefasciatus, principal vector of bancroftian filariasis from filaria endemic areas of northern India. Asian Pac J Trop Med. 2011;4(6):426-429. DOI: 10.1016/S1995-7645(11)60119-3.

[24] Lopez RP, Lima JBP, Martins AJ. Insecticide resistance in *C. quinquefasciatus* Say, 1823 in Brazil: a review. Parasit Vectors. 2109;12: 1-12. DOI:10.1186/s13071-019-3850-8.

[25] Norris L, Douglas N. Insecticide Resistance in *C. quinquefasciatus* Mosquitoes After the Introduction of Insecticide-Treated Bed Nets in Macha, Zambia. J Vector Ecol. 2011;36(2): 411-20. DOI: 10.1111/j.1948-7134.2011.00182.x.

[26] Jones CM, Liyanapathirana M, Agossa FR, Weetman D, Ranson H, Donnelly MJ, Wilding CS. Footprints of positive selection associated with a mutation (*N1575Y*) in the voltage-gated sodium channel of *A. gambiae*. Proc Natl Acad Sci U S A. 2012;109(17):6614-9. DOI: 10.1073/pnas.1201475109.

[27] Claborn DM. The biology and control of leishmaniasis vectors. J Glob Infect Dis. 2010;2(2):127-134. DOI: 10.4103/0974-777X.62866.

[28] Dhiman RC, Yadav RS. Insecticide resistance in phlebotomine sandflies in

Southeast Asia with emphasis on the Indian subcontinent. Infect Dis Poverty. 2016;5(1):106. DOI: 10.1186/s40249-016-0200-3.

[29] Balaska S, Fotakis EA, Chaskopoulou A, Vontas J. Chemical control and insecticide resistance status of sand fly vectors worldwide. PLoS Negl Trop Dis. 2021;15(8):e0009586. DOI: /10.1371.

[30] Alexander B, Maroli M. Control of Phlebotomine sand flies. Med Vet Entomol 2003;17: 1-18. DOI: 10.1046/j.1365-2915.2003.00420.x.

[31] Messenger LA, Shililu J, Irish SR, Anshebo GY, Tesfaye AG, Ye-Ebiyo Y, Chibsa S, Dengela D, Dissanayake G, Kebede E, Zemene E, Asale A, Yohannes M, Taffese HS, George K, Fornadel C, Seyoum A, Wirtz RA, Yewhalaw D. Insecticide resistance in *Anopheles arabiensis* from Ethiopia (2012-2016): a nationwide study for insecticide resistance monitoring. Malar J. 2017;18;16(1):469. DOI: 10.1186/s12936-017-2115-2.

[32] Mondal D. Elimination efforts in Bangladesh: lessons learned and challenges remaining. SPEAK India Delhi, India, 3-5 November 2016.

[33] Torr SJ, Chamisa A, Mangwiro TNC, Vale GA (2012) Where, When and Why Do Tsetse Contact Humans? Answers from Studies in a National Park of Zimbabwe. PLoS Negl Trop Dis 6(8):e1791. DOI: 10.1371.

[34] Leak SG, Peregrine AS, Mulatu W, Rowlands GJ, D'Ieteren G. Use of insecticide-impregnated targets for the control of tsetse flies (Glossina spp.) and trypanosomiasis occurring in cattle in an area of south-west Ethiopia with a high prevalence of drug-resistant trypanosomes. Trop Med Int Health. 1996;1(5):599-609. DOI: 10.1111/j.1365-3156.1996.tb00085.x.

[35] Rowlands GJ, Leak SG, Mulatu W, Nagda SM, Wilson A, d'Ieteren GD. Use of deltamethrin 'pour-on' insecticide for the control of cattle trypanosomosis in the presence of high tsetse invasion. Med Vet Entomol. 2001;15(1):87-96. DOI: 10.1046/j.1365-2915.2001.00272.x.

[36] Mougabure-Cueto G, Picollo MI. Insecticide resistance in vector Chagas disease: evolution, mechanisms and management. Acta Trop. 2015;149: 70-85. DOI: 10.1016/j.actatropica.2015.05.014.

[37] Zerba EN. Susceptibility and resistance to insecticides of Chagas disease vectors. Medicina (B Aires). 1999;59 Suppl 2:41-6.

[38] Pessoa GC, Trevizani NA, Dias LS, Bezerra CM, Melo BV, Diotaiut L. Toxicological profile of deltamethrin in *Triatoma brasiliensis* (Hemiptera: Reduviidae) in State of Ceara, Northeastern Brazil. Rev. Soc Bras Med Trop. 2015; 48: 39-43. DOI: 10.1590/0037-8682-0010-2015.

[39] Lardeux F, Depickère S, Duchon S, Chavez T. Insecticide resistance of *T. infestans* (Hemiptera, Reduviidae) vector of Chagas disease in Bolivia. Trop Med Int Health. 2010;15:1037-1048. DOI: 10.1111/j.1365-3156.2010.02573.x

[40] Acevedo MA, Prosper O, Lopiano K, Ruktanonchai N, Caughlin TT, Martcheva M, Osenberg CW, Smith DL. Spatial heterogeneity, host movement and mosquito-borne disease transmission. PLoS One. 2015;10(6):e0127552. DOI: 10.1371/journal.pone.0127552.

[41] David JP, Ismail HM, Chandor-Proust A, Paine MJ. Role of cytochrome P450s in insecticide resistance: impact on the control of mosquito-borne diseases and use of insecticides on Earth. Philos Trans R Soc Lond B Biol Sci. 2013;368(1612): 20120429.DOI: 10.1098/rstb.2012.0429.

[42] Rai P, Bharati M, Subba A, Saha D. Insecticide resistance mapping in the vector of lymphatic filariasis, *C. quinquefasciatus* Say from northern region of West Bengal, India. PLoS One. 2019;14(5):e0217706. DOI: 10.1371/journal.pone.0217706.

[43] Ibrahim SS, Ndula M, Riveron JM, Irving H, Wondji CS. The P450 *CYP6Z1* confers carbamate/pyrethroid cross-resistance in a major African malaria vector beside a novel carbamate-insensitive N485I *acetylcholinesterase-1* mutation. Mol Ecol. 2016;25(14):3436-52. DOI: 10.1111/mec.13673.

[44] Gong Y, Ting L, Yucheng F, Nannan L. The function of two P450s, CYP9M10 and CYP6AA7, in the permethrin resistance of *C. quinquefasciatus*. Scientific Reports. 2017;7: 587. DOI: 10.1038/s41598-017-00486-0.

[45] Talipouo A, Konstantinos M, Elysée N, Borel DT, Emmanouil F, Edmond K, Roland B, Sévilor K, Parfait AA, Vasileia B, Sofia B, Charles W, John V, Christophe AN. High insecticide resistance mediated by different mechanisms in *C. quinquefasciatus* populations from the city of Yaoundé, Cameroon. Scientific Reports. 2021;11.7322. DOI: 10.1038/s41598-021-86850-7.

[46] Sandeu MM, Mulamba C, Weedall GD, Wondji CS. A differential expression of pyrethroid resistance genes in the malaria vector *Anopheles funestus* across Uganda is associated with patterns of gene flow. PLoS One. 2020;15(11):e0240743. DOI: 10.1371/journal.pone.0240743.

[47] Zhou D, Liu X, Sun Y, Ma L, Shen B, Zhu C. Genomic Analysis of Detoxification Supergene Families in the Mosquito *Anopheles sinensis*. PLoS One. 2015;10(11):e0143387. DOI: 10.1371/journal.pone.0143387.t001.

[48] Yan L, Yang P, Jiang F, et al. Transcriptomic and phylogenetic analysis of *C. pipiens quinquefasciatus* for three detoxification gene families. BMC Genomics. 2012;13:609. DOI: 10.1186/1471-2164-13-609.

[49] Gueye OK, Tchouakui M, Dia AK, Faye MB, Ahmed AA, Wondji MJ, Nguiffo DNJ, Mugenzi LM, Tripet F, Konaté L, Diabate A, Dia I, Gaye O, Faye O, Niang EHAS, Wondji C. Insecticide Resistance Profiling of *Anopheles coluzzii* and *A. gambiae* Populations in the Southern Senegal: Role of Target Sites and Metabolic Resistance Mechanisms. Genes. 2020;11:1403. DOI: 10.3390/genes11121403.

[50] Riveron JM, Irving H, Ndula M, Barnes KG, Ibrahim SS, Paine M, Wondjii C. Directionally selected cytochrome P450 alleles are driving the spread of pyrethroid resistance in the major malaria vector *Anopheles funestus*. Proc Natl Acad Sci USA. 2013;110:252-257. DOI: 10.1073/pnas.1216705110.

[51] Ranson H, Abdallah H, Badolo A, Guelbeogo WM, Kerah-Hinzoumbe C, Yangalbe-Kalnone E, et al. Insecticide resistance in *A. gambiae*: data from the first year of a multicountry study highlight the extent of the problem. Malar J. 2009;8:299. DOI: 10.1186/1475-2875-8-299.

[52] Singh OP, Dykes CL, Das MK, Pradhan S, Bhatt RM, Agrawal OP, Adak T. Presence of two alternative kdr-like mutations, L1014F and L1014S, and a novel mutation, V1010L, in the voltage gated Na + channel of *Anopheles culicifacies* from Orissa, India. Malar J. 2010;9:146. DOI: 10.1186/1475-2875-9-146.

[53] Irving H, Wondji CS. Investigating knockdown resistance (kdr) mechanism against pyrethroids/DDT in the malaria vector *Anopheles funestus* across Africa.

BMC Genet. 2017;18(1):76. DOI: 10.1186/s12863-017-0539-x.

[54] Diouf El, Hadji NE, Badara S, Cheikh D, Mbaye D, Abdoulaye K, Ibrahima D, Ousmane F, Lassana K. Multiple insecticide resistance target sites in adult field strains of *An.gambiae* (s.l.) from southeastern Senegal. Parasit Vectors. 2020;13. DOI: 10.1186/s13071-020-04437-z.

[55] Ibrahim SS, Muhammad MM, Datti JA, Irving H, Kusimo MO, Tchapga W, et al. Temporal escalation of Pyrethroid Resistance in the major malaria vector *Anopheles coluzzii* from Sahelo-Sudanian Region of northern Nigeria. Sci Rep. 2019;14;9(1):7395. DOI: 10.1038/s41598-019-43634-4.

[56] Keïta M, Kané F, Thiero O, Traoré B, Zeukeng F, Sodio AB, Traoré SF, Djouaka R, Doumbia S, Sogoba N. Acetylcholinesterase (ace-1^R) target site mutation G119S and resistance to carbamates in *A. gambiae* (sensu lato) populations from Mali. Parasit Vectors. 2020;13(1):283. DOI: 10.1186/s13071-020-04150-x.

[57] Vannini L, Dunn WA, Reed TW, Willis JH: Changes in transcript abundance for cuticular proteins and other genes three hours after a blood meal in *A. gambiae*. Insect Biochem Mol Biol 2014;44:33-34. DOI: 10.1016/j.ibmb.2013.11.002

[58] Jacob RM, Silvie H, Kamga T, Magellan T, Wondji M, Micareme T, Irving H, Cuamba N, Maquina M, Paaijmans K, Wondji C. Escalation of pyrethroid resistance in the malaria vector *Anopheles funestus* induces a loss of efficacy of PBO-based insecticide-treated nets in Mozambique. J Infect Dis. 2019; 220. DOI: 10.1093/infdis/jiz139.

[59] Zhu F, Lavine L, O'Neal S, Lavine M, Foss C, Walsh D. Insecticide Resistance and Management Strategies in Urban Ecosystems. Insects. 2016;7(1):2. DOI: 10.3390/insects7010002.

[60] Saeidi Z, Vatandoost H, Akhavan AA, Yaghoobi-Ershadi MR, Rassi Y, Sahu SS, Thankachy S, Dash S, Swaminathan S, Kasinathan G, Purushothaman J. Multiple insecticide resistance in *Anopheles culicifacies* s.l. (Diptera: Culicidae) in east-central India. Pathog Glob Health. 2019;113(8):352-358. DOI: 10.1080/20477724.2019.1707398.

[61] Owusu HF, Jančáryová D, Malone D. *et al.* Comparability between insecticide resistance bioassays for mosquito vectors: time to review current methodology?. Parasit Vectors. 2015;8:357. DOI: 10.1186/s13071-015-0971-6.

[62] WHO. Test Procedures for Insecticide Resistance Monitoring in Malaria Vectors, Bio-Efficacy and Persistence of Insecticides on treated surfaces: report of the WHO informal consultation, Geneva, 28-30 September 1998. World Health Organization. https://apps.who.int/iris/handle/10665/64879 1998.

[63] Sternberg ED, Waite JL, Thomas MB. Evaluating the efficacy of biological and conventional insecticides with the new 'MCD bottle' bioassay. Malar J. 2014;13:499. DOI: 10.1186/1475-2875-13-499.

[64] Brogdon WG, Chan A. Guidelines for evaluating insecticide resistance in vectors using the CDC bottle bioassay/methods in *Anopheles* research. CDC Atlanta USA: CDC Technical Report; 2010. 28p.

[65] Yin J, Yamba F, Zheng C, Zhou S, Smith SJ, Wang L, Li H, Xia Z, Xiao N. Molecular Detection of Insecticide Resistance Mutations in *A. gambiae* from Sierra Leone Using Multiplex SNaPshot and Sequencing. Front Cell Infect Microbiol. 2021;11:666469. DOI: 10.3389/fcimb.2021.666469.

[66] Zhang C, Shi Q, Li T, Cheng P, Guo X, Song X, Gong M. Comparative proteomics reveals mechanisms that underlie insecticide resistance in *C. pipiens pallens* Coquillett. PLoS Negl Trop Dis. 2021;15(3):e0009237. DOI: 10.1371/journal.pntd.0009237

[67] Yared S, Gebressielasie A, Damodaran L, Bonnell V, Lopez K, Janies D, Carter TE. Insecticide resistance in *Anopheles stephensi* in Somali Region, eastern Ethiopia. Malar J. 2020;19(1):180. DOI: 10.1186/s12936-020-03252-2.

[68] Al-Koleeby Z, El Aboudi A, Assada M, Al-Hadi M, Abdalr Ahman M, Awash A, Ahmed AS, Mohamedi H, Al Jarbany J, Faraj C. The Current Insecticide Resistance in Main Malaria Vector *Anopheles arabiensis* in Yemen. J Trop Med. 2020;2020:5625019. DOI: 10.1155/2020/5625019.

[69] Marcombe S, Mathieu RB, Pocquet N, Riaz MA, Poupardin R, Sélior S, Darriet F, Reynaud S, Yébakima A, Corbel V, David JP, Chandre F. Insecticide resistance in the dengue vector *Aedes aegypti* from Martinique: distribution, mechanisms and relations with environmental factors. PLoS One. 2012;7(2):e30989. DOI: 10.1371/journal.pone.0030989.

[70] Matowo J, Kulkarni MA, Mosha FW, Oxborough RM, Kitau JA, Tenu F, Rowland M. Biochemical basis of permethrin resistance in *Anopheles arabiensis* from Lower Moshi, north-eastern Tanzania. Malar J. 2010;9:193. DOI: 10.1186/1475-2875-9-193.

[71] Orjuela LI, Morales JA, Ahumada ML, Rios JF, González JJ, Yañez J, Rosales A, Cabarcas DM, Venegas J, Yasnot MF, Quiñones ML. Insecticide Resistance and Its Intensity in Populations of Malaria Vectors in Colombia. Biomed Res Int. 2018;2018:9163543. DOI: 10.1155/2018/9163543.

[72] Soma DD, Zogo BM, Somé A, Tchiekoi BN, Hien DFS, Pooda HS, Coulibaly S, Gnambani JE, Ouari A, Mouline K, Dahounto A, Ouédraogo GA, Fournet F, Koffi AA, Pennetier C, Moiroux N, Dabiré RK. *Anopheles* bionomics, insecticide resistance and malaria transmission in southwest Burkina Faso: A pre-intervention study. PLoS One. 2020;15(8):e0236920. DOI: 10.1371/journal.pone.0236920.

[73] Main BJ, Everitt A, Cornel AJ, Hormozdiari F, Lanzaro GC. Genetic variation associated with increased insecticide resistance in the malaria mosquito, *Anopheles coluzzii*. Parasit Vectors. 2018;11(1):225. DOI: 10.1186/s13071-018-2817-5.

[74] Irving H, Wondji CS. Investigating knockdown resistance (kdr) mechanism against pyrethroids/DDT in the malaria vector *Anopheles funestus* across Africa. BMC Genet. 2017;18(1):76. DOI: 10.1186/s12863-017-0539-x.

[75] Irving H, Riveron JM, Ibrahim SS, Lobo NF, Wondji CS. Positional cloning of rp2 QTL associates the P450 genes CYP6Z1, CYP6Z3 and CYP6M7 with pyrethroid resistance in the malaria vector *Anopheles funestus*. Heredity (Edinb). 2012;109(6):383-92. DOI: 10.1038/hdy.2012.53.

[76] Nabet C, Chaline A, Franetich JF, Brossas JY, Shahmirian N, Silvie O, Tannier X, Piarroux R. Prediction of malaria transmission drivers in *Anopheles* mosquitoes using artificial intelligence coupled to MALDI-TOF mass spectrometry. Sci Rep. 2020;10(1):11379. DOI: 10.1038/s41598-020-68272-z.

[77] Sikulu MT, Monkman J, Dave KA, Hastie ML, Dale PE, Kitching RL, Killeen GF, Kay BH, Gorman JJ, Hugo LE. Proteomic changes occurring in the malaria mosquitoes *A. gambiae* and *Anopheles stephensi* during aging.

J Proteomics. 2015;126:234-44. DOI 10.1016/j.jprot.2015.06.008.

[78] Mohanty AK, Dey G, Kumar M, Sreenivasamurthy SK, Garg S, Prasad TSK, Kumar A. Proteome data of female *Anopheles stephensi* antennae. Data Brief. 2019;24:103911. DOI: 10.1016/j.dib.2019.103911.

[79] Vijay S, Rawal R, Kadian K, Raghavendra K, Sharma A. Annotated differentially expressed salivary proteins of susceptible and insecticide-resistant mosquitoes of *Anopheles stephensi*. PLoS One. 2015;10(3): e0119666. DOI: 10.1371/journal.pone.0119666.

[80] Chandrasiri PKGK, Fernando SD, De Silva BGDNK. Insecticide resistance and molecular characterization of knockdown resistance (kdr) in *C. quinquefasciatus* mosquitoes in Sri Lanka J Vector Ecol. 2020;45(2): 204-210. DOI: 10.1111/jvec.12391

[81] Zayed AB, Szumlas DE, Hanafi HA, Fryauff DJ, Mostafa AA, Allam KM, Brogdon WG. Use of bioassay and microplate assay to detect and measure insecticide resistance in field populations of *C. pipiens* from filariasis endemic areas of Egypt. J Am Mosq Control Assoc. 2006;22(3):473-82. DOI: 10.2987/8756-971X.

[82] Sarkar M, Borkotoki A, Baruah I, Bhattacharyya IK, Srivastava RB. Molecular analysis of knock down resistance (kdr) mutation and distribution of kdr genotypes in a wild population of *C. quinquefasciatus* from India. Trop Med Int Health. 2009;14(9):1097-104. DOI: 10.1111/j.1365-3156.2009.02323.x.

[83] Li CX, Dong YD, Song FL, Zhang XL, Gu WD, Zhao TY. Company amplification of estalpha2/estbeta2 and correlation between esterase gene copy number and resistance to insecticides in the field *C. pipiens pallens* strains collected from Beijing, China. J Med Entomol. 2009;46(3):539-45. DOI: 10.1603/033.046.0319.

[84] Tmimi FZ, Faraj C, Bkhache M, Mounaji K, Failloux AB, Sarih M. Insecticide resistance and target site mutations (G119S ace-1 and L1014F kdr) of *C. pipiens* in Morocco. Parasit Vectors. 2018;11(1):51. DOI: 10.1186/s13071-018-2625-y.

[85] Jones CM, Machin C, Mohammed K, Majambere S, Ali AS, Khatib BO, McHa J, Ranson H, Kelly-Hope LA. Insecticide resistance in *C. quinquefasciatus* from Zanzibar: implications for vector control programmes. Parasit Vectors. 2012;5:78. DOI: 10.1186/1756-3305-5-78.

[86] Silva Martins WF, Wilding CS, Isaacs AT, Rippon EJ, Megy K, Donnelly MJ. Transcriptomic analysis of insecticide resistance in the lymphatic filariasis vector *C. quinquefasciatus*. Sci Rep. 2019;9(1):11406. DOI: 10.1038/s41598-019-47850-w.

[87] Zhang C, Shi Q, Li T, Cheng P, Guo X, Song X, Gong M. Comparative proteomics reveals mechanisms that underlie insecticide resistance in *C. pipiens pallens* Coquillett. PLoS Negl Trop Dis. 2021;15(3):e0009237. DOI: 10.1371/journal.pntd.0009237.

[88] Pathirage DRK, Karunaratne SHPP, Senanayake SC, Karunaweera ND. Insecticide susceptibility of the sand fly leishmaniasis vector *Phlebotomus argentipes* in Sri Lanka. Parasit Vectors. 2020;13(1):246. DOI 10.1186/s13071-020-04117-y.

[89] Faraj C, Ouahabi S, Adlaoui el B, El Elkohli M, Lakraa L, El Rhazi M, Ameur B. Insecticide susceptibility status of *Phlebotomus (Paraphlebotomus) sergenti* and *Phlebotomus (Phlebotomus) papatasi* in endemic foci of cutaneous leishmaniasis in Morocco. Parasit Vectors. 2012;5:51. DOI: 10.1186/1756-3305-5-51.

[90] Rocha DA, Costa LMD, Pessoa GDC, Obara MT. Methods for detecting insecticide resistance in sand flies: A systematic review. Acta Trop. 2021;213:105747. DOI: 10.1016/j.actatropica.2020.105747.

[91] Denlinger DS, Creswell JA, Anderson JL, Reese CK, Bernhardt SA. Diagnostic doses and times for *Phlebotomus papatasi* and *Lutzomyia longipalpis* sand flies (Diptera: Psychodidae: Phlebotominae) using the CDC bottle bioassay to assess insecticide resistance. Parasit Vectors. 2016;9:212. DOI: 10.1186/s13071-016-1496-3.

[92] Fotakis EA, Giantsis IA, Demir S, Vontas JG, Chaskopoulou A. Detection of Pyrethroid Resistance Mutations in the Major Leishmaniasis Vector *Phlebotomus papatasi*. J Med Entomol. 2018;55(5):1225-1230. DOI: 10.1093/jme/tjy066.

[93] Balaska S, Fotakis EA, Chaskopoulou A, Vontas J. Chemical control and insecticide resistance status of sand fly vectors worldwide. PLoS Negl Trop Dis. 2021;15(8):e0009586. DOI: 10.1371/journal.pntd.0009586.

[94] Pessoa GC, Vinãs PA, Rosa AC, Diotaiuti L. History of insecticide resistance of Triatominae vectors. Rev. Soc Bras Med Trop. 2015;48(4):380-389. DOI: 10.1590/0037-8682-0081-2015.

[95] May-Concha I, Remón C, Mougabure-Cueto G. Behavioral response mediated by feces in *T. infestans* (Hemiptera: Reduviidae: triatominae) susceptible and resistant to deltamethrin. Acta Trop. 2020;206:105442. DOI: 10.1016/j.actatropica.2020.105442.

[96] Lobbia PA, Rodríguez C, Mougabure-Cueto G. Effect of nutritional state and dispersal on the reproductive efficiency in *T. infestans* (Klug, 1834) (Hemiptera: Reduviidae: Triatominae) susceptible and resistant to deltamethrin. Acta Trop. 2019;191:228-238. DOI: 10.1016/j.actatropica.2019.01.012.

[97] Traverso L, Lavore A, Sierra I, Palacio V, Martinez-Barnetche J, Latorre-Estivalis JM, Mougabure-Cueto G, Francini F, Lorenzo MG, Rodríguez MH, Ons S, Rivera-Pomar RV. Comparative and functional triatomine genomics reveals reductions and expansions in insecticide resistance-related gene families. PLoS Negl Trop Dis. 2017;11(2):e0005313. DOI: 10.1371/journal.pntd.0005313.

[98] Calderón-Fernández GM, Moriconi DE, Dulbecco AB, Juárez MP. Transcriptome Analysis of the *T. infestans* (Hemiptera: Reduviidae) Integument. J Med Entomol. 2017;54(6):1531-1542. DOI: 10.1093/jme/tjx151.

[99] Roca-Acevedo G, Picollo MI, Capriotti N, Sierra I, Santo-Orihuela PL. Examining Mechanisms of Pyrethroid Resistance in Eggs of Two Populations of the Chagas' Disease Vector *T. infestans* (Hemiptera: Reduviidae). J Med Entomol. 2015;52(5):987-92. DOI: 10.1093/jme/tjv078.

[100] Traverso L, Sierra I, Sterkel M, Francini F, Ons S. Neuropeptidomics in *T. infestans*. Comparative transcriptomic analysis among triatomines. J Physiol Paris. 2016 Oct;110(3 Pt A):83-98. DOI: 10.1016/j.jphysparis.2016.12.005.

[101] Marcombe S, Mathieu RB, Pocquet N, Riaz MA, Poupardin R, Sélior S, Darriet F, Reynaud S, Yébakima A, Corbel V, David JP, Chandre F. Insecticide resistance in the dengue vector *A. aegypti* from Martinique: distribution, mechanisms and relations with environmental factors. PLoS One. 2012;7(2):e30989. DOI: 10.1371/journal.pone.0030989.

[102] Ben-Dov E. *B. thuringiensis* subsp. *israelensis* and its dipteran-specific toxins. Toxins (Basel). 2014;6(4):1222-1243. DOI: 10.3390/toxins6041222.

[103] Zlotkin E, Fishman Y, Elazar M. AaIT: from neurotoxin to insecticide. Biochimie. 2000;82(9-10):869-881. DOI: 10.1016/s0300-9084(00)01177-9.

Chapter 17

Determination of Pesticides Residues in Bee Products: An Overview of the Current Analytical Methods

Enrique Mejías and Tatiana Garrido

Abstract

The presence of undesirable compounds in honey and other bee products may modify their biological attributes. Such molecules may be present because of different human activities (i.e., pollutants, pesticides) or because of veterinary treatments designed to control and prevent diseases that affect bees. The use of pesticides in agricultural crops has been related with negative effects with and acute damages for bees. The widespread agricultural use of neonicotinoids is a common exposure pathway for bees, and it may be an important factor in declining bee health. In 2013, the European Union has forbidden the use of three pesticides belonging to the neonicotinoids: Imidacloprid, Thiamethoxam, and Clothianidin after the analysis of several scientific results of some studies where those pesticides were involved in an increased death of bees.

Keywords: honey, beehives, pesticides residues, good agricultural practices, analytical methods

1. Introduction

Honey has been described as a natural sweet mixture produced by honeybees from the nectar of flowers or from living parts of plants. Bees combine this mixture with substances of their own, and then it is deposited, dehydrated, and stored in the honeycomb for further uses [1]. Honey is the most characterized bee product due to its nutritional value as a natural. Honey is composed of several carbohydrates, mainly fructose and glucose (85–95% of total sugars). Glucose has a lower degree of solubility than fructose. The ratio of glucose to fructose determines the liquid state of a given honey. Other types of sugar are present due to the union of two or more molecules of fructose or glucose as polysaccharides. Additionally, certain substances are available in honey, such as organic acids, amino acids, proteins, enzymes, lipids, flavonoids, and vitamins that are responsible for its biological properties including antioxidant or antibiotic activities [2].

Additionally, certain substances are available in honey, such as organic acids, amino acids, proteins, enzymes, lipids, flavonoids, and vitamins that are responsible for its biological properties including antioxidant or antibiotic activities [3, 4].

Melissopalynological analysis is used to establish whether a honey is unifloral or not. Unifloral honey has a higher market price because at least 45% of the pollen grains in its solids are from the same plant species. Therefore, the quality of a honey depends on the presence and concentration level of specific compounds and the botanical origin classification [5].

Honey can obtain the characteristics of plants whose pollen grains and nectar have been taken by bees. Thus, the biological properties are related to the plant species and its attributes [6]. Antioxidant activity is one of the observed biological properties of honey. The presence of enzymatic antioxidants (glucose oxidase, catalase) and non-enzymatic antioxidants (flavonoids, ascorbic acid, and phenolic acids) have been detected in many honeys [7, 8]. Several studies have looked to establish some relationship between phenolic compounds and the antioxidant properties of honey. An analysis of the phenolic compounds profile of unifloral Rhododendron honey produced in Turkey demonstrated that increased antioxidant activity was related to higher concentrations of those molecules. The same effect was observed for the antibacterial capabilities of honey samples [9, 10].

The identification of phenolic compounds includes many extraction techniques that permit the isolation of the phenolic fraction from the rest of the honey's components. Solid-Phase Extraction (SPE) procedures are recommended for cleaning the samples, followed by High-Performance Liquid Chromatography (HPLC) or Capillary Electrophoresis, CE [11]. Those techniques have been used to determine the chemical profiles of natural products from extracts obtained from complex organic matrices such as honey. Despite its high resolving power, high-performance liquid chromatography (HPLC) may present some limitations for the separation of molecules belonging to the same family, even when proper sample cleaning is performed to achieve better results. In the same way, capillary electrophoresis and the related technique, electrokinetic chromatography (EKC), in zone format (CZE) allow for the analysis of ionic and neutral compounds on the same column. The great advantage of this methodology is the amount of sample needed for each analysis; it requires only a few nanoliters of extract with a solvent waste of 1 mL–2 mL per assay [12]. Several research studies have focused on the identification of phenolic compounds in bee products. Samples of commercial propolis were studied using CE, and 15 polyphenols were separated with a buffer of sodium tetraborate 30 mM, pH 9.0, and under an applied voltage of 15 kV. Borate buffers form complexes with orthodihydroxyl groups on the flavonoid skeleton and facilitate separation. In the same study, three different extracts were produced (ethanolic, aqueous-ethanolic, and aqueous-glycolic extracts) to compare the levels of available analytes in each one. After this procedure, it was possible to establish a reproducible fingerprint of the polyphenolic profiles, the pattern of which depended on the nature of the extraction solvent [13].

By the way, the determination of the antioxidant capability of honey requires UV–Vis determinations. For instance, the colorimetric assays for the general quantification of phenolics is done by Folin–Ciocalteu reaction; assessment of radical scavenging using the reduction reaction of the radical 2,2-diphenyl-1-picrylhydrazyl (DPPH) is a very helpful method for this aim [3]; the assessment of antioxidant activity using the ferric reducing/antioxidant power assay (FRAP) [14, 15] and the oxygen radical absorbance capacity (ORAC) [16] has been frequently used.

2. Chilean honey

In 2020, according to figures obtained from Agricultural and Livestock Service of Chile, there are 8777 beekeepers who manage 1.241,504 beehives distributed throughout the country [17]. In that sense, 1991 tons were exported, and most of the honey was sent to European markets, with Germany being the main buyer, followed by Belgium. Also, China and the United Arab Emirates have also emerged as important buyers of Chilean honey. Chilean bee products have interesting biological properties that improve their natural potential as an attractive exportable nutritional food. Chemical characterization is necessary for certifying their natural attributes. Furthermore, chemical content analysis enables the fulfillment of international regulations for healthy and safe foods because those markets are very strict in terms of the food safety issues. Several studies on the potential properties of native unifloral honey have been conducted. The Ulmo Honey (Eucryphia cordifolia) demonstrated the greatest antibacterial power among the selected Chilean unifloral honey samples. The main identified compounds are gallic, caffeic, coumaric, and chlorogenic acids [18, 19].

3. Undesirable residues

The presence of undesirable compounds in honey occurs when beehives or plants are exposed to pollutants caused by human activities; in this case, the final composition of honey is modified, and the effectiveness of its biological activity changes. Recently, it has been demonstrated that honey has a specific chemical profile of inorganic elements related to the place where it was produced [20]. This information enables the certification of the geographical origin of a honey [21]. Similarly, studies showed that the inorganic content is not dependent on the botanical origin, but rather on the composition of the soils and water in the areas surrounding beehives, and other environmental conditions play an important role in this case [10, 22]. Also, honeys that contained metals in their composition showed a decreased antioxidant activity compared with control samples [23]. The same trend was found in bee pollen samples obtained from the same beehives [24]. Furthermore, it was possible to observe that the chemical behavior of phenolic compounds was modified due to the presence of metals, based on analyses by capillary electrophoresis with diode array detector (CE-DAD) [12].

4. Good agricultural practices (GAP)

According to the Food and Agricultural Organization of the United Nations (FAO), the current definition of Good Agricultural Practices (GAP) includes codes, standards, and regulations that have been developed in recent years by the food industry and producers' organizations but also governments and nongovernmental organizations, aiming to codify agricultural practices at farm level for a range of commodities. Their purpose varies from fulfillment of trade and government regulatory requirements (in particular with regard to food safety and quality), to more specific requirements of specialty or niche markets. The objective of these GAP codes, standards, and regulations includes, to a varying degree: ensuring safety and quality of produce in the food chain; capturing new market advantages by modifying supply chain governance; improving natural resources.

Currently, despite those findings, the standard methods for the measurement of those compounds still include an analysis by HPLC with mass spectrometry (MS). The efficiency of mass spectrometry and the optimization of chromatographic procedures have helped to decrease the experimental time for each analysis. Moreover, several antibiotics can be detected in just one chromatogram. This was the case in a study that was performed by selecting 11 honey samples from different botanical origins produced in Granada, Spain. After HPLC coupled to electrospray ionization (ESI) time-of-flight (TOF) mass analyzer, the following antibiotics were quantified with an average of MRL from 0.05 to 0.76 μg Kg^{-1} and a run time of approximately 11 minutes: chlortetracycline, demeclocycline, doxycycline, methacycline, minocycline, oxytetracycline, tetracycline, and rolitetracycline [25].

5. Xenobiotics

In several areas where agricultural activities occur, the possibility of finding pesticide-free areas has decreased. Because of this, an increase in the density of apiaries has been observed with the systematic appearance of bee diseases. The continuous exposure of bees to xenobiotics (agricultural pesticides and veterinary products) is responsible for the presence of these compounds in recycled waxes [26]. Although the information available is just related to reports from a limited group of countries, there is enough evidence that the presence of these compounds in the surrounding areas of beehives, as well as in the composition of products obtained from those apiaries, may have a long-term effect on the reproductive health of beekeepers, even farmer workers, or consumers [27].

6. Residues of pesticides: analytical methods

The use of pesticides in agriculture is allowed under strict regulations, but nowadays there is enough evidence about the negative effects of those products over bee health. In that way, the presence of pesticides in the honey content may be detected owing to direct contamination from beekeeping practices or by indirect contamination from environmental sources [28]. Since the mid-1990s, several beekeepers from different parts of United States and Europe reported high colony losses caused by a phenomenon known as Colony Collapse Disorder (CCD) [29]. One of the key factors of this disorder has been the use of neonicotinoids, a class of neurotoxic pesticides. Some studies have showed its effects by killing bees after expositions of these compounds, and there **is** evidence of damages at sublethal doses of neonicotinoids over their nervous systems affecting foraging abilities, navigation, learning communication, and memory. Also, the suppression of the immune systems of bees may be caused by neonicotinoids [30, 31]. In 2013, the European Food Safety Authority (EFSA) identified several risks posed to bees by three neonicotinoid insecticides: Clothianidin, Imidacloprid, and Thiamethoxam as seed treatment or as granules, with particular regard to their acute and chronic effects on bee colony survival and development. For this reason, the use of those products has been restricted by European Union. In addition, these restrictions are extensive for treated seeds with those pesticides (EFSA Journal 2013; 3066; 3067; 3068) [32]. It has been described that neonicotinoids are not the only group of insecticides with negative effects. The organophosphate and organochlorine also cause damage to the bees [33].

Among the methods used for the detection and identification of both insecticides and pesticides, the most useful technique is HPLC with mass spectrometry [34]. As previously indicated for pesticides, the critical step in those determinations is related to the pretreatments of samples before chromatographic analysis. One alternative methodology was developed for 13 pesticides detected in 40 samples of honeys from Poland. In this case, all the samples were subjected to liquid–liquid extraction process on a diatomaceous earth support. The main difficulty in this methodology was the matrix effects over the percentages of recoveries of pesticides (63–117%), when the samples were fortified for assessing the success of this extraction process [35]. At the present time, the analyses include the QuEChERS method followed by dispersive solid-phase extraction (d-SPE). This is simple sample preparation technique recommended for pesticides detection in a wide variety of food and agricultural products. Several studies have achieved satisfactory results in the determination of a long list of pesticides belonging to different classes such as organophosphates, triazoles, carbamates, dicarboximides, dinitroanilines, and neonicotinoids in honeybee bodies, honey, and bee pollen. The advantage of QuEChERS is the improvement of precision of measurements and percentages of recovery of pesticides after analysis without a matrix effect of samples affecting the reliability of results. This permits better sensitivity and lower detection limits for each pesticide [36–38].

The group of compounds corresponding to agronomic pesticides and veterinary products is wide and growing day by day. Despite this, there is a consensus on the analytical methods available for their detection and quantification. **Table 1** presents a list of susceptible compounds that can be identified in bee products with their main methodologies for their extraction, detection, and/or quantification (adapted from [39]).

N°	Pesticide	DL	QL	Extraction method		Chromatography method	
		ng g^{-1}	ng g^{-1}	SPE	QuEChERS	LC–MS/MS	GC–MS
1	2,4 D	0.05	0.1	√		√	
2	ABAMECTIN	0.005	0.01	√		√	
3	ACEPHATE	0.005	0.01	√		√	
4	ACEQUINOCYL	0.005	0.01	√		√	
5	ACETAMIPRID	0.005	0.01	√		√	
6	ACETOCHLOR	0.005	0.01		√		√
7	ACRINATHRIN	0.005	0.01		√		√
8	ALACHLOR	0.005	0.01		√		√
9	ALDICARB	0.005	0.01	√		√	
10	ALDICARB SULFONE	0.005	0.01	√		√	
11	ALDICARB SULFOXIDO	0.005	0.01	√		√	
12	ALDRIN	0.005	0.01		√		√
13	AMINOMETHYLPHOSPHONIC ACID	0.005	0.01	√		√	
14	AMITRAZ	0.005	0.01	√		√	
15	ATRAZINE	0.005	0.01		√		√
16	AZADIRACHTIN	0.005	0.01	√		√	
17	AZINPHOS ETHYL	0.005	0.01		√		√
18	AZINPHOS METHYL	0.005	0.01		√		√
19	AZOXYSTROBIN	0.05	0.1	√		√	
20	BAC C10 BENZALKONIUM CHLORIDE	0.005	0.01	√		√	
21	BAC C12 BENZALKONIUM CHLORIDE	0.005	0.01	√		√	
22	BAC C14 BENZALKONIUM CHLORIDE	0.005	0.01	√		√	
23	BENALAXYL	0.005	0.01	√		√	

N°	Pesticide	DL	QL	Extraction method		Chromatography method	
		ng g^{-1}	ng g^{-1}	SPE	QuEChERS	LC–MS/ MS	GC–MS
24	BENOMYL/CARBENDAZIM	0.005	0.01	√		√	
25	BENTAZON	0.005	0.01	√		√	
26	BHC ALPHA	0.005	0.01		√		√
27	BHC BETA	0.005	0.01		√		√
28	BHC DELTA	0.005	0.01		√		√
29	BIFENAZATE	0.005	0.01		√		√
30	BIFENTHRIN	0.005	0.01		√		√
31	BITERTANOL	0.005	0.01		√		√
32	BOSCALID	0.005	0.01	√		√	
33	BRODIFACOUM	0.005	0.01	√		√	
34	BROMACIL	0.005	0.01	√		√	
35	BROMADIOLONE	0.005	0.01		√		√
36	BROMOPHOS ETHYL	0.005	0.01		√		√
37	BROMOPHOS METHYL	0.005	0.01		√		√
38	BROMOPROPYLATE	0.005	0.01		√		√
39	BUPROFEZIN	0.005	0.01		√		√
40	CADUSAFOS	0.005	0.01		√		√
41	CAPTAFOL	0.005	0.01		√		√
42	CAPTAN	0.005	0.01		√		√
43	CARBARYL	0.005	0.01	√		√	
44	CARBENDAZIM	0.005	0.01	√		√	
45	CARBOFURAN	0.005	0.01	√		√	
46	CARBOPHENOTHION	0.005	0.01		√		√
47	CARTAP HCL	0.005	0.01	√		√	
48	CHLOFENTEZINE	0.005	0.01	√		√	
49	CHLORANTRANILIPROLE	0.005	0.01	√		√	
50	CHLORDANE CIS	0.005	0.01		√		√
51	CHLORDANE TRANS	0.005	0.01		√		√
52	CHLORDENE	0.005	0.01		√		√
53	CHLORFENAPYR	0.005	0.01		√		√
54	CHLORFENSON	0.005	0.01		√		√
55	CHLORFENVINPHOS	0.005	0.01		√		√
56	CHLOROBENZILATE	0.005	0.01		√		√
57	CHLOROTHALONIL	0.005	0.01		√		√
58	CHLORPYRIFOS ETHYL	0.005	0.01		√		√
59	CHLORPYRIFOS METHYL	0.005	0.01		√		√
60	CYHEXATIN/AZOCICLOTIN	0.005	0.01	√		√	
61	CLETODIM (EYZ)	0.005	0.01		√		√

N°	Pesticide	DL	QL	Extraction method		Chromatography method	
		ng g^{-1}	ng g^{-1}	SPE	QuEChERS	LC–MS/MS	GC–MS
62	CLOTHIANIDIN	0.005	0.01	√		√	
63	COUMAPHOS	0.005	0.01	√		√	
64	CYANAZINE	0.005	0.01		√		√
65	CYFLUTHRIN (**)	0.005	0.01		√		√
66	CYFLUTHRIN BETA	0.005	0.01		√		√
67	CYHALOTHRIN GAMMA	0.005	0.01		√		√
68	CYHALOTHRIN L	0.005	0.01		√		√
69	CYPERMETHRIN	0.005	0.01		√		√
70	CYPROCONAZOLE	0.005	0.01	√		√	
71	CYPRODINIL	0.005	0.01	√		√	
72	CYROMAZINE	0.005	0.01	√		√	
73	DDAC - DIDECYLDIMETHYLAMMONIUM CHLORIDE	0.005	0.01	√		√	
74	DDD op	0.005	0.01		√		√
75	DDD pp	0.005	0.01		√		√
76	DDE op	0.005	0.01		√		√
77	DDE pp	0.005	0.01		√		√
78	DDT op	0.005	0.01		√		√
79	DDT pp	0.005	0.01		√		√
80	DELTAMETHRIN	0.005	0.01		√		√
81	DEMETON-S	0.005	0.01		√		√
82	DIAZINON	0.005	0.01		√		√
83	DICHLOBENIL	0.005	0.01		√		√
84	DICHLOFLUANID	0.005	0.01		√		√
85	DICHLORVOS	0.005	0.01		√		√
86	DICLORAN	0.005	0.01		√		√
87	DICOFOL op (**)	0.005	0.01		√		√
88	DICROTOPHOS (**)	0.005	0.01		√		√
89	DIELDRIN	0.005	0.01		√		√
90	DIFENOCONAZOLE	0.005	0.01	√		√	
91	DIFLUBENZURON	0.005	0.01	√		√	
92	DIMETHENAMID	0.005	0.01		√		√
93	DIMETHOATE	0.005	0.01	√		√	
94	DIMETHOMORF	0.005	0.01	√		√	
95	DIPHENYLAMINE	0.005	0.01		√		√
96	DISULFOTON	0.005	0.01		√		√
97	DODINE	0.005	0.01	√		√	
98	EMAMECTIN BENZOATE	0.005	0.01	√		√	

N°	Pesticide	DL	QL	Extraction method		Chromatography method	
		ng g^{-1}	ng g^{-1}	SPE	QuEChERS	LC–MS/MS	GC–MS
99	ENDOSULFAN I	0.005	0.01		√		√
100	ENDOSULFAN II	0.005	0.01		√		√
101	ENDOSULFAN SULFATE	0.005	0.01		√		√
102	ENDRIN	0.005	0.01		√		√
103	EPTC	0.005	0.01		√		√
104	ESFENVALERATE/FENVALERATE	0.005	0.01	√		√	
105	ETHION	0.005	0.01		√		√
106	ETHOPROFOS	0.005	0.01	√		√	
107	ETOFENPROX	0.005	0.01	√		√	
108	FENAMIPHOS	0.005	0.01		√		√
109	FENARIMOL	0.005	0.01		√		√
110	FENAZAQUIN	0.005	0.01	√		√	
111	FENBUCONAZOLE	0.005	0.01		√		√
112	FENCLORPHOS	0.005	0.01		√		√
113	FENHEXAMID	0.005	0.01		√		√
114	FENITROTHION	0.005	0.01		√		√
115	FENOXYCARB	0.005	0.01	√		√	
116	FENPROPATHRIN	0.005	0.01		√		√
117	FENPROPIMORF	0.005	0.01	√		√	
118	FENPYROXIMATE	0.005	0.01	√		√	
119	FENTHION	0.005	0.01		√		√
120	FERBAM	0.005	0.01		√		√
121	FIPRONIL	0.005	0.01		√		√
122	FLOCOUMAFEN	0.005	0.01	√		√	
123	FLUAZINAM	0.005	0.01	√		√	
124	FLUDIOXINIL	0.005	0.01	√		√	
125	FLUFENOXURON (**)	0.005	0.01	√		√	
126	FLUMETRALIN	0.005	0.01		√		√
127	FLUQUINCONAZOLE	0.005	0.01		√		√
128	FLUSILAZOLE	0.005	0.01		√		√
129	FLUTRIAFOL (**)	0.005	0.01		√		√
130	FLUTOLANIL	0.005	0.01	√		√	
131	FLUVALINATE (**)	0.005	0.01		√		√
132	FOLPET	0.005	0.01		√		√
133	FONOFOS	0.005	0.01		√		√
134	FORCHLORFENURON	0.005	0.01	√		√	
135	FORMETANATE	0.005	0.01	√		√	
136	FORMOTHION (**)	0.005	0.01		√		√

Nº	Pesticide	DL	QL	Extraction method		Chromatography method	
		ng g^{-1}	ng g^{-1}	SPE	QuEChERS	LC–MS/MS	GC–MS
137	GLUFOSINATE AMONNIUM	0.005	0.01		√		√
138	GLYPHOSATE	0.005	0.01	√		√	
139	HALOXIFOP METHYL	0.005	0.01	√		√	
140	HEPTACHLOR	0.005	0.01		√		√
141	HEPTACHLOR EPOXIDE	0.005	0.01		√		√
142	HEPTENOPHOS	0.005	0.01		√		√
143	HEXACHLOROBENZENE	0.005	0.01		√		√
144	HEXACONAZOLE	0.005	0.01		√		√
145	HEXAZINONE (**)	0.005	0.01		√		√
146	HEXYTIAZOX	0.005	0.01	√		√	
147	IMAZALIL	0.005	0.01	√		√	
148	IMIDACLOPRID	0.005	0.01	√		√	
149	INDOXACARB	0.005	0.01	√		√	
150	IPRODIONE	0.005	0.01		√		√
151	ISOFENPHOS	0.005	0.01		√		√
152	KRESOXIM METHYL	0.005	0.01		√		√
153	LENACIL	0.005	0.01		√		√
154	LINDANE	0.005	0.01		√		√
155	LINURON	0.005	0.01	√		√	
156	LUFENURON	0.005	0.01	√		√	
157	MALATHION	0.005	0.01		√		√
158	MANDIPROPAMID	0.005	0.01	√		√	
159	METALAXYL	0.005	0.01		√		√
160	METAMITRON	0.005	0.01	√		√	
161	METAFLUMIZOLE	0.005	0.01	√		√	
162	METHAMIDOPHOS	0.005	0.01	√		√	
163	METHIDATHION	0.005	0.01		√		√
164	METHIOCARB	0.005	0.01	√		√	
165	METHOXYCHLOR	0.005	0.01		√		√
166	METHOXYFENOZIDE	0.005	0.01		√		√
167	METOLACHLOR	0.005	0.01	√		√	
168	METOMYL	0.005	0.01	√		√	
169	METRAFENONA	0.005	0.01	√		√	
170	METRIBUZIN	0.005	0.01	√		√	
171	MEVINPHOS	0.005	0.01		√		√
172	MIREX	0.005	0.01		√		√
173	MONOCROTOPHOS	0.005	0.01	√		√	
174	MYCLOBUTANIL	0.005	0.01		√		√

N°	Pesticide	DL	QL	Extraction method		Chromatography method	
		ng g^{-1}	ng g^{-1}	SPE	QuEChERS	LC–MS/MS	GC–MS
175	NAPROPAMIDE	0.005	0.01		√		√
176	NOVALURON	0.005	0.01	√		√	
177	NUARIMOL	0.005	0.01		√		√
178	OMETHOATE	0.005	0.01	√		√	
179	OXADIAZON	0.005	0.01		√		√
180	OXAMYL	0.005	0.01	√		√	
181	OXYFLUORFEN	0.005	0.01		√		√
182	PACLOBUTRAZOL	0.005	0.01		√		√
183	PARATHION ETHYL	0.005	0.01		√		√
184	PARATHION METHYL	0.005	0.01		√		√
185	PENCONAZOLE	0.005	0.01	√		√	√
186	PENDIMETHALIN	0.005	0.01		√		√
187	PERMETHRIN	0.005	0.01		√		√
188	PHORATE	0.005	0.01		√		√
189	PHOSALONE	0.005	0.01		√		√
190	PHOSMET	0.005	0.01	√		√	√
191	PHOSPHAMIDON	0.005	0.01		√		√
192	PIRAZOPHOS	0.005	0.01		√		√
193	PIRIMETHANIL	0.005	0.01	√		√	√
194	PIRIMICARB	0.005	0.01		√		√
195	PIRIMIPHOS ETHYL	0.005	0.01		√		√
196	PIRIMIPHOS METHYL	0.005	0.01		√		√
197	PROCHLORAZ	0.005	0.01	√		√	√
198	PROCYMIDONE	0.005	0.01		√		√
199	PROFENOFOS	0.005	0.01		√		√
200	PROPAMOCARB	0.005	0.01	√		√	√
201	PROPARGITE	0.005	0.01		√		√
202	PROPICONAZOLE	0.005	0.01		√		√
203	PROPOXUR	0.005	0.01	√		√	
204	PROPYZAMIDE	0.005	0.01	√		√	
205	PROTIOCONAZOLE	0.005	0.01	√		√	
206	PYMETROZIN	0.005	0.01	√		√	
207	PYRACLOSTROBIN	0.005	0.01	√		√	
208	PYRIDABEN	0.005	0.01		√		√
209	PYRIPROXYFEN	0.005	0.01		√		√
210	QUINALPHOS	0.005	0.01		√		√
211	QUINOMETHIONATE	0.005	0.01		√		√
212	QUINOXIFENO	0.005	0.01		√		√

Nº	Pesticide	DL	QL	Extraction method		Chromatography method	
		ng g^{-1}	ng g^{-1}	SPE	QuEChERS	LC–MS/MS	GC–MS
213	QUINTOZENE	0.005	0.01		√		√
214	ROTENONE	0.005	0.01		√		√
215	SIMAZINE	0.005	0.01		√		√
216	SPINETORAM	0.005	0.01	√		√	
217	SPINOSAD	0.005	0.01	√		√	
218	SPIRODICLOFEN	0.005	0.01	√		√	
219	SPIROTETRAMAT	0.005	0.01	√		√	
220	SULFUR (S8)	0.005	0.01		√		√
221	TEBUCONAZOLE	0.005	0.01	√	√		√
222	TEBUFENOZIDE	0.005	0.01	√		√	
223	TEFLUTHRIN	0.005	0.01	√		√	
224	TERBACIL	0.005	0.01		√		√
225	TETRACONAZOLE	0.005	0.01		√		√
226	TETRADIFON	0.005	0.01		√		√
227	THIABENDAZOLE	0.005	0.01	√		√	
228	THIACLOPRID	0.005	0.01	√		√	
229	THIAMETHOXAM	0.005	0.01	√		√	
230	THIDIAZURON (**)	0.005	0.01	√		√	
231	THIOCYCLAM HYDROGEN OXALATE	0.005	0.01	√		√	
232	THIOPHANATE METHYL	0.005	0.01	√		√	
233	TOLCLOFOS METHYL	0.005	0.01	√		√	
234	TOLYLFLUANID	0.005	0.01		√		√
235	TOXAPHENE (**)	0.005	0.01		√		√
236	TRIADIMEFON	0.005	0.01	√		√	
237	TRIADIMENOL	0.005	0.01	√		√	
238	TRIAZOPHOS	0.005	0.01	√		√	
239	TRICHLORFON (**)	0.005	0.01		√		√
240	TRIFLOXYSTROBIN	0.005	0.01		√		√
241	TRIFLUMIZOLE	0.005	0.01	√		√	
242	TRIFLUMORON	0.005	0.01	√		√	
243	TRIFLURALIN	0.005	0.01		√		√
244	TRIFORINE	0.005	0.01	√		√	
245	UNICONAZOLE	0.005	0.01	√		√	
246	VAMIDOTHION	0.005	0.01	√		√	
247	VINCLOZOLIN	0.005	0.01		√		√

Table 1.
List of pesticides analyzed in honey and beeswax samples. DL: Detection Limit; QL: Quantification Limit.

7. Conclusion

The identification and detection of pesticides in the final content of honey and beeswax could be useful for beekeepers for understanding one of the potential causes of bee death. In that term, this is helpful for making improvements in the regulations for beekeeping directed to take care and preserve bees and the production of honey.

Phenolic compounds are the main molecules involved in the biological activity of honey. These compounds and its biochemical properties may be affected due to the presences of residues such as pesticides. Likewise, a specific survey applied to beekeepers to obtain production data (tons per year, detection of decreased bee population, and presence of diseases such as nosema disease and varroa mite infestation) is useful for understanding the real impact of pesticides exposure for bees.

In that way, values for biological and or physicochemical activities, production data, and detection of pesticide joined to georeferentiation of selected study sites will allow us to build a map per region describing appropriate zones for apiculture development. Finally, it shall enhance the chances of beekeepers for increasing their production by protecting bees' health.

Acknowledgements

Funding by ANID—PAI/Inserción sector productivo, 1era conv. 2019, Grant number I7819010001.This section of your manuscript may also include funding information.

Conflict of interest

The author declares no conflict of interest.

Author details

Enrique Mejías[1]* and Tatiana Garrido[2]

1 Centro de Tecnologías Nucleares en Ecosistemas Vulnerables (CTNEV), División de Investigación y Aplicaciones Nucleares (DIAN), Comisión Chilena de Energía Nuclear (CCHEN), Santiago of Chile, Chile

2 Facultad de Ciencias Químicas y Farmacéuticas, Departamento de Química Inorgánica y Analítica, Universidad de Chile, Chile

*Address all correspondence to: enrique.mejias@cchen.cl

References

[1] Codex Stan 12-1981. Revised Codex Standard for Honey. Codex Stan 12–1981, Rev. 1 (1987), Rev. 2; 1981/2001; Amended 2019. Food and Agriculture Organization of the United Nations. https://www.fao.org/fao-who-codexalimentarius/sh-proxy/es/?lnk=1&url=https%253A%252F%252Fworkspace.fao.org%252Fsites%252Fcodex%252FStandards%252FCXS%2B12-1981%252FCXS_012e.pdf

[2] Missio da Silva P, Gauche C, Gonzaga LV, Oliveira-Costa AC. Honey: Chemical composition, stability and authenticity. Food Chemistry. 2016;**196**:309-323

[3] Meda A, Lamien C, et al. Determination of the total phenolic, flavonoid and proline contents in Burkina Fasan honeys, as well as their radical scavenging activity. Food Chemistry. 2005;**91**:571-577

[4] Sakač M, Jovanov P, Marić A, et al. Physicochemical properties and mineral content of honey samples from Vojvodina (Republic of Serbia). Food Chemistry. 2019;**276**:15-21

[5] Montenegro G, Gomez M, et al. Application of the Chilean Official Standard to designate the botanical origins of honey for the characterization of the apicultural production. Cien. Inv. Agr. 2008;**35**:181-190

[6] Alotibi I, Harakeh S, Al-Mamary M, et al. Floral markers and biological activity of Saudi honey. Saudi Journal of Biological Sciences. 2018;**25**:1369-1374

[7] Marguitas L, Stanciu O, Dezmirean D, Bobis O, Popescu O, et al. In Vitro antioxidant capacity of honeybee-collected pollen of selected floral origin harvested from Romania. Food Chemistry. 2009;**115**:878-883

[8] Gül A, Pehlivan T. Antioxidant activities of some monofloral honey types produced across Turkey. Saudi Journal of Biological Sciences. 2018;**25**:1056-1065

[9] Pasias IN, Kiriakou IK, Kaitatzis A, Koutelidakis AE, Proestos C. Effect of late harvest and floral origin on honey antibacterial properties and quality parameters. Food Chemistry. 2018;**242**:513-518

[10] Silici S, Dogan O, Tuzen M, Soylak M. Assessment of trace element levels in Rhododendron honeys of Black Sea Region, Turkey. Journal of Hazardous Materials. 2008;**156**:612-618

[11] Pyrzynska K, Biesaga M. Analysis of phenolic acids and flavonoids in honey. Trends of Analytical Chemistry. 2009;**28**(7):893-902

[12] Mejías E, Gómez CJ, Gareil P, Delaunay N, Montenegro G. Characterization of phenolic profile alterations in metal-polluted bee pollen via capillary electrophoresis. Cien. Inv. Agr. 2018;**45**(1):51-63

[13] Volpi N. Separation of flavonoids and phenolic acids from propolis by capillary zone electrophoresis. Electrophoresis. 2004;**25**:1872-1878

[14] Bolanos de la Torre A, Henderson T, et al. A universally calibrated microplate ferring reducing antioxidant power (FRAP) assay for foods and applications to Manuka honey. Food Chemistry. 2015;**174**:119-123

[15] Bertoncelj J, Dobersek U, Jamnik M, Golob T. Evaluation of the phenolic content, antioxidant activity and colour of Slovenian honey. Food Chemistry. 2007;**105**:822-828

[16] Serem JC, Bester MJ. Physicochemical properties, antioxidant activity and celular protective effects of honeys from southern Africa. Food Chemistry. 2012;**133**:1544-1550

[17] División de Protección Pecuaria-Servicio Agrícola Ganadero. Boletin Apícola No. 6. Gobierno de Chile; 2021. pp. 1-9 Available from: https://www.sag.gob.cl/sites/default/files/boletin_apicola_no6_version_corregida_08-21.pdf

[18] Velásquez P, Montenegro G, Leyton F, Ascar L, Ramirez O, Giordano A. Bioactive compounds and antibacterial properties of monofloral Ulmo honey. CYTA-Journal of Food. 2020;**18**(1): 11-19

[19] Viteri R, Zacconi F, Montenegro G, Giordano A. Bioactive compounds in Apis mellifera monofloral honeys. Journal of Food Science. 2021;**86**: 1552-1582

[20] Terrab A, Recamales AF, et al. Contribution to the study of avocado honeys by their mineral contents using inductively coupled plasma optical emisión spectrometry. Food Chemistry. 2005;**92**:305-309

[21] Di Bella G, Lo TV, Potorti AG, Bua GD, et al. Geographical discrimination of Italian honey by multi-element analysis with a chemometric approach. Journal of Food Composition and Analysis. 2015;**44**:25-35

[22] Montenegro G, Mejías E. Biological applications of honeys produced by Apis mellifera. Biological Research. 2013; **46**(4):341-345

[23] Mejías E, Garrido T. Analytical procedures for determining heavy metals in honey used as bioindicator of environmental pollution. In: En Arnaut V, editor. Honey Analysis. Rijeka: IntechOpen; 2017. pp. 311-324. DOI: 10.5772/66328

[24] Mejias E, Montenegro G. The antioxidant activity of Chilean honey and bee pollen produced in the Llaima Volcano's zones. Journal of Food Quality. 2012;**35**(5):315-322

[25] Carrasco Pancorbo A, Casado Terrones S, Segura Carretero A, et al. Reversed-phase high-performance liquid chromatography coupled to ultraviolet and electrospray time-of-flight mass spectrometry on-line detection for the separation of eight tetracyclines in honey samples. Journal of Chromatography A. 2008;**1195**: 107-116

[26] Harriet J, Campá J, Grajales M, Lheritier C, et al. Agricultural pesticides and veterinary substances in Uruguayan Beeswax. Chemosphere. 2017;**177**:77-83

[27] El-Nahhal Y. Pesticide residues in honey and their potential reproductive toxicity. Science of the Total Environment. 2020;**741**:139953

[28] Panseri S, Catalano A, Giorgi A, Arioli F, Procopio A, Britti D, et al. Occurrence of pesticide residues in Italian honey from different areas in relation to its potential contamination sources. Food Control. 2014;**38**:150-156

[29] Agricultural Research Service. Colony Collapse Disorder Progress Report. CCD Steering Committee–Agricultural Research Service. United States Department of Agriculture USDA; 2012. https://www.ars.usda.gov/is/br/ccd/ccdprogressreport2012.pdf

[30] Williamson SM, Wright GA. Exposure to multiple cholinergic pesticides impairs olfactory learning and memory in honeybees. The Journal of Experimental Biology. 2013;**216**: 1799-1807

[31] Alaux C, Brunet JL, Dussaubat C, et al. Interactions between Nosema microspores and neonicotinoid weaken honeybees (Apis mellifera). Environmental Microbiology. 2010; **12**(3):774-782

[32] European Food Safety Authority. Conclusion on Pesticide Peer Review: Conclusion on the peer review of the

pesticide risk assessment for bees for the active substance clothianidin, thiamethoxam and imidacloprid. EFSA. 2013;**11**(1):3066

[33] Fawzy E, Sanaa E, Nour E. Determining pesticide residues in honey and their potential risk to consumers. Pol. J. Environ. Stud. Vol. 2014;**23**(5): 1573-1580

[34] Gbylik-Sikorska M, Sniegocki T, et al. Determination of neonicotinoid insecticides and their metabolites in honey bee and honey by liquid chromatography tandem mass spectrometry. Journal of Chromatography B. 2015;**990**:132-140

[35] Kujawski MW, Namiesnik J. Levels of 13 multi-class pesticide residues in Polish honeys determined by LC-ESI-MS/MS. Food Control. 2011;**22**:914-919

[36] Souza Tette PA, Rocha GL, Abreu Glória MB, Fernandes C. Pesticides in honey: A review on chromatographic analytical methods. Talanta. 2016;**149**: 124-141

[37] Kasiotis K., Anagnostopoulos C., et al. 2014. Pesticide residues in honeybee, honey and bee pollen by LC-MS/MS screening: Reported death incidents in honeybees. Science of the Total Environment 485 – 486: 633-642.

[38] Chen J, Mullin C. Determination of nonylphenol ethoxylate and octylphenol ethoxylate surfactants in beehive samples by high performance liquid chromatography coupled to mass spectrometry. Food Chemistry. 2014; **158**:473-479

[39] Mejías E, Gómez C, Garrido T. Suitable Areas for Apiculture expansion determined by antioxidant power, chemical profiles, and pesticide residues in *Caldcluvia paniculata* Honey and Beeswax Samples. Insects. 2022;**13**:31

Chapter 18

Review of Insecticide Resistance and Its Underlying Mechanisms in *Tribolium castaneum*

U. Shamjana and Tony Grace

Abstract

The red flour beetle *Tribolium castaneum* has emerged as the genetically tractable model insect for population genetics, functional genomics, and evolutionary studies. This agricultural pest is notorious for its potential to severely damage stored products. *T. castaneum* has developed resistance to almost all insecticides. The reports of insecticide resistance from different parts of the world show that sustained insecticide usage has only aggravated the problem. As insecticides continue to be the mainstay of pest control programs, it is essential to identify the factors influencing insecticide resistance for implementing effective pest-management strategies. The development and progression of insecticide resistance in *T. castaneum* is thus an escalating global issue requiring immediate solutions. Several studies have investigated the multiple resistance mechanisms found in *T. castaneum*, such as reduced cuticular penetration, increased metabolic detoxification, and target-site insensitivity. The availability of Whole Genome Sequence and recent advances in Next Generation Sequencing technology has furthered a geneticist's grasp of resistance study in *Tribolium*. The strategic containment of this organism calls for an in-depth understanding of resistance development. The review mainly focuses on different kinds of resistance mechanisms and genes mediating insecticide resistance. Also, it exhaustively explores the *CYP450* gene superfamily in *Tribolium* to emphasize its role in governing resistance. The consolidated insights from this study will facilitate further research on identifying biological targets, thereby developing novel control strategies for effective insect control.

Keywords: Insecticide resistance, Resistance Mechanisms, Detoxification genes, *CYP450* gene superfamily, *Tribolium castaneum*

1. Introduction

The global population is expected to cross 9.1 billion by the year 2050 and food production is projected to rise to 70% to feed this growing population [1]. Many of the fastest-growing populations are in developing countries, several of which are already facing moderate or severe food insecurity and a shortfall in food supply. One in every six children suffer from hunger in developing countries [2] and the proportion of undernourishment has been steadily increasing since 2015 [3]. The increasing trend globally of food insecurity attests to the fact that severe food deprivation or hunger is a real threat, and this scenario nullifies the ambitious "zero hunger target" by 2030. The severity of "food insecurity" underscores the immense challenge in attaining safe, nutritious, and sufficient food for all people [3]. Tackling problems of food insecurity demand intensive food production. However, increasing food production alone will not be a viable solution to achieve the "zero hunger target" by 2030 or for meeting the growing demand for food.

The pre-harvest and post-harvest issues combined with insect infestation represent a very strong limitation in optimal food production, causing mass losses of grains. After harvest, food grains undergo a series of processes such as threshing, cleaning, drying, storage, processing, and transportation before it reaches the consumer. It has been identified that food losses in the post-harvesting chain start at the time of harvest and continue up to food marketing at the consumer's end [4–6]. Grain losses may also take place due to technical limitations such as inadequate stock management facilities, improper packaging, and insufficient infrastructure.

In many countries, 15% of food grains are lost during or after harvest [7]. The Food and Agricultural Organization (FAO) estimated post-harvest grain loss at 40% and cereal loss at 30% in India [8]. The post-harvest losses account for on-farm, processing, and storage loss. Studies attribute massive grain loss in developing countries to manual operations in different stages of harvesting, which causes 15% loss on the field, 13–20% loss at processing, and 15–25% storage loss [9]. Several studies show that insects are the main contributor to storage loss in the food supply chain [10–12], which accounts for 10–20% of storage loss [13].

A diverse community of stored product species are associated with different environments where farmers store grains and cereals; from farm bins to processing facilities, to feed mills, to flour mills, to retailer stores [14–16]. Among this complex pest system, 600 species from Coleoptera and 70 species from Lepidoptera can cause substantial losses by eroding the quality of grains [17]. Coleoptera is the largest order of insects with over 250,000 described species and contains in its fold some of the most notorious stored grain pests. In these, *T. castaneum* requires special attention because of the significant harm that it can have on stored products. It attacks a large variety of stored and processed commodities and is the most harmful insect in the pest complex for its ability to inflict severe damage on stored products. Curtailing *Tribolium* infestations in the supply chain would be one critical step that can help strengthen food quality, reduce storage loss, and improve food security.

2. *Tribolium castaneum* and its damage

The red flour beetle, *Tribolium castaneum*, is an important model organism and a common inhabitant of milled cereal products, stored flour, and fungus-infested grain [18–21]. *T. castaneum* causes severe damage in flour mills and wherever dried foods and cereal products are stored or processed. They rank among the most harmful pests inhabiting grain storage facilities and processing facilities [16, 22, 23]. These insects frequently leave storage locations and migrate across heterogeneous landscapes on a daily, seasonal, or irregular basis to find new mates and resources [24]. The movement of grains from producers to consumers generates a complex network of grain storage and transportation that facilitates the dispersal of pests and pathogens associated with the grain [25]. The dispersal of the *T. castaneum*, through transportation and storage networks, allows them to find suitable habitats where they feed and reproduce, ultimately exploiting the resource patches of grain.

The adult females of *T. castaneum* lay eggs on the flour, complete their life cycle, and deplete the nutritional quality of grains over time. In the case of serious infestation, the flour becomes adulterated with a pungent odor, diminished in nutritional and market value [26, 27]. In addition to direct feeding, *T. castaneum* contaminates the food products through molting and excretion, which makes the product commercially undesirable. Depending on the level of infestation, the grain can be rejected or downgraded [28]. Product deterioration can also result from the production of quinones secreted from glands on the thorax and abdomen [29–31] leading to significant loss of quality and economic loss. The customer demand for infestation-free flour/grain has increased widely, raising the stakes of *T. castaneum* management in grain storage facilities. In a way, consumer demand for infestation-free products has been a fillip to the use of insecticides such as organophosphates and pyrethroids during storage.

Thus, a wide variety of insecticides has been applied as a primary strategy for *Tribolium* control by targeting the insect's neurological sites, including voltage-gated ion channels and acetylcholine system, causing irreversible disruption of neurological function, resulting in insect mortality. It has brought down the infestation rate, ensured long-term protection of stored commodities, and is relatively convenient to apply [32, 33]. But the incessant application of insecticides in storage facilities has accelerated the development of insecticide resistance in *T. castaneum* and resulted in the formation of particular resistant alleles in succeeding generations. The occurrence of insecticide resistance in *T. castaneum* found in grains and cereals during storage and shipping was recorded in many countries. The first instance of insecticide resistance was reported in *Tribolium* between 1959 and the early 1960s [34, 35]. Halisack and Beeman [36] applied discriminating doses of malathion to *T. castaneum* populations collected from cereal storages in the US and detected 20-fold resistance in 31 of 36 *T. castaneum* populations. In Canada, 54 strains of *T. castaneum* showed resistance to malathion at an LC99.9 value of 0.012 mg a.i/cm^2 [37]. The populations of *T. castaneum* collected from flour mills in the USA were exposed to discriminating doses of malathion to measure their resistance status. Of 28 strains, 93% of the *T. castaneum* population tolerated the discriminating doses of malathion [38]. The resistance status of Egyptian populations of *T. castaneum* was studied using the filter paper bioassay method against three contact insecticides and populations of *T. castaneum* were found to be more resistant against pirimiphos-methyl [39]. *T. castaneum* resistance is extended to pyrethroid insecticides, which is one of the most widely-used classes of insecticides in food and fodder houses as it is effective on a wide range of insects, has high efficacy at the minimum dose, and low toxicity on mammals [40–42]. Cases of pyrethroid resistance have been detected in *T. castaneum* populations from Pilot-Scale Warehouses [43] and peanut storage warehouses [44]. Several cases of resistance have been reported in different populations of *T. castaneum* collected from different countries across the world such as Italy [45], United States of America [46–50], Africa [51], Serbia [52], Bangladesh [53], Philippines [54], Pakistan [55, 56], Iran [57] Australia [58]. The occurrence of insecticide resistance in *T. castaneum* has been reported against various fumigants-methyl bromide and phosphine [59–66], synthetic pyrethroids, e.g., cypermethrin, deltamethrin, cyfluthrin, fenvalerate, and permethrin [67, 68], organophosphates [47, 52, 69, 70].

In the Indian context, the first cases of insecticide resistance were reported in 1971 by Bhatia et al. [71] who found *T. castaneum* collected from the Food Corporation of India, Delhi, to be resistant to malathion. Since then, high frequencies of insecticide resistance were recorded in *T. castaneum* collected from different storage facilities across India. Saxena et al. [72] monitored the dicholorvos resistance status of 13 samples from warehouses of the Food Corporation of India located at Mirzapur and Allahabad. The results revealed that strains from Allahabad exhibited more than ten-fold resistance compared to the Mirzapur strain. The *T. castaneum* population collected from different types of storage premises in Punjab varied in malathion resistance and was measured at a maximum in the populations of beetles collected from a public warehouse in Ropar [73].

Similarly, malathion resistance level in Indian populations of *T. castaneum* collected from thirteen different seed centres was tested and high levels of resistance were found in the Coimbatore strain. Eleven strains differed in terms of resistance levels in the range of 1.18 to 24.53 folds [74]. Insecticide resistance in *T. castaneum* has been studied in most Indian states vis-à-vis different insecticides such as malathion [75], dichlorvos [72], deltamethrin [76, 77], cypermethrin [78]. This rapid increase of resistance against different insecticide classes in India jeopardizes effective pest management strategies. The situation has only worsened with the recurring use of the same insecticide in grain storage facilities, which exert strong selection pressure on *T. castaneum* and hence reduce the efficacy of insecticides. The foregoing results confirm that the development and progression of insecticide resistance in *T. castaneum* is widespread and requires immediate solutions. Since insecticides exist as the mainstay in pest control programs, identifying the factors influencing insecticide resistance is essential in devising new and effective pest management strategies. This review presents a comprehensive picture of different resistance mechanisms and genes governing insecticide resistance in *T. castaneum*.

3. Insecticide resistance mechanisms in *T. castaneum*

The emergence and spread of insecticide resistance in an insect population is a slow and gradual evolutionary process. Following the initial exposure to the insecticide, there is a latent period in which resistance genes are segregated and linked with other genes that contribute favorable conditions for resistance development. During the evolution of resistance under insecticide selection pressure, the target species show a noticeable increase of tolerance to the pesticide. In the next stage, insecticide resistance slowly develops, followed by a period of rapid development, during which many factors influence the selection of resistance to insecticides. Rapidly developing resistance results in explosive population growth of the pests in stored products that become almost impossible to control. It is challenging to detect the resistance mechanism because they emerge over evolutionary time. Many key factors such as intensive application of insecticides, control operations, mode of inheritance of resistance genes, change in fitness of individuals, and genetic background of insects influence resistance [79]. Despite species diversity and chemical diversity of insecticides, only three mechanisms are known to cause insecticide resistance in *T. castaneum:* i) Target site insensitivity, where changes in sensitivity of target site inhibit insecticide binding ii) Metabolic resistance, where the elevated quantity of enzymes lead to increased activities of metabolic detoxification iii) Lack of penetration, where cuticular thickening or cuticular modification prevents penetration of insecticides and render them bound to the target.

The advances in genomic research (e.g., transcriptomic sequencing and whole-genome sequencing) have made significant progress in understanding resistance mechanisms such as metabolic resistance, penetration resistance, and knockdown resistance in *T. castaneum.* An even more fascinating and rapidly advancing area of microbiome research that blends entomology with microbiology is the study of the potential of entire communities of bacteria, viruses, and fungi, that live within the insect hosts, to detoxify insecticides. Existing studies highlight candidate resistance mechanisms such as symbiont-mediated insecticide resistance in various insects and have documented the major bacterial taxa in the adaptation to detoxify xenobiotic compounds [80–83].

Researchers around the world have begun to evaluate the symbiotic associations in different pest populations, how they interact with their hosts and whether they have the potential to detoxify insecticides. Interestingly, bacterial symbionts have been involved in insecticide degradation and resistance development in some insect pests, weeds, and nematodes. There are a growing number of reports where pest resistance to insecticides is not only due to the mechanisms within the pest genome but also due to the organisms in the microbiome community [84]. However, the microbial communities inhabiting *T. castaneum* and the unique intricate connection between symbionts and insecticide resistance have not yet been investigated. Many fundamental questions about the microbial shifts in response to insecticides and the functions of particular microorganisms in mediating resistance in *T. castaneum* remain unresolved.

3.1 Target site insensitivity

Insecticides such as organophosphates, carbamates, and pyrethroids produce neurotoxicity by inhibiting the enzyme acetylcholine esterase associated with the central nervous system [85–88]. These insecticides also affect other target sites such as voltage-gated sodium channels (VGSC) and gamma aminobutyric acid (GABA) receptors in the insect nervous system [89]. The DDT and pyrethroid insecticides primarily target VGSC in the nervous system [90]. Several potential insecticides such as cyclodienes and fipronil bind to the GABA receptor and block the receptor function [91]. Most commonly used insecticides primarily target different receptors on the nervous system (**Figure 1**).

Insecticide-resistant insects perform normal neurological functions despite the presence of insecticide because they have evolved insensitive acetylcholine receptors which provide resistance to organophosphate and carbamate insecticides. The reduced sensitivity of acetylcholinesterase to OP and carbamate insecticides has been studied in many resistant insect species of agricultural and veterinary importance [92–96]. The reduced target site sensitivity is a result of altered insecticide target molecules. There are mainly four types of target site insensitivity mechanisms observed in various insect species. These include a) Altered Acetylcholinesterase (AChE) resistance mechanism, which provides resistance to organophosphates and carbamates b) Knockdown resistance (*kdr*) mechanism which confers resistance to DDT and pyrethroids c) Reduced GABA receptor sensitivity mechanism, which causes resistance to phenylpyrazoles and cyclodienes and d) Altered nAChRs, which provide resistance to neonicotinoids [89, 97, 98].

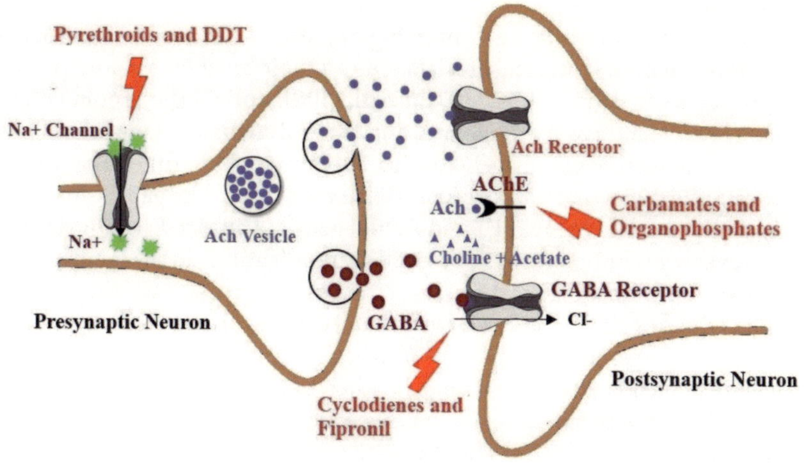

Figure 1.
Diagrammatic representation of pre and post synaptic neurons, showing the different target sites of most commonly used insecticide classes. Source: Adapted and modified from [85–91].

3.1.1 Altered acetylcholine esterase (AChE) resistance mechanism

Acetylcholinesterase (AchE) is a vital enzyme required for regulating the neurotransmitter acetylcholine (ACh). It terminates the synaptic transmission by hydrolyzing acetylcholine into acetate and choline at cholinergic synapses in insects [99]. The inhibition of AchE increases the concentration of the acetylcholine at the synaptic cleft, which leads to a prolonged binding of ACh to its postsynaptic receptor. The high quantity of acetylcholine at the postsynaptic receptor causes neuro-excitation and produces intoxication symptoms such as tremors, convulsions, and eventually paralysis-related death. This enzyme is a target site of organophosphates and carbamates insecticides, which are bound to a serine residue on the active site of AchEs and convert the AchEs into their non-functional form. This causes the accumulation of acetylcholine at the nerve endings and disrupts nerve activity, resulting in paralysis and the death of insects [100].

Several Organophosphorous compounds have been used to protect agricultural commodities from insect infestation. But most of the insects have developed resistance against these insecticides due to insensitive AchE. Target insensitivity of AchE to insecticides occurs through the mutations in the active site of "*Ace*" genes that encode acetylcholinesterase enzyme, which is the most common reason for conferring insecticide resistance. Because of the incessant application of insecticides, different mutations are induced in the *Ace* genes either singly or in combination, which reduces the sensitivity of AchE to the insecticides [101, 102]. The first major research in insects was conducted on *Drosophila melanogaster* that mapped the *Ace* locus at the molecular level and the genomic sequencing effort confirmed that *Ace* encodes the acetylcholinesterase enzyme [103]. The existence of the *Ace* gene and its genome structure has been identified in many insects such as *Drosophila melanogaster* [104], *Musca domestica* [105], *Anopheles gambiae* [106], *Lucilia cuprina* [107], *Tribolium castaneum* [108], *Pieris rapae* [109], *Sitobion avenae* [110], *Bemisia tabaci* [111], *Bombyx mori* [112], *Aphis gossypii* [113], *Plutella xylostella* [114], *Blattella germanica* [115], *Aedes aegypti* [116]. The gene sequence and genomic organization of *Ace* genes in different insects revealed that most insect species possess two *Ace* genes (*Ace* 1 and *Ace* 2) except *Drosophila melanogaster, M. domestica,* and *Lucilia cuprina.* The introduction of the point mutation in *Ace* genes through single amino acid substitution reduces the sensitivity of AChE to insecticides inhibition. These insensitive acetylcholinesterases impart resistance to carbamate and organophosphorus insecticides.

Recent evidence for resistance-conferring mutations in *Ace* genes has focussed on their involvement in insecticide resistance and their biochemical and physiological properties in different insects. Lu et al. [108] studied the genome organization, expression patterns, phylogenies, and three-dimensional models of two *Ace* genes in *T. castaneum* extensively to better understand the functional role of *Ace* genes and the molecular basis of insecticide resistance. The gene sequencing and comparative analysis of AChE1 and AchE2 genes (*Tcas Ace 1 & Tcas Ace 2*) in *Tribolium* revealed that both genes possess different features in the length of their genomic DNA, chromosome locations, and intron/exon organizations. Sequencing full-length cDNAs of AChE genes showed that AChE1 is distributed on chromosome 5 and AchE2 on chromosome 2. In addition, AChE1 consisted of one intron, whereas AchE2 consisted of five introns. Further, extensive protein simulation studies provided evidence that AChE1 has been associated with the hydrolysis of acetylcholine, whereas AChE2 has not been involved in the hydrolysis of acetylcholinesterase substrates. This novel finding prompted Lu et al. [108] to investigate the functional differences of two AChE genes in cholinergic, non-cholinergic neurotransmission, and insecticide resistance by gene-silencing in *T. castaneum*.

RNAi results of both *TcAce1* and *TcAce2* in *T. castaneum* larvae were consistent with the observations of protein modeling studies. Thus, the protein simulation studies of AchE coupled with RNAi experiments have proved that AChE1 is essential for cholinergic neurotransmission and is the target for anticholinesterase insecticides such as organophosphorous and carbamates, which disable the hydrolysis activity of AchE1and cause incapacitation. Whereas, AChE2 is not responsible for neurotransmission in *T. castaneum.* This study also suggested that genetic modifications of AchE1 are most likely responsible for the insensitivity of acetylcholinesterase to organophosphorus and carbamate insecticides. It is remarkable that target site insensitivity is due to different mutations (mainly point mutation) at the catalytic sites of AchEs and conferred resistance in *Drosophila melanogaster* [117], *M. domestica* [105], *Bactrocera oleae* [118], *Leptinotarsa decemlineata* [119], *Chilo auricilius* [120], *Apolygus lucorum* [121]. Sequencing of the gene encoding AchE has generated insights on different point mutations which causes the alteration of AchE genes and a decrease in the sensitivity of anti-cholinesterases insecticides inhibition. In addition, the efficacy of organophosphates and carbamates has been challenged by multiple mutations in the same AchEs of the insects [117, 122]. Thus, the point mutations and multiple mutations result in decreased hydrolytic efficiency of AchE and are associated with insecticide resistance.

3.1.2 Knockdown resistance (kdr) mechanism

Over the years, Pyrethroids have come to be the most sought-after class of insecticides for pest control in commercial and household environments because of their affordable and durable qualities [123]. However, their utility has been limited by the widespread development of insecticide resistance in many major pests. Pyrethroids are synthetic derivatives of pyrethrin, and the pyrethroids were classified into two groups namely class I and class II based on their physical characteristics and knockdown effect against insects. Class I pyrethroids contain a basic structure of cyclopropane carboxylic ester. These compounds include permethrin, resmethrin, phenothrin, bifenthrin, allethrin, tefluthrin, and tetramethrin. Class II pyrethroids contain a cyano group and these compounds include cypermethrin, deltamethrin, cyhalothrin, fenvalerate, cyfluthrin, fenpropathrin, flumethrin. The toxicity of pyrethroids was found to be 2250 times higher in insects than mammals due to their increased sodium channel sensitivity, lower body temperature, and smaller structure [124]. When an insect is intoxicated with non-cyano pyrethroids (class I), it produces strong excitatory action and tremors on the nervous system. The cyano pyrethroids trigger a quite different action, which includes salivation and choreoathetosis. It has been suggested that poisoning symptoms differ based on the cyano or non-cyano pyrethroids [125]. The pyrethrin and pyrethroid insecticides primarily target the VGSC in the nervous system. Pyrethroids and DDT produce their toxicity by binding onto the voltage-gated channels in axonal membranes, altering their gating properties, and the channels remain open for a long time. This causes a prolonged sodium influx, thereby depolarizing the axonal membrane and stimulating the neurons to produce repetitive discharges, finally resulting in paralysis [90, 126]. In insects, modification of voltage-gated sodium channel structure by point mutation or substitution causes insensitivity and reduces the binding affinity of the insecticides to protein.

Knockdown resistance (*kdr*) is one of the major mechanisms involved in resistance to all pyrethroids, pyrethrins, DDT, and its analogs [127]. The *kdr* resistance was first identified in the house fly [126]. Since then, *kdr* has been described in several insects against pyrethroids and an organochlorine class of insecticides [128]. Knockdown resistance occurs due to different point mutations in voltage-gated sodium channels. These sodium channels are composed of a larger α-subunit (260 kDa) and smaller β-subunits (30–40 kDa). The pore-forming α-subunit has four homologous domains (I – IV), and each domain possesses six transmembrane helices (S1 – S6). The domains are joined together to form a central aqueous pore and the pore is lined by S5, S6 linkers, and S5, S6 helices. In each domain, the S4 segment is involved in voltage sensing, and a positively charged amino acid residue is embedded in every third position [129, 130]. In mammals, the sodium channel encodes nine genes [131] whereas, in insects, the sodium channel encodes only a single gene known as *para* [132, 133]. However, *para* undergoes an extensive alternative splicing process to increase the heterogeneity and functional diversity of sodium channel [133, 134]. These distinct variants of the *para* sodium channel in insects produce different levels of sensitivity to pyrethroids and DDT. The different amino acid substitutions in the para sodium channel variants of nerve membranes have been demonstrated in *M. domestica* [135, 136], *Blatella germanica* [137, 138], *Ctenocephalides felis* [139], *Drosophila melanogaster* [140] that render the loss of sensitivity to pyrethroids. The secondary mutation designated as *super kdr* has been identified mostly in the domain II region of the sodium channel and reported in *B. tabaci* [141], *Haematobia irritans* [142], *Plutella xylostella* [143] which confers enhanced resistance to pyrethroids. The occurrence of the *kdr* mutations in voltage-gated sodium channels limits the efficacy of pyrethroids and it remains a threat to the control of *T. castaneum* [144]. This has been earlier reinforced by the functional expression studies of voltage-gated sodium channel paralytic A gene ($TcNa_v$) of *T. castaneum.* RNAi-induced knockdown results reveal that the $TcNa_v$ gene of *T. castaneum* is a potential candidate to target for the future control of *T. castaneum* and lends support for the use of RNAi as a viable method for controlling this insect [139]. The results of this study provide convincing evidence which shows that pyrethroid resistance in *T. castaneum* correlated with the presence of point mutations in the sodium channel *para* gene of insect nervous membrane. Thus, the identification of *kdr* mutations provides insights into the resistance mechanism in *T. castaneum* and has also proven critical for designing new insecticides for insect control.

3.1.3 Reduced GABA receptor sensitivity mechanism

The GABA ionotrophic receptor of the neuron membrane is formed by the oligomerization of five subunits around a central pore and each subunit possesses a large N terminal domain and four membrane-spanning domains (M1–M4). Several potential insecticides such as cyclodienes and fipronil stick to the M2 membrane-spanning a region of GABA receptor as competitive inhibitors, which prevent the chloride uptake of the GABA ion channel. The inhibition of GABA stimulated chloride uptake enhances the firing of nerve impulses in insects, which initiates lethal effects on insects [145]. However, the mutations or modifications in the molecular structure of the GABA ion channel influence the activity of insecticides. The mutated GABA receptor becomes insensitive to the insecticides at varying levels and this insensitivity has the potential to increase resistance in an insect species.

Such a resistant modification was first characterized in the GABA receptor subunit gene *Rdl* ('Resistance to Dieldrin gene') of *Drosophila* by using positional cloning approach [146, 147] and polymerase chain reaction [148]. The resistance-associated mutation studies in resistant *D. melanogaster* strains showed that the amino acid alanine^{302} in GABA gated chloride channel encoded by *Rdl* was replaced by serine. This single amino acid replacement was found to be associated with resistance in *D. melanogaster* and exhibited 4000 fold resistance against *cyclodiene* [149]. The cyclodiene resistance was conferred by replacing a single alanine at the second membrane-spanning region M2 with serine or glycine, which readily prevents direct binding of drugs and allosterically modifies the conformation of the Rdl receptor. Further studies on point mutation (Ala^{302} to Ser) in the GABA gated chloride channel gene *Rdl* in different insect orders, namely Diptera, Coleoptera, and Dictyoptera, conclude the fact that mutations associated with single base-pair replacement are highly conserved [150]. The remarkable conserved nature of single Ala > Ser mutation in the GABA receptor was confirmed in *D. simulans, T. castaneum,* and *B. germanica* by amplifying the resistance-associated *Rdl* sequences and the same replacement of Ala > Ser as found in *D. melanogaster* was observed. This finding raises the question as to whether this resistance-associated substitution arises once and then disseminates globally or if resistance arises independently in different populations. To address this, Andreev et al. [151] collected 141 strains of *Tribolium* globally and screened for dieldrin resistance. Of the 141 strains, 23 homozygous resistant strains and 6 susceptible strains were chosen for *Rdl* sequencing and mutational differences were compared among them. Phylogenetic analysis provided strong evidence of multiple independent origins of dieldrin-associated mutation, which suggests that resistance surfaced independently in 23 resistant *Tribolium* strains. *T. castaneum* genome sequencing effort facilitated the characterization of GABA gated ion channels and the post-translational modification of the *Rdl* gene. The post-translational modification such as alternative splicing have been identified in exons 3 and 6 of the *T.cas Rdl* gene and observed the three splice variants for exons 3. This information on *Rdl* isoforms of *T.cas* helped to investigate their contribution in modifying the tolerance to insecticides [152]. These results suggest that variant isoforms of the *Rdl* gene define insensitivity to cyclodiene and have gone on to develop resistance to cyclodiene in *T. castaneum* through alternative splicing of exon3 and exon6. This is a significant factor enhancing cyclodiene insensitive transcripts in *T. castaneum.*

In addition, genome sequencing advancements have facilitated the detection of multiple *Rdl* genes in Lepidopteran genomes [153, 154] and *Acyrthosiphon pisum* genome [155]. The multiple origins of the resistance-associated mutation have implications for understanding the evolution of resistance, the spread of resistance, and its management. Interestingly, the polymerase reaction amplified the same region in different resistant strains of insects *T. castaneum* [138], *D. simulans* [156], *B. germanica* [157] and confirmed that this single base pair substitution conferred the cyclodiene resistance. Another replacement at residue A^{296}S (equivalent to position 301 in *D. melanogaster*) was reported in *Anopheles arabiensis* and *A. gambiae* [158] was found to be associated with higher levels of dieldrin resistance. Similarly, *Rdl* mutation with the replacement of Ala^{302} with Serine conferred 237-fold fipronil resistance in *Nilaparvata lugens* [159]. The single base pair substitution (Ala to Ser/Gly/Asn) in the *Rdl* receptor at the site analogues to 301 in *Drosophila* has also been identified in *Laodelphax striatellus* [160, 161], *Anopheles funestus* [162]. These findings suggest that point mutation and post translational modification of the *Rdl* gene are significant evolutionary phenomena in insects and modulate insecticides' binding/sensitivity, which makes insects resistant to most natural and synthetic insecticides.

3.2 Metabolic resistance

The most common resistance mechanism in insects is the metabolic detoxification mechanism, enabling the insect to degrade or sequester the insecticides faster before releasing their toxic effect. This resistance mechanism allows insects to overproduce the enzymes mainly cytochrome *P450* monooxygenases (*CYP450s*), carboxylesterases (*CarEs*), and glutathione S transferases (*GSTs*), to thwart the toxic effects of insecticides. This advantage helps to evolve resistance in *T. castaneum* populations for all main classes of insecticides currently used for stored product pest control such as organophosphates, carbamates, and pyrethroids [163–165]. The comprehensive genomic and transcriptomic analysis has led to the identification of the key genes encoding detoxifying enzymes such as *CYP450s*, *GSTs*, and *CarEs*. These genes are being frequently associated, over the past few decades, with the rise of insecticide resistance in *T. castaneum* through overexpression, copy number variation or gene duplication, coding sequence mutations, or as a combined effect of these mechanisms.

Several studies showed that resistant insects possess generally higher levels of *P450* dependent monooxygenases, and have high catalytic activity towards the toxicant [166, 167]. Some studies showed that amplification of transferase gene exerts insecticide sequestration/detoxification of many different endogenous and xenobiotic substances including insecticides [168–170]. The studies of the mechanisms of metabolic resistance to carbamates, organophosphates, and pyrethroids have revealed the role of esterases (especially carboxylesterases) in resistant insect species [171–176]. These enzymes are capable of sequestering insecticide substrates through two principal mechanisms 1) Overexpression of one or more esterases and 2) Mutations in gene encoding esterase [171].

3.2.1 Cytochrome P450s (CYP450s)

Recent years have witnessed the rapid evolution of insecticide resistance due to their continuous exposure. However, the resistance mechanism in insects is not fully understood, and the evolution of resistance to insecticides in *T. castaneum* populations threatens the long-term future of the food storage system. The exposure of *T. castaneum* to insecticides triggers a complex defense response that includes genes that encode key Cytochrome *P450* monooxygenase detoxification enzymes. These metabolic systems are involved in the inactivation of xenobiotic compounds such as pesticides and drugs [177]. The availability of whole-genome sequence and well-functioning RNAi proves *T. castaneum* as a powerful model system for studying insecticide resistance and functional genetics [178]. Whole-genome sequencing of *T. castaneum* identified 133 functional *CYP* genes and 10 *CYP* pseudogenes [179]. These 143 genes belong to four clans (clan1, clan2, clan3 and mitochondrial clan), 26 families, and 59 subfamilies. Nine new families were identified including *CYP3* clan families *CYP345, 346, 347*, and *348* and mitochondrial family *CYP353*; *CYP4* clan families *CYP349, 350, 351*, and *352* and mitochondrial family *CYP353*. To identify the phylogenetic and evolutionary relationships of *T. castaneum CYPs* with *CYP*omes of other insects, four phylogenetic trees were constructed and a remarkable 1:1 orthology of *CYP2* and mitochondrial clans of *T. castaneum* with *D. melanogaster, A. gambiae,* and *A. mellifera* insect genomes was observed, suggesting functional conservation of these *CYPs* [180]. The number of *P450* genes in *T. castaneum* is much larger than *D. melanogaster* and *A. gambiae* but considerably lower than *C. quinquefasciatus* and *A. aegypti*. The large number of *CYP* genes in *T. castaneum* provides excellent protection against xenobiotics and other insecticides via an enzymatic detoxification mechanism [178]. Among 143 *CYP* genes, 99 *T. castaneum CYPs* were mapped on 9 chromosomes, 87 of which were located on six chromosomes LG3, LG4, LG5, LG6, LG8, and LG9. No *CYP* gene was mapped on the LG1 = X chromosome. The distribution and location of another 44 *CYPs* on the chromosome remain unknown. This confirms that several genes are under gene duplication events and that they descended from a common ancestral *P450* gene [180, 181].

Increased detoxification by cytochrome *P450s* has been considered to be the major mechanism involved in insecticide resistance of *T. castaneum*. *CYP450* gene

CYP6BQ9 showed 200-fold overexpression in the deltamethrin-resistant *QTC279* strain of *T. castaneum* and this upregulation suggests that *CYP6BQ9* has a significant impact on *T. castaneum* to metabolize deltamethrin [144]. Functional genomic and qRT-PCR based methods revealed that the high expression of *CYP6BQ9* in the brain might enhance the ability of the brain cells to catalyze deltamethrin and provide the defenses to protect the target site [144]. Additionally, RNAi-mediated knockdown of possible transcription factors was performed to understand the mechanism of the overexpression of the *CYP6BQ9* gene. Out of the 7 transcription factors tested, CncC and Maf transcription factors have been identified as key regulators for the activation of *CYP6BQ* genes and responsible for deltamethrin resistance in *T. castaneum* [182]. In another study, RNA sequencing, RNAi knockdown, and qRT-PCR data showed the involvement of CncC in the regulation of expression of multiple detoxification genes involved in phase I (*P450s*) and phase II (*GSTs*), and Phase III (ABC transporters) detoxification mechanisms in pyrethroid resistance strain of *T. castaneum* [183]. Both studies suggest that transcription factor CncC is required for the induction of genes coding for proteins involved in xenobiotic degradation. Many studies in flies and beetles have reported CncC regulation of expression of genes coding for proteins involved in phase I (*P450s*) and phase II (*GSTs*) detoxification mechanisms [184, 185].

CYP4BN6 and *CYP6BQ11* expression was induced in *T. castaneum* by dichlorvos and carbofuran and a higher level of expression of these two genes in late pupal and adult stages was detected. Furthermore, RNA interference (RNAi) mediated knockdown repressed the expression and increased the susceptibility of *Tribolium* to these two insecticides, suggesting that *CYP4BN6* and *CYP6BQ11* genes play an important role in developing resistance. More significantly, in addition to the findings mentioned above, expression of both *TcCYP4BN6* and *TcCYP6BQ11* was reduced by *latrophilin* (*lph*) gene knockout, indicating that these two *CYP* genes are controlled by the *lph* gene responsible for the susceptibility of the beetles to insecticides [186]. Cytochrome *P450s* are a supergene family of metabolic enzymes and the upregulation of *CYP* genes mediated by different insecticides has been extensively studied in *T. castaneum* [163]. Liang et al. [163] found that three (*CYP4G7, CYP4BR3,* and *CYP345A1)* out of the eight selected *CYP* genes (*CYP4G7*, *CYP4Q4*, *CYP4BR3*, *CYP12H1*, *CYP6BK11*, *CYP9D4*, *CYP9Z5*, and *CYP345A1*) showed high expression when the insects were exposed to four insecticides- cypermethrin, permethrin, cyhalothrin, lambda imidacloprid. Also, selected genes from *CYP6* and *CYP9* families did not exhibit any insecticide mediated overexpression in this study, although the genes from these families are known to confer resistance to a wide range of insecticides and metabolic detoxification [144, 182]. Specifically, they found that the upregulation of a specific gene can be influenced by insecticide concentration, developmental stage of insects, and exposure duration. They also suggested that the overexpression of *CYP* genes was affected by relatively low concentrations of insecticides, and increasing insecticide concentration did not show any significant upregulation, possibly due to increased toxic stress to the insects. In addition, tissue-specific expression patterns of *CYP* genes revealed that 7 out of 8 *CYP* genes were significantly upregulated in insect detoxification tissues including malpighian tubules, midgut, and fat bodies. The possible role of *CYP450* genes in phosphine resistant strain of *T. castaneum* has been studied previously and two *CYP* genes (*CYP4Q4* and *CYP4Q7*) are overexpressed in the midgut of permethrin resistant *T. castaneum* strain [187]. Two *CYP450* genes *CYP4Q4* and *CYP4Q7* identified in a pyrethroid-resistant strain of *T. castaneum* showed some level of upregulation which indicates that overexpression of *CYP450* genes is an important factor governing insecticide resistance in *Tribolium* [188]. The previous studies have shown that *T. castaneum* has developed insecticide resistance to 33 active ingredients [189] and genomic sequence analysis revealed an expansion of members of *CYP* families belonging to metabolic detoxification enzymes [179]. Zhu et al. [178] characterized the expression and induction of *CYP6BQ* gene cluster in deltamethrin resistant strain of *T. castaneum*, revealing that 10 out of these 12 genes were significantly upregulated in resistant strain than in the Lab-S susceptible strain. Moreover, the tissue-specific expression pattern of genes within the *CYP6BQ* cluster found that four genes (*CYP6BQ9, CYP6BQ5, CYP6BQ2, CYP6BQ4*) and three genes (*CYP6BQ11, CYP6BQ2, CYP6BQ4*) were significantly upregulated (>100 fold) in the tissues of the head and midgut respectively. All these studies have shown that overexpression of a specific *CYP* gene can be influenced by the type of insecticide, toxicity of insecticide, concentration of insecticide, exposure duration, and physiological status of insects.

3.2.2 Glutathione S tranferase (GSTs)

The glutathione S tranferase (*GSTs*) is a superfamily of multifunctional enzymes involved in insecticide resistance [190]. These enzymes metabolize the insecticides by conjugation reaction with reduced glutathione to hydrophobic xenobiotics and produce water-soluble metabolites that are easily excreted. Insect's *GSTs* were classified based on their location within the cell- cytosolic, microsomal, and mitochondrial [191, 192], and these *GSTs* are members of Delta, Epsilon, Sigma, Theta, Omega, and Zeta protein classes in arthropods [193]. Cytosolic *GSTs* possess a carboxyl (C)-terminal α-helical domain and an amino (N)-terminal α/β-domain joined by a variable linker region. The N terminal region is comprised of a highly conserved G site, which binds reduced GSH, and the C terminal domain consists of a highly variable H site that interacts with hydrophilic substrates. This hypervariability characteristic of the H site allows *GSTs* to metabolize various hydrophobic residues [194]. Sequencing the insect's genome provided an opportunity to identify and characterize the *GSTs* on a genome-wide scale [192, 195, 196]. This provides a platform for a better understanding of the evolution of insecticide resistance in arthropods. Using the genome sequence of *T. castaneum,* 36 putative cytosolic *GSTs* and 5 microsomal *GSTs* were discovered [192]. Among the 41 *GSTs*, thirty-eight *GSTs* were located on 4 chromosomes and the remaining three *GSTs* were mapped to other 3 of the 10 *T. castaneum* chromosomes. *T. castaneum* possesses the 3 Delta *GST* genes and 19 Epsilon *GSTs* gene, which were the fewer and higher *GST* genes than in *Drosophila, Anopheles, Apis, Bombyx, and Acyrthosiphon* [192, 197]. The expansion of the Epsilon class in *T. castaneum* indicates that they are frequently involved in high duplication events than the other four *GST* subclasses (Omega, Theta, Zeta, Sigma) and are fairly variable between different species and conserved within the species. And the four *GST* subclasses of Omega, Theta, Zeta, and Sigma of *T. castaneum* possess three, one, one, and seven genes respectively [192]. The previous studies reported that the Epsilon class of *GSTs* encoding enzymes are responsible for degrading certain insecticides such as DDT and pyrethroids in *Aedes aegypti* [198] and *Anopheles gambiae* [199]. The detoxification ability of the Epsilon class of *GSTs* in different insects suggested that *T. castaneum* maintains higher insecticide resistance and such tolerance may be due to the presence of expanded epsilon *GSTs* [192]. In addition to the epsilon class of *GST* mediated detoxification, the delta class of *GSTs* in *T. castaneum* was engaged in resisting poisonous chemicals and developing resistance to certain kinds of insecticides [200].

To gain insights on the regulatory, functional, and biological significance of *GST* delta 1 *T. castaneum* (*TcGSTd1*), Chen et al. [200] merged the RNA-sequencing technology and RNAi of control and RNAi treated larvae (*ds-TcGSTd1*) of *T. castaneum.* The results from this study established that *TcGSTd1* took part not only in the detoxification process but was also involved in insect fitness, survival, reproduction, and development. Further, Song et al. [201] conducted functional research on three deltas *GSTs* of *T. castaneum* (*TcGSTd1*, *TcGSTd2*, and *TcGSTd3*) to identify their role in insecticide degradation, metamorphosis, and physiology. The three delta *GSTs* of *T. castaneum* with their full-length sequences were identified and further characterized by cloning and sequencing. In this study, the expression levels of three delta *GSTs* were consistent across all developmental stages, implying that they may act as housekeeping genes and play an important role in the metamorphosis of *T. castaneum*. The expression profiling experiments revealed greater expression of *TcGSTd3* and *TcGSTd2* and lower expression of *TcGSTd1* after exposure to phoxim and lambda-cyhalothrin. Interestingly, the expression of *TcGSTd2* and *TcGSTd3* significantly increased by phoxim treatment than with the lambda-cyhalothrin. The results from this study imply that under elevated *TcGSTd2* and *TcGSTd3* activity conditions, *T. castaneum* can detoxify phoxim activation products, leading to resistance development. Similarly, elevated levels of *GSTs* have been reported to be associated with insecticide metabolism and producing resistance in many insects [171, 198, 202, 203]. In addition, resistance was induced by gene duplication within the structural *GST* genes which changes their substrate specificity [204]. Thus, the knowledge of *GST* mediated detoxification mechanism helps to detect resistance at an early stage, to remove the particular insecticide before the resistance alleles become fixed in the populations, and to design an effective molecule of insecticide.

3.2.3 Carboxyl Esterases (CarEs)

Carboxylesterases are ubiquitous enzymes involved in the detoxification of ester-containing xenobiotics. They are members of the esterase family of enzymes and have been isolated from all living organisms. As their name suggests, they are involved in hydrolysis reactions and convert the carboxyl esters into carboxylic acid and alcohol. Hydrolysis of the ester bond includes hydrolysis of a diverse range of phospho, thio, carboxylic, and other ester substrates. For carboxylesterases, the hydrolysis reaction is accomplished by 2 steps- first, the nucleophilic attack of oxygen of a serine residue on the carbonyl group of the substrate, removing the alcohol product, and generating relatively stable acyl enzymes. Second, a water molecule acts as an intermediary and makes a nucleophilic attack to remove the acid product of the reaction and produce the free enzyme. This reaction mechanism causes insecticide resistance in many insect species. As a key component of the detoxification mechanism, esterases have focused on the research of xenobiotic metabolism and resistance. The expression of carboxylesterases was significantly upregulated in the organophosphorous resistant *Aphis gossypii* strain than the susceptible strain [205]. Elevated carboxylesterase activity and carboxylesterase expression were identified in the pyrethroid-resistant strain of *Musca domestica* [206] and the tolerance to cypermethrin in *Musca domestica* was induced by high *CarE* enzyme activity. Similarly, the elevation of *CarE* activity in OP resistant and susceptible *Nilaparvata lugens strain* and the increased expression suggests that *CarE* mRNA was related to OP resistance in *Nilaparvata lugens* [207].

The occurrence of multiple mechanisms in an insect develops a very high level of resistance and in the case of *T. castaneum,* resistance was highest against pirimiphos-methyl and bifenthrin. It is interesting to observe that *T. castaneum* used two genetic strategies to adapt to these insecticides attack 1) a pool of Laccase2 enzyme ensured the protection by synthesizing the thicker cuticle which prevented the entry of insecticide into the insect body 2) a pool of esterases and lipases contributed the protection by hydrolysing or sequestering which rendered the insecticides ineffective [208]. Similarly, the functional role of two carboxylesterase genes of *T. castaneum* (*Tcest4* or *Tcest6*) were investigated by RNAi and identified their interaction with *Latrophilin* [165]. *Latrophilin (lph)* is an adhesion G-protein-coupled receptor that is essentially involved in the physiological process and cellular detoxification process although one member of *lph* existed in *T. castaneum* [209, 210]. The induction of *Tcest4* and *Tcest6* gene expression after treatment with carbofuran or dichlorvos insecticides revealed their detoxification ability and the RNAi of *Tcest4* and *Tcest6* further confirmed that it had a vital role in insecticide resistance. In addition, the study suggested that *lph* has a vital role in regulating the activity of *Tcest4* and *Tcest6* [165]. All these studies revealed the detoxification ability of carboxylesterases towards toxicants and the induction characteristics of some carboxylesterases could be used as biomarkers to assess the resistance against certain xenobiotics.

3.3 Reduced cuticular penetration resistance mechanism

Reduced penetration resistance is uncommon and little is known about its workings in insects. Reduced penetration is also called cuticle resistance that reduces the dose of the insecticide reaching into the insect's body and in all probability strongly associated with insecticide resistance. Normally contact insecticide penetrates through the insect cuticle and reaches the target site for action [172]. The cuticle is composed mainly of two different components, chitin, and cuticular protein, and the three functional layers of the cuticle consist of the outermost envelope, protein-rich epicuticle, and chitin-rich procuticle [211, 212]. Cuticular barriers develop resistance in insects by altering the cuticular thickness or by changing the cuticular composition [213, 214] or remodeling the cuticle by the high occurrence of cuticular proteins. The overexpression of laccases and ABC transporters has been reported to be involved in the compositional change of cuticle, which increases insects' tolerance to insecticides in the environment. Arkane et al. [215] revealed the association between the cuticle tanning and the expression profile of *T. castaneum* Laccase2 (*TcLac2*) from pupation to adult eclosion. This study unambiguously demonstrated that RNAi of *TcLac2* affected the cuticle tanning and produced the cuticle to be unpigmented. The dysfunctional *TcLac2* affected not only the adult and pupal cuticle but also the larval cuticle of *T. castaneum*. The suppressed *TcLac2* produces white and more flexible cuticles in pupal or newly molted adults of *T. castaneum*, facilitating the entry of insecticides into its body, and the high occurrence of expressed *TcLac2* mediates the compositional change of cuticle, which reduces the penetration of insecticides. Consistent with this interpretation, a pool of laccases was expressed highly in *T. castaneum* that served as protection against pirimiphos-methyl and bifenthrin [208]. Besides the overexpression of laccase, cuticular penetration resistance has been reported in combination with the upregulated activity of ABC transporters and *CYP450* genes [178, 208]. Typically, a combination of two or more mechanisms contributes significantly strong resistance than a single resistance mechanism [177]. In addition, several cuticular proteins (*TcCPR18*, *TcCPR4*, and *TcCPR27*) were identified from *T. castaneum* and found to be associated with the formation of the rigid cuticle [216]. However, their exact role in the cuticular penetration mechanism remains elusive. Thus, future functional characterization of cuticular proteins could provide a strong foundation for identifying the major players involved in the cuticular resistance mechanism, enabling the development of new resistance management strategies.

Researchers have been previously using bioassays, genetic and biochemical techniques to study the resistance mechanisms in *T. castaneum*. The application of whole-genome and transcriptome sequencing platforms provided a larger repository of gene resources for further investigations of resistance mechanisms in

T. castaneum. Known mechanisms that confer insecticide resistance in *T. castaneum* include 1) target site insensitivity 2) metabolic resistance 3) reduced cuticular penetration. These different forms of resistance mechanisms have been identified to act in compounding layers to degrade or sequester the insecticides faster before releasing their toxic effect. (**Figure 2**). In pest model beetle *T. castaneum*, different biochemical and molecular mechanisms were investigated against different insecticide classes (**Table 1**).

3.4 Symbiont mediated insecticide resistance mechanism

Recent years have seen a sharp increase in the study of insect microbiome, its crucial role in metabolic detoxification, and modulation of host immune responses. In some insect hosts, the symbiotic association appears to be causal for insecticide degradation, whereas, in others, studies suggest that it is mediated by physiological trade-offs [225]. In addition, the relationship between microbial community and insecticide resistance differs greatly and is context-dependent [225]. Several studies have established a causal connection between the fitness-enhancing symbionts and insecticide resistance in the bean bug, *Riptortus pedestris* [226], *Bactrocera dorsalis* [81],

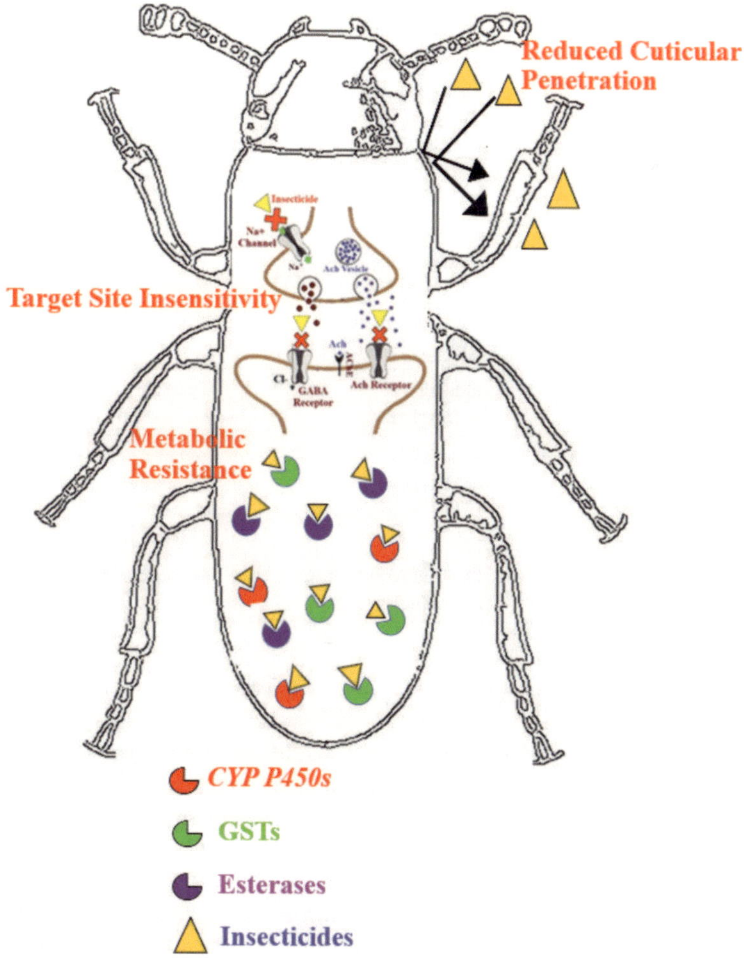

Figure 2.
Overview of types of resistance mechanisms in T. castaneum. Source: Adapted and modified from [108, 144, 151, 177, 215].

Sl. No.	Insecticide	Type of Resistance	Mechanism	References
1	Cyclodiene	Point mutations in the gene Resistance to dieldrin (*Rdl*)	Target site insensitivity	[138, 151]
2	Pirimiphos-methyl and bifenthrin	Elevated activity of lipases, esterases, and laccase2	Combination of reduced cuticular penetrance and metabolic detoxification	[208]
3	Dichlorvos, malathion, carbaryl, and carbofuran	Mutations in *AchE* gene	Reduced insensitivity of AChE	[108]
4	Malathion	Elevated activity of carboxylesterase	Metabolic detoxification	[217–220]
5	Malathion	Elevated activity of glutathione transferase	Metabolic detoxification	[221]
6	Deltamethrin	Elevated activity of *CYP6BQ9*	Metabolic detoxification	[144]
7	Phosphine	Elevated activity of *CYP346B1*, *CYP346B2*, and *CYP346B3*	Metabolic detoxification	[222, 223]
8	Phosphine	Elevated activity of *CYP4Q7*, *CYP4Q4*	Metabolic detoxification	[187]
9	Permethrin	Elevated activity of *CYP4Q4* and *CYP4Q7*	Metabolic detoxification	[187]
10	Cypermethrin, Permethrin and Cyhalothrin	Elevated activity of *CYP4G7* and *CYP345A1*	Metabolic detoxification	[163]
11	Phosphine	Elevated activity of *CYP345A1* and *CYP345A2*	Metabolic detoxification	[164]
12	Dichlorvos and carbofuran	Elevated activity of *CYP4BN6* and *CYP6BQ11*	Metabolic detoxification	[186]
13	Carbofuran or dichlorvos	Induction of carboxylesterase	Metabolic detoxification	[165]
14	Pirimiphos-methyl	Reduced cuticular penetration	Cuticular penetration resistance mechanism	[224]
15	Phoxim	Elevated activity of *GSTd2* and *GSTd3*	Metabolic detoxification	[201]

Table 1.
Studies performed in T. castaneum *showing different resistance mechanisms.*

Anopheles stephensi [227], *Lasioderma serricorne* [228], *Spodoptera frugiperda* [229], *Plutella xylostella* [230]. Among these studies, symbionts like *Burkholderia*, *Citrobacter freundii*, *Bacillus thuringiensis*, *Enterococcus mundtii,* and several gut bacteria provide physiological and evolutionary modifications to their insect host, thereby enhancing protection in a significant portion of insect pest taxa.

Increased access to a rapidly advancing metagenomic approach facilitated the understanding of the role of microbial communities in insecticide resistance, and several studies hint at this association. The initial research that isolated the bacteria and fungi from different life stages of *Tribolium confusum* (closely related to *T. castaneum*) dates back 60 years. It provided important clues on insect fecundity and pupation rate when grown in microbe-free flour. Momir Futo [231] studied the immune priming phenomenon (a form of immune memory) in *T. castaneum* and investigated its high survival response when orally exposed to bacterial components of *B. thuringiensis bv. Tenebrionis*. The evidence suggests that microbiota plays a significant role in the activation of immune priming and produces various immunological responses to enhance the protection of this species. A recent study systematically characterized the microbiome of *T. castaneum* across life stages and sexes and observed microbiome-mediated fitness benefits such as increased fecundity, increased offspring survival, and long lifespan. They also observed a correlation between wheat flour microbiome and host beetle microbiome [232]. All these studies have established different aspects of microbiome-derived fitness in *T. castaneum*. However, the microbial communities inhabiting *T. castaneum* and the unique intricate connection between symbionts and insecticide resistance have not been investigated thus far. A high-throughput metagenome sequencing approach is required to reveal the link between microbiota and insecticide resistance. Discovering this link and exploring the bacteria known to degrade insecticides within *T. castaneum* may offer novel insights into the unknown insecticide resistance mechanisms. Understanding the symbiont-associated resistance mechanism in *T. castaneum* has broader implications for developing integrative resistance management strategies.

4. Methods to monitor insecticide resistance in *T. castaneum*

Multiple reports indicate that *T. castaneum* has developed resistance to almost all insecticides [47, 51, 67, 72, 73]. The deleterious consequences of insecticide resistance in *Tribolium* have created a crisis in pest management programs. In the last decade, various research groups have identified several mechanisms involved in insecticide resistance in *T. castaneum* [39, 163, 164, 178, 208, 210]. They approached this problem at different levels and incorporated a diverse range of approaches like conventional toxicity bioassays, biochemical assays, gene expression, and functional genomic studies.

Different toxicity bioassays were used to measure resistance in the early stage in a cost-efficient manner. The conventional bioassay that is used to diagnose resistance involves collecting *T. castaneum* from the grain storage facilities and warehouses and rearing them until sufficient populations for testing. Mortality of larva, pupa, or adults is then estimated after exposure to series of concentrations of insecticide compounds. Subsequently, a Log dose probit assay is used to determine the LC_{50} or LC_{99} values of field-collected and susceptible populations. Then the resistance ratio was calculated using LC_{50} and LC_{95} values from *T. castaneum* field-collected populations compared with LC_{50} and LC_{95} values of the susceptible laboratory strain of *T. castaneum*. Different techniques such as topical application [233], residue exposure test [163], filter paper [39], diet incorporation method [234] are used to diagnose and determine the causes of pest control failure by insecticide selection pressures under field conditions.

The fully sequenced and annotated genome of *T. castaneum* has facilitated the identification of many genes involved in resistance [179]. Thus, further resistance research is centered on identifying genes involved in mediating insecticide resistance, particularly *CYP450s*, *GSTs*, and *CarEs*. Transcriptome profiling study identified differentially expressed miRNAs in four major life stages of *T. castaneum* and validated them by using real-time PCR experiments [235]. Several studies relied on different techniques such as Next-generation sequencing, RNA isolation, First-strand cDNA Synthesis, Real-Time PCR to evaluate the responses (upregulation or downregulation) of detoxifying genes against various insecticides in different tissues, developmental stages [163]. However, the resistance mechanism in insects was not fully understood. In this scenario, extensive genetic analysis is critical for understanding the function of genes involved in developing insecticide resistance which can then be targeted to suppress the further evolution of resistance. Thus, the RNAi technology complemented with expression studies to investigate the correlation between mRNA and the function of genes to rule out the role of these genes in resistance [144]. This gene silencing mechanism suppressed the target gene expression in *T. castaneum*, caused rapid and widespread mortality within the pest population. Here, different steps include target genes selection, isolation of *T. castaneum* at the proper stage for injection, establishing dsRNA production methodology, knockdown of target genes by injecting dsRNA directly into egg and larva and relative expression of knocked down gene-specific transcripts were employed to accomplish gene silencing [236]. To evaluate the specificity of dsRNA effects, quantitative PCR experiments were also carried out to check the expression of housekeeping genes such as actin or tubulin as a control [236]. RNAi-based gene silencing has the potential to down-regulate the expression of resistance-relevant genes and accelerate the discovery of gene function in *T. castaneum*. Thus, knock-down of upregulated resistance-relevant genes in *T. castaneum* is incredibly valuable in elucidating gene functions and provides information on the process that makes *T. castaneum* resistant or susceptible.

Resistance research on this beetle further improved by the most advanced approach Clustered regularly interspaced short palindromic repeats (CRISPR) system [237]. Recently, CRISPR is the best available method on vogue in order to explore the functional genes relevant to resistance in *T. castaneum* [238]. Considering the relevance of resistance inducing genes in *T. castaneum,* genome editing technology plays an important role in determining the functional genes involved in insecticide resistance. An enhanced conceptual understanding of Genome editing in *T. castaneum* will facilitate the application of CRISPR for dissection of gene function and fast-track the application of CRISPR to control these destructive pests [238].

Further, on the basis of gene expression, knockdown and genome editing studies, we could generate information on insecticide resistance levels of pest populations in our country. This information would offer a unique opportunity for overcoming or delaying resistance in *T. castaneum*. With the introduction of these advanced genetic technologies, the food security of our country could outstrip the population growth and ensure the supply of high-quality, safe, and economically stored grain products.

5. Conclusion

Insecticide resistance poses a major threat to global pest control efforts and elucidating the underlying mechanisms is critical for effective pest management. Most of the important pests of stored products have evolved resistance to commonly used insecticides the world over. *T. castaneum* is a pest of stored products that causes significant damage to cereal products, flour, grain, and rice bran. This omnivorous beetle has become so resistant to a range of insecticides that it can tolerate exposure to any insecticide. Many key factors such as intensive application of insecticides, control operations, mode of inheritance of resistance genes, change in fitness of individuals, and genetic background of *T. castaneum* influence the resistance.

This review attempts to address critical questions around how insecticide resistance emerges in *T. castaneum*, such as how many resistance mechanisms exist in a species genome? How do the different point mutations in ion channel receptors cause resistance? How does this mutation transfer globally? How many single or multiple mutations give rise to resistance? How many detoxifying genes are involved in resistance development? How do these genes provide resistance to all insecticides? How does their expression control resistance? What new mechanisms are still unexplored and what might be their role?

The availability of whole-genome sequence and applications of RNAi have made significant progress in understanding resistance mechanisms in *T. castaneum*. If *T. castaneum* comes into contact with an insecticide, the cuticle may be modified or thickened, eventually slowing down the penetration of insecticide molecules beyond the cuticular layer. If the insecticides enter the insect's body, *T. castaneum* can increase the expression of several genes from metabolic enzyme families (e.g., esterases, mixed-function oxidases, glutathione S-transferases) to detoxify the insecticidal effect. Eventually, if the insecticides enter the nervous system to act on the target sites, mutations are introduced into the active site of genes (e.g., *kdr* mutations, super *kdr* mutations), which can decrease the sensitivity of the target site to the insecticide. These are the three genetic modifications that reduce the lethal effects of an insecticide, thus developing pest resistance to organophosphates, pyrethroids, and neonicotinoids. Although metabolic detoxification and reduced target site insensitivity have been extensively studied, reduced cuticular penetrance mechanisms exist outside these paradigms. However, there are reasons beyond these factors that could shape resistance in insects, including microorganisms that enhance the degradation of insecticides. Many fundamental questions about the microbial shifts in response to insecticides and the functions of particular microorganisms in mediating resistance *in T. castaneum* remain unresolved.

The comprehensive genomic and transcriptomic analysis have improved the identification of the key genes encoding detoxifying enzymes such as CY*P450s*, *GSTs* and *CarEs*. The understanding of the detoxification mechanism responsible for insecticide resistance allows us to detect resistance at an early stage, remove the particular insecticide before the resistance alleles become fixed in the populations, and design an effective molecule of insecticide. Thus, understanding which insecticide is degraded by what genes is crucial to tackling the resistance problem. However, the contribution of most of the genes still needs to be confirmed by extensive genetic analysis such as RNAi and gene functional characterization. Knockdown of upregulated detoxifying genes in *T. castaneum* is immensely valuable in elucidating gene function and provides information on the factors that make *T. castaneum* resistant or susceptible. This double-stranded RNAi-mediated experiment paves the way for understanding the mechanisms causing resistance in *T. castaneum.*

T. castaneum resistance research has progressed effectively, from initial single-mutation study to multiple mutations in *Rdl* gene of GABA receptor, from examining the mutation in *Ace* gene to functional characterization of *Ace* gene of Acetylcholineesterase receptor, from sequencing the amino acid substitution of the *para* sodium channel to gene knockdown characterization, from single-gene sequencing to whole-genome analysis, from exploring transcriptional gene expression to functional expression analysis and from synergistic measures of metabolic detoxification to specific gene expression quantitation. The outcomes of these efforts have provided a clearer picture of molecular targets of different insecticides, complex resistance mechanisms, detoxifying genes, gene expressions, modification of target receptors, and have generated fresh insights for the development of targeted novel insecticides. Thus, this detailed review on complex resistance mechanisms and the genes involved in resistance will enhance the knowledge pool of all possible insecticide targets in *T. castaneum* and render greater selectivity in insecticide design, thereby improving the efficacy of insecticides. The consolidated insights from the literature review will provide much-needed insights as to what makes *T. castaneum* resistant to insecticides. The comprehensive overview will help successive research to initiate focussed monitoring of resistance. It contributes towards a deeper understanding of insecticide resistance and improved management of this destructive pest that threatens food storage, food safety, health, and economic security. Scientific efforts that make use of this pool of knowledge can lead to more sustainable agricultural practices.

Author details

U. Shamjana and Tony Grace*
Department of Genomic Science, Central University of Kerala, Kasaragod, Kerala, India

*Address all correspondence to: tonygrace@cukerala.ac.in

References

[1] United Nations. World Population Prospects: The 2017 Revision; United Nations, Department of Economic and Social Affairs, Population Division: New York, NY, USA [Internet]. 2017. Available from: https://www.un.org/development/desa/publications/world-population-prospects-the-2017-revision.html [Accessed: 2021-07-15].

[2] World Health Organization. World Health Statistics 2012: World Health Organization [Internet]. 2012 Available from: https://apps.who.int/iris/handle/10665/44844 [Accessed: 2021-08-08].

[3] FAO, IFAD, UNICEF, WFP and WHO. 2019. The State of Food Security and Nutrition in the World 2019. Safeguarding against economic slowdowns and downturns. Rome, FAO. Licence: CC BY-NC-SA 3.0 IGO

[4] Hodges RJ, Buzby JC, Bennett B. Postharvest losses and waste in developed and less developed countries: Opportunities to improve resource use. Journal of Agricultural Science, 2011;149 (S1):37-45. DOI:10.1017/S0021859610000936

[5] Mesterházy Á, Oláh J, Popp J. Losses in the grain supply chain: Causes and solutions. Sustainability. 2020;12(6): 2342. DOI:10.3390/su12062342

[6] Grolleaud M. Post-harvest losses: Discovering the full story. Overview of the phenomenon of losses during the post-harvest system. Rome, Italy: FAO, Agro Industries and Post-Harvest Management Service (AGSI). [Internet]. 2002. Available from: https://agris.fao.org/agris-search/search.do?recordID=XF2016055548 [Accessed: 2021-07-24]

[7] Parfitt J, Barthel M, Macnaughton S. Food waste within food supply chains: quantification and potential for change to 2050. Philosophical transactions of the royal society B: biological sciences. 2010; 365(1554):3065-3081. DOI:10.1098/rstb.2010.0126

[8] National Academy of Agricultural Sciences. National Academy of Agricultural Sciences Saving the Harvest: Reducing the Food Loss and Waste; Policy Brief No.5; National Academy of Agricultural Sciences: New Delhi, India. [Internet] 2019. Available from: http://naasindia.org/documents/Saving%20the%20Harvest.pdf [Accessed: 2021-07-12].

[9] Abass AB, Ndunguru G, Mamiro P, Alenkhe B, Mlingi N, Bekunda M. Post-harvest food losses in a maize-based farming system of semi-arid savannah area of Tanzania. Journal of stored products research. 2014;57:49-57. DOI:10.1016/j.jspr.2013.12.004

[10] Boxall RA. Damage and loss caused by the larger grain borer Prostephanus truncatus. Integrated Pest Management Reviews. 2002;7(2):105-21. DOI:10.1023/a:1026397115946

[11] Tapondjou LA, Adler CL, Bouda H, Fontem DA. Efficacy of powder and essential oil from Chenopodium ambrosioides leaves as post-harvest grain protectants against six-stored product beetles. Journal of stored products research. 2002;38(4):395-402. DOI:10.1016/s0022-474x(01)00044-3

[12] Kumar D, Kalita P. Reducing postharvest losses during storage of grain crops to strengthen food security in developing countries. Foods. 2017;6(1):8. DOI:10.3390/foods6010008

[13] Phillips TW, Throne JE. Biorational approaches to managing stored-product insects. Annual review of entomology. 2010;55:375-97. DOI:10.1146/annurev.ento.54.110807.090451

[14] Dowdy AK, McGaughey WH. Seasonal activity of stored-product insects in and around farm-stored wheat. Journal of Economic Entomology. 1994;87(5):1351-8. DOI:10.1093/jee/87.5.1351

[15] Arbogast RT, Kendra PE, Mankin RW, McGovern JE. Monitoring insect pests in retail stores by trapping and spatial analysis. Journal of economic entomology. 2000;93(5):1531-1542. DOI:10.1603/0022-0493-93.5.1531

[16] Campbell JF, Arbogast RT. Stored-product insects in a flour mill: population dynamics and response to fumigation treatments. Entomologia Experimentalis et Applicata. 2004;112(3):217-25. DOI:10.1111/j.0013-8703.2004.00197.x

[17] Rajendran S, Sriranjini V. Plant products as fumigants for stored-product insect control. Journal of stored products Research. 2008;44(2):126-35. DOI:10.1016/j.jspr.2007.08.003

[18] Park T. Beetles, competition, and populations. Science. 1962;138(3548): 1369-75. DOI:10.1126/science.138.3548.1369.

[19] Beeman RW. Distribution of the Medea factor M4 in populations of Tribolium castaneum (Herbst) in the United States. Journal of Stored Products Research. 2003;39(1):45-51. DOI:10.1016/s0022-474x(02)00016-4

[20] Campbell JF, Hagstrum DW. Patch exploitation by Tribolium castaneum: movement patterns, distribution, and oviposition. Journal of Stored Products Research. 2002;38(1):55-68. DOI:10.1016/s0022-474x(00)00042-4

[21] McKay T, Bowombe-Toko MP, Starkus LA, Arthur FH, Campbell JF. Monitoring of Tribolium castaneum (Coleoptera: Tenebrionidae) in rice mills using pheromone-baited traps. Journal of economic entomology. 2019;112(3):1454-62. DOI:10.1093/jee/toy422

[22] McKay T, White AL, Starkus LA, FH, Campbell JF. Seasonal patterns of stored-product insects at a rice mill. Journal of economic entomology. 2017;110(3):1366-1376. DOI:10.1093/jee/tox089

[23] Mills R, Pedersen J. A flour mill sanitation manual. Eagen Press; 1990. DOI:10.1002/food.19910350438

[24] Ridley AW, Hereward JP, Daglish GJ, Raghu S, Collins PJ, Walter GH. The spatiotemporal dynamics of Tribolium castaneum (Herbst): adult flight and gene flow. Molecular ecology. 2011;20(8):1635-1646. DOI:10.1111/j.1365-294x.2011.05049.x

[25] Hernandez Nopsa JF, Daglish GJ, Hagstrum DW, Leslie JF, Phillips TW, Scoglio C, Thomas-Sharma S, Walter GH, Garrett KA. Ecological networks in stored grain: Key postharvest nodes for emerging pests, pathogens, and mycotoxins. BioScience. 2015;65(10):985-1002. DOI:10.1093/biosci/biv122

[26] Payne NM. Some Effects of Triboliumon on Flour. Journal of Economic Entomology. 1925;18(5):737-744. DOI:10.1093/jee/18.5.737

[27] Atwal AS. Dhaliwal GS. Insect pests of stored grain and other Products. In: Agricultural pests of India and South-East Asia. Kalyani Publisher, New Delhi, India. 1976; 389-415. Available from: https://www.cabi.org/ISC/abstract/19790565441

[28] Pinniger DB. Food-baited traps; past, present and future. Journal of the Kansas Entomological Society. 1990:533-8. DOI:10.2307/25085219

[29] Ogden JC. Effect of components of conditioned medium on behavior in Tribolium confusum. Physiological

Zoology. 1969;42(3):266-74. DOI:10.1086/physzool.42.3.30155490

[30] Hodges RJ, Robinson R, Hall DR. Quinone contamination of dehusked rice by Tribolium castaneum (Herbst) (Coleoptera: Tenebrionidae). Journal of Stored Products Research. 1996; 32(1):31-7. DOI:10.1016/0022-474x(95)00036-7

[31] Villaverde ML, Juárez MP, Mijailovsky S. Detection of Tribolium castaneum (Herbst) volatile defensive secretions by solid phase microextraction–capillary gas chromatography (SPME-CGC). Journal of Stored Products Research. 2007;43(4):540-5. DOI:10.1016/j.jspr.2007.03.003

[32] Evans NJ. The effectiveness of various insecticides on some resistant beetle pests of stored products from Uganda. Journal of stored products research. 1985;21(2):105-9. DOI:10.1016/0022-474x(85)90030-x

[33] Collins PJ, Schlipalius DI. Insecticide resistance. In Recent advances in stored product protection. Berlin, Heidelberg: Springer. 2018; pp. 169-182. DOI:10.1007/978-3-662-56125-6_8

[34] Anonymous. Tolerances and exemptions from tolerances for pesticide chemicals in or on raw agricultural commodities. Federal Register 1958;23:6417.

[35] Kumar V, Morrison FO. Recording the susceptibility levels of current stored product pest populations to current insecticides. In: Proceedings of XII International Congress of Entomology; 1965; pp. 656-657

[36] Haliscak JP, Beeman RW. Status of malathion resistance in five genera of beetles infesting farm-stored corn, wheat, and oats in the United States. Journal of Economic Entomology. 1983;76(4):717-22. DOI:10.1093/jee/76.4.717

[37] White ND, Watters FL. Incidence of malathion resistance in Tribolium castaneum and Cryptolestes ferrugineus populations collected in Canada. In: Proceedings of the Third International Working Conference on Stored-Product Entomology. October 23-28, 1983, Kansas State University, Manhattan, Kansas USA; 1984; pp. 290-302. Kansas State University.

[38] Zettler LJ. Pesticide resistance in Tribolium castaneum and T. confusum (Coleoptera: Tenebrionidae) from flour mills in the United States. Journal of Economic Entomology. 1991;84(3):763-7. DOI:10.1093/jee/84.3.763

[39] Attia MA, Wahba TF, Shaarawy N, Moustafa FI, Guedes RN, Dewer Y. Stored grain pest prevalence and insecticide resistance in Egyptian populations of the red flour beetle Tribolium castaneum (Herbst) and the rice weevil Sitophilus oryzae (L.). Journal of Stored Products Research. 2020;87:101611. DOI:10.1016/j.jspr.2020.101611

[40] Hadaway AB. Toxicity of insecticides to tsetse flies. Bulletin of the World Health Organization. 1972;46(3):353. Available from: https://www.ncbi.nlm.nih.gov/pmc/articles/PMC2480747/

[41] Casida JE. Pyrethrum flowers and pyrethroid insecticides. Environmental health perspectives. 1980;34:189-202. DOI:10.1289/ehp.8034189

[42] Arthur FH. Grain protectants: current status and prospects for the future. Journal of Stored Products Research. 1996;32(4):293-302. DOI:10.1016/s0022-474x(96)00033-1

[43] Toews MD, Phillips TW, Payton ME. Estimating populations of grain beetles using probe traps in wheat-filled

concrete silos. Environmental Entomology. 2005;34(3):712-8. DOI:10.1603/0046-225x-34.3.712

[44] Halliday WR, Arthur FH, Zettler JL. Resistance status of red flour beetle (Coleoptera: Tenebrionidae) infesting stored peanuts in the southeastern United States. Journal of economic entomology. 1988;81(1):74-7. DOI:10.1093/jee/81.1.74

[45] Rossi E, Cosimi S, Loni A. Insecticide resistance in Italian populations of Tribolium flour beetles. Bulletin of insectology. 2010;63(2):251-8. Available from: https://www.cabdirect.org/cabdirect/abstract/20113005057

[46] Champ BR, Dyte CE. Report of the FAO global survey of pesticide susceptibility of stored grain pests. FAO; 1976. Available from: https://www.cabdirect.org/cabdirect/abstract/19776721377

[47] Arnaud L, Brostaux Y, Assie LK, Gaspar C, Haubruge E. Increased fecundity of malathion-specific resistant beetles in absence of insecticide pressure. Heredity. 2002;89(6):425-9. DOI:10.1038/sj.hdy.6800167

[48] Opit GP, Phillips TW, Aikins MJ, Hasan MM. Phosphine resistance in Tribolium castaneum and Rhyzopertha dominica from stored wheat in Oklahoma. Journal of Economic Entomology. 2012;105(4):1107-14. DOI:10.1603/EC12064

[49] Cato A. *Phosphine resistance in North American Tribolium castaneum (Herbst) (Coleoptera: Tenebrionidae)* (Doctoral dissertation, Kansas State University). Available from: http://krex.k-state.edu/dspace/handle/2097/20337. [Accessed: 2021-08-04].

[50] Stadler T, Subramanjam B, Ferrero AA. Monitoring for insecticide resistance in major stored product pests in Argentina: a review. Agriscientia. 2003;20:99-110.). Available from: http:/revistas.unc.edu.ar/index.php/agris/article/download/2838/2720/9844

[51] Pieterse AH, Schulten GG, Kuyken W. A study on insecticide resistance in Tribolium castaneum (Herbst)(Coleoptera, Tenebrionidae) in Malawi (Central Africa). Journal of Stored Products Research. 1972;8(3):183-91. DOI:10.1016/0022-474x(72)90038-0

[52] Andrić G, Pražić-Golić M, Kljajić P. Toxicity of several contact insecticides to Tribolium castaneum (Herbst) populations after selection with pirimiphos-methyl and deltamethrin. Pesticidi i fitomedicina. 2015;30(4):209-16. DOI: 10.2298/PIF1504209A

[53] Talukder F, Rahman A, hahjahan M. Malathion Resistance i n Tribolium castaneum (Coleoptera: Tenebrionida e) in Bangladesh. Biological Sciences - PJSIR [Internet]. 2007; 50(3):204-209. Available from: https://www.v2.pjsir.org/index.php/biological-sciences/article/view/966/541 [Accessed: 2021-07-30]

[54] Gibe AJG, Motoyama N. Malathion resistance in the red flour beetle, Tribolium castaneum Herbst (Coleoptera:Tenebrionidae), Asia Life Sciences. 2002;11:75-83.

[55] Saleem MA, Shakoori AR. Toxicity of malathion, permethrin and cypermethrin against resistant and susceptible strains of Tribolium castaneum (Herbst.). Pakistan Journal of Zoology (Pakistan). [Internet] 1989. Available from: https://agris.fao.org/agris-search/search.do?recordID=PK9001154. [Accessed: 2021-07-19].

[56] Wakil W, Kavallieratos NG, Usman M, Gulzar S, El-Shafie HA. Detection of phosphine resistance in field populations of four key

stored-grain insect pests in Pakistan. Insects. 2021;12(4):288. DOI:10.3390/insects12040288

[57] Javadzadeh M, Sheikhi-Garjan A, Hosseini-Gharalari A. Susceptibility of different populations of Tribolium confusum (Coleoptera: Tenebrionidae) to malathion (EC 57%) in flour mills of Iran. Acta Phytopathologica et Entomologica Hungarica. 2017;52(1): 111-5. DOI: 10.1556/038.52.2017.002

[58] Collins PJ. Inheritance of resistance to pyrethroid insecticides in Tribolium castaneum (Herbst). Journal of Stored Products Research. 1998;34(4):395-401. DOI:10.1016/s0022-474x(98)00020-4

[59] Nakakita H, Winks RG. Phosphine resistance in immature stages of a laboratory selected strain of Tribolium castaneum (Herbst)(Coleoptera: Tenebrionidae). Journal of Stored Products Research. 1981;17(2):43-52. DOI:10.1016/0022-474x(81)90016-3

[60] Mills KA. Resistance to the fumigant hydrogen phosphide in some stored-product species associated with repeated inadequate treatments. Mitteilungen der Deutschen Gesellschaft für allgemeine und angewandte Entomologie. [Internet] 1983;4(1/3):98-101. Available from: https://www.cabdirect.org/cabdirect/abstract/19840512864. [Accessed: 2021-07-29]

[61] Tyler PS, RWD Taylor, DP Rees. Insect resistance to phosphine fumigation in food warehouses in Bangladesh. International Pest Control [Internet] 1983;25:10-13. Available from: http://pascal-francis.inist.fr/vibad/index.php?action=getRecordDetail&idt=PASCALZOOLINEINRA83X0402224 [Accessed: 2021-07-22]

[62] Attia FI. Insecticide and fumigant resistance in insects of grain and stored-products in Australia. In: Proceedings of the 3rd International Working Conference on Stored-product Entomology. Manhattan, Kansas USA. [Internet] 1983;pp. 196-208. Available from: http://spiru.cgahr.ksu.edu/proj/iwcspp/pdf2/3/196.pdf [Accessed: 2021-08-02]

[63] Pacheco IA, Santori MR, Taylor RW. Survey of phosphine resistance in stored grain insect pests in the state of Sao Paulo. Coletanea do Instituto de Tecnologia de Alimentos. 1990;20(2): 144-54. Available from: www.cabdirect.org/cabdirect/abstract/19921161132

[64] Benhalima H, Chaudhry MQ, Mills KA, Price NR. Phosphine resistance in stored-product insects collected from various grain storage facilities in Morocco. Journal of Stored Products Research. 2004;40(3):241-9. DOI:10.1016/s0022-474x(03)00012-2

[65] Emery RN, Nayak MK, Holloway JC. Lessons learned from phosphine resistance monitoring in Australia. Stewart Postharvest Review. 2011;7(3). DOI:10.2212/spr.2011.3.8

[66] Lorini I, Collins PJ, Daglish GJ, Nayak MK, Pavic H. Detection and characterisation of strong resistance to phosphine in Brazilian Rhyzopertha dominica (F.)(Coleoptera: Bostrychidae). Pest Management Science: formerly Pesticide Science. 2007;63(4):358-64. DOI: 10.1002/ps.1344

[67] Collins PJ. A new resistance to pyrethroids in Tribolium castaneum (Herbst). Pesticide Science. 1990;28(1):101-15. Available from: https://www.cabdirect.org/cabdirect/abstract/19911146037

[68] Hasan M, Ahmed F, Ashraf K, Ahmad M. Survey of resistance against insecticides indifferent strains of Tribolium castaneum (Herbst.) collected from Bahawalpur Division. Pakistan Entomology. 1996;18: 41-2

[69] Hussain R, Ashfaq M, Saleem MA, Ahmed SO. Toxicity of some insecticides with novel modes of action against malathion-resistant and organophosphate-susceptible strains of Tribolium castaneum larvae. International Journal of Agriculture and Biology. 2005;7(5):768-772. DOI:1560-8530/2005/07-5-768-772

[70] Beeman RW. Inheritance and linkage of malathion resistance in the red flour beetle. Journal of Economic Entomology. 1983;76(4):737-40. DOI:10.1093/jee/76.4.737

[71] Bhatia SK, Pradhan S. Studies on resistance to insecticides in Tribolium castaneum (Herbst)—III. Selection of a strain resistant to lindane and its biological characteristics. Journal of Stored Products Research. 1971;6(4): 331-7. DOI:10.1016/0022-474x (71)90046-4

[72] Saxena JD, Bhatia SK, Sinha SR. Status of insecticide resistance in Tribolium castaneum (Herbst.) in India. IV: Resistance to phosphine. Bulletin of grain technology. 1991;29(3):148-51. Available from: https://pascal-francis.inist.fr/vibad/index.php?action=getRecordDetail&idt=4045123

[73] Dhaliwal BK, Chawla RF. Evaluation of Current Status of Malathion Resistance in Tribolium castaneum (Herbst)(Coleoptera: Tenebrionidae) in Punjab, India. Pesticide Research Journal. 1995;7(1):54-7. Available from: https://www.indianjournals.com/ijor.aspx?target=ijor:prj&volume=7&issue=1&article=008

[74] Srivastava, C., Sinha, S. R., and Singh, D. 2001. Susceptibility of red flour beetle Tribolium castaneum (Herbst) to malathion. Indian Journal of Entomology. 63(2): 176-178

[75] Bhatia SK, Saxena JD, Sinha SR. Status of insecticide resistance in Tribolium castaneum (Herbst) in India. I: Resistance to malathion. Bulletin of grain technology. 1990;28(3):250-4. Available from: https://pascal-francis.inist.fr/vibad/index.php?action=getRecordDetail&idt=5416534

[76] Padhee AK, Saxena JD, Sinha SR, Srivastava C. Selection for Resistance to Deltamethrin in Red-Flour Beetle, Tribolium castaneum (Herbst). Annals of Plant Protection Sciences. 2002;10(2):220-4. Available from: ttps://www.indianjournals.com/ijor.aspx?target=ijor:apps&volume=10&issue=2&article=011

[77] Singh S, Prakash S. Development of resistance in Tribolium castaneum, Herbst (Coleoptera: Tenebrionidae) towards deltamethrin in laboratory. International Journal of Scientific and Research Publications. 2013;3(8):1-4. Available from: http://www.ijsrp.org/e-journal.html

[78] Arshad A, Munawar A, Mastoi MI, Sohail S, Liang C. Evaluation of Phosphine and Cypermethrin resistance in field collected strains of Rhyzopertha Dominica (Coleoptera: Bostrichidae) and Tribolium castaneum (Coleoptera: Tenebrionidae) under laboratory conditions. Evaluation. 2019;4(5).

[79] Roush RT, McKenzie JA. Ecological genetics of insecticide and acaricide resistance. Annual review of entomology. 1987;32(1):361-80. DOI:10.1146/annurev.en.32.010187.

[80] Dada N, Sheth M, Liebman K, Pinto J, Lenhart A. Whole metagenome sequencing reveals links between mosquito microbiota and insecticide resistance in malaria vectors. Scientific reports. 2018;8(1):1-3. DOI:10.1038/s41598-018-20367-4

[81] Cheng D, Guo Z, Riegler M, Xi Z, Liang G, Xu Y. 2017. Gut symbiont enhances insecticide resistance in a significant pest, the orient fruit fly Bactrocera dorsalis (Hendel).

Microbiome. 2017;5:13. DOI:10.22194/JGIASS/4.3.753

[82] Wang YT, Shen RX, Xing D, Zhao CP, Gao HT, Wu JH, Zhang N, Zhang HD, Chen Y, Zhao TY, Li CX. Metagenome sequencing reveals the midgut microbiota makeup of Culex pipiens quinquefasciatus and its possible relationship with insecticide resistance. Frontiers in microbiology. 2021;12:228. DOI:10.3389/fmicb.2021.625539

[83] Xia X, Zheng D, Zhong H, Qin B, Gurr GM, Vasseur L, Lin H, Bai J, He W, You M. DNA sequencing reveals the midgut microbiota of diamondback moth, Plutella xylostella (L.) and a possible relationship with insecticide resistance. PloS one. 2013;8(7):e68852. DOI:10.1371/journal.pone.0068852

[84] Ceja-Navarro JA, Vega FE, Karaoz U, Hao Z, Jenkins S, Lim HC, Kosina P, Infante F, Northen TR, Brodie EL. Gut microbiota mediate caffeine detoxification in the primary insect pest of coffee. Nature communications. 2015;6(1):1-9. DOI:10.1038/ncomms8618

[85] Eto M. Zweig G. Organophosphorus Pesticides: Organic and Biological Chemistry. Cleveland, Ohio: CRC Press. Inc. 1979. DOI:10.1201/9781351075305

[86] Kuhr RJ, Dorough HW. Carbamate insecticides: chemistry, biochemistry, and toxicology. Cleveland, Ohio: CRC Press, Inc. 1976. Available from: https://www.cabdirect.org/cabdirect/abstract/19760536242

[87] Russell RJ, Claudianos C, Campbell PM, Horne I, Sutherland TD, Oakeshott JG. Two major classes of target site insensitivity mutations confer resistance to organophosphate and carbamate insecticides. Pesticide Biochemistry and Physiology. 2004; 79(3):84-93. DOI:10.1016/j.pestbp.2004.03.002

[88] Pang YP. Insect acetylcholinesterase as a target for effective and environmentally safe insecticides. Advances in Insect Physiology. 2014;46:435-94. DOI:10.1016/b978-0-12-417010-0.00006-9

[89] Heong KL, Tan KH, Garcia CPF, Liu Z, Lu Z. Research methods in toxicology and insecticide resistance monitoring of rice planthoppers. 2nd ed. International Rice Research Institute: Los Banos; 2013. Available from: https://ageconsearch.umn.edu/record/164481/files/ResearchToxicology.pdf

[90] Narahashi T, Frey JM, Ginsburg KS, Roy ML. Sodium and GABA-activated channels as the targets of pyrethroids and cyclodienes. Toxicology letters. 1992;64:429-36. DOI:10.1016/0378-4274(92)90216-7

[91] Casida JE. Insecticide action at the GABA-gated chloride channel: Recognition, progress, and prospects. Archives of insect biochemistry and physiology. 1993;22(1-2):13-23. DOI:10.1002/arch.940220104

[92] Fournier D, Mutero A. Modification of acetylcholinesterase as a mechanism of resistance to insecticides. Comparative Biochemistry and Physiology Part C: Pharmacology, Toxicology and Endocrinology. 1994;108(1):19-31. DOI:10.1016/1367-8280(94)90084-1

[93] Clark JM, Scott JG, Campos F, Bloomquist JR. Resistance to avermectins: extent, mechanisms, and management implications. Annual review of entomology. 1995;40(1):1-30. DOI:10.1146/annurev.en.40.010195.000245

[94] Tomita T, Hidoh O, Kono Y. Absence of protein polymorphism attributable to insecticide-insensitivity of acetylcholinesterase in the green rice leafhopper, Nephotettix cincticeps. Insect Biochemistry and Molecular

Biology. 2000;30(4):325-333. DOI:10.1016/s0965-1748(00)00006-0

[95] Gao JR, Zhu KY. Increased expression of an acetylcholinesterase gene may confer organophosphate resistance in the greenbug, Schizaphis graminum (Homoptera: Aphididae). Pesticide Biochemistry and Physiology. 2002;73(3):164-173. DOI:10.1016/s0048-3575(02)00105-0

[96] Weill M, Fort P, Berthomieu A, Dubois MP, Pasteur N, Raymond M. A novel acetylcholinesterase gene in mosquitoes codes for the insecticide target and is non–homologous to the ace gene Drosophila. Proceedings of the Royal Society of London. Series B: Biological Sciences. 2002;269(1504): 2007-16. DOI 10.1098/rspb.2002.2122

[97] Liu N. Insecticide resistance in mosquitoes: impact, mechanisms, and research directions. Annual review of entomology. 2015;60:537-559. DOI:10.1146/annurev-ento-010814-020828

[98] Gorbel V, N'Guessan R. Distribution, mechanisms, impact and management of insecticide resistance in malaria vectors: a pragmatic review. In: Manguin S, editor. Anopheles mosquitoes - new insights into malaria vectors. Rijeka: InTech; 2013. DOI:10.5772/56117

[99] Pitman RM. Transmitter substances in insects: a review. Comparative and general pharmacology. 1971;2(7):347-71. DOI:10.1016/0010-4035(71)90060-7

[100] Siegfried BD, Scharf ME. Mechanisms of organophosphate resistance in insects. In: Biochemical sites of insecticide action and resistance. 2001: pp. 269-291. Springer, Berlin, Heidelberg. DOI:10.1007/978-3-642-59549-3_13

[101] Toutant JP. Insect acetylcholinesterase: catalytic properties, tissue distribution and molecular forms. Progress in neurobiology. 1989;32(5):423-46. DOI:10.1016/0301-0082(89)90031-2

[102] Kono Y, Tomita T. Amino acid substitutions conferring insecticide insensitivity in Ace-paralogous acetylcholinesterase. Pesticide biochemistry and physiology. 2006;85(3):123-32. DOI:10.1016/j.pestbp.2005.12.002

[103] Hall LC, Spierer P. The Ace locus of Drosophila melanogaster: structural gene for acetylcholinesterase with an unusual 5′ leader. The EMBO Journal. 1986;5(11):2949-2954. DOI:10.1002/j.1460-2075.1986.tb04591.x

[104] Adams MD, Celniker SE, Holt RA, Evans CA, Gocayne JD, Amanatides PG, Scherer SE, Li PW, Hoskins RA, Galle RF, George RA. The genome sequence of Drosophila melanogaster. Science. 2000;287(5461):2185-2195. DOI:10.1126/science.287.5461.2185

[105] Walsh SB, Dolden TA, Moores GD, Kristensen M, Lewis T, Devonshire AL, Williamson MS. Identification and characterization of mutations in housefly (Musca domestica) acetylcholinesterase involved in insecticide resistance. Biochemical Journal. 2001;359(1):175-81. DOI:10.1042/0264-6021:3590175

[106] Weill M, Malcolm C, Chandre F, Mogensen K, Berthomieu A, Marquine M, Raymond M. The unique mutation in ace-1 giving high insecticide resistance is easily detectable in mosquito vectors. Insect molecular biology. 2004;13(1):1-7. DOI:10.1111/j.1365-2583.2004.00452.x

[107] Chen Z, Newcomb R, Forbes E, McKenzie J, Batterham P. The acetylcholinesterase gene and organophosphorus resistance in the Australian sheep blowfly, Lucilia cuprina. Insect biochemistry and

molecular biology. 2001;31(8):805-816. DOI:10.1016/s0965-1748(00)00186-7

[108] Lu Y, Park Y, Gao X, Zhang X, Yao J, Pang YP, Jiang H, Zhu KY. Cholinergic and non-cholinergic functions of two acetylcholinesterase genes revealed by gene-silencing in Tribolium castaneum. Scientific Reports. 2012;2(1):1-7. DOI:10.1038/srep00288

[109] Jiang XC, Jiang XY, Liu S. Molecular characterization and expression analysis of two acetylcholinesterase genes from the small white butterfly Pieris rapae (Lepidoptera: Pieridae). Journal of Insect Science. 2018;18(5):2. DOI:10.1093/jisesa/iey085

[110] Chen M, Han Z, Qiao X, Qu M. Resistance mechanisms and associated mutations in acetylcholinesterase genes in Sitobion avenae (Fabricius). Pesticide Biochemistry and Physiology. 2007;87(3):189-195. DOI:10.1016/j.pestbp.2006.07.009

[111] Alon M, Alon F, Nauen R, Morin S. Organophosphates' resistance in the B-biotype of Bemisia tabaci (Hemiptera: Aleyrodidae) is associated with a point mutation in an ace1-type acetylcholinesterase and overexpression of carboxylesterase. Insect Biochemistry and Molecular Biology. 2008;38(10):940-949. DOI:10.1016/j.ibmb.2008.07.007

[112] Seino A, Kazuma T, Tan AJ, Tanaka H, Kono Y, Mita K, Shiotsuki T. Analysis of two acetylcholinesterase genes in Bombyx mori. Pesticide Biochemistry and Physiology. 2007;88(1):92-101. DOI:10.1016/j.pestbp.2006.09.005

[113] Li F, Han ZJ. Two different genes encoding acetylcholinesterase existing in cotton aphid (Aphis gossypii). Genome. 2002;45(6):1134-41. DOI: 10.1139/G02-085

[114] Ni XY, Tomita T, Kasai S, Kono Y. cDNA and deduced protein sequence of acetylcholinesterase from the diamondback moth, Plutella xylostella (L.)(Lepidoptera: Plutellidae). Applied entomology and zoology. 2003;38(1):49-56. DOI:10.1303/aez.2003.49

[115] Kim JI, Jung CS, Koh YH, Lee SH. Molecular, biochemical and histochemical characterization of two acetylcholinesterase cDNAs from the German cockroach Blattella germanica. Insect molecular biology. 2006;15(4):513-522. DOI:10.1111/j.1365-2583.2006.00666.x

[116] Mori A, Lobo NF, deBruyn B, Severson DW. Molecular cloning and characterization of the complete acetylcholinesterase gene (Ace1) from the mosquito Aedes aegypti with implications for comparative genome analysis. Insect biochemistry and molecular biology. 2007;37(7):667-674. DOI:10.1016/j.ibmb.2007.03.014

[117] Mutero A, Pralavorio M, Bride JM, Fournier D. Resistance-associated point mutations in insecticide-insensitive acetylcholinesterase. Proceedings of the national academy of sciences. 1994;91(13):5922-5926. DOI:10.1073/pnas.91.13.5922

[118] Vontas JG, Cosmidis N, Loukas M, Tsakas S, Hejazi MJ, Ayoutanti A, Hemingway J. Altered acetylcholinesterase confers organophosphate resistance in the olive fruit fly Bactrocera oleae. Pesticide Biochemistry and Physiology. 2001;71(2):124-132. DOI:10.1006/pest.2001.2568

[119] Zhu KY, Lee SH, Clark JM. A point mutation of acetylcholinesterase associated with azinphosmethyl resistance and reduced fitness in Colorado potato beetle. Pesticide biochemistry and physiology. 1996;55(2):100-108. DOI:10.1006/pest.1996.0039

[120] Luo GH, Li XH, Zhang ZC, Liu BS, Huang SJ, Fang JC. Cloning of two Acetylcholinesterase genes and analysis of point mutations putatively associated with Triazophos resistance in Chilo auricilius (Lepidoptera: Pyralidae). Journal of economic entomology. 2015;108(3):1289-1297. DOI:10.1093/jee/tov086

[121] Wu S, Zuo K, Kang Z, Yang Y, Oakeshott JG, Wu Y. A point mutation in the acetylcholinesterase-1 gene is associated with chlorpyrifos resistance in the plant bug Apolygus lucorum. Insect biochemistry and molecular biology. 2015;65:75-82.

[122] Menozzi P, Shi MA, Lougarre A, Tang ZH, Fournier D. Mutations of acetylcholinesterase which confer insecticide resistance in Drosophila melanogaster populations. BMC evolutionary biology. 2004;4(1):1-7. DOI:10.1186/1471-2148-4-4

[123] Butler D. Mosquitoes score in chemical war. Nature. 2011;475:19. DOI:10.1038/475019a

[124] Bradberry SM, Cage SA, Proudfoot AT, Vale JA. Poisoning due to pyrethroids. Toxicological reviews. 2005;24(2):93-106. DOI:10.2165/00139709-200524020-00003

[125] Vijverberg HP, van den Bercken J. Neurotoxicological effects and the mode of action of pyrethroid insecticides. Critical Reviews in Toxicology. 1990;21(2):105-26. DOI: 10.3109/10408449009089875.

[126] Soderlund DM, Bloomquist JR. Molecular mechanisms of insecticide resistance. In: Roush RT, Tabashnik BE editors. Pesticide resistance in arthropods. Boston, MA: Springer. 1990. pp. 58-96. DOI:10.1007/978-1-4684-6429-0_4

[127] Soderlund DM, Knipple DC. The molecular biology of knockdown resistance to pyrethroid insecticides. Insect biochemistry and molecular biology. 2003;33(6):563-577. DOI:10.1016/s0965-1748(03)00023-7

[128] Coats JR. Mechanisms of toxic action and structure-activity relationships for organochlorine and synthetic pyrethroid insecticides. Environmental Health Perspectives. 1990;87:255-262. DOI:10.1289/ehp.9087255

[129] Catterall WA. From ionic currents to molecular mechanisms: the structure and function of voltage-gated sodium channels. Neuron. 2000;26(1):13-25. Available from: https://www.bio.fsu.edu/~dfadool/Zhenbo1.pdf

[130] Sato C. Ueno Y, Asal K, Takahashi K, Sato M, Engel A, and Fujiyoshi Y. The voltage-sensitive sodium channel is a bell-shaped molecule with several cavities. Nature. 2001;409:1047-1051. DOI:10.1038/35059098

[131] Goldin AL. Evolution of voltage-gated Na+ channels. Journal of Experimental Biology. 2002;205(5):575-584. Available from: https://pubmed.ncbi.nlm.nih.gov/11907047/

[132] Dong K. Progress in insect sodium channel research. In: Gilbert LI, Gill SS, editors. Insect pharmacology: Channels, receptors, toxins and enzymes. Academic Press. 2010; pp. 25-27.

[133] Loughney K, Kreber R, Ganetzky B. Molecular analysis of the para locus, a sodium channel gene in Drosophila. Cell. 1989;58(6):1143-1154. DOI:10.1016/0092-8674(89)90512-6

[134] Olson RO, Liu Z, Nomura Y, Song W, Dong K. Molecular and functional characterization of voltage-gated sodium channel variants from Drosophila melanogaster. Insect biochemistry and molecular biology.

2008;38(5):604-10. DOI:10.1016/j.ibmb.2008.01.003

[135] Williamson MS, Denholm I, Bell CA, Devonshire AL. Knockdown resistance (kdr) to DDT and pyrethroid insecticides maps to a sodium channel gene locus in the housefly (Musca domestica). Molecular and General Genetics MGG. 1993;240(1):17-22. DOI:10.1007/bf00276878

[136] Williamson MS, Martinez-Torres D, Hick CA, Devonshire AL. Identification of mutations in the housefly para-type sodium channel gene associated with knockdown resistance (kdr) to pyrethroid insecticides. Molecular and General Genetics MGG. 1996 Aug;252(1):51-60. DOI:10.1007/bf02173204

[137] Tan J, Liu Z, Tsai TD, Valles SM, Goldin AL, Dong K. Novel sodium channel gene mutations in Blattella germanica reduce the sensitivity of expressed channels to deltamethrin. Insect biochemistry and molecular biology. 2002;32(4):445-454. DOI:10.1016/s0965-1748(01)00122-9

[138] Miyazaki M, Matsumura F, Beeman RW. DNA sequence and site of mutation of the GABA receptor of cyclodiene-resistant red flour beetle, Tribolium castaneum. Comparative Biochemistry and Physiology Part B: Biochemistry and Molecular Biology. 1995;111(3):399-406. DOI:10.1016/0305-0491(95)00007-u

[139] Bass C, Schroeder I, Turberg A, Field LM, Williamson MS. Identification of mutations associated with pyrethroid resistance in the para-type sodium channel of the cat flea, Ctenocephalides felis. Insect biochemistry and molecular biology. 2004;34(12):1305-13. DOI:10.1016/j.ibmb.2004.09.002

[140] Vais H, Williamson MS, Goodson SJ, Devonshire AL, Warmke JW, Usherwood PN, Cohen CJ. Activation of Drosophila sodium channels promotes modification by deltamethrin: reductions in affinity caused by knock-down resistance mutations. The Journal of general physiology. 2000;115(3):305-18. DOI:10.1085/jgp.115.3.305

[141] Morin S, Williamson MS, Goodson SJ, Brown JK, Tabashnik BE, Dennehy TJ. Mutations in the Bemisia tabaci para sodium channel gene associated with resistance to a pyrethroid plus organophosphate mixture. Insect biochemistry and molecular biology. 2002;32(12):1781-1791. DOI:10.1016/s0965-1748(02)00137-6

[142] Guerrero FD, Jamroz RC, Kammlah D, Kunz SE. Toxicological and molecular characterization of pyrethroid-resistant horn flies, Haematobia irritans: identification of kdr and super-kdr point mutations. Insect biochemistry and molecular biology. 1997;27(8-9):745-755. DOI:10.1016/s0965-1748(97)00057-x

[143] Schuler TH, Martinez-Torres D, Thompson AJ, Denholm I, Devonshire AL, Duce IR, Williamson MS. Toxicological, electrophysiological, and molecular characterisation of knockdown resistance to pyrethroid insecticides in the diamondback moth, Plutella xylostella (L.). Pesticide Biochemistry and Physiology. 1998;59(3):169-82. DOI:10.1006/pest.1998.2320

[144] Zhu F, Parthasarathy R, Bai H, Woithe K, Kaussmann M, Nauen R, Harrison DA, Palli SR. A brain-specific cytochrome P450 responsible for the majority of deltamethrin resistance in the QTC279 strain of Tribolium castaneum. Proceedings of the National Academy of Sciences. 2010;107(19):8557-62. DOI:10.1073/pnas.1000059107

[145] Wafford KA, Lummis SC, Sattelle DB. Block of an insect central

nervous system GABA receptor by cyclodiene and cyclohexane insecticides. Proceedings of the Royal Society of London. B. Biological Sciences. 1989;237(1286):53-61. DOI:10.1098/rspb.1989.0036

[146] Ffrench-Constant RH, Roush RT, Mortlock D, Dively GP. Isolation of dieldrin resistance from field populations of Drosophila melanogaster (Diptera: Drosophilidae). Journal of economic entomology. 1990;83(5):1733-1737. DOI:10.1093/jee/83.5.1733

[147] Rocheleau TA, Steichen JC, Chalmers AE. A point mutation in a Drosophila GABA receptor confers insecticide resistance. Nature. 1993;363(6428):449-451. DOI:10.1038/363449a0

[148] Henderson JE, Knipple DC, Soderlund DM. PCR-based homology probing reveals a family of GABA receptor-like genes in Drosophila melanogaster. Insect biochemistry and molecular biology. 1994;24(4):363-371. DOI:10.1016/0965-1748(94)90029-9

[149] Steichen JC, Rocheleau TA, Aronstein K, Roush RT. A single-amino acid substitution in a gamma-aminobutyric acid subtype A receptor locus is associated with cyclodiene insecticide resistance in Drosophila populations. Proceedings of the National Academy of Sciences. 1993;90(5):1957-61. DOI:10.1073/pnas.90.5.1957

[150] Thompson M, Steichen JC, Ffrench-Constant RH. Conservation of cyclodiene insecticide resistance-associated mutations in insects. Insect molecular biology. 1993;2(3):149-154. DOI:10.1111/j.1365-2583.1993.tb00134.x

[151] Andreev D, Kreitman M, Phillips TW, Beeman RW. Multiple origins of cyclodiene insecticide resistance in Tribolium castaneum (Coleoptera: Tenebrionidae). Journal of Molecular Evolution. 1999;48(5):615-624. DOI:10.1007/pl00006504

[152] Jones AK, Sattelle DB. The cys-loop ligand-gated ion channel gene superfamily of the red flour beetle, Tribolium castaneum. BMC genomics. 2007;8(1):1-6. DOI:10.1186/1471-2164-8-327

[153] Yu LL, Cui YJ, Lang GJ, Zhang MY, Zhang CX. The ionotropic γ-aminobutyric acid receptor gene family of the silkworm, Bombyx mori. Genome. 2010;53(9):688-697. DOI:10.1139/G10-056

[154] Yuan G, Gao W, Yang Y, Wu Y. Molecular cloning, genomic structure, and genetic mapping of two Rdl-orthologous genes of GABA receptors in the diamondback moth, Plutella xylostella. Archives of Insect Biochemistry and Physiology: Published in Collaboration with the Entomological Society of America. 2010;74(2):81-90. DOI:10.1002/arch.20361

[155] Dale RP, Jones AK, Tamborindeguy C, Davies TG, Amey JS, Williamson S, Wolstenholme A, Field LM, Williamson MS, Walsh TK, Sattelle DB. Identification of ion channel genes in the Acyrthosiphon pisum genome. Insect molecular biology. 2010;19:141-53. DOI:10.1111/j.1365-2583.2009.00975.x

[156] Le Goff G, Hamon A, Bergé JB, Amichot M. Resistance to fipronil in Drosophila simulans: influence of two point mutations in the RDL GABA receptor subunit. Journal of neurochemistry. 2005;92(6):1295-305. DOI:10.1111/j.1471-4159.2004.02922.x

[157] Ang LH, Nazni WA, Kuah MK, Shu-Chien AC, Lee CY. Detection of the A302S Rdl mutation in fipronil bait-selected strains of the German cockroach (Dictyoptera: Blattellidae). Journal of economic entomology. 2013;106(5):2167-2176. DOI:10.1603/ec13119

[158] Du W, Awolola TS, Howell P, Koekemoer LL, Brooke BD, Benedict MQ, Coetzee M, Zheng L. Independent mutations in the Rdl locus confer dieldrin resistance to Anopheles gambiae and An. arabiensis. Insect molecular biology. 2005;14(2):179-83. DOI:10.1111/j.1365-2583.2005.00544.x

[159] Garrood WT, Zimmer CT, Gutbrod O, Lüke B, Williamson MS, Bass C, Nauen R, Davies TE. Influence of the RDL A301S mutation in the brown planthopper Nilaparvata lugens on the activity of phenylpyrazole insecticides. Pesticide biochemistry and physiology. 2017;142:1-8. DOI:10.1016/j.pestbp.2017.01.007

[160] Nakao T, Kawase A, Kinoshita A, Abe R, Hama M, Kawahara N, Hirase K. The A2′ N mutation of the RDL γ-aminobutyric acid receptor conferring fipronil resistance in Laodelphax striatellus (Hemiptera: Delphacidae). Journal of economic entomology. 2011;104(2):646-652. DOI:10.1603/ec10391

[161] Wei Q, Wu SF, Gao CF. Molecular characterization and expression pattern of three GABA receptor-like subunits in the small brown planthopper Laodelphax striatellus (Hemiptera: Delphacidae). Pesticide biochemistry and physiology. 2017;136:34-40. DOI:10.1016/j.pestbp.2016.08.007

[162] Wondji CS, Dabire RK, Tukur Z, Irving H, Djouaka R, Morgan JC. Identification and distribution of a GABA receptor mutation conferring dieldrin resistance in the malaria vector Anopheles funestus in Africa. Insect biochemistry and molecular biology. 2011;41(7):484-491. DOI:10.1016/j.ibmb.2011.03.012

[163] Liang X, Xiao D, He Y, Yao J, Zhu G, Zhu KY. CYP-mediated up-regulation of cytochrome P450 genes in the red flour beetle (Tribolium castaneum). International journal of molecular sciences. 2015;16(1):2078-2098. DOI:10.3390/ijms16012078

[164] Huang Y, Li F, Liu M, Wang Y, Shen F, Tang P. Susceptibility of Tribolium castaneum to phosphine in China and functions of cytochrome P450s in phosphine resistance. Journal of Pest Science. 2019;92(3):1239-1248. DOI:10.1007/s10340-019-01088-7

[165] Wei L, Gao S, Xiong W, Liu J, Mao J, Lu Y, Song X, Li B. Latrophilin mediates insecticides susceptibility and fecundity through two carboxylesterases, esterase4 and esterase6, in Tribolium castaneum. Bulletin of entomological research. 2019;109(4):534-543. DOI:10.1017/s0007485318000895

[166] Gilbert LI, Iatrou K, Gill SS. Comprehensive molecular insect science. Oxford: Elsevier; 2005;1-77. Available from: https://agris.fao.org/agris-search/search.do?recordID=US201300103376

[167] Feyereisen R. Evolution of insect P450. Biochemical Society Transactions. 2006;34(Pt 6):1252-1255. DOI: 10.1042/BST0341252.

[168] Mouches C, Pasteur N, Berge JB, Hyrien O, Raymond M, de Saint Vincent BR, De Silvestri M, Georghiou GP. Amplification of an esterase gene is responsible for insecticide resistance in a California Culex mosquito. Science. 1986;233(4765):778-780. DOI:10.1126/science.3755546

[169] Field LM, Devonshire AL, Forde BG. Molecular evidence that insecticide resistance in peach-potato aphids (Myzus persicae Sulz.) results from amplification of an esterase gene. Biochemical journal. 1988;251(1):309-312. DOI: 10.1042/bj2510309

[170] Sanil D, Shetty V, Shetty NJ. Differential expression of glutathione

s-transferase enzyme in different life stages of various insecticide-resistant strains of Anopheles stephensi: A malaria vector. Journal of vector borne diseases. 2014;51(2):97. http://www.mrcindia.org/journal/issues/512097.pdf

[171] Hemingway J, Hawkes NJ, McCarroll L, Ranson H. The molecular basis of insecticide resistance in mosquitoes. Insect biochemistry and molecular biology. 2004;34(7):653-665. DOI:10.1016/j.ibmb.2004.03.018

[172] Liu N, Zhu F, Xu Q, Pridgeon JW, Gao X. Behavioral change, physiological modification, and metabolic detoxification: mechanisms of insecticide resistance. Acta Entomologica Sinica. 2006;49:671-679. Available from: https://www.airitilibrary.com/Publication/alDetailedMesh?docid=04546296-200608-49-4-671-679-a

[173] Li X, Schuler MA, Berenbaum MR. Molecular mechanisms of metabolic resistance to synthetic and natural xenobiotics. Acta Entomologica Sinica. 2007;52:231-253. DOI:10.1146/annurev.ento.51.110104.151104

[174] Lilly DG, Dang K, Webb CE, Doggett SL. Evidence for metabolic pyrethroid resistance in the common bed bug (Hemiptera: Cimicidae). Journal of economic entomology. 2016;109(3):1364-1368. DOI:10.1093/jee/tow041

[175] Bass C, Puinean AM, Zimmer CT, Denholm I, Field LM, Foster SP, Gutbrod O, Nauen R, Slater R, Williamson MS. The evolution of insecticide resistance in the peach potato aphid, Myzus persicae. Insect biochemistry and molecular biology. 2014;51:41-51. DOI:10.1016/j.ibmb.2014.05.003

[176] Punta M, Coggill PC, Eberhardt RY, Mistry J, Tate J, Boursnell C, Pang N, Forslund K, Ceric G, Clements J, Heger A. The Pfam protein families database. Nucleic acids research. 2012;40(D1):D290-301. DOI:10.1093/nar/gkr1065

[177] Scott JG. Cytochromes P450 and insecticide resistance. Insect biochemistry and molecular biology. 1999;29(9):757-77. DOI:10.1016/s0965-1748(99)00038-7

[178] Zhu F, Moural TW, Shah K, Palli SR. Integrated analysis of cytochrome P450 gene superfamily in the red flour beetle, Tribolium castaneum. BMC genomics. 2013;14(1):1-2. DOI:10.1186/1471-2164-14-174

[179] Richards S, Gibbs RA, Weinstock GM, Brown SJ, Denell R, Beeman RW, Gibbs R, Bucher G, Friedrich M, Grimmelikhuijzen C, Klingler M. The genome of the model beetle and pest Tribolium castaneum. Nature. 2008;452(7190):949-955. DOI:10.1038/nature06784

[180] Feyereisen R. Insect CYP genes and P450 enzymes. In: Gilbert LI. editor. Insect molecular biology and biochemistry. Oxford: Elsevier, Academic Press. 2012: pp 236-316. DOI:10.1016/b978-0-12-384747-8.10008-x

[181] Nebert DW, Nelson DR, Feyereisen R. Evolution of the cytochrome P450 genes. Xenobiotica. 1989;19(10):1149-1160. DOI:10.3109/00498258909043167

[182] Kalsi M, Palli SR. Transcription factors, CncC and Maf, regulate expression of CYP6BQ genes responsible for deltamethrin resistance in Tribolium castaneum. Insect Biochemistry and Molecular Biology. 2015;65:47-56. DOI:10.1016/j.ibmb.2015.08.002

[183] Kalsi M, Palli SR. Transcription factor cap n collar C regulates multiple

cytochrome P450 genes conferring adaptation to potato plant allelochemicals and resistance to imidacloprid in Leptinotarsa decemlineata (Say). Insect biochemistry and molecular biology. 2017;83:1-2. DOI:10.1016/j.ibmb.2017.02.002

[184] Sykiotis GP, Bohmann D. Stress-activated cap'n'collar transcription factors in aging and human disease. Science signaling. 2010;3(112):re3-re3. DOI:10.1126/scisignal.3112re3

[185] Misra JR, Horner MA, Lam G, Thummel CS. Transcriptional regulation of xenobiotic detoxification in Drosophila. Genes & development. 2011;25(17):1796-1806. DOI:10.1101/gad.17280911

[186] Xiong W, Gao S, Mao J, Wei L, Xie J, Liu J, Bi J, Song X, Li B. CYP4BN6 and CYP6BQ11 mediate insecticide susceptibility and their expression is regulated by Latrophilin in Tribolium castaneum. Pest management science. 2019;75(10):2744-2755. DOI:10.1002/ps.5384

[187] Lu Y, Liu F, Wang X, Wang Z. Cytochrome P450-mediated detoxification involves in phosphine resistance mechanism in the red flour beetle of Tribolium castaneum. Archives of Applied Science Research. 2015;7(9):49-58. DOI: 10.1080/0972060X.2016.1215263

[188] Assie LK, Francis F, Gengler N, Haubruge E. Response and genetic analysis of malathion-specific resistant Tribolium castaneum (Herbst) in relation to population density. Journal of stored products research. 2007;43(1):33-44. DOI:10.1016/j.jspr.2004.12.001

[189] Whalon ME, Mota-Sanchez D, Hollingworth RM. Analysis of Global pesticide resistance in arthropods. In: Whalon ME, Mota-Sanchez D, Hollingworth RM. editors. Global pesticide resistance in arthropods. Cambridge, MA: CABI. 2008; pp 5-31. http://sherekashmir.informaticspublishing.com/722/1/9781845933531.pdf

[190] Hayes JD, Wolf CR. Role of glutathione transferase in drug resistance. In: Sies H, Ketterer B. editors. Glutathione Conjugation: Mechanisms and Biological Significance. London: Academic Press.1988; pp. 315-355.

[191] Ladner JE, Parsons JF, Rife CL, Gilliland GL, Armstrong RN. Parallel evolutionary pathways for glutathione transferases: structure and mechanism of the mitochondrial class kappa enzyme rGSTK1-1. Biochemistry. 2004;43(2):352-61. DOI:10.1021/bi035832z

[192] Shi H, Pei L, Gu S, Zhu S, Wang Y, Zhang Y, Li B. Glutathione S-transferase (GST) genes in the red flour beetle, Tribolium castaneum, and comparative analysis with five additional insects. Genomics. 2012;100(5):327-335. DOI:10.1016/j.ygeno.2012.07.010

[193] Friedman R. Genomic organization of the glutathione S-transferase family in insects. Molecular phylogenetics and evolution. 2011;61(3):924-932. DOI:10.1016/j.ympev.2011.08.027

[194] Board PG, Menon D. Glutathione transferases, regulators of cellular metabolism and physiology. Biochimica et biophysica acta (bba)-general subjects. 2013;1830(5):3267-3288. DOI:10.1016/j.bbagen.2012.11.019

[195] Zhou S, Lien YC, Shuvaeva T, DeBolt K, Feinstein SI, Fisher AB. Functional interaction of glutathione S-transferase pi and peroxiredoxin 6 in intact cells. The international journal of biochemistry & cell biology. 2013;45(2):401-7. DOI:10.1016/j.biocel.2012.11.005

[196] Han JB, Li GQ, Wan PJ, Zhu TT, Meng QW. Identification of glutathione

S-transferase genes in Leptinotarsa decemlineata and their expression patterns under stress of three insecticides. Pesticide Biochemistry and Physiology. 2016;133:26-34. DOI:10.1016/j.pestbp.2016.03.008

[197] Schama R, Pedrini N, Juárez MP, Nelson DR, Torres AQ, Valle D, Mesquita RD. Rhodnius prolixus supergene families of enzymes potentially associated with insecticide resistance. Insect biochemistry and molecular biology. 2016;69:91-104. DOI:10.1016/j.ibmb.2015.06.005

[198] Lumjuan N, Rajatileka S, Changsom D, Wicheer J, Leelapat P, Prapanthadara LA, Somboon P, Lycett G, Ranson H. The role of the Aedes aegypti Epsilon glutathione transferases in conferring resistance to DDT and pyrethroid insecticides. Insect biochemistry and molecular biology. 2011;41(3):203-209. DOI:10.1016/j.ibmb.2010.12.005

[199] Ortelli F, Rossiter LC, Vontas J, Ranson H, Hemingway J. Heterologous expression of four glutathione transferase genes genetically linked to a major insecticide-resistance locus from the malaria vector Anopheles gambiae. Biochemical Journal. 2003;373(3):957-963. DOI:10.1042/BJ20030169

[200] Chen X, Xiong W, Li C, Gao S, Song X, Wu W, Li B. Comparative RNA-sequencing profiling reveals novel Delta-class glutathione S-transferases relative genes expression patterns in Tribolium castaneum. Gene. 2016;593(1):13-20. DOI:10.1016/j.gene.2016.08.013

[201] Song X, Pei L, Zhang Y, Chen X, Zhong Q, Ji Y, Tang J, Feng F, Li B. Functional diversification of three delta-class glutathione S-transferases involved in development and detoxification in Tribolium castaneum. Insect molecular biology. 2020;29(3):320-36. DOI:10.1111/imb.12637

[202] Ranson H, Rossiter L, Ortelli F, Jensen B, Wang X, Roth CW, Collins FH, Hemingway J. Identification of a novel class of insect glutathione S-transferases involved in resistance to DDT in the malaria vector Anopheles gambiae. Biochemical Journal. 2001;359(2):295-304. DOI:10.1042/0264-6021:3590295

[203] Enayati AA, Ranson H, Hemingway J. Insect glutathione transferases and insecticide resistance. Insect molecular biology. 2005;14(1):3-8. DOI:10.1111/j.1365-2583.2004.00529.x

[204] Zhou ZH, Syvanen M. A complex glutathione transferase gene family in the housefly Musca domestica. Molecular and General Genetics MGG. 1997;256(2):187-194. DOI:10.1007/s004380050560

[205] Cao CW, Zhang J, Gao XW, Liang P, Guo HL. Overexpression of carboxylesterase gene associated with organophosphorous insecticide resistance in cotton aphids, Aphis gossypii (Glover). Pesticide biochemistry and physiology. 2008;90(3):175-180. DOI:10.1016/j.pestbp.2007.11.004

[206] Zhang Z, Xie Y, Wang Y, Lin Z, Wang L, Li G. Toxicities of monoterpenes against housefly, Musca domestica L.(Diptera: Muscidae). Environmental Science and Pollution Research. 2017;24(31):24708-13. DOI:10.1007/s11356-017-0219-4

[207] Small GJ, Hemingway J. Molecular characterization of the amplified carboxylesterase gene associated with organophosphorus insecticide resistance in the brown planthopper, Nilaparvata lugens. Insect molecular biology. 2000;9(6):647-653. DOI:10.1046/j.1365-2583.2000.00229.x

[208] Julio AH, Gigliolli AA, Cardoso KA, Drosdoski SD, Kulza RA, Seixas FA, Ruvolo-Takasusuki MC, de Souza CG, Lapenta AS. Multiple resistance to pirimiphos-methyl and bifenthrin in Tribolium castaneum involves the activity of lipases, esterases, and laccase2. Comparative Biochemistry and Physiology Part C: Toxicology & Pharmacology. 2017;195:27-43. DOI:10.1016/j.cbpc.2017.01.011

[209] Boucard AA, Maxeiner S, Südhof TC. Latrophilins function as heterophilic cell-adhesion molecules by binding to teneurins: regulation by alternative splicing. Journal of Biological Chemistry. 2014;289(1):387-402. DOI:10.1074/jbc.M113.504779

[210] Gao S, Xiong W, Wei L, Liu J, Liu X, Xie J, Song X, Bi J, Li B. Transcriptome profiling analysis reveals the role of latrophilin in controlling development, reproduction and insecticide susceptibility in Tribolium castaneum. Genetica. 2018;146(3):287-302. DOI:10.1007/s10709-018-0020-4

[211] Locke M. The Wigglesworth Lecture: Insects for studying fundamental problems in biology. Journal of Insect Physiology. 2001;47 (4-5):495-507. DOI: 10.1016/s0022-1910(00)00123-2

[212] Moussian B. Recent advances in understanding mechanisms of insect cuticle differentiation. Insect biochemistry and molecular biology. 2010;40(5):363-375. DOI: 10.1016/j.ibmb.2010.03.003

[213] Koganemaru R, Miller DM, Adelman ZN. Robust cuticular penetration resistance in the common bed bug (Cimex lectularius L.) correlates with increased steady-state transcript levels of CPR-type cuticle protein genes. Pesticide biochemistry and physiology. 2013;106(3):190-197. DOI:10.1016/j.pestbp.2013.01.001

[214] Lilly DG, Latham SL, Webb CE, Doggett SL. Cuticle thickening in a pyrethroid-resistant strain of the common bed bug, Cimex lectularius L.(Hemiptera: Cimicidae). Plos one. 2016;11(4):e0153302. DOI:10.1371/journal.pone.0153302

[215] Arakane Y, Muthukrishnan S, Beeman RW, Kanost MR, Kramer KJ. Laccase 2 is the phenoloxidase gene required for beetle cuticle tanning. Proceedings of the National Academy of Sciences. 2005;102(32):11337-11342. DOI:10.1073/pnas.0504982102

[216] Noh MY, Muthukrishnan S, Kramer KJ, Arakane Y. Tribolium castaneum RR-1 cuticular protein TcCPR4 is required for formation of pore canals in rigid cuticle. PLoS genetics. 2015;11(2):e1004963. DOI:10.1371/journal.pgen.1004963

[217] Haubruge E, Amichot M, Cuany A, Berge JB, Arnaud L. Purification and characterization of a carboxylesterase involved in malathion-specific resistance from Tribolium castaneum (Coleoptera: Tenebrionidae). Insect biochemistry and molecular biology. 2002;32(9):1181-1190. DOI:10.1016/s0965-1748(02)00054-1

[218] Darvishzadeh A, Sharifian I. Effect of spinosad and malathion on esterase enzyme activities of Tribolium castaneum (Coleoptera: Tenebrionidae). Journal of Entomology and Zoology Studies. 2015;3(2):351-354. DOI: 10.13140/RG.2.1.1025.6481

[219] Bughio FM, Wilkins RM. Influence of malathion resistance status on survival and growth of Tribolium castaneum (Coleoptera: Tenebrionidae), when fed on flour from insect-resistant and susceptible grain rice cultivars. Journal of Stored Products Research. 2004;40(1):65-75. DOI:10.1016/s0022-474x(02)00077-2

[220] Wool D, Front L. Esterase variation in Tribolium confusum (Coleoptera: Tenebrionidae): genetic analysis of interstrain crosses in relation to malathion resistance. Journal of Stored Products Research. 2003;39(2):237-249. DOI:10.1016/s0022-474x(01)00057-1

[221] Wool D, Noiman S, Manheim O, Cohen E. Malathion resistance in Tribolium strains and their hybrids: inheritance patterns and possible enzymatic mechanisms (Coleoptera, Tenebrionidae). Biochemical genetics. 1982;20(7-8):621-636. DOI:10.1007/bf00483961

[222] Oppert B, Guedes RN, Aikins MJ, Perkin L, Chen Z, Phillips TW, Zhu KY, Opit GP, Hoon K, Sun Y, Meredith G. Genes related to mitochondrial functions are differentially expressed in phosphine-resistant and-susceptible Tribolium castaneum. BMC genomics. 2015;16(1):1-0. DOI:10.1186/s12864-015-2121-0

[223] Wang K, Liu M, Wang Y, Song W, Tang P. Identification and functional analysis of cytochrome P450 CYP346 family genes associated with phosphine resistance in Tribolium castaneum. Pesticide Biochemistry and Physiology. 2020;168:104622. DOI:10.1016/j.pestbp.2020.104622

[224] Walter CM, Price NR. The uptake and penetration of pirimiphos-methyl into susceptible and resistant strains of the rust red flour beetle—Tribolium castaneum, herbst (coleoptera: tenebrionidae). Comparative Biochemistry and Physiology Part C: Comparative Pharmacology. 1989;94(2):419-423. DOI:10.1016/0742-8413(89)90091-1

[225] Pietri JE, Liang D. The links between insect symbionts and insecticide resistance: causal relationships and physiological tradeoffs. Annals of the Entomological Society of America. 2018;111(3):92-97. DOI:10.1093/aesa/say009

[226] Kikuchi Y, Hayatsu M, Hosokawa T, Nagayama A, Tago K, Fukatsu T. Symbiont-mediated insecticide resistance. Proceedings of the National Academy of Sciences. 2012;109(22):8618-8622. DOI:10.1073/pnas.1200231109

[227] Patil CD, Borase HP, Salunke BK, Patil SV. Alteration in Bacillus thuringiensis toxicity by curing gut flora: novel approach for mosquito resistance management. Parasitology research. 2013;112(9):3283-3288. DOI:10.1007/s00436-013-3507-z

[228] Shen SK, Dowd PF. Detoxification spectrum of the cigarette beetle symbiont Symbiotaphrina kochii in culture. Entomologia Experimentalis et Applicata. 1991;60(1):51-59. DOI:10.1111/j.1570-7458.1991.tb01522.x

[229] Almeida VT, Ramos VM, Saqueti MB, Gorni PH, Pacheco CA, de Leão RM. Bioactivity of ethanolic extracts of Euphorbia pulcherrima on Spodoptera frugiperda (JE Smith) (Lepidoptera: Noctuidae). African Journal of biotechnology. 2017;16(13):615-622. DOI:10.5897/ajb2017.15972

[230] Li W, Jin D, Shi C, Li F. Midgut bacteria in deltamethrin-resistant, deltamethrin-susceptible, and field-caught populations of Plutella xylostella, and phenomics of the predominant midgut bacterium Enterococcus mundtii. Scientific reports. 2017;7(1):1-3. DOI:10.1038/s41598-017-02138-9

[231] Futo M, Armitage SA, Kurtz J. Microbiota plays a role in oral immune priming in Tribolium castaneum. Frontiers in microbiology. 2016;6:1383. DOI:10.3389/fmicb.2015.01383

[232] Agarwal A, Agashe D. The red flour beetle Tribolium castaneum: A

model for host-microbiome interactions. PloS one. 2020;15(10):e0239051. DOI:10.1371/journal.pone.0239051

[233] Arthur FH, Zettler LJ. Malathion resistance in *Tribolium castaneum* (Coleoptera: Tenebrionidae): Differences between discriminating concentrations by topical application and residual mortality on treated surfaces. Journal of economic entomology. 1991;84(3):721-6. DOI:10.1093/jee/84.3.721

[234] AS AA, Thangapandiyan S. Comparative bioassay of silver nanoparticles and malathion on infestation of red flour beetle, *Tribolium castaneum*. The Journal of Basic and Applied Zoology. 2019;80(1):1-0. DOI:10.1186/s41936-019-0124-0

[235] Perkin LC, Oppert B. Gene expression in *Tribolium castaneum* life stages: Identifying a species-specific target for pest control applications. PeerJ. 2019;7:e6946. DOI:10.7717/peerj.6946

[236] Miller SC, Miyata K, Brown SJ, Tomoyasu Y. Dissecting systemic RNA interference in the red flour beetle *Tribolium castaneum*: parameters affecting the efficiency of RNAi. PloS one. 2012;7(10):e47431. DOI:10.1371/journal.pone.0047431

[237] Homem RA, Davies TG. An overview of functional genomic tools in deciphering insecticide resistance. Current opinion in insect science. 2018;27:103-10. DOI:10.1016/j.cois.2018.04.004

[238] Gilles AF, Schinko JB, Averof M. Efficient CRISPR-mediated gene targeting and transgene replacement in the beetle *Tribolium castaneum*. Development. 2015;142(16):2832-2839. DOI:10.1242/dev.125054

Chapter 19

Insecticide Resistance in Whiteflies *Bemisia tabaci* (Gennadius): Current Global Status

Biswajit Patra and Tapan Kumar Hath

Abstract

The whitefly, *Bemisia tabaci* (Gennadius) is a polyphagous pest causing considerable yield loss to many crops around the globe. It is a phloem feeder and transmits several viral diseases as well. It has great genetic diversity and is considered a complex of biotypes. Despite the adoption of several available control strategies, management by chemical pesticides has still been the first choice for the farmers to protect their crops. However, prolonged use of chemical pesticides has ultimately accelerated the development of multifold resistance against various groups of insecticides in different parts of the world. The status of development of insecticide resistance against different groups of insecticides by this pest, mechanisms of resistance, cross-resistance, role of detoxifying enzymes, and management issues have been discussed in this chapter.

Keywords: *Bemisia tabaci*, biotype, insecticide resistance, cross-resistance, IRM

1. Introduction

Whitefly, *B. tabaci* (Gennadius, 1889) (Hemiptera: Aleyrodidae) is a pest of global significance. It is a serious pest of vegetable, field, and ornamental crops [1–3]. This notorious pest shares global distribution as an important pest in field as well as in greenhouse production systems [2, 4]. Both the nymph and adult cause severe economic loss to the growers by direct sucking sap from the phloem and thereby reducing the yield. They also cause indirect damage by transmitting the virus [5] and excreting honeydew on leaves. As a result of honeydew secretion, black sooty mold develops that impairs photosynthesis ability of the infested plants. *B. tabaci* is considered as complex of biotypes [4, 6, 7] and composed of at least 40 morphologically indistinguishable species [8–11]. These biotypes/species are mainly differentiated based on biochemical or molecular polymorphism markers. There are mainly two types of biotypes *viz.,* biotype B and biotype Q. Biotype B is considered to be originated from the Middle East–Asia Minor region (Middle East Asia Minor 1—MEAM1 group) [9] whereas biotype Q possibly originated in the Iberian Peninsula Mediterranean—MED group) [12, 13].

The various biotype of this tiny fly causes significant economic loss. Henneberry and Faust [14] reported approximately 10 billion US dollars (USD) economic loss

IntechOpen

during the years 1980 to 2000 due to whitefly infestation. They also revealed that there were about 300 USD economic losses due to the infestation of whitefly in different bean crops during 1991. Cotton growers in Arizona, California, and Texas spent 154 million USD during 1994–1998 to control the whitefly [15]. This pest is listed as one of the top 100 invasive species of the world by IUCN. Due to the severity of infestation and polyphagous in nature, farmers largely depend on the chemical management of this pest. As a result of the extensive application of synthetic insecticides, *B. tabaci* has developed multifold resistance to a wide range of insecticides.

2. Insecticide resistance in *B. tabaci*

Insecticide resistance is one of the important threats in the changing agricultural scenario. It has been increasing at an alarming rate since the introduction of synthetic insecticides. The first case of insecticide resistance was documented by A.L. Melander in 1914. He reported that the San Jose scale was resistant to lime sulfur. Since then many pests have developed various degrees of resistance against various insecticides. The list is ever-growing.

In Turkey, the whitefly population (Biotype B) showed 20–310-fold resistance to OPs [16]. In India, Asia 1 whiteflies showed high resistance to OPs such as acephate and triazophos [17]. The population from China showed a low level of resistance to chlorpyriphos, dichlorvos, and carbosulfan (carbamate) [18, 19].

This pest has also developed various degrees of resistance against synthetic pyrethroids and neonicotinoids. The magnitude of resistance varies from region to region, and it mainly depends on the frequency of insecticide use. B biotype population from northwestern China [20] and Cyprus [21] showed very high resistance against cypermethrin and bifenthrin. Neonicotinoids were introduced as one of the most important chemicals against whitefly and they also performed well due to their systemic and translaminar properties and high residual activity [22–24]. However, due to their frequent and extensive use, resistance against these chemicals has been reported from different corners of the world. The first report of neonicotinoid resistance was published in 1996, describing the low efficacy of imidacloprid against *B. tabaci* [25]. Low-to-moderate levels of resistance to imidacloprid and thiamethoxam were reported from Brazil [26], whereas high levels of resistance were detected from Florida [27]. The same biotype of the pest showed different degrees of resistance to the same class of insecticide. Biotype Q in Israel showed a high level of resistance to thiamethoxam but a moderate level of resistance against imidacloprid and acetamiprid [28]. Control failure of whitefly with neonicotinoid has been reported from Pakistan also. It is due to neonicotinoid resistance in *B. tabaci* [29]. Naveen et al. [30] reported a high degree of resistance against neonicotinoids from India (Asia I and Asia II-1). Neonicotinoid resistance has also been reported from different parts of China both in biotype B and Q [18, 19, 31]. Biotype Q population of southeastern Spain showed 1–7-fold resistance (low level) against spiromesifen [32] whereas 8–32-fold resistance has been reported from India [17]. Astonishingly, several field populations from Spain showed more than 10,000-fold resistance against spiromesifen [33]. Insect growth regulators (pyriproxyfen and buprofezin) are also proving vulnerable to resistance by *B. tabaci* [17, 34–36]. These two chemicals, that is, pyriproxyfen and buprofezin act primarily against immature stages of whiteflies. They have distinct modes of action. Therefore, the chance of cross-resistance is very low. Buprofezin is a chitin synthesis inhibitor and results in nymphal death during ecdysis [37], whereas pyriproxyfen is a juvenile hormone mimic inter-rupting nymphal and pupal

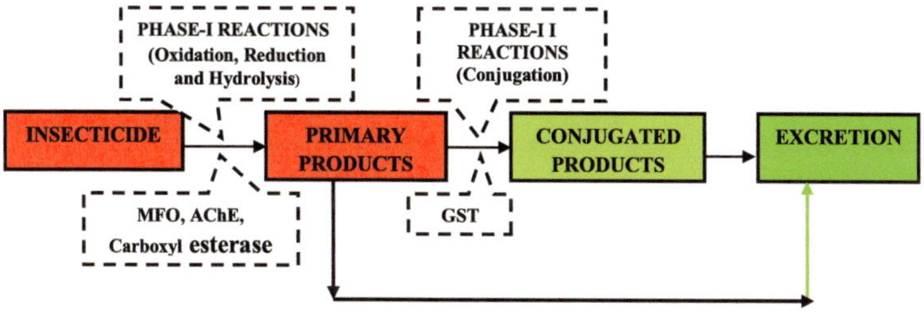

Figure 1.
A general pathway for insecticide detoxification.

development. It also suppresses egg hatching by direct exposure of eggs or transovarially *via* the treatment of adult females [38]. Resistance to buprofezin was first detected in the Netherlands and thereafter from Spain and Israel [39, 40]. Commercial introduction of pyriproxyfen for management of whitefly was done in the year 1991 in Israel. Within 1 year of its introduction, whitefly developed about 550-fold resistance at LC_{50} and it was reported from a rose greenhouse that had previously been sprayed only three times with this chemical [40, 41].

Resistance in *B. tabaci* is known to be multi-factorial based on multiple mechanisms. Enhanced detoxification and modifications to acteyl-cholinesterase (AChE), GABA-gated chloride-ion channel, and voltage-sensitive sodium channel are important mechanisms. Pesticide detoxifying enzymes play an important role in reducing the susceptibility of insecticides. Insecticides are generally hydrophobic in nature. Metabolism converts water-insoluble (apolar) or fat-soluble (lipophilic) insecticides into polar compounds or less lipophilic compounds. Conversion of apolar substances to less lipophilic or polar metabolites takes place by two reactions, that is, phase I (primary) and phase II (secondary) reactions. The phase I metabolites are sometimes polar enough to be excreted but are usually further converted to water-soluble conjugates by phase II reactions. A general insecticide detoxification pathway is as follows (**Figure 1**).

3. Resistance detection methods

Several standard methods are there to detect resistance development. At the field level, resistance development is suspected when control failure occurs. There are various reasons for the control failure of pests at the field level. Pesticide resistance is one of the important reasons for control failure. Once control failure is observed in the field, the population is collected and tested in the laboratory for confirmation. In the laboratory leaf-dip bioassays, biochemical assays, or molecular assays are conducted to know the susceptibility status of the population.

3.1 Leaf-dip bioassays

Several bioassay procedures are there to detect the resistance status but the leaf-dip bioassays are usually followed in the case of whitefly. In leaf-dip bioassay serial dilution of the tested insecticides are prepared from the commercial formulation of insecticide. Leaves are collected from the unsprayed field and washed under tap water. Then the washed leaves are air-dried under normal room temperature. Leaf discs of appropriate diameter are prepared and dipped in the insecticidal solution and agitated slightly for 5 to 10 seconds for complete wetting. Sometimes whole leaves are dipped in the insecticidal solution. Then the treated leaves or discs are dried for a few minutes until the surface liquid gets dried. In case of control, leaves or discs are dipped in distilled water. The leaves are kept on a Petri plate or in an appropriate container with a perforated lid or covered with a muslin cloth. Then 20 to 30 adult whiteflies are released in the Petri plates. The tests were carried out and maintained at 24 to 25°C temperature. Mortality is observed 48–72 hours after the treatment depending on the type of insecticides tested. Adults showing no signs of movement are considered as dead. Three replicates are usually carried out for each concentration of each insecticide and also for controls. Percentage of mortality of nymph for each concentration of test insecticide and control was calculated using the following formula:

$$\text{Percent mortality} = \frac{\text{Number of dead whiteflies}}{\text{Total number of whiteflies treated}} \times 100 \tag{1}$$

Then corrected percent mortality was calculated using Abbott's formula [42] based on the control mortality, if any. The corrected mortality data of each test insecticide of each location were subjected to probit analysis based on Finney [43]. Now, the resistance ratio (RR) or resistance factor (RF) is calculated using the following formula:

$$\text{Resistance Ratio (RR)} = \frac{\text{LC50 of a field population}}{\text{LC50 of the most susceptible population}} \tag{2}$$

3.2 Biochemical assays

Biochemical assays are frequently used to characterize resistance mechanisms but they can also be used to detect resistance status. In biochemical assays, the activity of detoxifying enzymes *viz.*, general esterases (GEs), glutathione S-transferases (GSTs), and cytochrome P-450 dependent mixed-function oxidases are estimated using spectrophotometer or microplate reader and compared in different populations. In resistant populations, the activity of pesticide-degrading enzymes is overproduced resulting in efficient degradation of the pesticides.

3.3 Molecular assays

Molecular assays detect genes or mutations involved in resistance. There are different types of nucleic acid-based assays for resistance study. Quantitative PCR, reverse transcription plus quantitative PCR, whole transcriptome sequencing, etc., are some of the important molecular assay techniques. These techniques require highly sophisticated and costly instruments and reagents.

4. Geographical distribution of insecticide resistance in whitefly

The whitefly has developed multifold resistance against various groups of insecticides. Approximately 650 cases of insecticide resistance have been reported in the genus *Bemisia*, and it has developed resistance against more than 60 active ingredients [44]. Among two biotypes, MED (biotype Q) is considered to be more resistant to insecticides than MEAM1 (biotype B) [28, 35, 45, 46]. In the case of India, the species Asia II-7 has developed more resistance to insecticides than Asia I and Asia II-1 (**Table 1**) [30].

Country	Insecticides	Level of resistance or resistance ratio (RR)	References
Brazil	Imidacloprid Acetamiprid Thiamethoxam Chlorpyrifos Endosulfan	209–2972 150–556 346–4662 97–109 88–284	[26]
	Azadirachtin, cartap, chlorantraniliprole, diafenthiuron, imidacloprid, lambda-cyhalothrin, spiromesifen	Low to very high (MEAM 1)	[47]
China (NW)	Cypermethrin and Bifenthrin	Very high	[20]
	Imidacloprid	Low to medium	
	Pyriproxyfen	Medium to high	
	Abamectin	None	
China, SE	pymetrozine(biotype B and Q) imidacloprid, thiamethoxam (B, Q), nitenpyram (B, Q)	Low	[46]
	cyantraniliprole (B, Q)	Medium to high	
	Alpha-cypermethrin (B)	Very high	[31]
	Alpha-cypermethrin (Q)	Medium to very high	
	imidacloprid, thiamethoxam (B, Q)	Medium to very high	
	Spinosad (B, Q)	Low	
	fipronil (B, Q)	L to M	
China, E	Dichlorvos (Q)	Low	[19]
	cypermethrin(Q)	Low	
	imidacloprid, nitenpyram(Q)	Low to high	
China(B biotype)	Cypermethrin	2100–6200	[18]
	Bifenthrin	1000–2200	[31]
	Imidacloprid	28–1900	
	Thiamethoxam	29–1200	
	Alpha-cypermethrin	22–610	
China (Q biotype)	Acetamiprid	20–33	
	Imidacloprid	24–1900	
	Thiamethoxam	29–1200	
	Alpha-cypermethrin	22–610	[18]
Germany	Imidacloprid Acetamiprid Thiamethoxam	>100 (RR)	[48]

Country	Insecticides	Level of resistance or resistance ratio (RR)	References
Greece	α-cypermethrin	Medium to very high	[49]
	Imidacloprid	Low to very high	
	Bifenthrin	1–23	[50]
	Alpha-cypermethrin	30–600	[49]
	Imidacloprid	38–1958	
Cyprus	imidacloprid,	High to very high	[21]
	Thiamethoxam	High to very high	
	Acetamiprid	Low	
	Bifenthrin	High to very high	
India	Acephate	High to very high	[17]
	dinotefuran	Low to high	
	Spiromesifen	Low to high	
	Pyriproxyfen	Low to medium	
	Flonicamid	Low to high	
	Cypermethrin	26–136	[30]
	Deltamethrin	12–76	[51, 52]
	Imidacloprid	3–18	
	Thiamethoxam	3–7	
Israel	Thiamethoxam (Q)	Very high	[28]
	Imidacloprid (B, Q)	Low to high	
	Acetamiprid (Q)	Low to high	
	Thiamethoxam (Q biotype)	624–2434	[53]
	Acetamiprid (Q biotype)	4–110	[28]
	Imidacloprid (Q biotype)	2–83	[20]
Italy	Imidacloprid	>100	[48]
	Acetamiprid		
	Thiamethoxam		
Netherland	Buprofezin	47	[39]
Pakistan	Neonicotinoids	Low to medium	[34]
	Buprofezin	Low to medium	
	Diafenthiuron	Low to high	[29]
	Dimethoate	2–4931	[34]
	Deltamethrin	2–1115	
Spain	Azadirachtin	Low to medium	[32]
	Buprofezin	Medium to very high	
	Imidacloprid	Low to medium	
	Spiromesifen	Low	

Country	Insecticides	Level of resistance or resistance ratio (RR)	References
	Spiromesifen	Low to very high	[33]
	Spirotetramat	Low to high	
	Buprofezin	11–59 (RR)	[54]
	Imidacloprid	>100	[48]
	Acetamiprid	>100	[55]
	Thiamethoxam	>100	
	Pyradabin	0.9–9	
	Pyriproxyfen	0.7–15	
	Spiromesifen	1–7	
	Pymetrozine	Highly resistant	
Turkey	Pyrethroids and OPs	Medium to very high	[16]
	Imidacloprid	Medium to very high	
	Thiacloprid	High to very high	
	Thiamethoxam	Low to very high	
	Thiamethoxam	Low to high	[56]
	Acetamiprid	Low to medium	
	Bifenthrin (B biotype)	190–360	[16]
	Fenpropathrin	57–290	[56]
	Acetamiprid	4–30	[56]
	Thiamethoxam (B biotype)	9–32	[57]
USA (California, Arizona)	Neonicotinoids	High to very high	[58]
USA, Florida	Neonicotinoids	Low	[59]
	Bifenthrin	Low	
	Buprofezin	Low	
USA(A)	Cypermethrin	24	[60]
	Bifenthrin	7	[61]
	Etofenprox	110	[62]
	Imidacloprid (B biotype)	24–120	[63]
	Thiamethoxam	25	[63]
	Buprofezin (Q biotype)	˃1000	[64]
Egypt	Carbosulfan	20–80	[65]
	Cypermethrin	20–80	[66]
	Lambdacyhalothrin	528	
Guatemala	Imidacloprid	58–78	[67]
Sudan	Dimethoate	>454	[68]

Table 1.
Status of insecticide resistance to different classes of insecticides in whitefly.

5. Consequence of insecticide resistance

Control failure of a pest is a very common phenomenon in the case of resistance development. As a result of the control failure, farmers usually increase the frequency of application along with the higher dose of the chemical. Sometimes farmers reduce the interval between two consecutive sprays. All these incidences ultimately worsen the resistance scenario. When the chemical is no more effective to control the pest, then the ineffective pesticide is replaced with a new one provided that a suitable one is available. Initially, the new one also gives good efficacy. However, after repeated and prolonged use, the new one also becomes ineffective. This cycle continues and the pests develop multiple resistance. This scenario has been repeated in various agroecosystems of the world and the effect is enormous. Moreover, under the present scenario, it will be increasingly difficult to design, develop, and introduce new pesticides to solve the problems. Even there is a chance of development of resistance prior to the introduction of a new chemical. Hence, every care should be taken to delay and combat the resistance issue.

The development of pesticide resistance leads to control failure of the pests and thereby increases the cost of production due to higher requirements of the chemicals and frequent application costs. Control failure resulting from the resistance development leads to more use of the chemicals that ultimately deteriorates the quality of the produces. The pesticide residues in the product may be harmful to the health of the consumers and thereby society as a whole. In addition to that pest, resurgence may occur due to disruption of the pest-defender ratio. The secondary pests may attain the status of major pest due to harmful effects on the available natural enemies. As whitefly acts as a vector of various virus diseases of plants, the viral diseases of the plants may increase that may result in huge economic loss to the farmers.

6. Integrated resistance management (IRM) strategy

The biological characteristics like migratory ability and polyphagous nature of whitefly promote resistance development. These characteristics cannot be controlled directly. Therefore, multipronged strategies need to be adopted to manage the problem. Insecticide resistance management (IRM) strategies need to be followed to delay and combat the problem. Excessive use of chemical insecticide for the management of whitefly is the main cause of insecticide resistance development. So, we need to use insecticide as last resort. The frequency of insecticide use and thereby the degree of selection pressure is the main driving force for the development of insecticide resistance. Emphasis has to be given in the rotation of insecticides having different modes of action (MOA) based on IRAC's MoA classification scheme [69]. Insecticides having similar modes of action should not be used frequently. Moreover, the information on the cross-resistance phenomenon needs to be considered. The chance of multiple resistance development is also there in case of extreme selection pressure [50]. Chemicals having different modes of action are enlisted in **Table 2**. These insecticides may be used in rotation programs for the management of whitefly. The availability and use of an insecticide in a specific crop vary in different countries and it is regulated by law. Hence, it is important to check whether an insecticide is approved for use in a particular crop or not before recommending it. The mode of action, dose, and waiting period of chemicals that are approved for management of whitefly in India are as follows [70]. The cross-resistance information needs to be taken into account before recommendation. The insecticides should be used as per the label claim and should be selected based on the local recommendation or local efficacy and selectivity.

Crop	Insecticides	Mode of action	Dose (g a.i./ha)	PHI (days)
Cotton	Acetamiprid 20.00% SP	Nicotinic acetylcholine receptor (nAChR) competitive modulator	20	15
	Afidopyropen 50 g/L DC	Chordotonal organ TRPV channel modulators	50	1
	Bifenthrin 10EC	Sodium channel modulators	80	15
	Buprofezin 25%SC	Inhibitor chitin biosynthesis, type 1	250	20
	Cholorpyriphos 20%EC	Acetylcholinesterase (AChE) inhibitor	250	—
	Clothianidin 50WDG	Nicotinic acetylcholine receptor (nAChR) competitive modulator	20–25	20
	Diafenthiuron 47.80% SC	Mitochondrial ATP synthesis inhibitotor	239	30
	Diafenthiuron 50.00% WP	Mitochondrial ATP synthesis inhibitotor	300	21
	Dinotefuran 20SG	Nicotinic acetylcholine receptor (nAChR) competitive modulator	25–30	15
	Fipronil 05.00% SC	GABA-gated chloride channel blocker	75–100	6
	Fenpropathrin 30.00% EC	Sodium channel modulator	75–100	14
	Flonicamid 50.00% WG	Chordotonal organ modulator-undefined target site.	75	25
	Imidacloprid 17.80% SL	Nicotinic acetylcholine receptor (nAChR) competitive modulator	20–25	40
	Pyriproxyfen 10.00% EC	Juvenile hormone mimic	100	31
	Pyridaben 20.00% w/w WP	Mitochondrial electron transport inhibitor (complex –I)	100	28
	Spiromesifen 22.90% SC	Inhibitor of acetyl CoA carboxylase (lipid synthesis inhibitor)	144	10
	Thiacloprid 21.70% SC	Nicotinic acetylcholine receptor (nAChR) competitive modulator	120–144	52
	Thiamethoxam 25.00% WG	Nicotinic acetylcholine receptor (nAChR) competitive modulator	50	21
	Acephate 50.00% + Imidacloprid 01.80% SP	Acetylcholinesterase (AChE) inhibitor + (nAChR) competitive modulator	518	40
	Buprofezin 15.00% + Acephate 35.00% w/w WP	Inhibitor chitin biosynthesis, type 1 + Acetylcholinesterase (AChE) inhibitor	187.5 + 437.5	—
	Chlorpyrifos 50.00% + Cypermethrin 05.00% EC	Acetylcholinesterase (AChE) inhibitor + Sodium channel modulators	500 + 50	15
	Diafenthiuron 47.00% + Bifenthrin 09.40% w/w SC	Mitochondrial ATP synthesis inhibitotor + Sodium channel modulators	293.75 + 58.7	30
	Pyriproxyfen 05.00% + Fenpropathrin 15.00% EC	Juvenile hormone mimic + Sodium channel modulator	60 + 60	19
	Pyriproxyfen 10.00% + Bifenthrin10.00% w/w EC	Juvenile hormone mimic + Sodium channel modulator	60 + 60	19

Crop	Insecticides	Mode of action	Dose (g a.i./ha)	PHI (days)
	Fipronil 04.00% + Acetamiprid 04.00% w/w SC	GABA-gated chloride channel blocker + Nicotinic acetylcholine receptor (nAChR) competitive modulator	40 + 40	30
Brinjal	Afidopyropen 50 g/L DC	Chordotonal organ TRPV channel modulators	50	25
	Diafenthiuron 50.00% WP	Mitochondrial ATP synthesis inhibitor	300	3
	Fenpropathrin 30.00% EC	Sodium channel modulator	75–100	10
	Pyriproxyfen 10.00% EC	Juvenile hormone mimic	50	7
	Thiamethoxam 25.00% WG	Nicotinic acetylcholine receptor (nAChR) competitive modulator	50	3
Tomato	Carbofuran 3%CG	Acetylcholinesterase (AChE) inhibitor	1200	—
	Cyantraniliprole 10.26% OD	Ryanodine receptor modulator	90	3
	Diafenthiuron 50.00% WP	Mitochondrial ATP synthesis inhibitor	300	5
	Imidacloprid 17.80% SL	Nicotinic acetylcholine receptor (nAChR) competitive modulator	30–35	3
	Spiromesifen 22.90% SC	Inhibitor of acetyl CoA carboxylase (lipid synthesis inhibitor)	150	3
	Thiamethoxam 25.00% WG	Nicotinic acetylcholine receptor (nAChR) competitive modulator	50	5
Okra	Diafenthiuron 50.00% WP	Mitochondrial ATP synthesis inhibitor	300	5
	Fenpropathrin 30.00% EC	Sodium channel modulator	75–100	7
	Flupyradifurone 17.09% w/w SL	Nicotinic acetylcholine receptor (nAChR) competitive modulator	250	3
	Pyriproxyfen 10.00% EC	Juvenile hormone mimic	50	7
	Thiamethoxam 25.00% WG	Nicotinic acetylcholine receptor (nAChR) competitive modulator	25	5
	Tolfenpyrad 15.00% EC	Mitochondrial electron transport inhibitor (complex –I)	150	3
	Buprofezin 15.00% + Acephate 35.00% w/w WP	Inhibitor chitin biosynthesis, type 1 + Acetylcholinesterase (AChE) inhibitor	187.5 + 437.5	7

Table 2.
Insecticides with different modes of action for rotation program in whitefly management.

Spraying of mixture formulation of insecticides is also an important tactic for the management of whitefly. The basis of this approach is that individual insecticides in the mixture formulation have different modes of action and they lack cross-resistance [71, 72]. However, it has been found that the frequent and injudicious use of synergized insecticides lead to the development of high degree of resistance to both the chemicals. Despite several controversies over the use of mixture formulation of insecticide, the insecticide mixtures are still popular among farmers.

The chemical insecticide should be used as a last resort for the management of whitefly. Insecticides' use may be regulated in such a way that the full diversity of available chemicals is exploited instead of over-reliance on a single chemical for a long time. The Israeli strategy introduced in 1987 to preserve susceptibility to insecticides by optimizing and restricting their use to a single treatment per year [40, 41, 72] may be followed. The use of broad-spectrum insecticides should be avoided to minimize nontarget toxicity. This will conserve the natural enemies. The status of resistance should be monitored at regular intervals.

IPM strategies must be emphasized to combat the resistance problem. Some of the IPM strategies for management of whitefly are mass trapping and monitoring of whitefly using yellow sticky traps, use of entomopathogenic fungi (EPF), augmentative release of natural enemies, etc.

7. Biological control of whitefly using natural enemies

About 115 species of whitefly parasitoid belonging to 23 genera in five families (Aphelinidae, Azotidae, Signiphoridae, Encrytidae, and Platygastridae) have been reported [73]. The two most important genera of whitefly parasitoid are *Encarsia* and *Eretmocerus*. These two parasitoids significantly reduced the population of whitefly [74–76]. Apart from these two parasitoids, about 150 predators of whitefly have been reported from different parts of the world. Some of the important predators are ladybird beetles, predaceous bugs, lacewings, and spiders. It has also been reported that *Euseious scutalis* and *Typhlodromips swirskii* can significantly reduce the whitefly population on a single plant.

8. Biological control of whitefly using EPF

Entomopathogenic fungi or EPF are an important group of biological control agents that play a key role in the natural mortality of whitefly populations. They infect directly through the cuticle. *Beauveria bassiana* and *Metarhizium anisopliae* are some of the most commonly used microbial insecticides for the management of whitefly [77, 78]. In addition to these two microbials, *Aschersonia, Isaria (Paecilomyces)*, and *Lecanicillium (Verticillium)* are also used for the management of whitefly. Moreover, it has recently been reported that *Clonostachys rosea* has a pathogenic effect on the fourth instar nymphal and adult stages of the whitefly [79].

9. Conclusion

Modern agriculture is highly dependent on pesticide-based pest control and resistance development is inevitable. Therefore, resistance management strategies need to be adopted to maintain the efficacy of pesticides for successful pest management. For this, resistance diagnosis of whitefly populations needs to be done at regular intervals. Resistance is an evolutionary process. Hence, it needs to be managed wisely. The main reason for resistance development is over-reliance on pesticides. Therefore, pesticides should be used as a last resort and other nonchemical methods of pest management need to be emphasized. These will reduce the selection pressure on whitefly and thereby delay the resistance development.

Author details

Biswajit Patra[1] and Tapan Kumar Hath[2*]

1 Regional Research Station (Hill Zone), Uttar Banga Krishi Viswavidyalaya, Kalimpong, West Bengal, India

2 Department of Agricultural Entomology, Uttar Banga Krishi Viswavidyalaya, Cooch Behar, West Bengal, India

*Address all correspondence to: tapanhath@gmail.com

References

[1] Byrne DN, Bellows TS Jr. Whitefly biology. Annual Review of Entomology. 1991;**36**:431-457

[2] Oliveira MRV, Henneberry TJ, Anderson P. History, current status, and collaborative research projects for *Bemisia tabaci*. Crop Protection. 2001; **20**:709-723

[3] Stansly PA, Naranjo SE. Bemisia: Bionomics and Management of a Global Pest. Dordrecht: Springer; 2010

[4] Perring TM. The *Bemisia tabaci* species complex. Crop Protection. 2001; **20**(9):725-737

[5] Jones DR. Plant viruses transmitted by whiteflies. The European Journal of Plant Pathology. 2003;**109**:195-219

[6] Brown JK, Frohlich DR, Rosell RC. The sweetpotato or silverleaf whiteflies: Biotypes of *Bemisia tabaci* or a species complex? Annual Review of Entomology. 1995;**40**(1):511-534

[7] Xue M, Pang YH, Li QL, Liu TX. Effects of four host plants on susceptibility of *Spodoptera litura* (Lepidoptera: Noctuidae) larvae on five insecticides and activities of detoxification esterases. Pest Management Science. 2010;**66**(12): 1273-1279

[8] Boykin LM, De Barro PJ. A practical guide to identifying members of the *Bemisia tabaci* species complex: And other morphologically identical species. Frontiers in Ecology and Evolution. 2014;**2**:45. DOI: 10.3389/fevo.2014.00045

[9] De Barro PJ, Liu SS, Boykin LM, Dinsdale AB. *Bemisia tabaci*: A statement of species status. Annual Review of Entomology. 2011;**56**: 1-19

[10] Dinsdale A, Cook L, Riginos C, Buckley YM, Barro PD. Refined global analysis of *Bemisia tabaci* (Hemiptera: Sternorrhyncha: Aleyrodoidea: Aleyrodidae) mitochondrial cytochrome oxidase 1 to identify species level genetic boundaries. Annals of Entomological Society of America. 2010;**103**(2):196-208

[11] Hu J, De Barro PJ, Zhao H, Wang J, Nardi F, Liu S-S. An extensive field survey combined with a phylogenetic analysis reveals rapid and widespread invasion of two alien whiteflies in China. PLoS One. 2011;**6**(1):e16061. DOI: 10.1371/journal.pone.0016061

[12] Boykin LM, Shatters RG Jr, Rosell RC, McKenzie CL, Bagnall RA, De Barro PJ, et al. Global relationships of *Bemisia tabaci* (Hemiptera: Aleyrodidae) revealed using Bayesian analysis of mitochondrial COI DNA sequences. Molecular Phylogenetic Evolution. 2007;**44**(3):1306-1319

[13] Horowitz AR, Denholm I, Gorman K, Cenis JL, Kontsedalov S, Ishaaya I. Biotype Q of *Bemisia tabaci* identified in Israel. Phytoparasitica. 2003;**31**(1): 94-98

[14] Henneberry TJ, Faust RM. Introduction. In: Gould J et al., editors. Classical biological control of *Bemisia tabaci* in the United States. Berlin: Springer; 2008

[15] Ellsworth PC, Tronstad R, Leser J, Goodell PB, Godfrey LD, Henneberry TJ, et al. Sticky Cotton Sources and Solutions. IPM Series No. 13. The University of Arizona Cooperative Extension. Publ. #AZ1156. Tucson, p 4. 1999. Available from: http://ag.arizona.edu/crops/cotton/insects/wf/stickycss.pdf

[16] Erdogan C, Moores GD, Gurkan MO, Gorman KJ, Denholm I. Insecticide resistance and biotype status of

populations of the tobacco whitefly *Bemisia tabaci* (Hemiptera: Aleyrodidae) from Turkey. Crop Protection. 2008;**27**: 600-605

[17] Roy D, Bhattacharjee T, Biswas A, et al. Resistance monitoring for conventional and new chemistry insecticides on *Bemisia tabaci* genetic group Asia-I in major vegetable crops from India. Phytoparasitica. 2019;**47**: 55-66

[18] Wang S, Zhang Y, Yang X, Xie W, Wu Q. Resistance monitoring for eight insecticides on the sweetpotato whitefly (Hemiptera:Aleyrodidae) in China. Journal of Economic Entomology. 2017; **110**:660-666

[19] Yuan L, Wang SL, Zhou JC, Du YZ, Zhang YJ, Wang JJ. Status of insecticide resistance and associated mutations in Q-biotype of whitefly, *Bemisia tabaci*, from eastern China. Crop Protection. 2012;**31**:67-71

[20] Ma D, Gorman KJ, Devine GJ, Luo W, Denholm I. The biotype and insecticide-resistance status of whiteflies, *Bemisia tabaci* (Hemiptera: Aleyrodidae), invading cropping systems in Xinjiang Uygur Autonomous Region, northwestern China. Crop Protection. 2007;**26**:612-617

[21] Vassiliou V, Emmanouilidou M, Perrakis A, Morou E, Vontas J, Tsagkarakou A, et al. Insecticide resistance in *Bemisia tabaci* from Cyprus. Insect Science. 2011;**18**:30-39

[22] Elbert A, Nauen R, Leicht W. Imidacloprid, a novel chloronicotinyl insecticide: Biological activity and agricultural importance. In: Ishaaya I, Degheele D, editors. Insecticides with Novel Modes of Action: Mechanism and Application. Berlin: Springer; 1998. pp. 50-73

[23] Horowitz AR, Mendelson Z, Weintraub PG, Ishaaya I. Comparative toxicity of foliar and systemic applications of two chloronicotinyl insecticides, acetamiprid and imidacloprid, against the cotton whitefly, *Bemisia tabaci*. Bulletin of Entomological Research. 1998;**88**: 437-442

[24] Takahashi H, Mitsui J, Takausa N, Matsud M, Yoneda H, Suszuki J, et al. NI-25, a new type of systemic and broad spectrum insecticide. In: Proceedings, Brighton Crop Protection Conference, Pests and Diseases, 1992 Brighton, November 23–26, 1992. Farnham, Surrey, UK: British Crop Protection Council; 1992. pp. 89-96

[25] Cahill M, Gorman K, Day S, Denholm I, Elbert A, Nauen R. Baseline determination and 523 detection of resistance to imidacloprid in *Bemisia tabaci* (Homoptera: Aleyrodidae). Bulletin 524 of Entomological Research. 1996;**86**:343-349

[26] Silva LD, Omoto C, Bleicher E, Dourado PM. Monitoring the susceptibility to insecticides in *Bemisia tabaci* (Gennadius) (Hemiptera: Aleyrodidae) populations from Brazil. Neotropical Entomology. 2009;**38**: 116-125

[27] Schuster DJ, Mann RS, Toapanta M, Cordero R, Thompson S, Cyman S, et al. Monitoring neonicotinoid resistance in biotype B of *Bemisia tabaci* in Florida. Pest Management Science. 2010;**66**: 186-195

[28] Kontsedalov S, Abu-Moch F, Lebedev G, Czosnek H, Horowitz AR, Ghanim M. *Bemisia tabaci* biotype dynamics and resistance to insecticides in Israel during the years 2008–2010. Journal of Integrative Agriculture. 2012; **11**:312-320

[29] Ahmad M, Khan RA. Field-evolved resistance of *Bemisia tabaci* (Hemiptera: Aleyrodidae) to carbodiimide and neonicotinoids in Pakistan. Journal of

Economic Entomology. 2017;**110**: 1235-1242

[30] Naveen NC, Chaubey R, Kumar D, Rebijith KB, Rajagopal R, Subrahmanyam B, et al. Insecticide resistance status in the whitefly, *Bemisia tabaci* genetic groups Asia-I, Asia-II-1 and Asia-II-7 on the Indian subcontinent. Scientific Reports. 2017;7: 40634. DOI: 10.1038/srep40634

[31] Wang Z, Yan H, Yang Y, Wu Y. Biotype and insecticide resistance status of the whitefly *Bemisia tabaci* from China. Pest Management Science. 2010; **66**:1360-1366

[32] Fernandez E, Gravalos C, Javier Haro P, Cifuentes D, Bielza P. Insecticide resistance status of *Bemisia tabaci* Q-biotype in southeastern Spain. Pest Management Science. 2009;**66**: 885-891

[33] Bielza P, Moreno I, Belando A, Grávalos C, Izquierdo J, Nauen R. Spiromesifen and spirotetramat resistance in field populations of *Bemisia tabaci* Gennadius in Spain. Pest Management Science. 2019;**75**:45-52

[34] Basit M, Saeed S, Saleem MA, Denholm I, Shah M. Detection of resistance, cross-resistance, and stability of resistance to new chemistry insecticides in *Bemisia tabaci* (Homoptera: Aleyrodidae). Journal of Economic Entomology. 2013;**106**: 1414-1422

[35] Horowitz AR, Ishaaya I. Dynamics of biotypes B and Q of the whitefly *Bemisia tabaci* and its impact on insecticide resistance. Pest Management Science. 2014;**70**:1568-1572

[36] Li X, Degain BA, Harpold VS, Marçon PG, Nichols RL, Fournier AJ, et al. Baseline susceptibilities of B- and Q-biotype *Bemisia tabaci* to anthranilic diamides in Arizona. Pest Management Science. 2012;68(1):83-91.

[37] Uchida M, Asai T, Sugimoto T. Inhibition of cuticle deposition and chiton synthesis by a new insect growth regulator, buprofezin, in JfVilaparvatalug ensS tal. Pesticide Biochemistry Physiology. 1985;**27**:71-75

[38] Ishaaya I, Horowitz AR. Novel phenoxy juvenile hormone analog (pyriproxyfen) suppresses embryogenesis and adult emergence of sweet potato whitefly (Homoptera: Aleyrodidae). Journal of Economic Entomology. 1992;**85**:2113-2117

[39] Cahill M, Jarvis W, Gorman K, Denholm I. Resolution of baseline responses and documentation of resistance to buprofezin in *Bemisia tabaci* (Homoptera: Aleyrodidae). Bulletin of Entomological Research. 1996;**86**:117-122

[40] Horowitz AR, Ishaaya I. Managing resistance to insect growth regulators in the sweetpotato whitefly (Homoptera: Aleyrodidae). Journal of Economic Entomology. 1994;**87**:866-871

[41] Horowitz AR, Forer G, Ishaaya I. Managing resistance in *Bemisia tabaci* in Israel with particular emphasis on cotton. Pesticide Science. 1994;**42**: 113-122

[42] Abbott WS. A method of computing the effectiveness of an insecticide. Journal of Economic Entomology. 1925; **18(2)**:265-267.

[43] Finney DJ. Probit Analysis: A Statistical Treatment of the Sigmoid Curve. London: Cambridge University Press; 1971. p. 333

[44] Mota-Sanchez D, Wise JC. The Arthropod Pesticide Resistance Database. East Lansing, MI: Michigan State University; 2019. Available from: http://www.pesticideresistance.org

[45] Luo C, Jones CM, Zhang F, Denholm I, Gorman K. Insecticide

resistance in *Bemisia tabaci* biotype Q (Hemiptera: Aleyrodidae) from China. Crop Protection. 2010;**29**:429-434

[46] Yao FL, Zheng Y, Huang XY, Ding XL, et al. Dynamics of *Bemisia tabaci* biotypes and insecticide resistance in Fujian province in China during 2005–2014. Scientific Reports. 2017;7:40803. DOI: 10.1038/srep41409

[47] Dângelo RAC, Michereff-Filho M, Campos MR, da Silva PS, Guedes RNC. Insecticide resistance and control failure likelihood of the whitefly *Bemisia tabaci* (MEAM1; B biotype): A neotropical scenario. The Annals of Applied Biology. 2018;**172**:88-99

[48] Nauen R, Stumpf N, Elbert A. Toxicological and mechanistic studies on neonicotinoid cross resistance in Q-type *Bemisia tabaci* (Hemiptera: Aleyrodidae). Pest Management Science. 2002;**58**:868-875

[49] Roditakis E, Grispou M, Morou E, Kristoffersen JB, Roditakis N, Nauen R, et al. Current status of insecticide resistance in Q biotype *Bemisia tabaci* populations from Crete. Pest Management Science. 2009;**65**:313-322

[50] Roditakis E, Roditakis NE, Tsagkarakou A. Insecticide resistance in Bemisia tabaci (Homoptera: Aleyrodidae) populations from Crete. Pest Management Science. 2005;**61**:577-582

[51] Kranthi KR, Jadhav DR, Wanjari RR, Shakir Ali S, Russell D. Carbamate and organophosphate resistance in cotton pests in India, 1995 to 1999. Bulletin of Entomological Research. 2001;**91**:37-46

[52] Shadmany M, Omar D, Muhamad R. Biotype and insecticide resistance status of *Bemisia tabaci* populations from Peninsular Malaysia. Journal of Applied Entomology. 2015;**139**:67-75

[53] Horowitz AR, Kontsedalov S, Khasdan V, Ishaaya I. Biotypes B and Q of Bemisia tabaci and their relevance to neonicotinoid and pyriproxyfen resistance. Archives of Insect Biochemistry and Physiology. 2005;**58**: 216-225

[54] Fernández E, Grávalos C, Haro PJ, Cifuentes D, Bielza P. Insecticide resistance status of Bemisia tabaci Q biotype in south eastern Spain. Pest Management Science. 2009;**65**:885-891

[55] Elbert A, Nauen R. Resistance of *Bemisia tabaci* (Homoptera: Aleyrodidae) to insecticides in southern Spain with special reference to neonicotinoids. Pest Management Science. 2000;**56**:60-64

[56] Şahin İ, İkten C. Neonicotinoid resistance in Bemisia tabaci (Genn., 1889) (Hemiptera: Aleyrodidae) populations from Antalya, Turkey. Turkish Journal of Entomology. 2017; **41**(2):169-175

[57] Horowitz AR, Ishaaya I. Insect Pest Management: Field and Protected Crops. New York: Springer Science & Business Media; 2004

[58] Castle SJ, Prabhaker N. Monitoring changes in *Bemisia tabaci* (Hemiptera: Aleyrodidae) susceptibility to neonicotinoid insecticides in Arizona and California. Journal of Economic Entomology. 2013;**106**:1404-1413

[59] Caballero R, Cyman S, Schuster DJ. Monitoring insecticide resistance in biotype B of *Bemisia tabaci* (Hemiptera: Aleyrodidae) in Florida. Florida Entomologist. 2013;**96**:1243-1256

[60] Cahill M, Byrne FJ, Gorman K, Denholm I, Devonshire AL. Pyrethroid and organophosphate resistance in the tobacco whitefly *Bemisia tabaci* (Homoptera: Aleyrodidae). Bulletin of Entomological Research. 1995;**85**:181-187

[61] Prabhaker N, Castle S, Henneberry TJ, Toscano NC. Assessment of cross-

resistance potential to neonicotinoid insecticides in Bemisia tabaci (Hemiptera: Aleyrodidae). Bulletin of Entomological Research. 2005;**95**: 535-543

[62] Dennehy TJ, Degain BA, Harpold VS, Zaborac M, Morin S, Fabrick JA, et al. Extraordinary resistance to insecticides reveals exotic Q biotype of Bemisia tabaci in the New World. Journal of Economic Entomology. 2010; **103**:2174-2186

[63] Omer AD, Johnson MW, Tabashnik BE, Costa HS, Ullman DE. Sweetpotato whitefly resistance to insecticides in Hawaii: Intra-island variation is related to insecticide use. Entomologia Experimentalis et Applicata. 1993;**67**: 173-182

[64] Pérez CJ, Alvarado P, Narváez C, Miranda F, Hernández L, Vanegas H, et al. Assessment of insecticide resistance in five insect pests attacking field and vegetable crops in Nicaragua. Journal of Economic Entomology. 2000; **93**:1779-1787

[65] Kady HEL, Devine GJ. Insecticide resistance in Egyptian populations of the cotton whitefly, *Bemisia tabaci* (Hemiptera: Aleyrodidae). Pest Management Science. 2003;**59**: 865-871

[66] Farghaly SF. Biochemical monitoring for insecticides resistance in whitefly *Bemisia tabaci*. American-Eurasian Journal of Agriculture & Environmental Sciences. 2010;**8**:383-389

[67] Byrne FJ, Castle S, Prabhaker N, Toscano NC. Biochemical study of resistance to imidacloprid in B biotype *Bemisia tabaci* from Guatemala. Pest Management Science. 2003;**59**:347-352

[68] Abdeldaffie EYA, Elhag EA, Bashir NHH. Resistance in the cotton whitefly, *Bemisia tabaci* (Genn.), to insecticide recently introduced into Sudan Gezira. International Journal of Pest Management. 1987;**33**:283-286

[69] Sparks TC, Nauen R. IRAC: Mode of action classification and insecticide resistance management. Pesticide Biochemistry and Physiology. 2015;**121**: 122-128

[70] Anonymous. Major Uses of Pesticides. Central Insecticide Board & Registration Committee, Ministry of Agriculture & Farmers Welfare, Government of India; 2021. pp. 1-53 Available from: http://ppqs.gov.in/sites/default/files/major_use_of_pesticide_insecticide_as_on_30.11.2021.pdf

[71] Denholm I, Rowland M. Tactics for managing pesticide resistance in arthropods: Theory and practice. Annual Review of Entomology. 1992;**37**: 91-112

[72] Horowitz AR, Ishaaya I. Chemical control of Bemisia: management and application. In: Gerling D, Mayer RT, editors. Bemisia 1995: Taxonomy, Biology, Damage, Control and Management. Andover, Hants: Intercept; 1996. pp. 537-556

[73] Lahey A, Lahey Z, Stansly P. An updated list of parasitoid Hymenoptera reared from the Bemisia tabaci species complex (Hemiptera: Aleyrodidae). Florida Entomologist. 2020;**98**: 456-463

[74] Kedar SC, Saini RK, Kumaranag KM, Sharma SS. Record of natural enemies of whitefly, *Bemisia tabci* (Gennadius) (Hemiptera: Aleyridae) in some cultivated crops in Haryana. Journal of Biopesticides. 2014;7:57-59

[75] Kidane D, Yang N, Wan F. Evaluation of a banker planned system for biological control of *B. tabaci* (Hemiptera: Aleyrodidae) on tomato, using two aphelinid parasitoids under field-cage conditions. Biocontrol Science and Technology. 2018;**28**:1054-1073

[76] Xu H, Yang N, Wan F. Competitive interactions between parasitoids provide new insight into host suppression. PLoS One. 2013;**8**:e82003

[77] Al-Deghari MA. Bioassay evaluation of the entomopathogenic fungi, *Beauveria bassina* Vuellemin against eggs and nymphs of *Bemisia tabaci* Gennadius (Homoptera:Aleyrodidae). Pakistan Journal of Biological Science. 2008;**11**:1551-1560

[78] Perring TM, Stansly PA, Liu TX, Smith HA, Andreason SA. Whiteflies: Biology, ecology and management. In: Wakil W, Brust GE, Perring TM, editors. Sustainable Management of Arthropod Pests of Tomato. Cambridge, MA, USA; Amsterdam, The Netherlands: Academic Press; 2018. pp. 73-110

[79] Anwar W, Ali S, Nawaz K, Iftikhar S, Javed MA, Hashem A, et al. Entomopathogenic fungus *Clonostachys rosea* as a biocontrol agent against whitefly*(Bemisia tabaci)*. Biocontrol Science and Technology. 2018;**28**: 750-760

Chapter 20

Insect Resistance to Neonicotinoids - Current Status, Mechanism and Management Strategies

Shrawan Kumar Sahani, Vikas Kumar and Subhajit Pal

Abstract

Pesticides are any substance used for controlling, preventing, destroying, repelling, or mitigating of pests. Neonicotinoids have been the most commonly used insecticide since the early 1990s, current market share of more than 25% of total global insecticide sales. Neonicotinoid insecticides are highly selective agonists of insect nicotinic acetylcholine receptors (nAChRs) that exhibit physicochemical properties, rendering them more useful over other classes of insecticides. This includes having a wide range of application techniques and efficacy in controlling sucking and biting insects. Although neonicotinoids are applied as foliar insecticides with possible direct exposure risks to honeybees, a large part of neonicotinoid use consists of seed coating or root drench application. There are three major detoxification enzymes involved in the development of resistance against insecticides viz., cytochrome P450 monooxygenases, carboxylesterases, and glutathione S-transferases. The repeatedly used use of compounds of the same active ingredients and application of excessive organophosphates (OPs) and pyrethroids in *Bemisia tabaci*. Resistance to insecticides resulting in loss of efficacy of many older insecticides has placed excessive pressure on novel products. One of the major limitations to resistance management is the occurrence of cross-resistance. This review briefly summarizes the current status of neonicotinoid resistance, the biochemical and mechanisms involved, and the implications for resistance management.

Keywords: insect resistance, neonicotinoids, overuse, nicotinic acetylcholine receptors, management strategies

1. Introduction

Chemical control remains the most important and widely used strategy against noxious insect pests around the world. It is used to kill, harm, repel, or mitigate one or more species of an insect by disrupting the nervous system and damaging their exoskeletons. Insecticides have not only controlled insects, but it also used to control diseases carrier agents and helps in the economy and social benefits through better health and increase food production [1, 2]. After the introduction of neonicotinoids in the 1990s, most widely used against the sap-feeding insect. Among these imidacloprid is the most widely used insecticide in the world. Neonicotinoids

IntechOpen

currently account for approximately 25% of the total insecticide market and are increasing in use as they replace the organophosphate (OP) and carbamate insecticides, causing less toxicity in birds and mammals than insects. Overwhelming evidence has risen over the past decade regarding potentially harmful risks to humans, nontarget insects, aquatic invertebrates, and side effects on the natural environment following usage of specific classes of insecticides [3, 4].

There is various kind of factors that helps the occurrence and initial successful establishment of neonicotinoids to control, mitigate the especially soft body insect pests. At that time there was no known pesticide resistance in target pests, mainly because of recently synthesized nicotine contain plants, their physicochemical properties included many advantages such as selectivity, target-specific, less residual effect on soil, and metabolism rate fast over previous generations of insecticides (i.e., organophosphates, carbamates, pyrethroids, etc.) they shared an assumed reduced operator and consumer risk [5, 6]. But after some time, due to large and indiscriminate use of the same mode of action insecticides have been responsible for developed resurgence and insecticide tolerance ability increased. The first report of neonicotinoid resistance was published in 1996, describing the low efficacy of imidacloprid against Spanish greenhouse populations of cotton whitefly. There are three major detoxification enzymes involved in the development of resistance against insecticides viz., cytochrome P450 monooxygenases, carboxylesterases, and glutathione S-transferases [1, 7].

Several field problems such as poor selection of chemicals and substandard application practices exacerbated the control failures of insecticides against *Bemisia tabaci* in India. The repeated use of compounds of the same active ingredients and application of excessive organophosphates, carbamate, pyrethroids, and neonicotinoids against insect pests cause the development of resistance. Resistance to insecticides resulting in loss of efficacy of many older insecticides has placed excessive pressure on novel products. One of the major limitations to resistance management is the occurrence of cross-resistance [8].

2. Landmark to development of neonicotinoids and their mode of action

- In 1970, first-time nithiazine was precursor by Henry Feuer, a reputed chemist at Purdue University [4, 9].

- Shell (oil refinery company) researchers found in screening that this precursor showed insecticide activity on insect pest management and refined it to develop nithiazine.

- In 1984, the mode of action of nithiazine was found to be as a postsynaptic acetylcholine receptor same as nicotine.

- In 1985, Bayer patented imidacloprid as the first commercial neonicotinoid, and till 1990 used at large scale.

- The early 2000s, two other neonicotinoids, clothianidin, and thiamethoxam, entered the market, which is drastically changing the thinking of people.

- During 2013, virtually all corn planted in the United States was treated with one of these two insecticides. Thiamethoxam among the neonicotinoids has less residual effect and persistence in nature.

- Beginning of 2014, about a third of US soybean acreage was planted with neonicotinoid-treated seeds, usually imidacloprid or thiamethoxam [3].
- Electrophysiological studies on identified cockroach neurons, and the binding to cockroach nervous system membrane.

2.1 Neonicotinoid groups Vs Older groups

Neonicotinoids have been the most commonly used insecticides since the early 1990s as an alternative to older organophosphate and carbamate insecticides. Neonicotinoids are insecticides that exhibit physicochemical properties, rendering them more useful over other classes of insecticides [10]. Over the last few years, neonicotinoids have been combined with pyrethroids in formulated products, and with diatomaceous earth (e.g., Alpine dust insecticide, with dinotefuran) for the control of insect pests. Moderate to high levels of tolerance/resistance to various neonicotinoids showed by insect pests. Romero and Anderson reported that resistance to neonicotinoids may likely be conferred by the increased enzymatic activities found in these populations. Those findings showed that tolerance or even resistance to neonicotinoids is now present in field and storage pest populations [1, 11].

Neonicotinoids show high acute toxicity to honeybees. The neonicotinoid family includes imidacloprid, clothianidin, and thiamethoxam (the latter is metabolized to clothianidin in the plant and the insect). Recently, imidacloprid has been replaced by thiamethoxam and clothianidin in some parts of the world. To date, neonicotinoids have proved the development of resistance, such as *Myzus persicae* and *Phorodon humuli*. The effects of imidacloprid on *Nilaparvata lugens*, tebufenozide on *Plutella xylostella* and *Spodoptera exigua,* thiamethoxam on *Bemisia tabaci,* trichlorphon on *Bactrocera dorsalis*, imidacloprid on *Spodoptera litura*, and emamectin benzoate on *Chrysoperla carnea* have been reported [12].

The first report of neonicotinoid resistance was published in 1996, describing the low efficacy of imidacloprid against Spanish greenhouse populations of cotton whitefly. Later-generation, show stronger resistance (up to 17-fold in the first 15 generations, but >80-fold resistance after 24 generations, which has been confirmed in some populations of the whitefly (*Bemisia tabaci*) and the Colorado potato beetle (*Leptinotarsa decemlineata*) [13]. Although neonicotinoids are applied as foliar insecticides with possible direct exposure risks to honeybees, a large part of neonicotinoid use consists of seed coating or root drench application [14, 15].

2.2 Mode of action of neonicotinoid

All neonicotinoids act on the insect central nervous system as agonists of the postsynaptic nicotinic acetylcholine receptors (nAChRs). Neonicotinoids act as agonists on the postsynaptic insect nicotinic acetylcholine receptors (nAChRs), biodegradable substituents which have a much higher affinity on insects than mammals [16, 17]. Neonicotinoid insecticides are highly toxic to many invertebrates, including honey bees, bumblebees, and solitary bees. The neonicotinoids (including imidacloprid, dinotefuran, clothianidin, and thiamethoxam) are nitro functional group ($-NO_2$) instead of a cyano functional group ($-C{=}N$) in their molecular structure. This slight difference in their molecular structure affects the toxicity level of neonicotinoids, which bind to an insect receptor site. The nitro-group neonicotinoids are much more toxic to bees than the cyano-group neonics, which include acetamiprid and thiacloprid [11, 13, 18].

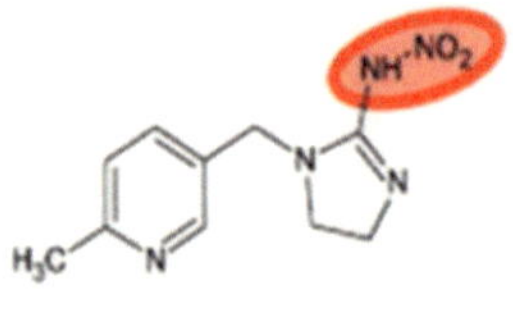

Imidacloprid

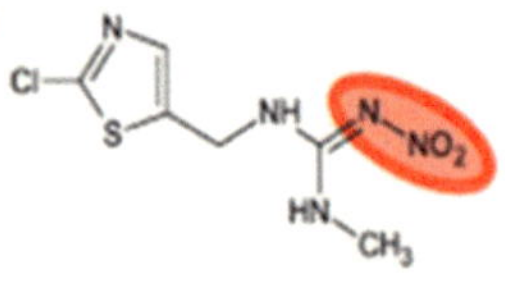

Clothianidin

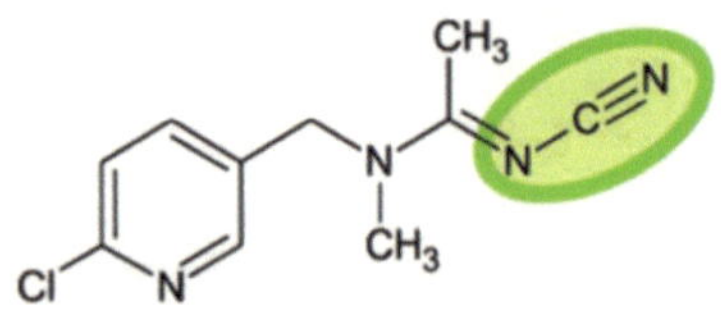

Acetamiprid

3. Global growth, status, and uses of neonicotinoid insecticide

During 1990, the global insecticide market was dominated by carbamates, organophosphates, and pyrethroids. In 2008, one-quarter of the insecticide market was neonicotinoid to 27% in 2010 and nearly 30% in 2012. The Overuse of chemical products in different spheres of life not only brings benefits for humanity but also presents a large number of threats against the environment and in consequence to human health. The present graph indicates the maximum use of neonicotinoids insecticide and its application in different countries. Here, thiamethoxam shares maximum contribution in the market (37.6) followed by imidacloprid (33.5), clothianidin (14.7), acetamiprid (7.2), thiacloprid (3.8), dinotefuran (2.9), nitenpyram (0.3), and the area

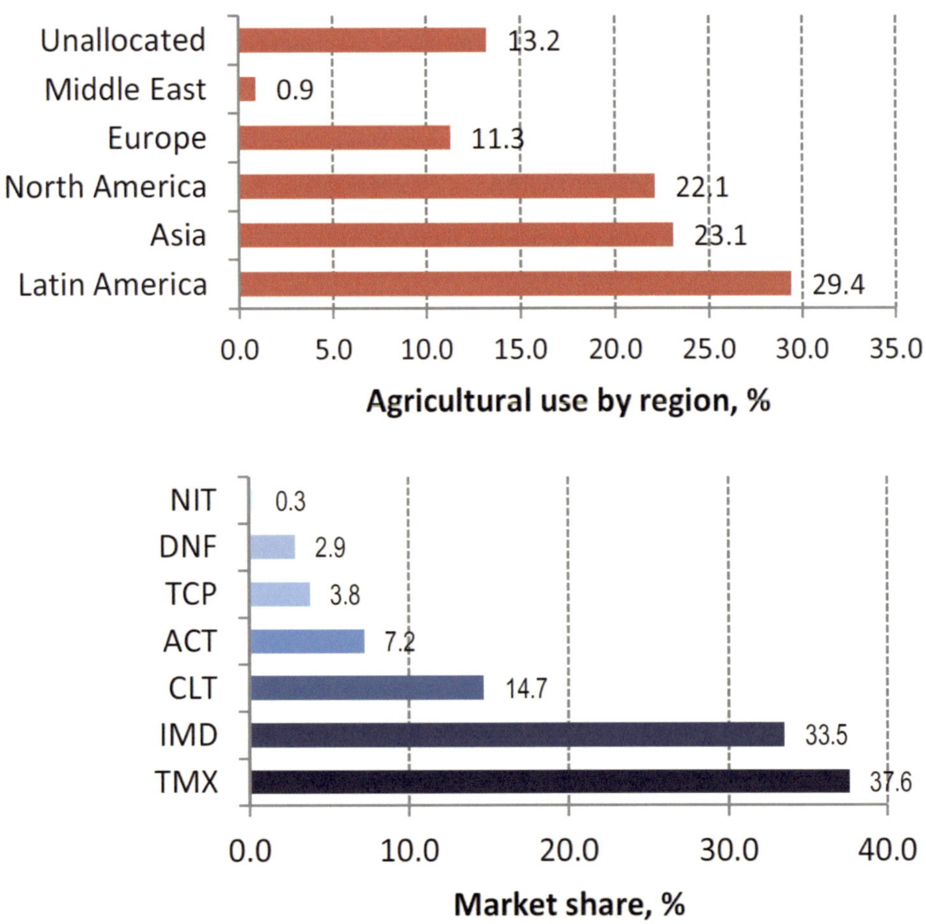

Figure 1.
Agricultural use by region and market share of individual neonicotinoids in percent. Abbreviations: TMX (thiamethoxam), IMD (imidacloprid), CLT (clothianidin), ACT (acetamiprid), TCP (thiacloprid), DNF (dinotefuran), NIT (nitenpyram).

uses maximum in agriculture production in Latin America (29.4), Asia (23.1), and North America (22.1) followed by others unallocated areas (**Figure 1**).

3.1 The use of neonicotinoids covers four major domains

The uses of neonicotinoid are to protect crops and ornamentals against polyphagous insects and mites, urban pest control to target harmful organisms such as cockroaches, ants, termites, wasps, flies, etc., apart from the agricultural uses, it is also applicable in veterinary sciences to reduce the chances of occurrence (fleas, ticks on pet animals) and fish farming infestations. In agriculture as well as horticulture crop, neonicotinoid can be functional in many different ways such as foliar spraying, seed dressing, soil drenching, furrow application, trunk injections in trees, mixing with irrigation water, drenching of flower, soil treatment, granular application, dipping of seedlings, bulbs and application with a brush on the stems of fruit trees. Seed and soil applications denote approximately 60% of their uses globally [16, 19]. The usage of neonicotinoid insecticides has grown considerably since the forerunner of this group, it is first introduced among neonicotinoids in the year 1991 followed by acetamiprid and thiamethoxam. Till now, seven insecticides be in the right place to this chemical class are available to farmers all over the world and classified as Group A within the Insecticide Resistance Action Committee (IRAC) and Mode of Action Classification Scheme. All neonicotinoids are agonists of insect nicotinic acetylcholine receptors [18]. In 1941, the first case documented by insects demonstrated resistance to an inorganic resistance; 1987—first reported in First reported on tobacco in vaeck, and later on tomato in Bischoff district. Insect resistance genetically modified crops (primarily cotton and maize), are toxic to certain insects. They are often called Bt crops because the introduction genes were originally identified in a bacterial species called *Bacillus thuringiensis* (**Figure 2**).

3.2 Insecticide resistance to neonicotinoid

Resistance is quickly developed due to the selection of highly effective compounds with kill or mitigate to insect, long residual effect, and is regular use of single biochemical target site. The toxicant is converted into a nontoxicant form in the body of an insect by various enzymes. All these enzymatic changes are carried forward and transmitted through genes. Resistance in B- and Q-type has been noticed in *Bemisia tabaci* to enhanced oxidative detoxification of neonicotinoids due to overexpression of monooxygenases. No evidence for target-site resistance has been found in whiteflies [18, 19]. Biotic and abiotic degradation processes contribute to the environmental persistence of neonicotinoids. The half-life of neonicotinoids varies depending on physiochemical

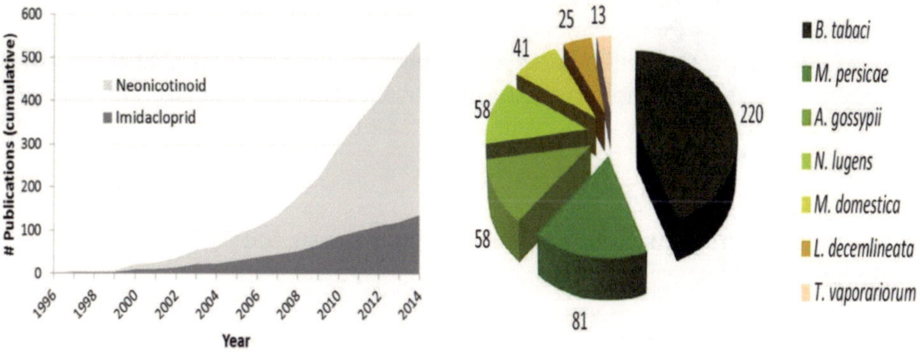

Figure 2.
Several reported cases of neonicotinoid resistance up to 2014. Only those pests with >10 (fold) reported cases are shown.

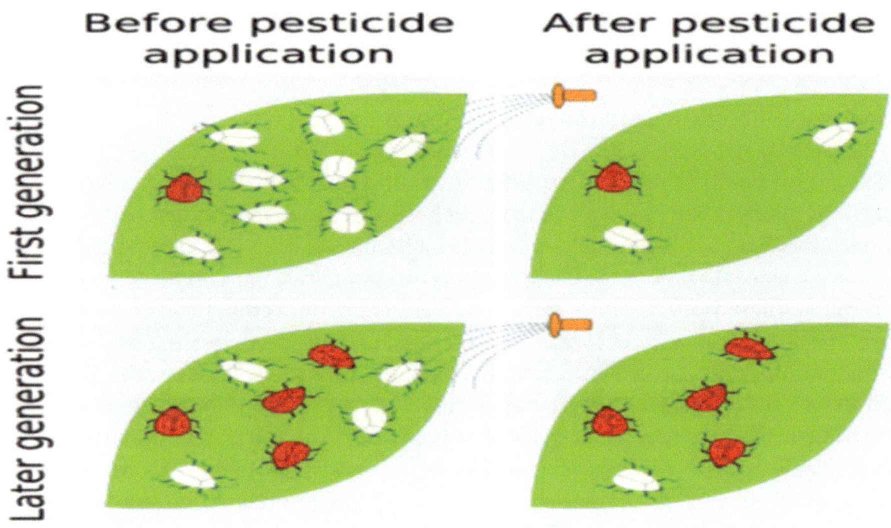

Figure 3.
Pesticide resistance can build up in the pest population when a change in the genetic characteristic of the pest population is inherited from one generation to the next.

conditions (i.e., organic matter content, soil texture, residence time) before undergoing complete degradation. The development of resistance is a complex and dynamic process and depends upon many factors. As per the IRAC, resistance is well-defined as a heritable change in the sensitivity of a pest population, that is reflected in the constant failure of a product to attain the expected level of control when used according to the label endorsement for that pest species [20]. The environmental protection agency (EPA) is divided into two classes of toxicity agents i.e., II and III. Accumulation (increase the level of pesticides) of those pesticides into the soil affects the pollen quality of sprayed plants, especially due to their toxic effects, against pollinators the consequences of the occurrence of these insecticides have been discussed [10, 21]. It is determined that the transfer of vertical gene (a particular gene) from the microorganism, higher plants, animals into a host plant for crop improvement and researches are called transgenic plant or insects, virus, fungus resistance plant. It gives us facts for future research. Mutation (can build up in the pest population when a change in the genetic characteristic of the pest population is inherited from one generation to the next) in nAChR subunits and in most cases, metabolism is also responsible for the development of resistance (**Figure 3**). The brown planthopper, *Nilaparvata lugens*, was selected with imidacloprid treatment at a sublethal dose to obtain resistant mutation. Both *Bemisia tabaci* (sweet potato whitefly) and *Trialeurodes vaporariorum* (greenhouse whitefly) have been shown to have a high possibility for resistance development and characterize some of the main targets for which IRAC specific strategies have been developed [8, 22]. Global resistance management guiding principles were designed by the Neonicotinoid Working Group of the Insecticide Resistance Action Committee and are based on guidelines published and updated earlier [9].

4. Mechanism of resistance and factors that influence resistance development

4.1 Resistance mechanisms

The various mechanisms that enable insects to resist the action of insecticides can be grouped into several categories:

4.1.1 Single resistance

Resistance to Dichloro diphenyl trichloroethane (DDT) amounts to resistance to several DDT analogs such as methoxychlor, but not to hexa chloro cyclohexane (HCH). Due to excessive and continuous use of the insecticides.

4.1.2 Cross-resistance

It occurs when resistance to one insecticide or within a group. Eg: organophosphate insecticides, fungicides etc.

4.1.3 Multiple resistance

It involves multiple, independent resistance mechanisms, which often lead to resistance to chemicals from different families (i.e., organophosphate and carbamate insecticides) [23].

4.1.4 Metabolic resistance

It plays one of the significant roles, in the development of resistance which helps to change the activity of enzyme systems that all insects possess to help them detoxify naturally occurring foreign materials. Enzymes are classified viz., esterases, monooxygenases, and glutathione S-transferases typically fulfill this function. These enzyme systems are often enhanced in resistant insect strains enabling them to metabolize or degrade insecticides before they can exert a toxic effect. Metabolic resistance appliances have been noticed in whitefly, aphid, and Colorado potato beetle populations for all major classes of insecticides, currently also used for soft body insect control including neonicotinoids insecticide [24].

4.1.5 Target site resistance

Insecticides generally perform on a specific site within the insect, especially within the nervous system (e.g., OP, carbamate, and pyrethroid insecticides). The site of action can be reformed in resistant strains of insects such that the insecticide no longer binds effectively. This results in the insects being unaffected, or less affected, by the action of insecticide than susceptible insects.

4.1.6 Reduced penetration

Changes in the insect cuticle or digestive tract linings that avoid or slow the absorption or diffusion of insecticides can be found in some strains of resistant insects. This resistance mechanism can affect a broad range of insecticides. Examples of reduced penetration mechanisms are limited and are often considered a contributing factor to reduced susceptibility.

4.1.7 Behavioral resistance

This resistance illustrates any adjustment in insect behavior that helps to avoid the lethal effects of insecticides [22]. Insecticide resistance in mosquitoes is not always based on biochemical mechanisms such as metabolic detoxification or target site mutations, but may also be conferred by behavioral changes in response to prolonged exposure to an insecticide.

Mechanism	Species
Resistance to neonicotinoids	
Enhanced expression of CYP6G1	*Drosophila melanogaster*
Enhanced expression of CYP6AY1, CYP6ER1, CYP4CE1, and CYP6CW1	*Nilaparvata lugens*
Enhanced expression of glutathione *S*-transferase (MdGST) and galactosyltransferase (MdGT1)	*Musca domestica*
Mutation of Dα1/Dβ2 subunits	*Drosophila melanogaster*
Deletion of Dα1 subunit	
Y151 mutation in Nlα1/Nlα3 subunits	*Nilaparvata lugens*
Reduced expression of Nlα8 nicotinic acetylcholine receptor (nAChR) subunit	
R81T mutation in Mpβ1 subunit	*Myzus persicae*
	Aphis gossypii
Tolerance to neonicotinoids	
Reduced sensitivity to thiacloprid and acetamiprid due to metabolism by CYP9Q3	*Apis mellifera*
Reduced sensitivity to thiacloprid and acetamiprid due to metabolism by CYP9Q4 and CYP9Q6	*Bombus terrestris*
Reduced sensitivity to thiacloprid due to metabolism by CYP9BU1 and CYP9BU2	*Osmia bicornis*

Table 1.
Mechanisms of neonicotinoid resistance and tolerance in insects.

4.1.8 Genetic basis of resistance

It occurs naturally, genetic mutations allow a small proportion of the population to resist and endure the effects of the insecticide. This occurs due to continually using the same insecticide and horizontal genes. Resistance insects will reproduce and the genetic changes that confer resistance are transferred from parents to offspring so that eventually they become numerous within the population (**Table 1**).

5. Factors influence in the development of resistance

5.1 Frequency of application

However, usually pesticide or management measures are used, one of the important factors that influence resistance development. With every use, an advantage is given to the resistance insects inside a population. The speed of increase of resistance in any population can usually be faster within the presence of a lower applicable value.

5.2 Dosage and persistence of effect

The period of pesticide persistence remains effective, additionally referred to as its persistence, relies upon the chemical science of the pesticide, the short of formulation, and also the application rate. The product that gives a persistent impact equally gives continuous selection pressure to multiple species. As an example, an area spray can persist for short time and can choose solely against one generation of mosquitoes [22]. In addition, a residual wall application or a bed net treatment can persist for months or years providing choice pressure against several generations of an equivalent insect. It is so vital to frequently follow manufacturer and United Nations agency recommendations once victimization such pesticides.

5.3 Rate of reproduction

As we have got identified that usually, insects that have a short life cycle, and high reproductive rate are possible to develop resistance earlier than species that have a lower reproductive rate, as any resistance generation will quickly unfold throughout the population. The homopterans insect has an associate in nursing account for pesticide resistance and is considered by a comparatively short life cycle and high fecundity, with female laying a huge number of eggs throughout their reproductive life. However, the tse-tse fly have shorter life cycle and fecundity comparatively less than hemipterans and comparatively low rate of reproduction, females produce in total fewer approximately 10 offspring.

5.4 Population isolation

With vectors of sickness, the goal is common to get rid of all or the bulk of the population, but the larger the choice pressure that is placed on a population, the quicker status could also be lost. Immigration of people possessing susceptible genes from untreated areas can beneficially dilute and contend with the resistance genes within the overall population. Associate in nursing early step during a vector management program should therefore to be to estimate the importance of immigration of untreated insects. As an example, associate in nursing island wherever the whole space was treated would have the next risk of developing resistance as few untreated mosquitoes would be part of the treated population. The hazard of pesticide resistance rising out to be measured once designing resistance management ways [25, 26]. Awareness of and coordination with neighboring vector management programs and agricultural activities should be excited, so the regional impact on the target population is deliberated.

- Prolonged exposure to a single insecticide
- High selection pressure
- Large coverage area
- Insects multiplying by asexual means
- The selection at every stage of the insect life cycle

6. Neonicotinoid resistance management

- Always use products at the recommended label rates and spray intervals with the appropriate application equipment.
- Rotation of insecticide group against the rapid selection of resistant population.
- Use suitable integrated pest management (IPM) approaches.

- Neonicotinoids are used against different pests in the same cultivars.

Repeatable uses of different chemistry of neonicotinoids against more than one pest species in the same crop are less susceptible but need at the local level, to take into account the pest populations dynamics, overlapping of the various species, their relative importance, and each species' potential risk for developing resistance [2].

- Do not control a multigeneration pest exclusively with neonicotinoids.

- The use of nonspecific products helps to prevent the development of resistance.

- Plan to use neonicotinoid insecticides in such a way that they do not affect the beneficial organisms.

- Good agricultural practices should be applied alongside physical and biological pest control methods.

- Judicious use of insecticides (need-based and recommended dose).

- The use of insecticide synergists.

- Window system of pesticide application.

- Area-wide management.

- Crop pest host management.

- Monitor problematic pest populations to detect first shifts insensitivity.

6.1 Alternative prospects

- Use other synthetic or naturally occurring chemical insecticides

- Biological control with microorganisms

- Biological control through farming practices

- Use of semiochemicals for mass trapping, mating disruption, repulsion, antifeeding effects, push-and-pull or attract-and-kill techniques

- Use other techniques like physical and mechanical methods to minimize insecticidal loads

- Genetically improved plant varieties

- We used four criteria to rank the alternatives to neonicotinoids—efficacy (E), applicability (A), durability (D), and practicability (P)

7. Summary

The widespread use of synthetic insecticides has given rise to the serious problem of insecticide resistance all over the world. The problem of insecticide resistance is growing in magnitude is no doubt steadily diminishing the choice of effective insecticides for vector control. The frequent change in insecticides involves a substantial increase in cost. The practice with neonicotinoid develops harmful possible impacts on nontarget species and the environment worldwide. This review provides a beneficial means for categorizing regions that may need improved development of best management practices (BMPs) to mitigate the adverse consequences associated with extensive use of insecticides in surface and groundwater. Pesticides must be used judiciously in an IPM program to preserve cost-effective pesticides and maintain susceptible individuals in a pest population. The recent finding that nAChR subunit composition can be switched in insects exposed to sublethal concentrations of neonicotinoids is of considerable interest. To manage pest species effectively while minimizing conditions that lead to the onset of resistance, we need to know how messenger ribonucleic acids (mRNAs) encoding, nAChR subunits, and their associated proteins, as well as enzymes involved in metabolism, are dynamically modified. The challenge of optimizing and implementing such tactics for specific pests depends on a suite of ecological, genetic, operational, and socioeconomic.

Author details

Shrawan Kumar Sahani[1]*, Vikas Kumar[1] and Subhajit Pal[2]

1 Department of Entomology, Bihar Agricultural University, Bhagalpur, Bihar, India

2 Department of Agricultural Entomology, Visva-Bharati University, Birbhum, West Bengal, India

*Address all correspondence to: sahanishrawan@gmail.com

References

[1] Buszewski B, Bukowska M, Ligor M, Staneczko-Baranowska I. A holistic study of neonicotinoids neuroactive insecticides—Properties, applications, occurrence, and analysis. Environmental Science and Pollution Research. 2019;**26**:34723-34740

[2] Gorman K. Report of resistance to the neonictonoid insecticide imidacloprid in *Trialeurodes vaporariorum*. Pest Management Science. 2007;**63**:555

[3] Insecticide Resistance Action Committee (IRAC). Prevention and Management of Insecticide Resistance in Vectors of Public Health Importance. 2nd ed; 2011

[4] Pisa LW, Amaral-Rogers V, Belzunces LP, Bonmatin JM, Downs CA, Goulson D, et al. Effects of neonicotinoids and fipronil on non-target invertebrates. Environmental Science and Pollution Research. 2015;**22**:68-102

[5] Korrat RAA, Ahmed SA, Badr NF. The potential side effects of certain insecticide formulations on the green lacewing, *Chrysoperla carnea* (Stephens). Journal of Plant Protection and Pathology. 2019;**10**:605-612

[6] Tomizawa M, Casida JE. Neonicotinoid insecticide toxicology: Mechanisms of selective action. Annual Review of Pharmacology and Toxicology. 2005;**45**:247-268

[7] Dang K, Doggett SL, Singham GV, Lee CY. Insecticide resistance and resistance mechanisms in bed bugs, *Cimex* spp. (Hemiptera: Cimicidae). Parasites & Vectors. 2017;**10**:1-31

[8] Horowitz AR, Kontsedalov S, Denholm I, Ishaaya I. Dynamics of insecticide resistance in *Bemisia tabaci*: A case study with the insect growth regulator pyriproxyfen. Pest Management Science. 2002;**58**:1096-1100

[9] Nauen R, Denholm I. Resistance of insect pests to neonicotinoid insecticides: Current status and future prospects. Archives of Insect Biochemistry and Physiology. 2005;**58**:200-215

[10] Bose A, Murugan M, Shanthi M, Ramalingam J, Mini ML. Prevalence of neonicotinoid resistance in cotton aphid *Aphis gossypii*. Indian Journal of Entomology. 2020;**82**:110-114

[11] Bonmatin JM, Giorio C, Girolami V, Goulson D, Kreutzweiser DP, Krupke C, et al. Literature review: Global neonicotinoid insecticide occurrence in aquatic environments. Water. 2020;**12**:33-88

[12] IRAC. Mode of Action Classification Scheme. Version 9.4. Issued: March 2020

[13] Bass C, Denholm I, Williamson MS, Nauen R. The global status of insect resistance to neonicotinoid insecticides. Pesticide Biochemistry and Physiology. 2015;**121**:78-87

[14] Jeschke P, Nauen R. Chapter 3: Neonicotinoid insecticides. In: Gilbert LI, Gill S, editors. Insect Control: Biological and Synthetic Agents. Vol. 62. London, England: Academic Press; 2010. ISBN 9780123814500

[15] Matsuda K, Ihara, Sattelle DB. Neonicotinoid insecticides: Molecular targets, resistance, and toxicity. Annual Review of Pharmacology and Toxicology. 2020;**60**:241-255

[16] Marzaro M, Mitchell EAD. Environmental fate and exposure; neonicotinoids and fipronil. Environmental Science and Pollution Research. 2015;**22**:35-67

[17] Satar G, Ulusoy MR, Nauen R, Dong K. Neonicotinoid insecticide resistance among populations of *Bemisia tabaci* in the Mediterranean region of

Turkey. Bulletin of Insectology. 2018;**71**:171-177

[18] Jeschke P, Nauen R. Neonicotinoids from zero to hero in insecticide chemistry. Pest Management Science. 2008;**64**:1084-1098

[19] Rauch N, Nauen R. Biochemical markers linked to neonicotinoid cross-resistance in *Bemisia tabaci*. Archives of Insect Biochemistry and Physiology. 2003;**54**:165

[20] Elbert A, Bailo-Schleiermacher I, Bruggen KU, Nauen R, Rogers D, Steffens R, et al. Bayer crop science guidelines on resistance management for neonicotinoids. Pflanzenschutz-Nachrichten Bayer. 2005;**58**:4-32

[21] Zhang C, Tian D, Yi X, Zhang T, Ruan J, Wu R, et al. Occurrence, distribution and seasonal variation of five neonicotinoid insecticides in surface water and sediment of the Pearl Rivers, South China. Chemosphere. 2019:437-446

[22] Simon-Delso N, Amaral-Rogers V, Belzunces LP, Bonmatin JM, Chagnon M, Downs C, et al. Systemic insecticides (neonicotinoids and fipronil): Trends, uses, mode of action and metabolites. Environmental Science and Pollution Research. 2015;**22**:5-34

[23] Radolinski J, Wu J, Xia K, Hession WC, Stewart RD. Plants mediate precipitation-driven transport of a neonicotinoid pesticide. Chemosphere. 2019;**222**:445-452

[24] Zota AR, Calafat AM, Woodruff TJ. Temporal trends in phthalate exposures: Findings from the National Health and Nutrition Examination Survey, 2001-2010. Environmental Health Perspectives. 2014;**122**:235-241

[25] Zeng T, Arnold WA. Pesticide photolysis in prairie potholes: Probing photosensitized processes. Environmental Science & Technology. 2013;**47**:6735-6745

[26] Zhu F, Lavine L, O'Neal S, Lavine M, Foss C, Walsh D. Insecticide resistance and management strategies in urban ecosystems. Insects. 2016;7:2

Section 8

Nano-Biopesticides

Chapter 21

Nano-Biopesticides as an Emerging Technology for Pest Management

Abu Hazafa, Muhammad Murad, Muhammad Umer Masood, Shahid Bilal, Muhammad Nasir Khan, Qasim Farooq, Muhammad Omer Iqbal, Muhammad Shakir and Muhammad Naeem

Abstract

With an increasing world population, the demand for quality food is rising. To meet safe food demand, it is necessary to double or maybe triple agriculture production. Annually, almost 25% of the world crop is destroyed due to pests. During the past few decades, different pesticides, including chemical, synthetic, biological, and botanical have been adopted to achieve adequate results against pests for agriculture interests and plant safety. Globally, more than 200,000 people died every year due to direct chemical and synthetic pesticides poisoning. But these pesticides did not achieve the desired results due to delivery problems, less stability, low biodegradability, less specificity, and high cost. To overcome these problems, the rapidly emerging field of nanotechnology is considered an important achievement in the agriculture sectors in order to improve pest mortality rates and crop production. The nano-biopesticides attained special attention against the insect pests due to their small size (1-100 nm), large surface area, high stability, cost-effectiveness, fever toxicity, and easy field application. The current chapter highlights the relevance of nano-biopesticides for pest insect management on several crops of agricultural concern. The mechanisms of action, delivery, and environmental sustainability of nano-biopesticides are also discussed in the present chapter.

Keywords: Pest management, nanotechnology, Nano-biopesticides, environmental sustainability, crop yield, agriculture

1. Introduction

The entire world population is around 7.7 billion, which is growing steadily. One of the main predicaments is the lack of quality food for human beings due to environmental biotic and abiotic problems such as weeds, pests, and diseases [1]. Over 65,000 kinds of pests are recorded, including weeds, arthropods, and fungi or are also regarded primarily as plant pathogens [2]. The recent evidence recommended that pests prompted an 8-10% loss in wheat crops, 20% in sugar, 25% in rice, 30% in pulses, 35% in oilseeds, and 50% in cotton. The estimated annual crop loss caused by pests and diseases is USD 2000 billion. Therefore, different pesticidal technologies should be extended in these circumstances, particularly in developing countries, to subdue these food predicaments [1]. For the last several years, pest management in industrialized counties has depended on the application of pesticides. Hence, the application of pesticides was raised above 1900% within the 1940s-1980s. According to a calculation, today, 2.3 billion kg of pesticides have been applied annually, making up to $ 58.5 billion of the global exchange [2, 3].

Every year, almost 25% of the world's crop production is destroyed by pests [4]. Many types of pests including *Acalitus vaccinia* (Blueberry bud mite), *Acrobasis vaccinia* (Cranberry fruit worm), *Acrosternum hilare* (Green stink bug), *Agrotis ipsilon* (Black cutworm), *Altica Sylvia* (Blueberry flea beetle), *Aphis gossypii* (Cotton aphid), and *Bemisia tabaci* (Sweet potato whitefly) are detrimental to crop production due to their huge nutritional needs [5]. Thus, the challenge is to enhance the resistance of crops against pests without disturbing the crop yields. According to recent advances, the use of synthetic pesticides has increased to kill pests for better crop production [5, 6]. Pesticides are substances or a mixture of substances that are used to kill, resist, and repel pests. The total consumption of pesticides in developed countries is about 3000 g/ha [7]. Synthetic pesticides have received much attention due to their broad spectrum of insect control and ability to kill pests in the agroecosystem. Plants and their secondary metabolites, including alkaloids, organic acids, and glycoalkaloids are considered promising sources of plant effecting pests (known as biopesticides) [8].

The pesticides are divided into chemical, biological, synthetic, microbial, biopesticides, biochemical, and plant-incorporated pesticides. Chemical pesticides are delivered to plants either directly for seed treatment and weed control or indirectly through spraying the chemical on plants. Some chemical pesticides show good pesticidal activity, but they exert negative impacts both on human health and the environment; for example, methyl bromide has been reported as a good pesticide over the last 40 years against soil-borne pathogens, pests, and nematodes in many crops like tomato, melon, pepper, and strawberry. But later on, due to its ozone depletion negativity, it was banned in 2015 following the Montreal Protocol. Moreover, some other chemicals like chloropicrin and dazomet are restricted in some areas due to their concern about food safety and human health [9].

Biopesticides, often known as biological pesticides, are insecticides derived from microorganisms or natural substances. Biopesticides are divided into three categories: microbial biopesticides, botanical biopesticides, and plant-incorporated protectants [10]. As an alternative to conventional insecticidal methods, biopesticides have recently gained much attention due to their potential target specificity, fewer harmful side effects, capacity to disintegrate fast, and high efficacy. Several substances have been investigated as biopesticides in recent years, including *Clitoria ternatea* extract, oxymatrine (an alkaloid component), stilbenes in grape cane, *Talaromyces flavus* strains (SAY-Y-94-01), and olive mill oil [11, 12]. The usage of biopesticides, which represent less of a hazard to the environment and human health than synthetic pesticides, should be done with caution. Certain products have been licensed for usage as biopesticides, although they still pose health concerns [13, 14]. However, in this regard, nanotechnology has gained special attention and has become a novel field during the last couple of years due to its multidisciplinary applications in pharmacology, agriculture, pest management, and parasitology fields [15]. Nano-particles are the most rapidly expanding area of nanotechnology, which provides the solution to many environmental problems due to their eco-friendly behavior and cost-effectiveness, small size (1-100 nm), and large surface area. According to emerging evidence, nanotechnology has been proven to be an effective tool for the formulation of new nanocomposites against pests and improving crop varieties [16].

The nano-biopesticides have superiority over the biopesticides and conventional techniques for many reasons, including environmentally friendly behavior, desired results within a few hours after applications, biodegradability, easy delivery to plants, and release slowly from the vector [15]. Furthermore, their small size makes them an effective carrier when combined with pesticides that can easily enter the plants. Another advantage of nano-biopesticides is that they did not have an adverse effect on soil microorganisms and phototoxicity of Ag-based nano-particles was suppressed by nano-coating them with biocompatible polyvinyl pyrrole compounds [17]. The nano-biopesticides can be synthesized by following two ways: either by extracting the biological active pesticidal compound (APC) from plants and blended it with nano-particles and inserted it into a suitable polymer that acts as a supporting material, or APC secrete the metallic salt with bind with nano-particles (NPs) that hemolyze and merge into an appropriate polymer. The APC integrated with NPs and merged into a compatible vector including micelles, liposomes, nanosphere, polymer, and nanofiber. These ingredients were used as a spray to kill the insect pests for food protection [18].

The accumulative data revealed that nano-biopesticides contain secondary plant metabolites and their mediated metal oxide nanomaterials. It was found that biopesticides have gained importance over chemical pesticides during the past few decades due to their eco-friendly behavior, high efficiency, and fewer side effects. The evidence reported that recently much research had been carried out on nano-biopesticides; either pests are attaining chemical pesticide resistance, or a small number of insecticides have expired due to severe environmental and human concerns. However, this situation demands novel plant-based pesticides on the nanoscale to formulate the nano-biopesticides for pest management. Recently, different biopesticides have been reported against different pests such as *Bacillus firmus* and *Bacillus sphaericus,* which are used against diamondback moths, while *Trichoderma harzianum* and *Trichoderma viride* are used against root rots and wilts, *Beauveria bassiana* against mango hoppers, neem-based nano-biopesticides against whitefly, and *Bacillus thuringiensis* against *israelensiss* but they have not shown an effective impression on pests. However, future studies are needed to overcome and improve nano-particle release rates and delivery [19–21]. Therefore, the present chapter demonstrated the importance of nano-biopesticides for pest insect management on different crops for agricultural interest to overtake these situations. Moreover, the present chapter explains the mechanism of action, delivery, and environmental sustainability of nano-biopesticides.

2. Pesticides

Agricultural output has been increased dramatically in the early 20th century, especially in the United States, to keep up with the rapidly growing human population. During the last century, the world's population has tripled from 1.5 billion in 1900 to 6.1 billion in 2000. The world's population has grown by one billion people in the past decade, and the UN predicts it could reach 9.4 to 10 billion by 2050 if current growth rates continue [22]. A lateral increase in food production was needed during the 20th century to keep up with the increase in the world population. This was accomplished via the use of fertilizers and other agricultural inputs throughout the twentieth century. Bio-fertilizers (such as guano) were first used in the late 1800s; inorganic phosphate fertilizers (such as urea) were first used in the early 1900s and have steadily gained in favor ever since [23]. Phosphates helps to increase the crop diversity and yields and aided in the unprecedented "green revolution" for agricultural productivity. This caused a tenfold increase in grain production per unit surface area of agricultural land, leading to a global food surplus [24]. An increasing global population and increased phosphate production were shown to be positively linked throughout the 20th century, with an R2 of 0.97 for the period 1900–1988 [25].

Synthetic crop preservation agents were introduced to the market in the 1940s, which increased food output. It grew from 0.2 million tons in 1950 to over 5 million tons in 2000, up from 0.2 million tons the year before. Between 1950 and 2000, pesticide production grew by about 11%, from 0.2 million tons to over 5 million tons. Crop preservation chemicals, sometimes known as pesticides, are composed of various composites, including growth regulators, neonicotinoids, organochlorines, pyrethroids, organophosphates, carbamates, and more recently, biopesticides. This wide spectrum of chemicals and insecticides has been developed throughout history to protect crops from pests and diseases. All sorts of pesticide sales grew; however, herbicides were the group that extended the most, accompanied by fungicides and insecticides. Pesticide application has suffered owing to lack of global uniformity, high cost of chemicals, human resources, and the vast diversity of pests present in each climatic or geographical area. Using FAO data, it was found that the mean pesticide application rates per hectare of arable land ranged from 6.5 to 60 kg/ha, with the greatest mean values occurring in Asia and a few South American nations. Unlike Western Europe and North America, Asia has not witnessed a rise in the usage of herbicides in both urban and agricultural areas. Compared to the widespread use of insecticides, herbicide usage in Asia has remained relatively low in recent years, according to World Bank and International Food Policy Research Institute data [26].

Throughout the twentieth century, ancient synthetic pesticides intended for agricultural pest control, such as DDT, were often used to treat human parasites and animal ticks. So, for example, DDT was designed to be used in agricultural pest control. Despite of being prohibited, it is nevertheless extensively used as a food prophylactic for various fish in South Asia, and to control home pests and malaria vectors globally, albeit seldom illegally [27]. Pesticides have been administered in agricultural settings for decades, employing techniques ranging from truck and aircraft spraying to old-fashioned field worker spraying. Studies on the effects and toxicity of manufactured chemicals on human health and well-being have shown that individuals report euphoria after pesticide application. This research included peasants, farmworkers, and their families following pesticide treatments. They previously discovered that unintentional poisoning affects about 355,000 individuals annually and is linked to high susceptibility and poor chemical management. They also discovered that increased sensitivity and poor management of hazardous substances are closely linked to such toxicities [28]. The research was conducted to assess the number of pesticides in the environment that killed various animals. Among the animals that resided there were fish, birds, bees, amphibians, and tiny mammals. It was also noted how much they were killed and how they were slaughtered [29].

Following the introduction of synthetic chemicals into the environment, it was only a short amount of time before it was thought that crop protection pesticides were causing disease both locally and internationally [26]. Many believe that sprayed on-crop DDT is deported into water bodies, quickly converted into DDE, and bio-accumulated in aquatic food systems before being reintroduced into the environment and ultimately reaching people. To manufacture endosulfan at this time, a rigorous and scientific decision-making procedure is undertaken. Additionally, this strategy includes scientific research to enhance food production, food safety, and environmental security in addition to the other objectives listed above [26].

Common use of synthetic pesticides inhibits the development of plant pathogen strains resistant to these chemicals, causing the reemergence of illnesses in the environment. Pesticides are being used more often by farmers, which is good [30]. Synthetic pesticides include active ingredients that are absorbed and retained by plants after application. People suffer chronic health issues due to the high concentration of harmful chemical deposits in these crops cultivated for human use [31]. Synthetic pesticides include active ingredients that are absorbed and retained by plants after application. People are suffering from chronic health issues due to the high concentration of harmful chemical deposits in these crops cultivated for human use [32].

3. Biopesticides

The need for biopesticides has been increased significantly in recent years, particularly in developing countries, due to restrictions placed on the use of some synthetic pesticides, such as organophthaloids, organochlorines, carbamates, and organophosphates, among other things. Synthetic pesticides are not only harmful to pests and diseases at the time of application, but they also have the potential to contaminate plant crops, posing a threat to human health, animal welfare, and environmental health. Synthetic pesticides are used to control pests and diseases in agriculture. In agriculture, synthetic pesticides are used to manage pests and illnesses that are introduced via the soil. As reported by the Environmental Protection Agency, synthetic pesticides are also harmful to both people and animals. They are also bad for the health of the ecosystem. When it comes to biochemistry, chemical pesticides are characterized by alterations in the signaling system, inhibition of enzymes, pH shifts, disruption of electrolytic balance, osmotic and membrane breakdown, pH gradients across membranes, and other characteristics. They also generate free radicals and other toxic compounds, which have the potential to damage proteins and DNA, as well as cause tissue degeneration, among other undesirable effects [14]. A wide range of diseases has been linked to the use of synthetic pesticides, including Parkinson's disease and neurotoxicity, type 2 diabetes, endocrine disruption, many cancers, and obesity, among others. Parkinson's disease is the most well-known of these disorders. It has been shown that the use of synthetic pesticides is linked with the development of these diseases, which may be due in part to the mechanisms of action of these chemicals, as well as the increasing exposure of individuals to these chemicals over time [33–35]. Despite the fact that it is regrettable, the majority of pesticides now in use are being phased out at a rapid rate, which is a good trend in the industry. On the other hand, pesticides that are still in use continue to accumulate in the human body with every meal that is eaten. In addition, employees who have been exposed to pesticides have been observed to get drunk as a result of the pesticides they have been exposed to over the course of their shift [36]. Natural pesticides offer many benefits over synthetic pesticides, the most significant of which is that they are less harmful to the environment and human health. However, this does not mean that they should be utilized recklessly or without consideration for the repercussions of their actions. Even if certain products have been authorized for use as biopesticides, it is conceivable that they may cause health issues among members of the general population. Large quantities of copper, which is an essential nutrient in the diets of both mammals and plants, have the potential to be poisonous to both humans and animals and hazardous to aquatic life if eaten over an extended period of time. There is also concern about toxic plant species, microalgae, and algae such as *Microcystis aeruginosa*, *Chrysanthemum* spp., *Gracilaria coronopifolia*, and others that are harmful because their appearance may be similar to that of hazardous compounds such as cyanide, among other things, and they are difficult to identify [13, 14].

As the name implies, biopesticides are pesticides that include active ingredients formed by microorganisms or natural materials rather than synthetic chemicals. They are used to control insects in a variety of circumstances and are referred to as "biopesticides." Pesticides derived from plants are divided into three categories: (a) microbial biopesticides, which are microorganisms that are effective against diseases and insects; (b) botanical biopesticides; and (c) plant-incorporated protectants. Microbial biopesticides are microorganisms that are effective against diseases and insects. Microbial biopesticides are microorganisms that have been shown to be efficient against many illnesses and insects in the field. A microbial biopesticide is a bacterium that is effective against a wide range of diseases and insect species, including fungi [10].

3.1 Microbial biopesticides

The presence of fungus is associated with insect damage. *Metarhizium anisopliae* and *Beauveria bassiana* are two forms of entomopathogenic fungi that may be found in the environment and are both harmful to insects. According to the World Health Organization, as soon as *B. bassiana* spores come into contact with the body of an insect host, they begin to develop, penetrate the cuticle, and multiply within the insect host, eventually killing the insect host and spreading to other insects in the surrounding area. *B. bassiana* is a fungus that can cause death in insects. As the body ages, it produces a white mold that spreads new spores into the surrounding environment, leading the environment to become more contaminated as a result of the pollution. A host insect becomes infected when the spores of the fungus *M. anisopliae* come into touch with the insect's body, causing the spores to sprout and the hyphae that emerge to pierce the insect's cuticle, causing the insect to succumb to the infection. It then begins to spread throughout the insect's body, resulting in the insect being infected within a few days after first becoming exposed to the fungus. In order to prevent the development of soilborne diseases, microorganisms such as *Pseudomonas* and *Trichoderma* have been extensively employed as biopesticides for many years, and they are still being utilized in this capacity today [37]. According to the University of California, Berkeley, a filamentous fungus such as *Trichoderma* may be found growing on organic materials such as rotting wood, soil and other organic materials. Many Trichoderma species, including *T. virens, T. viride*, and *T. harzianum*, have been found to have strong biocontrol capability. Many other competing methods for resources have been discovered to have biocontrol potential, including mycoparasitism, which is caused by the release of cell wall-degrading enzymes such as proteases, chitinases, and glucanases, among other things. The antibiotic compounds heptelidic acid and harzianic acid, as well as alamethicins, glisoprenins, tricholin and antibiotics, 6-pentylpyrone, peptaibols, viridin, and massoilactone, can all cause antibiosis. Heptelidic acid and harzianic acid are two of the most commonly encountered antibiotic compounds. Infections caused by bacteria are treated using the antibiotic heptelidic acid, which is produced by bacteria and used to treat infections caused by other bacteria [38].

Trichoderma is effective against a wide range of pathogenic fungi, including *Candida albicans*, *Phytophagthora*, *Fusarium*, *Sclerotia*, and other pathogenic fungi, in addition to *Candida albicans*. Cotton crops are affected by *Fusarium* wilt disease, caused by the fungus *Fusarium* sp., while other crops, such as maize, are affected by *Rhizoctonia* sp. and *Pythium* sp. As a consequence of using this technique, the development of cucumber resistance to the anthracnose disease caused by the fungus *Colletotrichum* sp. was aided. *Pseudomonas aeruginosa* weakened infections are characterized by the release of various derivatives and antibiotics such as pyoluteorin (Plt), 2,4-diacetylaminoglucinol (DAPG), phenazine-1-pyrrolnitrin (Prn), carboxylic acid (PCA), or the development of systemic resistance (ISR) [39]. When it comes to developing microbial biopesticide formulations, microorganisms such as algae, bacteria, and fungus must be incorporated if the usage of these pesticides is to become more generally accepted. According to the International Biopesticide Trade Association, the biopesticide industry is experiencing an outbreak of bacteria, particularly among *Bacillus thuringiensis* species, which are commonly used to control insect infestations in plantations and are now being transported across multiple countries [40]. Whenever parasites eat this bacterium, it creates a toxic endotoxin that attaches itself to the stomach of the insect and causes holes to develop, resulting in anion imbalances in the insect's body, insensitivity of the insect's digestive system, and eventually, the insect's death. According to industry standards, these pesticides are usually regarded as less toxic to birds, mammals, and non-target insects than conventional insecticides, and as a result, they are believed to be less damaging to the environment. Microalgae as biopesticides, despite this, have been proven to be helpful in the prevention and control of the spread of a wide variety of plant-borne diseases. Many studies have demonstrated this bacterium's ability to produce a diverse range of bio-compounds, including terpenes and growth regulators, as well as phenolic chemicals and other molecules. These studies have all demonstrated this bacterium's ability to produce these compounds, as well as its potential to produce other molecules. Terpenes, growth regulators, phenolic compounds, and other molecules are among the substances studied [41, 42].

3.2 Biochemical biopesticides

Non-toxic biochemical pesticides are natural insecticides produced by animals, plants, and insects. They do not damage the creatures that produce them. They are employed to manage pests without killing them. These chemicals may assist in growth and development by attracting or repelling pests (pheromones) and acting as plant growth regulators (PGR). It's difficult to tell whether a biopesticide is hazardous since so few countries have committees to test metabolites.

As a consequence, evaluating a biopesticide's safety is difficult [43]. Since their discovery, Auxin-type PGRs have been hailed as one of the most effective herbicides and biological control agents on the market. And for a good reason. It is generally recognized as one of the most efficient herbicides and biological control agents on the market today. Consider the difference in action selectivity between marijuana and PGR. Marijuana has a more selective effect, perhaps due to its fast detoxification process. Low concentrations of these chemicals promote cell elongation, biofertilizer activity, cell division, and cell growth. Dense doses cause weeds to get intoxicated and exhibit developmental abnormalities such as impaired respiration, carbon absorption, and transpiration. In the end, these anomalies harm weeds' circulatory systems and membranes, leading to their demise [14].

3.3 Botanical biopesticides

When applied to crops, pesticides (chemical compounds and plant extracts) are used to prevent the growth of pests (including insects) of various types. Pesticides are used to limit, halt, or otherwise manage pests of many kinds, including insects. Some ways in which plant security may be achieved include the utilization of a variety of secondary metabolites produced from plant sources such as essential oils, phenolics, and terpenes, among other things [44]. The non-persistency of essential oils in the environment, along with the fact that they are non-toxic to animals, has led to their being widely regarded as one of the most efficient agricultural pesticides presently available. As acaricides and insecticides, these compounds have the potential to be utilized in the environment, where they may also be used to inhibit the growth of fungus and bacteria. When essential oils are applied to plant cultures, the anti-oxidant properties of the oils protect the plants from pro-oxidants found in proteins and DNA, which cause cytotoxicity, the formation of reactive oxygen species, as well as the breakdown of cell membranes and organelles in the microorganisms that infect the plants [45]. However, the effectiveness of a biological pesticide can be affected by several factors, including the mist of the substance harvested, the method of extraction used to obtain this type of biopesticide, and the age of the plant from which the oil will be collected. The toxicity of a biological pesticide can also be affected by several factors, including the phenological age of the plant from which the oil will be collected. Although agricultural pesticides have many advantages, their use has been restricted for a variety of reasons, including their inability to maintain stability over time, the complexity of the extracted combination, extraction techniques, or formulation of the active component, as well as difficulties encountered during the purification process [46].

There are a number of plants that have been recognized as intrinsic sources of agricultural pesticides, as described in **Table 1**. The pests that are targeted by the insecticides contained in those plants are also included in the table. The ethanolic plant extracts of ginger (*Zingiber officinale*), turmeric (*Curcuma longa*), pepper (*Capsicum frutescens*), lemon (*Citrus limon*), and garlic (*Allium sativum*) have been shown to significantly inhibit the growth of *Fusarium oxysporum* sp., *Alternaria*

Plant	Host	Target pest	Reference
Allium sativum	Human and animal sp., *Oryza* sp., *Gossypium hirsutum*, Stored grain products; *Vigna unguiculate, Brassica oleracea*	*Colletotrichum* sp., *Bacillus subtilis*, *Salmonella senflenberg*, *Staphylococcus aureus*, *Staphylococcus epidermidis*, *Sitotroga cerealella*, *Spodoptera littorals*, *Tenebrio molitor*, *Callosobruchus maculatus*, *Plutella xylostella*, *Brevicoryne brassicae*	[47–49]
Azadirachta indica	*Vigna unguiculate*, *G. hirsutum*, *Solanum tuberosum*, *Triticum* sp., *Brassica* sp., *Prunus salicina*, *Lycopersicon esculentum*, *Capsicum chinense*, *Cardamomum sp.*	*Aphis craccivora*, *Aphis gossypii*, *Amrasca devastans*, *Myzus persicae*, *Sitobion avenae*, *Lipaphis erysimi*, *Bemisia tabaci*, *Sciothrips cardamomi*, *Rhizopus sp.*, *Aspergillus* sp., *Monilinia fructicola*, *Trichothecium roseum*, *Pythium aphanidermatum*, *Alternaria alternata*, *Helminthosporium* sp., *Vibrio cholerae*, *B. subtilis*, *Meloidogyne javanica*, *Meloidogyne ingognita*	[50–52]
Tagetes spp.	*Gladiolus grandifloras*, *Leucadendron*, Human and animals sp., *B. oleracea*	*Fusarium oxysporum*, *Klebsiella pneumoniae*, *B. brassicae*, *Plutella xylostella*, *Mamestra brassicae*, *Meloidogyne incognita*	[53, 54]

Plant	Host	Target pest	Reference
Thymus vulgaris	*Gallus gallus domesticus*, *Triticum aestivum*, *Solanum lycopersicum*, *Citrus aurantium*, *Cajanus cajan*, *G. hirsutum*	*Xanthomonas vesicatoria*, *Escherichia coli*, *Salmonella typhimurium*, *Diaphorina citri*, *Megalurothrips sjostedti*, *Eloidogyne incognita*, *Helicotylenchus dihystera*, *Pratylenchus brachyurus*	[55–57]
Cinnamomum zeylanicum	Zea mays, *Pinus densiflora*	*Botrytis cinereal*, *Penicillium expansum*, *Aspergillus oryzae*, *Fusarium solani*, *E. coli*, *S. aureus*, *B. subtilis*, *S. typhimurium*, *Bursaphelenchus xylophilus*, *Meloidogyne* sp.	[58, 59]
Curcuma longa	*T. aestivum*, *Prunus persica*, *B. oleracea*, *S. lycopersicum*, Human sp., Animal sp.	*Tribolium castaneum*, *Bactrocera zonata*, *Trichoplusia ni*, *Alternaria solani*, *Streptococcus pyogenes*, *Streptococcus mutants*, *Ralstonia solanacearum*, *E. coli*, *Listeria monocytogenes*, *B. subtilis*	[60, 61]
Zingiber officinale	*S. lycopersicum*, *Arachis hypogaea*, *Coffea* sp., *Mangifera indica*, *Oyza sativa*, *B. oleracea*, *Clarias gariepinus*	*Aspergillus parasiticus*, *E. coli*, *Salmonella typhi*, *T. castaneum*, *Drosicha mangiferae*, *Trichoplusia binotalis*, *Necrobial rufipes*, *Dermestes maculatu*	[62, 63]

Table 1.
The potential plant compounds as botanical pesticides and respective target pests.

solani, *Rhizoctonia solani*, *Pythium ultimum*, and *Lycopersicium* sp. [64]. Regarding growth inhibition, turmeric (*Curcuma longa)* has been shown to be the most effective herb, with results against *Alternaria solani* reaching up to 73 percent effectiveness. *In vitro* studies have demonstrated that the herbs *Rosmarinus officinalis, Rhus coriaria*, and *Eucalyptus globulus*, among others, effectively inhibit the development of the pathogen *Pseudomonas syringae* tomato [65]. A study conducted in a greenhouse showed that the *Eucalyptus globulus* tree was very efficient in reducing the bacterial specks of tomato (*Pseudominas syringae* p.v. tomato) to a degree of as much as 65% when cultivated in a greenhouse. In one research, when juvenile root-knot nematodes (*Meloidogyne* sp.) were exposed to extracts of the *Nerium oleander* at a 5 percent concentration and the extracts were applied topically, the mortality of the worms increased. When treated with extracts of *A. sativum*, *Eucalyptus* sp., *Azadiractha indica*, *Cinnamomum versicolor*, *Zingiber officinale*, and *Nerium oleander* at a concentration of 10 percent, the mortality rate of insects on second-stage juveniles ranges between 65 and 100 percent, and when treated with a concentration of 20 percent, the mortality rate ranges between 65 and 100 percent [66].

4. Limitations of biopesticides

Because of a number of factors, biopesticides are not widely utilized as a pest and disease management alternative, despite the fact that they offer many benefits, including the preservation of the environment and the safety of food for human consumption. For the component compounds to be effective in field settings, high dosages of the compounds are required [67]. The emerging evidence revealed that the biopesticides isolated from plants have to face more challenges regarding activity because they are extracted from plants that also contain several other bioactive compounds that could change their chemical properties. Moreover, the utilization of organic compounds as a solvent for the extraction of pesticides is involved in environmental pollution through their disposal. It was also found that biopesticides have a short shelf life that is associated with a high biodegradability rate. In addition to botanical pesticides, microbial pesticides could prove to be better pesticides for a limited type of pest in the field, but they only showed activity against one type of pest, that is one of the biggest disadvantages of microbial pesticides. Furthermore, other environmental factors such as desiccation, heat, light, and UV reduce the activity of microbial pesticides, resulting in continuous crop destruction [68].

The number of bioactive compounds present in plants and the kind of habitat in which they develop is influenced by the environment in which they are grown. Furthermore, the diversity of plants and their differences have an impact on the amount and kind of active chemicals contained in them, resulting in differences in how they respond to illnesses [69]. The quality of plant extracts, on the other hand, varies depending on the extraction method employed. It may be difficult to get the appropriate active and inert components ratios during the formulation process in certain instances. Aside from that, there are no established processes for preparation or assessment of efficacy, especially in field situations when time is of the essence [70]. However, although *in vitro* studies provide positive results, field outcomes are often inconsistent, in part due to the short shelf life of source materials and, in certain instances, the low quality of source materials and preparation methods. In order to use predatory biopesticides effectively, it is essential to do a thorough assessment of the host crops and their dispersal capacities. A manual application may be prohibitively expensive on small acreages because of the time commitment required to guarantee adequate crop coverage and exposure length. In order for products to be registered, data on chemistry, toxicity, packaging, and formulation must be supplied, which is not always the case in the pharmaceutical business [71].

5. Nanotechnology

From 1959 to 1960, developments in nanotechnology and nanoscience have been made to explore the synthesis and role of nano-particles prior to using them for different biomedical applications. Norio Taniguchi, a professor at Tokyo University of Science, made several successful attempts to synthesize nanometer-sized semiconductors in 1974. Later, it laid the foundation for research to perform experimentation on different types of nano-particles and nanocomposites. Nano-particles are found naturally in plants such as algae in the form of superoxide nano-particles and insects in the form of nanostructures. Nano-particles can be synthesized through physical, chemical, and biological methods [72].

Nano-particles fabricated via physical, chemical, and biological methods are classified by their chemical composition, Nanoparticles in the form of metals such as Cu, Fe, Zinc, Au and in the form of oxides such as ZnO, CuO, AlO, in the form of semiconductors such as ZnS, CdS, ZnSe, carbon-based nano-particles in the form of graphene, diamond, fullerenes, in the form of silicates such as nano clays, in the form of nano-particles based on dendrite with long chains of fibers [73]. Different nano-particles are divided into different dimensions on the basis of their application in different biomaterials. The one-dimensional object possesses thin layers and fine surfaces. Second-dimensional possesses the wires with excellent flexibility and long tubes. Third-dimensional materials can be synthesized from metal oxides through physical and biological methods. These dimensions of the nano-particles have different applications in the fields of agriculture, medical, pharmaceuticals, pest management, and different industrial sectors [72].

5.1 Nano-biopesticides

Nano-biopesticides are attractive due to their tiny size, high surface-area-to-volume ratio, stability, enhanced efficacy, better solubility, mobility, and decreased toxicity. Nano-biopesticides are also suggested because of their low toxicity (see **Figure 1**). Chemical pesticides are directly applied to plants can possess toxins released by air into the food chain and cause environmental issues. To control these issues, pesticides with formulations of nano-particles such as micelles and nano-composites reduce the chances of both environmental and health issues. Similarly, clay-based nanotubes deliver pesticides to control pests [75].

Like nano-fertilizers, nano-biopesticides are contained in carriers that enable for regulated release of active ingredients to accomplish desired effects in a given environment. Stiffness and penetrability are two properties enhanced by adding nano-biopesticides to biopolymers. Crystallinity, thermal stability, solubility, and

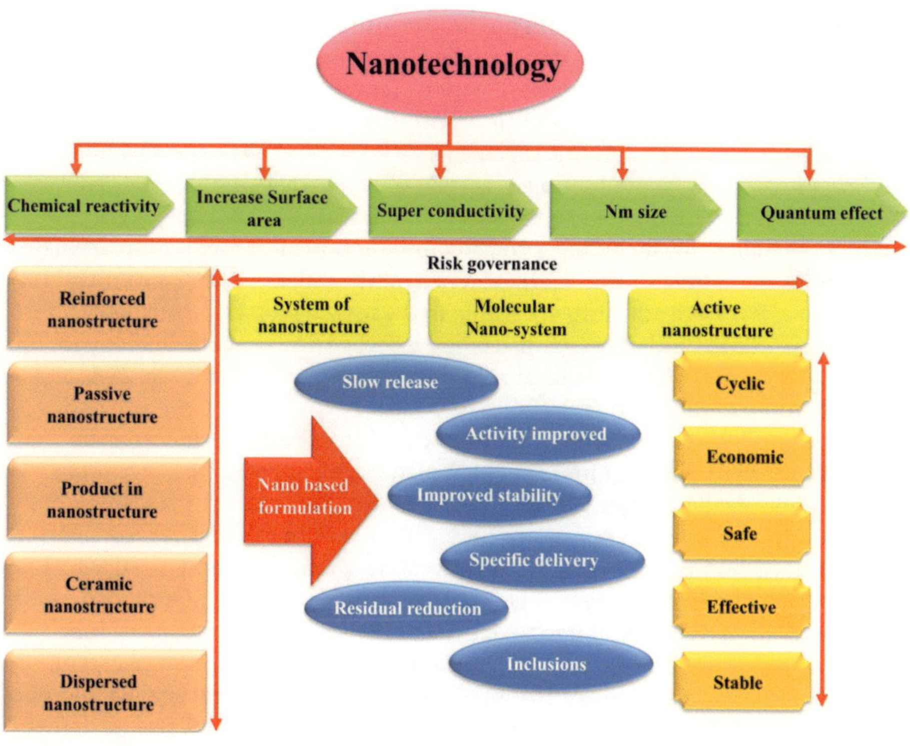

Figure 1.
The importance of nanotechnology for the formulation of nano-based biopesticides. This figure is reproduced from Lade et al. [74].

biodegradability are also enhanced [76, 77]. When nanomaterials were applied to the soils, nano-biopesticides containing nanomaterials resulted in the growth of mutualistic microorganisms that promote the pants' activities [17]. Sometimes, toxicity can be induced by coatings of silver-based nano-particles that could be reversed by biocompatible coatings, thus increasing the chances of seed germination in plants. Recently, nano-emulsions, nano-encapsulates, nanocontainers, and nano-cages have been reported as some nano-pesticide delivery techniques with different functionalities for plant protection [77].

Further research shows that cationic polymers may bind to polyanionic surfaces of bacteria, disrupt cell membranes, and kill pests. In agriculture, plants may be treated with biopesticides, such as nano-biopesticides, which can decrease microbial resistance, whereas chemicals applied directly to plants are unable to suppress a wide range of bacterial growth. Tertiary ammonium groups may be found in nano-particles as lengthy amino acid chains. Depending on their structure, these groups may attack various pests and illnesses, including bacteria. Because of their high activity in a wide variety of environmental and chemical conditions, polymers with quaternary ammonium groups in their chains are widely used [78]. Many polymers with this characteristic have been found and researched throughout time. For example, amphiphilic copolymers, functionalized cationic polycarbonates, poly(amidoamines), polyethylenimine, poly(methyl methacrylates), amino celluloses and chlorinated cellulose acetates are now available [78–80].

Essential oils (EOs) are highly volatile secondary metabolites found in many higher plants and flowers and certain fruits and vegetables. In addition to their traditional uses in medicine and cosmetics, a new study indicates they represent a major natural source of ecologically friendly pesticides. Essential oils are often used to treat gardening pants to keep insects and bees out of the garden. Invertebrates become neurotoxic when their nervous systems are suppressed of GABA and acetylcholine esterase (ACE) [81]. This 2007 research evaluated the anti-pest effects of plant extracts, essential oils, their purified components, and plant-based nano-formulations, as well as their modes of action. Temperature, light, and oxygen supply all have an impact on the EO's integrity. Researchers found that encapsulating flaxseed in gelatin and Arabic capsules may improve effectiveness by up to 84 percent, preventing the production of certain oxidants that stimulate the growth of some insects [82]. Sagiri *et al.* [83] summarized many techniques for encapsulating vegetable oils, including associated production processes, antimicrobial applications, insecticide/pesticide/pest repellant formulations, and antimicrobial applications. Purslane mustard, according to the manufacturer, is efficient against Sitophilus granaries and other grain-feeding insects. Using nanotechnology to control weevils and other pests improved their efficiency by about 7%, 21%, and 98%, according to prior research [84].

A variety of plants with nano-emulsions of ECs can be used to control the larval infections of different insects. These plants are *Tagetes minuta, Ageratum conyzoides* and *Achillea fragrantissima* used to control the growth of *Callosobruchus maculatus*. These nano-emulsions can be applied to kill or inhibit the growth of eggs and larvae in the form of fumes with treatment ranges from 16.1–40.5 μL/L air and 4.5–243 μL/L air, respectively [85]. The encapsulation of EOs can be performed through the use of nano-particles composed of liposomes and solid lipids. Encapsulation of EOs can be carried out through inverse gelation and oil emulsion [86]. Encapsulation of EOs through liposomes is helpful against microbes by protecting the cell membrane from the effects of EOs [87]. However, different types of nano-particles are used for pest management. Microparticles are also used to stabilize the effects of EOs. The encapsulation of carvacrol was performed with a diameter of 0.5 μm with the cell wall of *Saccharomyces cerevisiae* to control the larvae of *Rhipicephalus microplus* and LC50 formulations of about 0.71 mg/mL. The cell wall of *Saccharomyces cerevisiae* appears more helpful in maintaining the low volatility of encapsulated carvacrol that maintains the acaricidal activity up to 60 hours [88].

5.1.1 In vitro *nano-biopesticides bioassay*

Nano-biopesticides can be tested against a specific pest in order to check their efficiency before applying them in different crops. Nano-biopesticides can be synthesized through the active pesticidal compounds and combinations of different nanomaterials such as zinc oxides, silver oxides, and aluminum oxides [89]. The toxicity of nano-biopesticides can be measured through the minimum inhibitory concentration that employs the agar well diffusion method. Filter paper is usually coated with the outer surface of nano-biopesticides, and oral feeding directly applies to the target pest. The concentration of dead and alive pests can be precisely measured after 40 days of feeding [90].

5.1.1.1 *Pupicidal activity*

The pupicidal activity of nano-biopesticides is helpful in preventing the attack of pupae of different insect groups. It can be measured after applying the nano-biopesticides applied to the pupae of the target insects. This activity strongly measures the mortality rate after one day, which depends on the concentrations of nano-biopesticides. The work of Sivapriyajothi *et al.* [91] shows that nano-particles from the extraction of *Leucas aspera* show pupicidal activity in the concentration values (LC 50 and LC 90) for killing the larvae of mosquito vectors such as *Aedes aegypti*.

5.1.1.2 *Larvicidal activity*

The larvicidal activity of nano-biopesticides can be measured by the leaf disc method by introducing them into the leaf, and concentrations of larvae can be determined after 96 hours. Some plants show larvicidal activity, such as leaf extract of *Ambrosia arborescens* which is helpful for larvae of some pests in concentrations such as LC50 = 0.28 ppm and LC90 = 0.43 ppm [92].

5.1.1.3 *Anti-feeding activity*

The anti-feeding activity of nano-biopesticides can be measured by applying them to the leaf disc of pest food. The one-third-instar larva is introduced to the leaf, and the condemnations of leaf eaten by larvae can be measured every 24 hours. Anti-feeding activity has been observed about 92.4% in *Helicoverpa armigera* by applying silver-based nano-biopesticides [93].

5.1.2 In vivo *nano-biopesticide treatment*

Nano-biopesticides can be applied to plants in the right concentration in order to protect them from seasonal diseases. These concentrations (LC 50 and LC 90) aid in the identification of specific larvae, insets, and bee attacks. Nano-biopesticides are also applied in changing environments such as temperatures, humidity, and environmental stresses. In these conditions, nano-biopesticides are directly applied in the form of sprays to protect the plants from pest attacks. Therefore, the use of nano-biopesticides has become the most effective method in controlling the attack of animal vectors and disease-transmitting pests.

5.1.3 Activity against stored grain pests

Pests of stored grains are among the most difficult to manage in an agricultural system because of their large size [94]. Recently, it has been shown that alumina, silica, SiO2, zinc, and silver nano-particles have a substantial anti-pest effect against a range of pests when combined with other chemicals [95]. According to the researchers, when sprayed on plants or crops, nano-emulsions have been shown to be efficient in deterring the attack of attack insects that cause harm to grains that have been stored for extended periods of time. The researchers discovered that nano-biopesticide emulsions effectively prevented the spread of the *Tribolium castaneum* fungus in one of the case studies they conducted [96]. In this study, oil/water emulsions of *P. anisum* essential oil (14 percent, v/v of the total coarse emulsion), ethanol (3 percent, v/v), and Tween 80 (3 percent, v/v), which together represented 20 percent (v/v) of the total coarse emulsion, as well as various components, were used. The emulsions were properly mixed and kept at 86°C for 1 hour. Separating the mixture from the water was accomplished by centrifuging it at 10,000 g for 15 minutes (which made up 80 percent of the mixture). A technique known as photon correlation spectroscopy was used to evaluate a variety of characteristics such as conductivity, zeta potential of the emulsion, and polydispersity index. Contact and digestion techniques on the insects in the medium were used to assess the insecticidal activity of the *P. anisum* emulsified with nano-particles. During the testing against the insects in the medium, the researchers discovered that the *P. anisum* emulsion incorporating nano-particles proved to be very effective. Different concentrations of nano-emulsions can be prepared by acetone in 50 mL bottles containing the 20 g mash grains. The positive and negative controls as emulsions of *P. anisum*. Different groups of 20 beetles were properly made in the sex ratio of 1:1 and then transferred to 50 mL bottles containing the treated and control grains, and each concentration was recorded as three replicates. The mortality rate was carefully recorded after treatments with 3-day intervals till 12 days, while on the other hand, the mortality rate with essential oil containing the beetles was recorded at intervals of 48, 72, and 96 hours. Those insets that were survived and were attached to grain were removed. F1 progeny were recorded 60 days after insect infestation to avoid generation overlaps, as elaborated by Tofel *et al.* [97]. Standard procedures were used to determine the LC 50 values that were used to control the attack of battles and to understand the morphological changes associated with the nano-emulsions.

6. Methods for nano-suspensions preparations

Creating nanosuspensions may be accomplished using two distinct approaches, which are referred to as the bottom-up approach and top-down technology. The bottom-up approach is the more traditional way of creating nanosuspensions. In order to achieve top-down drug particle reduction, a number of techniques such as high-pressure homogenization and media milling are used. Following the bottom-up approach, pesticides (that are to be converted into nanosuspension) are solubilized in a suitable organic solvent and precipitated with the aid of a suitable stabilizer that has been dissolved in an antisolvent as a result of this solubilization and precipitation (often water). Methods such as precipitation, microemulsion, and melt emulsification, to name a few, are among the most often used in this method, and they are described in more detail below [98, 99]. The following are some of the most important methods for the production of nanosuspensions, which are described below.

6.1 High pressure homogenization method

The advantage of this method is the production of pesticides that are poorly soluble in water via the use of high-pressure homogenization. Successful completion of this procedure depends on completing three essential steps: In the first stage, a finely powdered medication is dispersed in a suitable stabilizer solution, resulting in a pre-suspension that is then subjected to further treatment after being stabilized. The pre-suspension is homogenized at a low pressure throughout the following procedure to guarantee consistency. Finally, but certainly not least, it is homogenized at high pressure for about 10 to 25 cycles, or until the desired size is achieved. Despite this, this method is only suitable for the production of highly concentrated nanosuspension formulations rather than diluted nanosuspension formulations since the pesticides must be micronized before they can be delivered to the field [99, 100].

6.2 Precipitation and homogenization methods

It has been shown that when exposed to high temperatures, precipitated plant extract nano-particles may crystallize and transform into microparticles. Greater energy pressures are thus required to homogenize them in order to avoid the development of microparticles. Because of their crystalline structure, these particles, which may be completely amorphous, completely crystalline, or slightly amorphous in nature, may cause bioavailability and long-term stability problems when used in pesticide formulations. It is necessary to homogenize the precipitated nanosuspension before it can be used to maintain the particle size achieved during precipitation. This method also has the benefit of being able to be used to produce pesticides that have low solubility in both organic and aqueous solutions, which is advantageous in both cases [101].

6.3 Media milling method

In this method, the plant extracts are exposed to an ultra-fine grinding medium, which results in the production of extract particles of a nanometer or smaller diameter. As a consequence of the contact of extracted particles with the milling medium, higher energy shear forces are produced throughout the milling process. This provides the required energy input to induce the microparticles to burst into nano-particles during the operation. For many days, milling material, which may consist of extract, a stabilizer, and water or another appropriate buffer, is rotated at a faster speed than the rest of the milling chamber and spun at a slower speed than the rest of the milling chamber [102].

6.4 Precipitation method

When the plant extract is dissolved in an organic solvent of choice, it is dried, which is referred to as precipitation. The surfactant is mixed with water (antisolvent), which also includes surfactant, to create a cohesive combination in order to achieve cohesiveness in the final organic phase of the reaction (aqueous phase). It is feasible to oversaturate the plant extract by adding the prepared organic phase to the aqueous phase in a fast manner (organic solvent to antisolvent). As a consequence, ultrafine particles are produced in large quantities (crystalline or amorphous). This process involves, among other things, the creation of nuclei as well as the growth of crystals, depending on the temperature. A high nucleation rate combined with a slow crystal development rate is required to do this since a stable solution with a smaller particle size than is presently accessible cannot be achieved without doing so [103].

7. Delivery of nano-biopesticides

Any nano-effective formulation in real-world applications depends on effective distribution. Environmentally friendly use of water, fertilizers, and pesticides is possible using nano-sensors and smart delivery systems (see **Figure 2**). Using satellite pictures of their fields in combination may allow farm managers to identify agricultural pests and collect evidence of stress caused by high heat, floods, or drought. Nanomaterials and GPS will be combined with satellite images of fields to produce a more realistic environmental model. Using this technology, farmers can now change agricultural inputs automatically. So, nano-sensors in the field may be able to detect plant viruses and soil nutrients, allowing for more precise crop management. Pesticide use and contamination will be minimized when slow-release nano-biopesticides contained in nano-particles are delivered to their targets [105]. Another alternative is to utilize a nano-barcode, a new technology that may be used to check the quality of agricultural products. Cornell University researchers used supermarket barcodes to create a low-cost, efficient, quick, and simple method for decoding and detecting diseases and illnesses. The technique was developed using grocery barcodes. These tiny probes or nano-barcodes may be scanned with a microscope using self-folding branching DNA constructs. It is feasible to detect a disease biomarker on agricultural goods or on the farm using a fluorescent color ratio. Because nano-barcodes and pathogen biomarkers are so compatible, any fluorescent-based device capable of detecting infection or illness should be able to recognize them. This continuing study's goal is to create a portable on-site detector that non-experts may utilize [106]. Auxins, plant hormones, are important in root development and seedling establishment

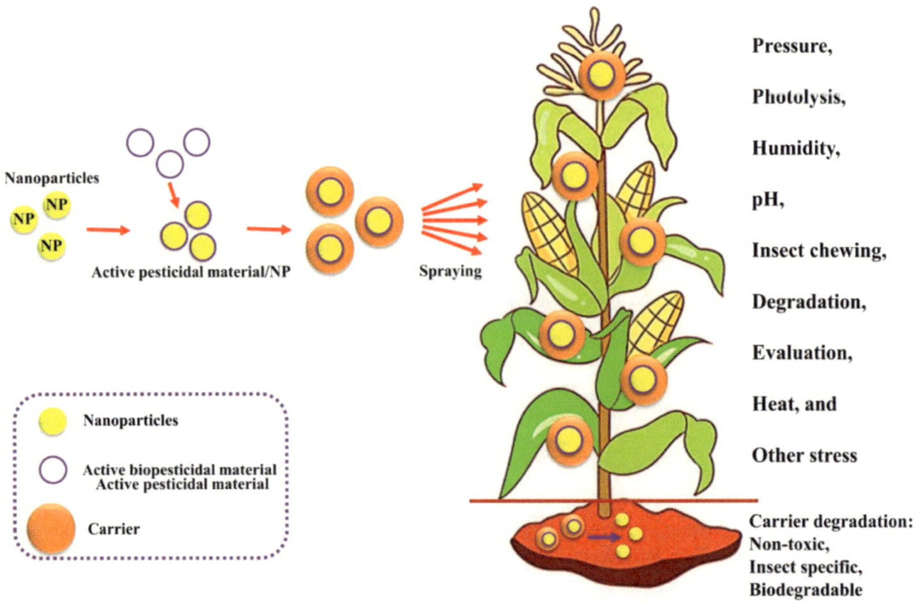

Figure 2.
The schematic diagram of delivery of nano-biopesticides to crop for pest management. This figure is reproduced from Lade and Gogle [104].

in both young and mature plants. Purdue University researchers have created an auxin-detecting nano-sensor that may be used to detect it in the environment. The interaction of auxin with biosensors produces a signal that can be monitored and used to detect the amount of auxin present at different locations along the root's length. Another method is to use mathematics to see whether neighboring cells absorb or release auxin at different rates. This advances auxin research by allowing scientists to better understand how plant roots adapt to their surroundings. This study's findings may help improve agricultural research in the future [107].

Using a micro- or nano-emulsion may enhance nano-biopesticide solubility, kinetic stability, optical transparency, and bioavailability while decreasing emulsion size and viscosity [108]. Despite not being intended for agricultural usage, a nano-permethrin formulation free of artificial polymers and stabilized with natural plant surfactants was shown to be an efficient larvicide. Developing nano-particles that act as a coating or protective layer for conventional nano-biopesticides and fertilizers may also be a future research topic. According to the National Science Foundation, nano-clay materials provide high aspect ratio interaction surfaces for encapsulating "agrochemicals such as fertilizers, plant growth stimulants, and insecticides" [109]. Incorporating silver nano-particles into electrospun polyacrylonitrile fibers is intriguing due to the possible antibacterial characteristics. This method may be used to entrap an active biopesticide or a nano-biopesticide for use in soil-applied pesticides or insecticides. To kill the soilborne bug, an electrospun nanofibrous mat loaded with nano-biopesticides is electrospun into the soil and subsequently removed [110].

8. Mechanism of action of biopesticides

Biopesticides have a variety of distinct modes of action that are distinct from one another and may be used in various settings, including agriculture. Through a variety of mechanisms, including parasitism, antibiosis, and predation, among others, microorganisms generate pesticides that are harmful to humans and animals. Botanical pesticides have been shown to be very effective since they kill insects while also interfering with the development of diseases. Prey is killed as a result of the attack by being parasitized or poisoned, which leads them to die as a result of the attack. Pests are attracted to the treatment area as a consequence of the application of the treatment, which results in the pests being killed or sterilized (see **Figure 3**). Extracts from plants belonging to the Asteraceae family have been reported to inhibit hyphal growth and induce structural modifications in the mycelia of plant pathogenic fungi [112]. *Asteraceae* plants contain compounds such as flavonoids, coumarin alkaloids, and terpenoids, leading to absolute fungal toxicity. Some compounds lead to changes in the cell wall as well as the morphology of cellular organelles [113]. As a result, when bioactive chemicals come into contact with fungal cell membranes, they may induce partitioning and penetration, which will allow the contents of the cell to escape via a hole created by this partitioning. In addition, it has been shown that the separation of the cytoplasmic membrane induced by plant bioactive substances results in the destruction of intracellular components and the growth of cells, ultimately resulting in the death of the cells [114].

There are different types of biopesticides, including sabadilla, pyrethrum, azadirachtin, and fluoroacetate that show different mechanisms of action against pests. For example, the alkaloid toxin of sabadilla significantly caused the loss of nerve cell membrane mechanism by affecting the nerve cell membrane of insects. It was found that sabadilla could kill most insects immediately after its use, but a few could survive up to few days in a state of paralysis before dying [115]. In addition,

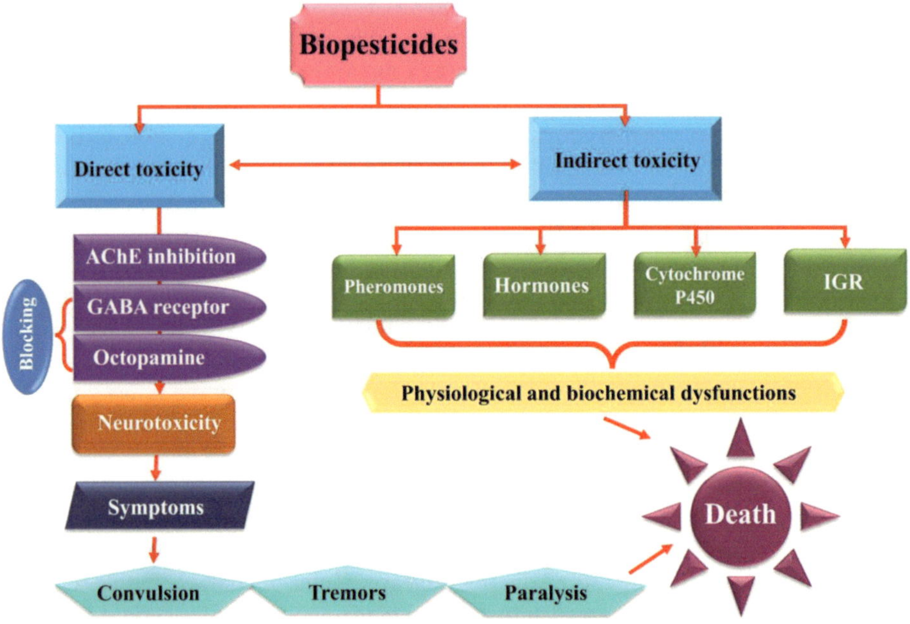

Figure 3.
The general mechanism of action of nano-biopesticides for pest insect management. This figure is reproduced from Mossa, [111].

the emerging evidence revealed that a low dose of pyrethrins significantly causes the immediate death of insects. For humans and warm-blooded animals, pyrethrins are not toxic. Allergic responses to humans, however, are frequent. It may cause a rash, and inhaling the dust can lead to headaches and illness. By altering the process of sodium and potassium ion exchanges in insect nerve fibers, pyrethrins exert their deadly effects by inhibiting the normal transmission of nerve impulses. The insecticides containing pyrethrin work very quickly and produce paralysis in the insects very quickly. But many insects can swiftly metabolize (break down) pyrethrins in spite of their acute toxicity. However, piperonyl butoxide (PBO) and pyrethrin could be used as combined therapy against these insects [116].

A recently reported study revealed that administration of azadirachtin to third instar larvae significantly reduces food consumption compared to control [117]. But, its antifeedant activity surely depends on the insect species and dose concentration [118]. It was reported that the inhibition of feeding behaviors after azadirachtin dose from stimulation of deterrent receptors was coupled with sugar receptors that lead to food restriction, starvation, and bad nutrition [119]. Recently, various studies have demonstrated the weight loss behavior of azadirachtin in different insects, including *Spodoptera eridania*, *Periplaneta americana*, *Drosophila melanogaster,* and *Helicoverpa armigera* [117, 120]. The pesticide mechanism of action of fluoroacetate is well known; it was reported that after ingestion of fluoroacetate by insects, it converted into fluoroacetyl-CoA and after that, into fluorocitric acid. However, the structure analogue (fluorocitric acid) to citric acid blocked the activity of an enzyme that was involved in the conversion of citric acid to cis-aconitic acid resulting in the energy production method being stopped. Due to the accumulation of citric acid inside the cell, the concentration of α-ketoglutaric acid, calcium, and glutamic acid reduced that resulting affected the nervous system of the insect because the nervous system is very sensitive to these acids, especially glutamic acid which is an essential neurotransmitter [115]. When bugs are exposed to insecticides such as allicin, which may be found in garlic bulbs (*A. sativum*), they suffocate and die. When applied to insects, allicin acts by interfering with the neurotransmitter receptors in their nervous systems. Suffocation is caused by substances such as allicin. Phytotoxins and terpenoids biochemically interact with insects via hydrophobic and ionic interactions. In addition, a large number of proteins are targeted and destroyed, resulting in physiological failure and degeneration. Plant extracts and essential oils include a range of compounds that may interact with an insect's nervous system and coordination, resulting in the insect's death as a consequence of the disruption produced by this contact [121].

9. Environmental sustainability of nano-biopesticides

Nano-biopesticides are eco-friendly, possess biodegradation properties, and are transported to the different parts of plants. Due to their bioavailability in the plant system, they are helpful in understanding the interactions and behavior of different pests that tack on crops. Spraying silver nano-particles with combinations of aloe vera extract and silver nitrate is helpful to control the growth of pests such as *H. armegera*. Formulations of biopesticides through nanotechnology help to control the harmful pests that attack crops [122]. Nano-biopesticide applications in plants not only control pests but also control environmental pollution by reducing the risks of accumulations of toxic metals in plants, which are helpful in cleaning the environment. For example, the delivery of zinc-based nano-biopesticides to roots protects the plants as they exhibit high efficiency compared to the toxic chemicals directly applied to plants. Nano-biopesticides are also helpful in the bioremediation process by degrading toxic compounds [123, 124].

Nano-biopesticides are biodegradable and transported to the different tissues of pants. Some studies have shown that soil applications of nano-biopesticides under optimum conditions are helpful for the degradation of toxic metabolites that are produced in plants. These metabolites cause the accumulation of toxic metals. It leads to an increase in the chances of death of plant tissues. On the other hand, traditionally used chemicals also increase the chances of death of plant tissues due to

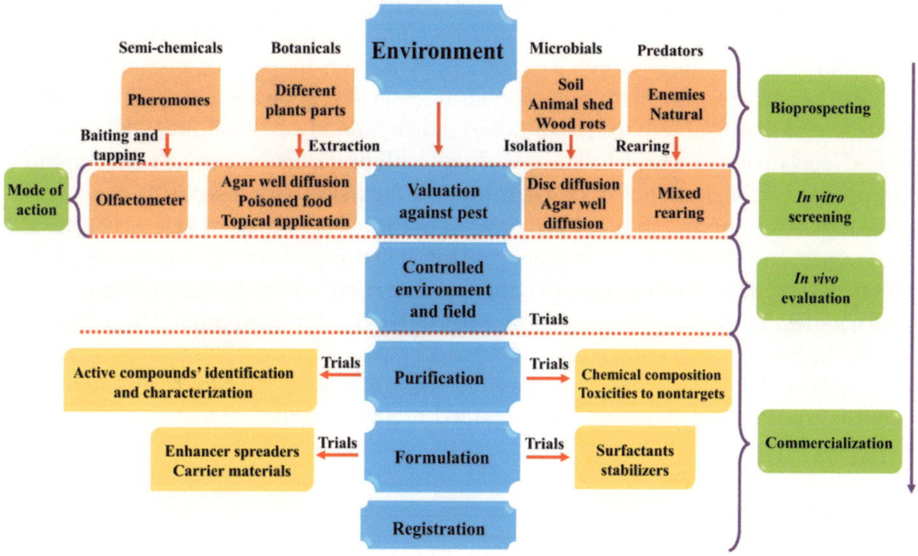

Figure 4.
The general process of formulation of nano-fertilizer to commercialization. This figure is reproduced from Lengai and Muthomi [128].

cellular toxicity in some cells. Therefore, the use of nano-biopesticides in environmental applications is much more reliable than other chemical compounds [125]. Nano-biopesticides reach the soil by activating the microbial activities that increase the chances of useful bacterial activities in plants such as mycorrhizal association.

Nano-biopesticides play an important role in maintaining environmental sustainability by replacing traditionally used chemicals in the form of sprays. The use of nano-biopesticides to control the pests also maintains the ecological chain. Nano-biopesticides for land conservation ensure the maximum yields and maintain the farming system. So, nano-biopesticides are also helpful for improving soil quality and increasing food yields under different cultivations. Other applications are found in crop protection by controlling pests and other animals such as bees and birds through sustainable development [126, 127]. The representation of the process of formulation of nano-biopesticides to fully commercialization is presented in **Figure 4**.

10. Future perspectives

Nano-biopesticides are used in the control of pests in order to prevent their action in agriculture sectors. These bio-pesticides will be helpful in targeting the different pests in more effective ways by reducing the chemical compounds in order to make profitable and environmentally friendly production. Due to unclear molecular mechanisms and sites of action to the target of the action, research progress for pest control in agriculture is slow [129]. Recent studies show that applications of nano-biopesticides are effective in controlling pests by replacing the traditionally used chemical compounds. These nano-biopesticides have fewer side effects as compared to directly applied chemical compounds. Nano-biopesticides have great potential to release active ingredients that are helpful in maintaining the different problems associated with agricultural systems, such as eutrophication. Although nano-biopesticides are widely used in different crops to control pests, their utilization in humans and animals remains unclear as they have entered into the food chain. More study is needed to characterize and formulate newly developed nano-biopesticides for controlling the different varieties of pests by ensuring no side effects on humans through the food chain [130].

As the world population increases rapidly, the feeding of humans will reach approximately 9 billion by 2050. It requires lots of nano-biopesticides to kill the pests and for the storage of food for long periods of time. It will be an emerging approach towards pest management that maintains environmental sustainability with fewer toxic effects on human health. The use of nano-biopesticides is also helpful in maintaining the nutrient balance in crops, minimize the risks to food security, and accumulating hazardous materials [131]. Nano-biopesticides have been extensively used in the agricultural fields for pest management or arthropod attack, but they possess chemical formulations that contain nano-particles that lead to toxicity concerns and health issues. These nano-biopesticides need to be standardized internationally to reduce their toxic effects on crops and the food chain. The use of nano-biopesticides in agriculture looks promising, but more research is needed in order to understand their toxic nature and monitor their application time to soils [132].

11. Conclusions

Approximatively, 25% of the world's food yield is destroyed each year by the attack of pests. According to recent studies, using synthetic pesticides has been related to an increase in some illnesses, including Parkinson's disease, neurotoxicity, type 2 diabetes, endocrine disruption, various malignancies, and even obesity. Insecticides produced from microorganisms or natural compounds are known as biopesticides. Due to their eco-friendliness, great efficacy, and few side effects, nano-biopesticides have gained in popularity over conventional pesticides over recent years. Biologically active pesticide compounds (APCs) may be produced in two ways: either by extracting APCs from plants and combining them with nano-particles or by inserting them into a polymer. As a result of their nano-size, high surface area/volume ratio, durability, enhanced effectiveness, greater solubility, mobility, and low toxicity, nano-biopesticides are superior to chemical pesticides. Biopesticides inhibit pathogen's growth by altering their cellular structures and morphology and exhibit neurotoxicity on insects. As a result, nano-biopesticides are environmentally benign and have biodegradation characteristics; they assist in cleaning the environment by reducing the danger of harmful metal buildup in plants. However, the use of nano-sensors and nano-based smart delivery systems could help in the efficient use of agricultural, natural resources such as water, nutrients, and chemicals through precision farming. Moreover, it is recommended to use a nano-barcode, which is a novel method to monitor the quality of agricultural products.

Acknowledgements

Hereby, we extend our gratitude to A.Q Research Group, Pakistan for reviewing the article and providing helpful comments.

Funding statement

This research received no specific grant from any funding agency in the public, commercial or not-for-profit sectors.

Conflict of interest

The authors declared no conflict of interest.

Research involving human participants and/or animals

This article does not contain any studies with human participants or animals performed by any of the authors.

Informed consent

For this type of study informed consent is not required.

Author details

Abu Hazafa[1*], Muhammad Murad[2], Muhammad Umer Masood[2], Shahid Bilal[3], Muhammad Nasir Khan[4], Qasim Farooq[5], Muhammad Omer Iqbal[6], Muhammad Shakir[7] and Muhammad Naeem[8]

1 Faculty of Sciences, Department of Biochemistry, University of Agriculture, Faisalabad, Pakistan

2 Faculty of Agriculture, Centre of Agricultural Biochemistry and Biotechnology, University of Agriculture, Faisalabad, Pakistan

3 Faculty of Agriculture, Department of Agronomy, University of Agriculture, Faisalabad, Pakistan

4 Pir Mehr Ali Shah Arid Agriculture University, Rawalpindi, Pakistan

5 Department of Botany, Ecology and Plant Physiology, University of Córdoba, Rabanales Campus, Spain

6 Shandong Provincial Key Laboratory of Glycoscience and Glycoengineering, School of Medicine and Pharmacy, Ocean University of China, Qingdao, China

7 School of Life Sciences, Northeast Normal University, China

8 College of Life Science, Hebei Normal University, Shijiazhuang, China

*Address all correspondence to: abuhuzaifavirk@gmail.com; abu.9093270@pern.onmicrosoft.com

References

[1] Sinha K, Ghosh J, Sil PC. New pesticides: A cutting-edge view of contributions from nanotechnology for the development of sustainable agricultural pest control. New Pesticides of Soil Sensers: Elsevier; 2017. pp. 47-79

[2] Berini F, Katz C, Gruzdev N, Casartelli M, Tettamanti G, Marinelli F. Microbial and viral chitinases: Attractive biopesticides for integrated pest management. Biotechnology Advances. 2018;**36**(3):818-838

[3] Atwood D, Paisley-Jones C. Pesticides industry sales and usage. 2008-2012 Market Estimates. 2017

[4] Rajendran G, Islam R. Plant clinics in Bangladesh: Are farmers losing less and feeding more. CABI Case Studies. 2017;**19**:145-152

[5] Douglas AE. Strategies for enhanced crop resistance to insect pests. Annual Review of Plant Biology. 2018;**69**: 637-660

[6] Chowański S, Adamski Z, Marciniak P, Rosiński G, Büyükgüzel E, Büyükgüzel K, et al. A review of bioinsecticidal activity of *Solanaceae* alkaloids. Toxin. 2016;**8**(3):60-69

[7] Khater HF. Prospects of botanical biopesticides in insect pest management. Pharmacologia. 2012;**3**(12):641-656

[8] Ibanez S, Gallet C, Després L. Plant insecticidal toxins in ecological networks. Toxin. 2012;**4**(4):228-243

[9] Huang X, Zhao J, Zhou X, Han Y, Zhang J, Cai Z. How green alternatives to chemical pesticides are environmentally friendly and more efficient. European Journal of Soil Science. 2019;**70**(3):518-529

[10] Mathew IL, Singh D, Singh R, Tripathi C. Bacillus thuringiensis: The biocontrol agent in a food web perspective. Biolife. 2014;**2**(1):348-362

[11] Koul O. Nanobiopesticides: An introduction. Nano-Biopesticides Today and Future Perspectives: Elsevier; 2019. pp. 1-15

[12] Damalas CA, Koutroubas SD. Current status and recent developments in biopesticide use. Agriculture. 2018;**8**(1):1-6

[13] Kumar A, Dhananjaya PS, Tyagi M. Biotoxic cyanobacterial metabolites exhibiting pesticidal and mosquito larvicidal activities. Journal of Microbiology and Biotechnology. 2003;**13**(1):50-56

[14] Costa JAV, Freitas BCB, Cruz CG, Silveira J, Morais MG. Potential of microalgae as biopesticides to contribute to sustainable agriculture and environmental development. Journal of Environmental Science and Health. Part. B. 2019;**54**(5):366-375

[15] Sayed AM, Kim S, Behle RW. Characterisation of silver nanoparticles synthesised by *Bacillus thuringiensis* as a nanobiopesticide for insect pest control. Biocontrol Science and Technology. 2017;**27**(11):1308-1326

[16] Chaloupka K, Malam Y, Seifalian AM. Nanosilver as a new generation of nanoproduct in biomedical applications. Trends in Biotechnology. 2010;**28**(11):580-588

[17] Lu W, Senapati D, Wang S, Tovmachenko O, Singh AK, Yu H, et al. Effect of surface coating on the toxicity of silver nanomaterials on human skin keratinocytes. Chemical Physics Letters. 2010;**487**(1-3):92-96

[18] Ragaei M, A-kH S. Nanotechnology for insect pest control. International Journal of Environmental Science and Technology. 2014;**3**(2):528-545

[19] Giongo AMM, Vendramim JD, Forim MR. Evaluation of neem-based nanoformulations as alternative to control fall armyworm. Ciencia e Agrotecnologia. 2016;**40**(1):26-36

[20] Mandal BK. Silver nanoparticles: Potential as insecticidal and microbial biopesticides. Nano-Biopesticides Today and Future Perspectives. London, United Kingdom: Elsevier; 2019. p. 281-302

[21] Mishra RK, Bohra A, Kamaal N, Kumar K, Gandhi K, Sujayanand G, et al. Utilization of biopesticides as sustainable solutions for management of pests in legume crops: Achievements and prospects. Egyptian Journal of Biological Pest Control. 2018;**28**(1):1-11

[22] Desa U. United nations department of economic and social affairs, population division. world population prospects: The 2015 revision, key findings and advance tables. Technical Report: Working Paper No ESA/P/ WP 2412015

[23] Gilland B. Nitrogen, phosphorus, carbon and population. Science Progress. 2015;**98**(4):379-390

[24] Carvalho FP. Agriculture, pesticides, food security and food safety. Environmental Science & Policy. 2006;**9**(7-8):685-692

[25] Hendrix CS. Applying hubbert curves and linearization to rock phosphate. Peterson Institute for International Economics. 2011;**4**(11-18):1-12

[26] Carvalho F. Pesticides, environment, and food safety. Food and Energy Security. 2017;**6**:48-60

[27] Taylor MD, Klaine SJ, Carvalho FP, Barcelo D, Everaarts J. Pesticide Residues in Coastal Tropical Ecosystems: Distribution, Fate and Effects. London, UK: CRC Press; 2002

[28] Alavanja MC, Bonner MR. Occupational pesticide exposures and cancer risk: A review. Journal of Toxicology and Environmental Health, Part B. 2012;**15**(4):238-263

[29] Paoli D, Giannandrea F, Gallo M, Turci R, Cattaruzza M, Lombardo F, et al. Exposure to polychlorinated biphenyls and hexachlorobenzene, semen quality and testicular cancer risk. Journal of Endocrinological Investigation. 2015;**38**(7):745-752

[30] Halimatunsadiah A, Norida M, Omar D, Kamarulzaman N. Application of pesticide in pest management: The case of lowland vegetable growers. International Food Research Journal. 2016;**23**(1):85-94

[31] Jantasorn A, Moungsrimuangdee B, Dethoup T. In vitro antifungal activity evaluation of five plant extracts against five plant pathogenic fungi causing rice and economic crop diseases. Journal of Biopesticides. 2016;**9**(1):1-7

[32] Dey K, Choudhury P, Dutta B. Impact of pesticide use on the health of farmers: A study in *Barak valley*, Assam (India). Journal of Environmental Chemistry and Ecotoxicology. 2013;**5**(10):269-277

[33] Starling AP, Umbach DM, Kamel F, Long S, Sandler DP, Hoppin JA. Pesticide use and incident diabetes among wives of farmers in the Agricultural Health Study. Occupational and Environmental Medicine. 2014;**71**(9):629-635

[34] Goodson W, Lowe L, Carpenter D, Gilbertson M, Manaf AA. Assessing the carcinogenic potential of low-dose exposures to chemical mixtures in the environment: The challenge ahead. Carcinogenesis. 2015;**36**:S254-SS96

[35] de Araujo JS, Delgado IF, Paumgartten FJ. Glyphosate and adverse pregnancy outcomes, a systematic

review of observational studies. BMC Public Health. 2016;**16**(1):472-484

[36] Attaullah M, Yousuf MJ, Shaukat S, Anjum SI, Ansari MJ, Buneri ID, et al. Serum organochlorine pesticides residues and risk of cancer: A case-control study. Saudi Journal of Biological Sciences. 2018;**25**(7): 1284-1290

[37] Kumar VV. Biofertilizers and biopesticides in sustainable agriculture. Role of rhizospheric microbes in soil: Singapore: Springer; 2018. p. 377-398

[38] Kumar VV. Plant growth-promoting microorganisms: Interaction with plants and soil. Plant, soil and microbes: Cham: Springer; 2016. p. 1-16

[39] Weller DM. Pseudomonas biocontrol agents of soilborne pathogens: Looking back over 30 years. Phytopathology. 2007;**97**(2):250-256

[40] Mnif I, Ghribi D. Potential of bacterial derived biopesticides in pest management. Crop Protection. 2015;**77**:52-64

[41] Priya M, Gurung N, Mukherjee K, Bose S. Microalgae in removal of heavy metal and organic pollutants from soil. In: Microbial biodegradation and bioremediation. Elsevier; 2014. pp. 519-537

[42] Stirk W, Bálint P, Tarkowská D, Novák O, Maróti G, Ljung K, et al. Effect of light on growth and endogenous hormones in Chlorella minutissima (Trebouxiophyceae). London, UK: Plant Physiology and Biochemistry. 2014;**79**:66-76

[43] Gupta S, Dikshit A. Biopesticides: An ecofriendly approach for pest control. Journal of Biopesticides. 2010;**3**(Special Issue):186-194

[44] Schuster C, Konstantinidou-Doltsinis S, Schmitt A. Glycyrrhiza glabra extract protects plants against important phytopathogenic fungi. Communications in Agricultural and Applied Biological Sciences. 2010;**75**(4):531-540

[45] Ntalli NG, Menkissoglu-Spiroudi U. Pesticides of botanical origin: A promising tool in plant protection. In: Pesticides-Formulations, Effects, Fate. Croatia: InTech; 2011. pp. 1-23

[46] Bardin M, Ajouz S, Comby M, Lopez-Ferber M, Graillot B, Siegwart M, et al. Is the efficacy of biological control against plant diseases likely to be more durable than that of chemical pesticides? Frontiers in Plant Science. 2015;**6**:566-576

[47] Aala F, Yusuf UK, Nulit R, Rezaie S. Inhibitory effect of allicin and garlic extracts on growth of cultured hyphae. Iranian Journal of Basic Medical Sciences. 2014;**17**(3):150-161

[48] Khalegi M, Salahi Eshlaghi B, Ghotaslou R. Antibacterial effects of Azerbaijan Garlic (*Allium sativum*). Journal of Zoonotic Diseases. 2017;**2**(1): 30-34

[49] Suleiman EA, Abdallah WB. *In vitro* activity of garlic (*Allium sativum*) on some pathogenic fungi. European Journal of Medicinal Plants. 2014;**10**: 1240-1250

[50] Biswas G. Comparative effectiveness of neem extracts and synthetic organic insecticide against mustard aphid. Bangladesh Journal of Agricultural Research. 2013;**38**(2):181-187

[51] Castillo-Sánchez LE, Jiménez-Osornio JJ, Delgado-Herrera MA, Candelaria-Martínez B, Sandoval-Gío J. Effects of the hexanic extract of neem *Azadirachta indica* against adult whitefly *Bemisia tabaci*. Journal of Entomology and Zoology Studies. 2015;**3**:95-99

[52] Raut RR, Sawant AR, Jamge BB. Antimicrobial activity of *Azadirachta indica* (Neem) against pathogenic

microorganisms. Journal of Academia and Industrial Research. 2014;**3**(7): 327-329

[53] Bissa S, Bohra A. Antibacterial potential of pot marigold. Journal of Microbiology and Antimicrobials. 2011;**3**(3):51-54

[54] Granja EM, Benitez SR, Sanjuanello D. Effect of antagonists and plant extracts in the control of *Protea wilt* (F. oxysporum). American Journal of Plant Sciences. 2014;**5**(21):3203-3219

[55] Abtew A, Subramanian S, Cheseto X, Kreiter S, Garzia GT, Martin T. Repellency of plant extracts against the legume flower thrips *Megalurothrips sjostedti* (Thysanoptera: Thripidae). Insect. 2015;**6**(3):608-625

[56] Karaca G, Bilginturan M, Olgunsoy P. Effects of some plant essential oils against fungi on wheat seeds. Indian Journal of Pharmaceutical Education and Research. 2017;**3**:385-388

[57] Witkowska D, Sowińska J, Żebrowska J, Mituniewicz E. The antifungal properties of peppermint and thyme essential oils misted in broiler houses. Brazilian Journal of Poultry Science. 2016;**18**(4):629-638

[58] Nikkhah M, Hashemi M, Najafi MBH, Farhoosh R. Synergistic effects of some essential oils against fungal spoilage on pear fruit. International Journal of Food Microbiology. 2017;**257**:285-294

[59] Zhang Y, Liu X, Wang Y, Jiang P, Quek S. Antibacterial activity and mechanism of cinnamon essential oil against Escherichia coli and Staphylococcus aureus. Food Control. 2016;**59**:282-289

[60] Ali S, Sagheer M, Hassan M, Abbas M, Hafeez F, Farooq M, et al. Insecticidal activity of turmeric (*Curcuma longa*) and garlic (*Allium sativum*) extracts against red flour beetle, Tribolium castaneum: A safe alternative to insecticides in stored commodities. Journal of Entomology and Zoology Studies. 2014;**2**(3):201-205

[61] de Souza Tavares W, Akhtar Y, Gonçalves GLP, Zanuncio JC, Isman MB. Turmeric powder and its derivatives from *Curcuma longa* rhizomes: insecticidal effects on cabbage looper and the role of synergists. Scientific Reports. 2016;**6**:34093-34099

[62] Al-Rahmah A, Mostafa A, Abdel-Megeed A, Yakout S, Hussein S. Fungicidal activities of certain methanolic plant extracts against tomato phytopathogenic fungi. African Journal of Microbiology Research. 2013;**7**(6):517-524

[63] Rizvi SAH, Hussain S, Rehman SU, Jaffar S, Rehman MFU. Efficacy of ecofriendly botanical extracts of Ginger (Zingiber officinale), Garlic (Allium sativum) and Tobacco (Nicotiana tabacum L) for the control of cabbage looper (Trichoplusia binotalis) under agro ecological conditions of Peshawar. Pakistan Journal of Entomology and Zoology Studies. 2016;**4**(1):88-90

[64] Muthomi JW, Lengai GM, Wagacha MJ, Narla RD. *In vitro* activity of plant extracts against some important plant pathogenic fungi of tomato. Australian Journal of Crop Science. 2017;**11**(6):683-688

[65] Bastas KK. Determination of antibacterial efficacies of plant extracts on tomato bacterial speck disease. The Journal of Turkish Phytopathology. 2015;**44**(1-2-3):1-10

[66] Salim HA, Salman IS, Majeed II, Hussein HH. Evaluation of some plant extracts for their nematicidal properties against root-knot nematode, Meloidogyne sp. Journal of Genetic and Environmental Resources Conservation. 2016;**3**:241-244

[67] Shiberu T, Getu E. Assessment of selected botanical extracts against

Liriomyza species (Diptera: Agromyzidae) on tomato under glasshouse condition. International Journal of Fauna and Biological Studies. 2016;**1**:87-90

[68] Essiedu JA, Adepoju FO, Ivantsova MN, editors. Benefits and limitations in using biopesticides: A review. In: AIP Conference Proceedings. AIP Publishing LLC; 2020

[69] Sales MDC, Costa HB, Fernandes PMB, Ventura JA, Meira DD. Antifungal activity of plant extracts with potential to control plant pathogens in pineapple. Asian Pacific Journal of Tropical Biomedicine. 2016;**6**(1):26-31

[70] Akinrinnola O, Okunlola A. Effectiveness of botanical formulations in vegetable production and bio-diversity preservation in Ondo State. United States: Nigerian Journal of Horticultural Science. 2014;**6**(1):6-13

[71] Lanzoni A, Martelli R, Pezzi F. Mechanical release of Phytoseiulus persimilis and Amblyseius swirskii on protected crops. Bull Insectology. 2017;**2**:245-250

[72] Lade BD, Gogle DP. Nano-biopesticides: Synthesis and applications in plant safety. In: Nanobiotechnology Applications in Plant Protection. Cham: Springer; 2019. pp. 169-189

[73] Bogutska K, Sklyarov YP, Prylutskyy YI. Zinc and zinc nanoparticles: Biological role and application in biomedicine. Ukrainica Bioorganica Acta. 2013;**1**:9-16

[74] Lade BD, Gogle DP, Lade DB, Moon GM, Nandeshwar SB, Kumbhare SD. Nanobiopesticide formulations: Application strategies today and future perspectives. In: Nano-Biopesticides Today and Future Perspectives. London, UK: Elsevier; 2019. pp. 179-206

[75] Sasson Y, Levy-Ruso G, Toledano O, Ishaaya I. Nanosuspensions: Emerging novel agrochemical formulations. In: Insecticides Design Using Advanced Technologies. Berlin, Heidelberg: Springer; 2007. pp. 1-39

[76] Manjunatha S, Biradar D, Aladakatti YR. Nanotechnology and its applications in agriculture: A review. Journal of Farm Science. 2016;**29**(1):1-13

[77] Bouwmeester H, Dekkers S, Noordam MY, Hagens WI, Bulder AS, De Heer C, et al. Review of health safety aspects of nanotechnologies in food production. Regulatory Toxicology and Pharmacology. 2009;**53**(1):52-62

[78] Ng VWL, Tan JPK, Leong J, Voo ZX, Hedrick JL, Yang YY. Antimicrobial polycarbonates: Investigating the impact of nitrogen-containing heterocycles as quaternizing agents. Macromolecules. 2014;**47**(4):1285-1291

[79] Bansal R, Pathak R, Jha D, Kumar P, Gautam HK. Enhanced antimicrobial activity of amine-phosphonium (NP) hybrid polymers against gram-negative and gram-positive bacteria. International Journal of Polymeric Materials and Polymeric Biomaterials. 2015;**64**(2):84-89

[80] Dong Q, Cai Q, Gao Y, Zhang S, Gao G, Harnoode C, et al. Synthesis and bactericidal evaluation of imide N-halamine-loaded PMMA nanoparticles. New Journal of Chemistry. 2015;**39**(3):1783-1791

[81] Regnault-Roger C, Vincent C, Arnason JT. Essential oils in insect control: Low-risk products in a high-stakes world. Annual Review of Entomology. 2012;**57**:405-424

[82] Liu J, Pu H, Liu S, Kan J, Jin C. Synthesis, characterization, bioactivity and potential application of phenolic acid grafted chitosan: A review. Carbohydrate Polymers. 2017;**174**: 999-1017

[83] Sagiri SS, Anis A, Pal K. Review on encapsulation of vegetable oils: Strategies, preparation methods, and applications. Polymer-Plastics Technology and Engineering. 2016;**55**(3):291-311

[84] Sabbour MM, El-Aziz SEA. Efficacy of three essential oils and their nano-particles against Sitophilus granarius under laboratory and store conditions. Journal of the Entomological Research. 2016;**40**(3):229-234

[85] Nenaah GE, Ibrahim SI, Al-Assiuty BA. Chemical composition, insecticidal activity and persistence of three Asteraceae essential oils and their nanoemulsions against Callosobruchus maculatus (F.). Journal of Stored Products Research. 2015;**61**:9-16

[86] Lopez M, Maudhuit A, Pascual-Villalobos M, Poncelet D. Development of formulations to improve the controlled-release of linalool to be applied as an insecticide. Journal of Agricultural and Food Chemistry. 2012;**60**(5):1187-1192

[87] Sherry M, Charcosset C, Fessi H, Greige-Gerges H. Essential oils encapsulated in liposomes: A review. Journal of Liposome Research. 2013;**23**(4):268-275

[88] da Silva LA, Maciel AP, Mendonça CJS, Junior LMC. Use of encapsulated carvacrol with yeast cell walls to control resistant strains of *Rhipicephalus microplus* (Acari: Ixodidae). Industrial Crops and Products. 2017;**108**:190-194

[89] Kamaraj C, Rajakumar G, Rahuman AA, Velayutham K, Bagavan A, Zahir AA, et al. Feeding deterrent activity of synthesized silver nanoparticles using *Manilkara zapota* leaf extract against the house fly, *Musca domestica* (Diptera: Muscidae). Parasitology Research. 2012;**111**(6):2439-2448

[90] Lade BD, Patil AS. Silver nano fabrication using leaf disc of Passiflora foetida Linn. Applied Nanoscience. 2017;**7**(5):181-192

[91] Sivapriyajothi S, Kumar PM, Kovendan K, Subramaniam J, Murugan K. Larvicidal and pupicidal activity of synthesized silver nanoparticles using Leucas aspera leaf extract against mosquito vectors, Aedes aegypti and Anopheles stephensi. Journal of Entomological and Acarological Research. 2014;**46**(2):77-84

[92] Morejon B, Pilaquinga F, Domenech F, Ganchala D, Debut A, Neira M. Larvicidal activity of silver nanoparticles synthesized using extracts of *Ambrosia arborescens* (Asteraceae) to control *Aedes aegypti L.*(Diptera: Culicidae). J of Nanotechnolg. 2018;**2018**. DOI: 10.1155/2018/6917938

[93] Siva C, Kumar MS. Pesticidal activity of eco-friendly synthesized silver nanoparticles using *Aristolochia indica* extract against *Helicoverpa armigera* Hubner (Lepidoptera: Noctuidae). International Journal of Advanced Science and Technology Research. 2015;**2**:197-226

[94] Dimetry N, Hussein H. Role of nanotechnology in agriculture with special reference to pest control. International Journal of PharmTech Research. 2016;**9**:121-144

[95] Stadler T, Buteler M, Valdez SR, Gitto JG. Particulate nanoinsecticides: A new concept in insect pest management. Insecticides: Agriculture Toxicology: IntechOpenRijeka; 2018. pp83-pp101

[96] Hashem AS, Awadalla SS, Zayed GM, Maggi F, Benelli G. *Pimpinella anisum* essential oil nanoemulsions against *Tribolium castaneum*—insecticidal activity and mode of action. Environmental Science and Pollution Research. 2018;**25**(19):18802-18812

[97] Tofel KH, Nukenine EN, Stähler M, Adler C. Degradation of azadirachtin A on treated maize and cowpea and the

persistence of Azadirachta indica seed oil on Callosobruchus maculatus and Sitophilus zeamais. Journal of Stored Products Research. 2016;**69**:207-212

[98] Ali HS, York P, Ali AM, Blagden N. Hydrocortisone nanosuspensions for ophthalmic delivery: a comparative study between microfluidic nanoprecipitation and wet milling. Journal of Controlled Release. 2011;**149**(2):175-181

[99] Bhowmik D, Harish G, Duraivel S, Kumar BP, Raghuvanshi V, Kumar KS. Nanosuspension-A novel approaches in drug delivery system. The Pharma Innovation. 2013;**1** (12, Part A):50

[100] Patil S, Rane B, Bakliwal S, Pawar S. Nano suspension: At a glance. International Journal of Pharmaceutical Sciences. 2011;**3**:947-960

[101] Soumya M, Gupta S, Jain R, Mazumder R. Solubility enhancement of poorly water soluble drug by using nano-suspension technology. International Journal of Research and Development in Pharmacy & Life Sciences. 2013;**2**(6):642-649

[102] Savjani KT, Gajjar AK, Savjani JK. Drug solubility: Importance and enhancement techniques. International Scholarly Research Notices. 2012;**2012**. DOI: 10.5402/2012/195727

[103] Kavitha V, Neethu C, Dineshkumar B, Krishnakumar K, John A. Nanosuspension formulation: An improved drug delivery system. Nanoscience and Nanotechnology: An International Journal. 2014;**4**:1-5

[104] Lade BD, Gogle DP. Nano-biopesticides: Synthesis and applications in plant safety. In: Nanobiotechnology Applications in Plant Protection. Cham: Springer; 2019. pp. 169-189

[105] Lade BD, Gogle DP, Lade DB, Moon GM, Nandeshwar SB, Kumbhare SD. Nanobiopesticide formulations: Application strategies today and future perspectives. In: Nano-Biopesticides Today and Future Perspectives. London, United Kingdom: Elsevier; 2019. pp. 179-206

[106] Li Y, Cu YTH, Luo D. Multiplexed detection of pathogen DNA with DNA-based fluorescence nanobarcodes. Nature Biotechnology. 2005;**23**(7):885-889

[107] McLamore ES, Diggs A, Calvo Marzal P, Shi J, Blakeslee JJ, Peer WA, et al. Non-invasive quantification of endogenous root auxin transport using an integrated flux microsensor technique. The Plant Journal. 2010;**63**(6):1004-1016

[108] Xu L, Liu Y, Bai R, Chen C. Applications and toxicological issues surrounding nanotechnology in the food industry. Pure and Applied Chemistry. 2010;**82**(2):349-372

[109] Ghormade V, Deshpande MV, Paknikar KM. Perspectives for nano-biotechnology enabled protection and nutrition of plants. Biotechnology Advances. 2011;**29**(6):792-803

[110] Schiffman JD, Schauer CL. A review: electrospinning of biopolymer nanofibers and their applications. Polymer Reviews. 2008;**48**(2):317-352

[111] Mossa A-TH. Green pesticides: Essential oils as biopesticides in insect-pest management. Journal of Environmental Science and Technology. 2016;**9**(5):354

[112] Nuzhat T, Vidyasagar G. Antifungal investigations on plant essential oils. A review. International Journal of Pharmacy and Pharmaceutical Sciences. 2014;**2**:19-28

[113] da Silva IF, de Oliveira RG, Soares IM, da Costa AT, Ascêncio SD, de Oliveira Martins DT. Evaluation of acute toxicity, antibacterial activity, and mode of action of the hydroethanolic

extract of *Piper umbellatum* L. Journal of Ethnopharmacology. 2014;**151**(1):137-143

[114] El-Wakeil NE. Botanical pesticides and their mode of action. Gesunde Pflanzen. 2013;**65**(4):125-149

[115] Oguh C, Okpaka C, Ubani C, Okekeaji U, Joseph P, Amadi E. Natural pesticides (biopesticides) and uses in pest management-a critical review. Asian Journal of Biotechnology and Genetic Engineering. 2019:1-18

[116] Rattan RS. Mechanism of action of insecticidal secondary metabolites of plant origin. Crop Protection. 2010;**29**(9):913-920

[117] Bezzar-Bendjazia R, Kilani-Morakchi S, Maroua F, Aribi N. Azadirachtin induced larval avoidance and antifeeding by disruption of food intake and digestive enzymes in Drosophila melanogaster (Diptera: Drosophilidae). Pesticide Biochemistry and Physiology. 2017;**143**:135-140

[118] Shannag HK, Capinera JL, Freihat NM. Effects of neem-based insecticides on consumption and utilization of food in larvae of Spodoptera eridania (Lepidoptera: Noctuidae). Journal of Insect Science. 2015;**15**(1):152

[119] Mordue A, Morgan E, Nisbet A, Gilbert L, Gill S. Azadirachtin, a natural product in insect control. In: Insect Control: Biological and Synthetic Agents. London, UK: Elsevier; 2010. pp. 185-197

[120] Ahmad S, Ansari MS, Muslim M. Toxic effects of neem based insecticides on the fitness of Helicoverpa armigera (Hübner). Crop Protection. 2015;**68**: 72-78

[121] Rodino S, Butu A, Butu M, Cornea CP. In vitro efficacy of some plant extracts against damping off disease of tomatoes. Journal of International Science Public Agricultural Food. 2014;**2**:240-244

[122] Bhattacharyya A, Duraisamy P, Govindarajan M, Buhroo AA, Prasad R. Nano-biofungicides: emerging trend in insect pest control. In: Advances and Applications Through Fungal Nanobiotechnology. Cham: Springer; 2016. pp. 307-319

[123] Srivastava R, Awasthi K, Tripathi D. Nanotechnology towards sustainable agriculture. Nanotechnology. 2018;**6**(6):155-158

[124] Patil SS, Shedbalkar UU, Truskewycz A, Chopade BA, Ball AS. Nanoparticles for environmental clean-up: A review of potential risks and emerging solutions. Environmental Technology and Innovation. 2016;**5**:10-21

[125] Khayatnezhad M, Nasehi F. Industrial pesticides and a methods assessment for the reduction of associated risks: A Review. Advancements in Life Sciences. 2021;**8**(2):202-210

[126] Grillo R, Fraceto LF, Amorim MJ, Scott-Fordsmand JJ, Schoonjans R, Chaudhry Q. Ecotoxicological and regulatory aspects of environmental sustainability of nanopesticides. Journal of Hazardous Materials. 2020:124148

[127] Chhipa H. Nanopesticide: Current status and future possibilities. Agricultural Research & Technology. 2017;**5**(1):1-4

[128] Lengai GM, Muthomi JW. Biopesticides and their role in sustainable agricultural production. Journal of Biosciences and Medicines. 2018;**6**(06):7

[129] Deka B, Babu A, Baruah C, Barthakur M. Nanopesticides: A systematic review of their prospects

with special reference to tea pest management. Frontiers in Nutrition. 2021;**8**:686131

[130] Deshpande MV. Nano-Biopesticides Today and Future Perspectives. London, UK: Elsevier; 2019.

[131] Seiber JN, Coats J, Duke SO, Gross AD. Pest Management with Biopesticides. Frontiers of Agricultural Science and Engineering. 2018;**5**(3):295-300

[132] Athanassiou C, Kavallieratos N, Benelli G, Losic D, Rani PU, Desneux N. Nanoparticles for pest control: Current status and future perspectives. Journal of Pest Science. 2018;**91**(1):1-15